Lecture Notes in Networks and Systems 987

The series "Lecture Notes in Networks and Systems" publishes the latest developments in Networks and Systems—quickly, informally and with high quality. Original research reported in proceedings and post-proceedings represents the core of LNNS.

Volumes published in LNNS embrace all aspects and subfields of, as well as new challenges in, Networks and Systems.

The series contains proceedings and edited volumes in systems and networks, spanning the areas of Cyber-Physical Systems, Autonomous Systems, Sensor Networks, Control Systems, Energy Systems, Automotive Systems, Biological Systems, Vehicular Networking and Connected Vehicles, Aerospace Systems, Automation, Manufacturing, Smart Grids, Nonlinear Systems, Power Systems, Robotics, Social Systems, Economic Systems and other. Of particular value to both the contributors and the readership are the short publication timeframe and the world-wide distribution and exposure which enable both a wide and rapid dissemination of research output.

The series covers the theory, applications, and perspectives on the state of the art and future developments relevant to systems and networks, decision making, control, complex processes and related areas, as embedded in the fields of interdisciplinary and applied sciences, engineering, computer science, physics, economics, social, and life sciences, as well as the paradigms and methodologies behind them.

Indexed by SCOPUS, INSPEC, WTI Frankfurt eG, zbMATH, SCImago.

All books published in the series are submitted for consideration in Web of Science.

For proposals from Asia please contact Aninda Bose (aninda.bose@springer.com).

Álvaro Rocha · Hojjat Adeli ·
Gintautas Dzemyda · Fernando Moreira ·
Aneta Poniszewska-Marańda
Editors

Good Practices and New Perspectives in Information Systems and Technologies

WorldCIST 2024, Volume 3

Editors
Álvaro Rocha
ISEG
Universidade de Lisboa
Lisbon, Portugal

Hojjat Adeli
College of Engineering
The Ohio State University
Columbus, OH, USA

Gintautas Dzemyda
Institute of Data Science and Digital Technologies
Vilnius University
Vilnius, Lithuania

Fernando Moreira
DCT
Universidade Portucalense
Porto, Portugal

Aneta Poniszewska-Marańda
Institute of Information Technology
Lodz University of Technology
Łódz, Poland

ISSN 2367-3370 ISSN 2367-3389 (electronic)
Lecture Notes in Networks and Systems
ISBN 978-3-031-60220-7 ISBN 978-3-031-60221-4 (eBook)
https://doi.org/10.1007/978-3-031-60221-4

This Springer imprint is published by the registered company Springer Nature Switzerland AG
The registered company address is: Gewerbestrasse 11, 6330 Cham, Switzerland

Preface

This book contains a selection of papers accepted for presentation and discussion at the 2024 World Conference on Information Systems and Technologies (WorldCIST'24). This conference had the scientific support of the Lodz University of Technology, Information and Technology Management Association (ITMA), IEEE Systems, Man, and Cybernetics Society (IEEE SMC), Iberian Association for Information Systems and Technologies (AISTI), and Global Institute for IT Management (GIIM). It took place in Lodz city, Poland, 26–28 March 2024.

The World Conference on Information Systems and Technologies (WorldCIST) is a global forum for researchers and practitioners to present and discuss recent results and innovations, current trends, professional experiences, and challenges of modern Information Systems and Technologies research, technological development, and applications. One of its main aims is to strengthen the drive toward a holistic symbiosis between academy, society, and industry. WorldCIST'24 is built on the successes of: WorldCIST'13 held at Olhão, Algarve, Portugal; WorldCIST'14 held at Funchal, Madeira, Portugal; WorldCIST'15 held at São Miguel, Azores, Portugal; WorldCIST'16 held at Recife, Pernambuco, Brazil; WorldCIST'17 held at Porto Santo, Madeira, Portugal; WorldCIST'18 held at Naples, Italy; WorldCIST'19 held at La Toja, Spain; WorldCIST'20 held at Budva, Montenegro; WorldCIST'21 held at Terceira Island, Portugal; WorldCIST'22 held at Budva, Montenegro; and WorldCIST'23, which took place at Pisa, Italy.

The Program Committee of WorldCIST'24 was composed of a multidisciplinary group of 328 experts and those who are intimately concerned with Information Systems and Technologies. They have had the responsibility for evaluating, in a 'blind review' process, the papers received for each of the main themes proposed for the conference: A) Information and Knowledge Management; B) Organizational Models and Information Systems; C) Software and Systems Modeling; D) Software Systems, Architectures, Applications and Tools; E) Multimedia Systems and Applications; F) Computer Networks, Mobility and Pervasive Systems; G) Intelligent and Decision Support Systems; H) Big Data Analytics and Applications; I) Human-Computer Interaction; J) Ethics, Computers & Security; K) Health Informatics; L) Information Technologies in Education; M) Information Technologies in Radiocommunications; and N) Technologies for Biomedical Applications.

The conference also included workshop sessions taking place in parallel with the conference ones. Workshop sessions covered themes such as: ICT for Auditing & Accounting; Open Learning and Inclusive Education Through Information and Communication Technology; Digital Marketing and Communication, Technologies, and Applications; Advances in Deep Learning Methods and Evolutionary Computing for Health Care; Data Mining and Machine Learning in Smart Cities: The role of the technologies in the research of the migrations; Artificial Intelligence Models and Artifacts for Business Intelligence Applications; AI in Education; Environmental data analytics; Forest-Inspired

Computational Intelligence Methods and Applications; Railway Operations, Modeling and Safety; Technology Management in the Electrical Generation Industry: Capacity Building through Knowledge, Resources and Networks; Data Privacy and Protection in Modern Technologies; Strategies and Challenges in Modern NLP: From Argumentation to Ethical Deployment; and Enabling Software Engineering Practices Via Last Development Trends.

WorldCIST'24 and its workshops received about 400 contributions from 47 countries around the world. The papers accepted for oral presentation and discussion at the conference are published by Springer (this book) in six volumes and will be submitted for indexing by WoS, Scopus, EI-Compendex, DBLP, and/or Google Scholar, among others. Extended versions of selected best papers will be published in special or regular issues of leading and relevant journals, mainly JCR/SCI/SSCI and Scopus/EI-Compendex indexed journals.

We acknowledge all of those that contributed to the staging of WorldCIST'24 (authors, committees, workshop organizers, and sponsors). We deeply appreciate their involvement and support that was crucial for the success of WorldCIST'24.

March 2024

Álvaro Rocha
Hojjat Adeli
Gintautas Dzemyda
Fernando Moreira
Aneta Poniszewska-Marańda

Organization

Conference

Honorary Chair

Hojjat Adeli	The Ohio State University, USA

General Chair

Álvaro Rocha	ISEG, University of Lisbon, Portugal

Co-chairs

Gintautas Dzemyda	Vilnius University, Lithuania
Sandra Costanzo	University of Calabria, Italy

Workshops Chair

Fernando Moreira	Portucalense University, Portugal

Local Organizing Committee

Bożena Borowska	Lodz University of Technology, Poland
Łukasz Chomątek	Lodz University of Technology, Poland
Joanna Ochelska-Mierzejewska	Lodz University of Technology, Poland
Aneta Poniszewska-Marańda	Lodz University of Technology, Poland

Advisory Committee

Ana Maria Correia (Chair)	University of Sheffield, UK
Brandon Randolph-Seng	Texas A&M University, USA

Chris Kimble	KEDGE Business School & MRM, UM2, Montpellier, France
Damian Niwiński	University of Warsaw, Poland
Eugene Spafford	Purdue University, USA
Florin Gheorghe Filip	Romanian Academy, Romania
Janusz Kacprzyk	Polish Academy of Sciences, Poland
João Tavares	University of Porto, Portugal
Jon Hall	The Open University, UK
John MacIntyre	University of Sunderland, UK
Karl Stroetmann	Empirica Communication & Technology Research, Germany
Marjan Mernik	University of Maribor, Slovenia
Miguel-Angel Sicilia	University of Alcalá, Spain
Mirjana Ivanovic	University of Novi Sad, Serbia
Paulo Novais	University of Minho, Portugal
Sami Habib	Kuwait University, Kuwait
Wim Van Grembergen	University of Antwerp, Belgium

Program Committee Co-chairs

Adam Wojciechowski	Lodz University of Technology, Poland
Aneta Poniszewska-Marańda	Lodz University of Technology, Poland

Program Committee

Abderrahmane Ez-zahout	Mohammed V University, Morocco
Adriana Peña Pérez Negrón	Universidad de Guadalajara, Mexico
Adriani Besimi	South East European University, North Macedonia
Agostinho Sousa Pinto	Polytechnic of Porto, Portugal
Ahmed El Oualkadi	Abdelmalek Essaadi University, Morocco
Akex Rabasa	University Miguel Hernandez, Spain
Alanio de Lima	UFC, Brazil
Alba Córdoba-Cabús	University of Malaga, Spain
Alberto Freitas	FMUP, University of Porto, Portugal
Aleksandra Labus	University of Belgrade, Serbia
Alessio De Santo	HE-ARC, Switzerland
Alexandru Vulpe	University Politechnica of Bucharest, Romania
Ali Idri	ENSIAS, University Mohamed V, Morocco
Alicia García-Holgado	University of Salamanca, Spain

Christos Chrysoulas	London South Bank University, UK
Christos Chrysoulas	Edinburgh Napier University, UK
Ciro Martins	University of Aveiro, Portugal
Claudio Sapateiro	Polytechnic of Setúbal, Portugal
Cosmin Striletchi	Technical University of Cluj-Napoca, Romania
Costin Badica	University of Craiova, Romania
Cristian García Bauza	PLADEMA-UNICEN-CONICET, Argentina
Cristina Caridade	Polytechnic of Coimbra, Portugal
Danish Jamil	Malaysia University of Science and Technology, Malaysia
David Cortés-Polo	University of Extremadura, Spain
David Kelly	University College London, UK
Daria Bylieva	Peter the Great St. Petersburg Polytechnic University, Russia
Dayana Spagnuelo	Vrije Universiteit Amsterdam, Netherlands
Dhouha Jaziri	University of Sousse, Tunisia
Dmitry Frolov	HSE University, Russia
Dulce Mourato	ISTEC - Higher Advanced Technologies Institute Lisbon, Portugal
Edita Butrime	Lithuanian University of Health Sciences, Lithuania
Edna Dias Canedo	University of Brasilia, Brazil
Egils Ginters	Riga Technical University, Latvia
Ekaterina Isaeva	Perm State University, Russia
Eliana Leite	University of Minho, Portugal
Enrique Pelaez	ESPOL University, Ecuador
Eriks Sneiders	Stockholm University, Sweden; Esteban Castellanos ESPE, Ecuador
Fatima Azzahra Amazal	Ibn Zohr University, Morocco
Fernando Bobillo	University of Zaragoza, Spain
Fernando Molina-Granja	National University of Chimborazo, Ecuador
Fernando Moreira	Portucalense University, Portugal
Fernando Ribeiro	Polytechnic Castelo Branco, Portugal
Filipe Caldeira	Polytechnic of Viseu, Portugal
Filippo Neri	University of Naples, Italy
Firat Bestepe	Republic of Turkey Ministry of Development, Turkey
Francesco Bianconi	Università degli Studi di Perugia, Italy
Francisco García-Peñalvo	University of Salamanca, Spain
Francisco Valverde	Universidad Central del Ecuador, Ecuador
Frederico Branco	University of Trás-os-Montes e Alto Douro, Portugal
Galim Vakhitov	Kazan Federal University, Russia

Gayo Diallo	University of Bordeaux, France
Gabriel Pestana	Polytechnic Institute of Setubal, Portugal
Gema Bello-Orgaz	Universidad Politecnica de Madrid, Spain
George Suciu	BEIA Consult International, Romania
Ghani Albaali	Princess Sumaya University for Technology, Jordan
Gian Piero Zarri	University Paris-Sorbonne, France
Giovanni Buonanno	University of Calabria, Italy
Gonçalo Paiva Dias	University of Aveiro, Portugal
Goreti Marreiros	ISEP/GECAD, Portugal
Habiba Drias	University of Science and Technology Houari Boumediene, Algeria
Hafed Zarzour	University of Souk Ahras, Algeria
Haji Gul	City University of Science and Information Technology, Pakistan
Hakima Benali Mellah	Cerist, Algeria
Hamid Alasadi	Basra University, Iraq
Hatem Ben Sta	University of Tunis at El Manar, Tunisia
Hector Fernando Gomez Alvarado	Universidad Tecnica de Ambato, Ecuador
Hector Menendez	King's College London, UK
Hélder Gomes	University of Aveiro, Portugal
Helia Guerra	University of the Azores, Portugal
Henrique da Mota Silveira	University of Campinas (UNICAMP), Brazil
Henrique S. Mamede	University Aberta, Portugal
Henrique Vicente	University of Évora, Portugal
Hicham Gueddah	University Mohammed V in Rabat, Morocco
Hing Kai Chan	University of Nottingham Ningbo China, China
Igor Aguilar Alonso	Universidad Nacional Tecnológica de Lima Sur, Peru
Inês Domingues	University of Coimbra, Portugal
Isabel Lopes	Polytechnic of Bragança, Portugal
Isabel Pedrosa	Coimbra Business School - ISCAC, Portugal
Isaías Martins	University of Leon, Spain
Issam Moghrabi	Gulf University for Science and Technology, Kuwait
Ivan Armuelles Voinov	University of Panama, Panama
Ivan Dunđer	University of Zagreb, Croatia
Ivone Amorim	University of Porto, Portugal
Jaime Diaz	University of La Frontera, Chile
Jan Egger	IKIM, Germany
Jan Kubicek	Technical University of Ostrava, Czech Republic
Jeimi Cano	Universidad de los Andes, Colombia

Jesús Gallardo Casero	University of Zaragoza, Spain
Jezreel Mejia	CIMAT, Unidad Zacatecas, Mexico
Jikai Li	The College of New Jersey, USA
Jinzhi Lu	KTH-Royal Institute of Technology, Sweden
Joao Carlos Silva	IPCA, Portugal
João Manuel R. S. Tavares	University of Porto, FEUP, Portugal
João Paulo Pereira	Polytechnic of Bragança, Portugal
João Reis	University of Aveiro, Portugal
João Reis	University of Lisbon, Portugal
João Rodrigues	University of the Algarve, Portugal
João Vidal de Carvalho	Polytechnic of Porto, Portugal
Joaquin Nicolas Ros	University of Murcia, Spain
John W. Castro	University de Atacama, Chile
Jorge Barbosa	Polytechnic of Coimbra, Portugal
Jorge Buele	Technical University of Ambato, Ecuador; Jorge Gomes University of Lisbon, Portugal
Jorge Oliveira e Sá	University of Minho, Portugal
José Braga de Vasconcelos	Universidade Lusófona, Portugal
Jose M. Parente de Oliveira	Aeronautics Institute of Technology, Brazil
José Machado	University of Minho, Portugal
José Paulo Lousado	Polytechnic of Viseu, Portugal
Jose Quiroga	University of Oviedo, Spain
Jose Silvestre Silva	Academia Military, Portugal
Jose Torres	University Fernando Pessoa, Portugal
Juan M. Santos	University of Vigo, Spain
Juan Manuel Carrillo de Gea	University of Murcia, Spain
Juan Pablo Damato	UNCPBA-CONICET, Argentina
Kalinka Kaloyanova	Sofia University, Bulgaria
Kamran Shaukat	The University of Newcastle, Australia
Katerina Zdravkova	University Ss. Cyril and Methodius, North Macedonia
Khawla Tadist	Morocco
Khalid Benali	LORIA - University of Lorraine, France
Khalid Nafil	Mohammed V University in Rabat, Morocco
Korhan Gunel	Adnan Menderes University, Turkey
Krzysztof Wolk	Polish-Japanese Academy of Information Technology, Poland
Kuan Yew Wong	Universiti Teknologi Malaysia (UTM), Malaysia
Kwanghoon Kim	Kyonggi University, South Korea
Laila Cheikhi	Mohammed V University in Rabat, Morocco
Laura Varela-Candamio	Universidade da Coruña, Spain
Laurentiu Boicescu	E.T.T.I. U.P.B., Romania

Lbtissam Abnane	ENSIAS, Morocco
Lia-Anca Hangan	Technical University of Cluj-Napoca, Romania
Ligia Martinez	CECAR, Colombia
Lila Rao-Graham	University of the West Indies, Jamaica
Liliana Ivone Pereira	Polytechnic of Cávado and Ave, Portugal
Łukasz Tomczyk	Pedagogical University of Cracow, Poland
Luis Alvarez Sabucedo	University of Vigo, Spain
Luís Filipe Barbosa	University of Trás-os-Montes e Alto Douro
Luis Mendes Gomes	University of the Azores, Portugal
Luis Pinto Ferreira	Polytechnic of Porto, Portugal
Luis Roseiro	Polytechnic of Coimbra, Portugal
Luis Silva Rodrigues	Polytencic of Porto, Portugal
Mahdieh Zakizadeh	MOP, Iran
Maksim Goman	JKU, Austria
Manal el Bajta	ENSIAS, Morocco
Manuel Antonio Fernández-Villacañas Marín	Technical University of Madrid, Spain
Manuel Ignacio Ayala Chauvin	University Indoamerica, Ecuador
Manuel Silva	Polytechnic of Porto and INESC TEC, Portugal
Manuel Tupia	Pontifical Catholic University of Peru, Peru
Manuel Au-Yong-Oliveira	University of Aveiro, Portugal
Marcelo Mendonça Teixeira	Universidade de Pernambuco, Brazil
Marciele Bernardes	University of Minho, Brazil
Marco Ronchetti	Universita' di Trento, Italy
Mareca María Pilar	Universidad Politécnica de Madrid, Spain
Marek Kvet	Zilinska Univerzita v Ziline, Slovakia
Maria João Ferreira	Universidade Portucalense, Portugal
Maria José Sousa	University of Coimbra, Portugal
María Teresa García-Álvarez	University of A Coruna, Spain
Maria Sokhn	University of Applied Sciences of Western Switzerland, Switzerland
Marijana Despotovic-Zrakic	Faculty Organizational Science, Serbia
Marilio Cardoso	Polytechnic of Porto, Portugal
Mário Antunes	Polytechnic of Leiria & CRACS INESC TEC, Portugal
Marisa Maximiano	Polytechnic Institute of Leiria, Portugal
Marisol Garcia-Valls	Polytechnic University of Valencia, Spain
Maristela Holanda	University of Brasilia, Brazil
Marius Vochin	E.T.T.I. U.P.B., Romania
Martin Henkel	Stockholm University, Sweden
Martín López Nores	University of Vigo, Spain
Martin Zelm	INTEROP-VLab, Belgium

Mazyar Zand	MOP, Iran
Mawloud Mosbah	University 20 Août 1955 of Skikda, Algeria
Michal Adamczak	Poznan School of Logistics, Poland
Michal Kvet	University of Zilina, Slovakia
Miguel Garcia	University of Oviedo, Spain
Mircea Georgescu	Al. I. Cuza University of Iasi, Romania
Mirna Muñoz	Centro de Investigación en Matemáticas A.C., Mexico
Mohamed Hosni	ENSIAS, Morocco
Monica Leba	University of Petrosani, Romania
Nadesda Abbas	UBO, Chile
Narasimha Rao Vajjhala	University of New York Tirana, Tirana
Narjes Benameur	Laboratory of Biophysics and Medical Technologies of Tunis, Tunisia
Natalia Grafeeva	Saint Petersburg University, Russia
Natalia Miloslavskaya	National Research Nuclear University MEPhI, Russia
Naveed Ahmed	University of Sharjah, United Arab Emirates
Neeraj Gupta	KIET group of institutions Ghaziabad, India
Nelson Rocha	University of Aveiro, Portugal
Nikola S. Nikolov	University of Limerick, Ireland
Nicolas de Araujo Moreira	Federal University of Ceara, Brazil
Nikolai Prokopyev	Kazan Federal University, Russia
Niranjan S. K.	JSS Science and Technology University, India
Noemi Emanuela Cazzaniga	Politecnico di Milano, Italy
Noureddine Kerzazi	Polytechnique Montréal, Canada
Nuno Melão	Polytechnic of Viseu, Portugal
Nuno Octávio Fernandes	Polytechnic of Castelo Branco, Portugal
Nuno Pombo	University of Beira Interior, Portugal
Olga Kurasova	Vilnius University, Lithuania
Olimpiu Stoicuta	University of Petrosani, Romania
Patricia Quesado	Polytechnic of Cávado and Ave, Portugal
Patricia Zachman	Universidad Nacional del Chaco Austral, Argentina
Paula Serdeira Azevedo	University of Algarve, Portugal
Paula Dias	Polytechnic of Guarda, Portugal
Paulo Alejandro Quezada Sarmiento	University of the Basque Country, Spain
Paulo Maio	Polytechnic of Porto, ISEP, Portugal
Paulvanna Nayaki Marimuthu	Kuwait University, Kuwait
Paweł Karczmarek	The John Paul II Catholic University of Lublin, Poland

Pedro Rangel Henriques	University of Minho, Portugal
Pedro Sobral	University Fernando Pessoa, Portugal
Pedro Sousa	University of Minho, Portugal
Philipp Jordan	University of Hawaii at Manoa, USA
Piotr Kulczycki	Systems Research Institute, Polish Academy of Sciences, Poland
Prabhat Mahanti	University of New Brunswick, Canada
Rabia Azzi	Bordeaux University, France
Radu-Emil Precup	Politehnica University of Timisoara, Romania
Rafael Caldeirinha	Polytechnic of Leiria, Portugal
Raghuraman Rangarajan	Sequoia AT, Portugal
Radhakrishna Bhat	Manipal Institute of Technology, India
Raiani Ali	Hamad Bin Khalifa University, Qatar
Ramadan Elaiess	University of Benghazi, Libya
Ramayah T.	Universiti Sains Malaysia, Malaysia
Ramazy Mahmoudi	University of Monastir, Tunisia
Ramiro Gonçalves	University of Trás-os-Montes e Alto Douro & INESC TEC, Portugal
Ramon Alcarria	Universidad Politécnica de Madrid, Spain
Ramon Fabregat Gesa	University of Girona, Spain
Ramy Rahimi	Chungnam National University, South Korea
Reiko Hishiyama	Waseda University, Japan
Renata Maria Maracho	Federal University of Minas Gerais, Brazil
Renato Toasa	Israel Technological University, Ecuador
Reyes Juárez Ramírez	Universidad Autonoma de Baja California, Mexico
Rocío González-Sánchez	Rey Juan Carlos University, Spain
Rodrigo Franklin Frogeri	University Center of Minas Gerais South, Brazil
Ruben Pereira	ISCTE, Portugal
Rui Alexandre Castanho	WSB University, Poland
Rui S. Moreira	UFP & INESC TEC & LIACC, Portugal
Rustam Burnashev	Kazan Federal University, Russia
Saeed Salah	Al-Quds University, Palestine
Said Achchab	Mohammed V University in Rabat, Morocco
Sajid Anwar	Institute of Management Sciences Peshawar, Pakistan
Sami Habib	Kuwait University, Kuwait
Samuel Sepulveda	University of La Frontera, Chile
Sara Luis Dias	Polytechnic of Cávado and Ave, Portugal
Sandra Costanzo	University of Calabria, Italy
Sandra Patricia Cano Mazuera	University of San Buenaventura Cali, Colombia
Sassi Sassi	FSJEGJ, Tunisia

Seppo Sirkemaa	University of Turku, Finland
Sergio Correia	Polytechnic of Portalegre, Portugal
Shahnawaz Talpur	Mehran University of Engineering & Technology Jamshoro, Pakistan
Shakti Kundu	Manipal University Jaipur, Rajasthan, India
Shashi Kant Gupta	Eudoxia Research University, USA
Silviu Vert	Politehnica University of Timisoara, Romania
Simona Mirela Riurean	University of Petrosani, Romania
Slawomir Zolkiewski	Silesian University of Technology, Poland
Solange Rito Lima	University of Minho, Portugal
Sonia Morgado	ISCPSI, Portugal
Sonia Sobral	Portucalense University, Portugal
Sorin Zoican	Polytechnic University of Bucharest, Romania
Souraya Hamida	Batna 2 University, Algeria
Stalin Figueroa	University of Alcala, Spain
Sümeyya Ilkin	Kocaeli University, Turkey
Syed Asim Ali	University of Karachi, Pakistan
Syed Nasirin	Universiti Malaysia Sabah, Malaysia
Tatiana Antipova	Institute of Certified Specialists, Russia
Tatianna Rosal	University of Trás-os-Montes e Alto Douro, Portugal
Tero Kokkonen	JAMK University of Applied Sciences, Finland
The Thanh Van	HCMC University of Food Industry, Vietnam
Thomas Weber	EPFL, Switzerland
Timothy Asiedu	TIM Technology Services Ltd., Ghana
Tom Sander	New College of Humanities, Germany
Tomasz Kisielewicz	Warsaw University of Technology
Tomaž Klobučar	Jozef Stefan Institute, Slovenia
Toshihiko Kato	University of Electro-communications, Japan
Tuomo Sipola	Jamk University of Applied Sciences, Finland
Tzung-Pei Hong	National University of Kaohsiung, Taiwan
Valentim Realinho	Polytechnic of Portalegre, Portugal
Valentina Colla	Scuola Superiore Sant'Anna, Italy
Valerio Stallone	ZHAW, Switzerland
Verónica Vasconcelos	Polytechnic of Coimbra, Portugal
Vicenzo Iannino	Scuola Superiore Sant'Anna, Italy
Vitor Gonçalves	Polytechnic of Bragança, Portugal
Victor Alves	University of Minho, Portugal
Victor Georgiev	Kazan Federal University, Russia
Victor Hugo Medina Garcia	Universidad Distrital Francisco José de Caldas, Colombia
Victor Kaptelinin	Umeå University, Sweden

Viktor Medvedev	Vilnius University, Lithuania
Vincenza Carchiolo	University of Catania, Italy
Waqas Bangyal	University of Gujrat, Pakistan
Wolf Zimmermann	Martin Luther University Halle-Wittenberg, Germany
Yadira Quiñonez	Autonomous University of Sinaloa, Mexico
Yair Wiseman	Bar-Ilan University, Israel
Yassine Drias	University of Algiers, Algeria
Yuhua Li	Cardiff University, UK
Yuwei Lin	University of Roehampton, UK
Zbigniew Suraj	University of Rzeszow, Poland
Zorica Bogdanovic	University of Belgrade, Serbia

Contents

Information and Knowledge Management

Improving Change Management in Large Engineering Projects: A Case Study

João Santos[1], Anabela Tereso[2](✉), and Cláudio Santos[3]

[1] Master's in Industrial Engineering and Management, University of Minho, Guimarães, Portugal
pg47297@alunos.uminho.pt

[2] ALGORITMI Research Centre/LASI, University of Minho, Guimarães, Portugal
anabelat@dps.uminho.pt

[3] Production and Systems Department, University of Minho, Guimarães, Portugal
claudio.santos@dps.uminho.pt

Abstract. Engineering change management is an essential component in the development of products, requiring many resources and time. A usual problem is the lack of standardized activities and the inexistence of activity tracking, usually resulting in increased time to complete and costs. This research aimed to identify and propose improvements in these topics within a manufacturing company. The first proposal consists of a revised standard for the activities in each Engineering Change Request phase, suggesting when each one of the activities should be started, as well as the advisable action for the Project Manager (PM). Then, the second proposal is related to the Engineering Change Request deadlines compliance, with two possible approaches. Approach 1 suggests the dates tracking in an Excel file, while approach 2 suggests using Power BI dashboards. Both proposals are expected to improve the normal flow of activities, resulting in higher compliance with the planned deadlines and ultimately reducing cost and the implementation time of the change.

Keywords: Engineering Change Management · Engineering Change Request · Project Management · Large Engineering Projects

1 Introduction

In today's increasingly competitive world, competitiveness has become a key factor in an organization's success. The growth of market competition, combined with the exceptional needs of consumers, contribute to a higher level of business renewal. For industries, this renewal usually means engineering changes to their products [1]. Organizations need to be constantly prepared to improve and update existing products [2]. These changes may be due to new product requirements requested by the customer, or due to new manufacturing requirements resulting from the company's investigation [3]. A change may involve any modification to the form, fit and function of the product and may affect the interactions and dependencies of the different parts of the product [4].

Á. Rocha et al. (Eds.): WorldCIST 2024, LNNS 987, pp. 3–12, 2024.
https://doi.org/10.1007/978-3-031-60221-4_1

Currently, the manufacturing industry is dealing with multiple Engineering Change Management (ECM) systems and processes to create and communicate changes. These systems include multiple formats and definitions, leading to increased confusion, cost and overall inefficiency in the system [5].

By addressing the challenges associated with change management, organizations can improve their ability to respond quickly to market demands, reduce costs, and improve overall productivity [3]. In this sense, this research aims to contribute to the development of more robust and streamlined processes for managing engineering changes in the manufacturing industry. Through a comprehensive analysis of activities and tools, this work seeks to improve the visibility and analysis of the ECM process, with a particular focus on addressing the challenges of handling a large number of ECRs. The goal is to develop methods and tools that enable efficient and rapid assessment of multiple ECRs, allowing stakeholders to more easily gain insight into the status and progress of ECRs. Achieving this goal is expected to improve decision making, streamline the ECM process, and increase overall organizational agility in managing engineering changes.

2 Literature Review

An Engineering Change (EC) can be defined as a change to components, drawings or software that has been performed during the product design process. The size and nature of the change, and even the number of people and duration of the application, can vary depending on the project and the various factors involved [4, 6]. So, ECM is responsible for organizing and controlling all the activities that make up the EC process [4].

ECM is a component of project management, and its implementation is an important step to achieve success in the projects as it assumes a relevant role, being responsible for keeping all the ECs under control [6, 8]. The activities of ECM include forecasting possible changes, identifying changes that have already occurred, planning preventive measures, and coordinating the changes [6].

The changes can be triggered by various stakeholders, including customers, suppliers, internal company departments, government bodies, and even market drivers [9]. In fact, change can not only be triggered by, but can also affect all the major business functions, illustrating its relevance in the project context [4]. Communication between all those involved in the change is of crucial importance to explore topics as potential risks, benefits, and requirements of the change [10].

2.1 Engineering Change Process

Over the years, various authors have proposed different definitions of EC process, each dividing the process into a different number of phases [4, 20]. However, it can be proposed a generic engineering change process (Fig. 1) based on the many models proposed by other authors [2]. This generic process covers the complete life cycle of ECs, from the trigger given by the need for change to the review of the process [11].

As shown in Fig. 1, EC process encompasses steps such as change identification, impact assessment, approval, implementation, and post-implementation evaluation. Following these steps or guidelines is essential to ensure the organization and effectiveness of the ECM process.

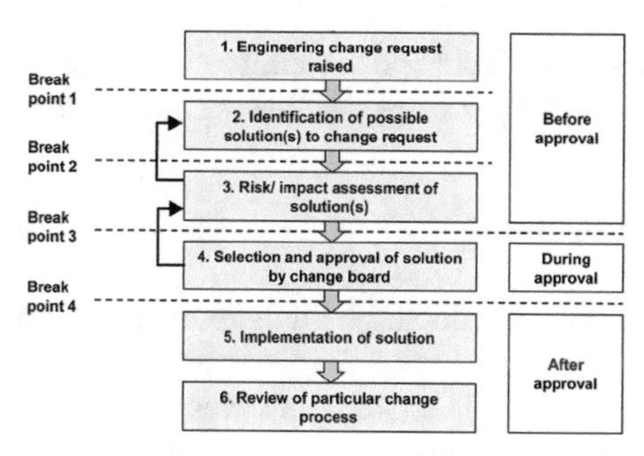

Fig. 1. Generic engineering change process [11]

In summary, although there is no universal best practice for the ECM, steps or guidelines should be followed to ensure a structured and successful ECM process.

2.2 Impacts of Change Management

The identification of the change impact is extremely important for the engineering change process. In general, changes have an impact on planning, scheduling and project costs [2, 4]. The scope of the change is usually greater than originally planned. The impact of adding an isolated part or component to a product is relatively easy to determine at the beginning of the design process [9]. However, at a later stage, the insertion of a component may conflict with others or even be incompatible, requiring rework, material waste, more people involved, and resulting in costs and delays [4, 9].

The change may also affect other products (e.g. other products in the same product family), processes (e.g. the manufacturing process) and companies (e.g. products with the same suppliers or partners) [2]. This propagation can create a snowball effect, and in the worst case, affect the whole system, increasing costs and delaying schedules [13].

The project manager is responsible for assessing the impact of the changes [7]. In order to do that it is recommended that the project team uses supporting tools and methods to make more accurate estimations [9].

2.3 Tools and Techniques

The change process can really be facilitated if it uses tools and techniques (e.g. Materials Requirements Planning, Failure Mode and Effect Analysis and Value Analysis) [14]. PDM systems are essential components for managing product data and activities throughout the product lifecycle [12]. PDM focuses on the product creation phase and links Enterprise Resource Planning (ERP) systems. Therefore, all PDM systems allow the coordination of ECM processes and the documentation and management of ECRs (usually well documented and stored in industrial databases) [19]. However, not all organizations have access to the recommended tools for various reasons, such as lack of awareness of their existence, failure to meet user requirements, high difficulty and time-consuming to use, among others [14]. Furthermore, despite the importance of the tools, it is vital that the organization clearly understands the process and how the changes are linked [4]. Despite the existence of numerous tools that provide ECM data, there is a weakness in the way this data is presented, forcing project managers to perform a step-by-step analysis of each ECR instead of having access to a comprehensive view.

3 Research Methodology

A case study methodology was adopted with the purpose of evaluating the opportunities for improvement in the engineering change management practices in a large engineering project environment. This strategy involves an empirical investigation of the case in question with the aim of formulating theoretical propositions [15, 16]. The primary researcher was integrated into the project management team of a large engineering project and, as the project progressed, investigated how the engineering change management process could be improved to implement changes more effectively and efficiently. Deductive approach and multi-method qualitative research were used to develop this research. Information was collected in several ways including company documentation, unstructured interviews, and direct and participant observation.

4 Case Study Analysis

The company under study has a project-oriented culture and considers project management as a core competence with high relevance in the product development process. The phases of project life cycle are aligned with the Process Groups (Initiate; Plan; Execute; Monitor and Control; and Close) [17]. In addition, the different phases are separated by milestones, following a model that can be understood as an adaptation of the Stage Gate model [18]. Parallel to the phases and milestones, there are different sample phases (named A, B, C and D) depending on the samples maturity level. Ultimately, once all these phases have been completed, the project moves on to series production, where large quantities of the already validated product are produced.

ECM is present in all the project phases. However, the ECR process only starts to be followed in D sample phase, since it works already as a preparation for the series production. In the early phases, the production can be considered residual and therefore the consequences of changes do not have the same impact as in the series production stage. The ECR process is used to ensure that all the information is reported to the appropriate people and that all data is documented, minimizing the risk of a wrong change implementation.

In the company under study, the ECR process (Fig. 2) is defined by five phases (Pre-clarification, Planning, Processing and Validation, Customer, and Implementation) with a review at the end of each one.

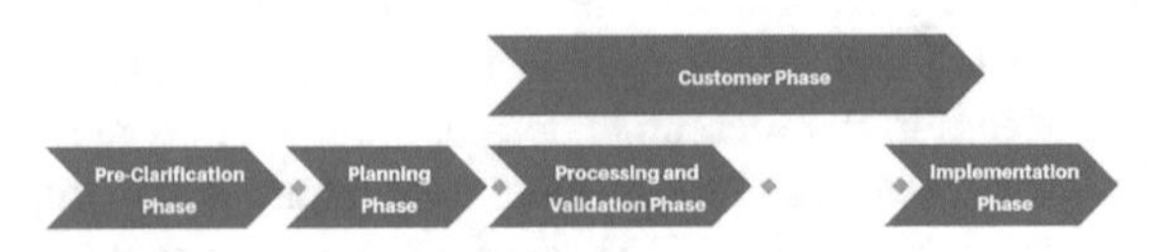

Fig. 2. ECR process overview

The process is initiated when some person (change initiator) creates the request, with a given title, the designated manager of the initiator (normally the project manager), and a brief description of the change. After these activities had been performed in the first phase, the ECR moves to the manager of the initiator responsibility that evaluates the request and, if it makes sense, approves the ECR for further processing, defining the person (change representative) that should follow the ECR creation and add all the documents, details and participants. After all the details (products affected, checklists, costs, requirements, participants, timeline, etc.) being added in the planning phase, the team (people from the different domains) perform a revision of the request and approves it. In processing and validation phase there is a production of prototype products that can help to evaluate the real impact of the change. After all the team approves again, the ECR waits until customer approval, entering in implementation after that.

Even though this process is already defined, there are several points that are left to the project manager to decide if it is necessary or not, at what time should be done and who should be involved in the process. Another important topic is the tracking of ECRs. In this context, the tracking can be understood as the controlling and monitoring of the status, the required activities and the next steps of the different ECRs. The project manager has to check each one individually to know what the blocker of the ECR implementation is. This leads to several delays on the changes implementation because there is not a regular control of the ECR status. In the series production, this topic assumes a vital importance since it is important to introduce the changes in a faster way.

5 Proposals for Improvement

The main objective of this chapter is to present and explain the proposals developed during the study. The proposals are divided into two main groups, the first being a standard for ECM activities in each of the ECR process phases, and the second being the application of a control mechanism for tracking ECRs. In order to validate the improvements, the proposals were presented to the main stakeholders. The suggestions for improvement were readily accepted as they were considered to add value. However, there are proposals that require a restructuring of the organizational tools, so the full implementation was postponed to future work due to time constraints.

5.1 Review of Standard Activities in Each ECR Phase

Although the activities are defined for each phase, there are some tasks that can be improved and reallocated in terms of the timing of a particular activity. It is necessary for the project manager to know when it is advisable to start the task, otherwise, the task will only be started to be developed when the need appears and then depending on the lead time of the task it can lead to unnecessary delays in the execution of the ECR.

In order to respond to this problem, it is necessary to adapt the timing as well as the way of working in order to ensure that the project manager is better supported in the correct respect of deadlines. With a well-defined process, the PM will have a better organization of the topics and be more efficient in the ECR implementation. The activities recommended in each phase are described separately in the following subsections.

Preliminary Agreement Phase

As the first phase of the ECR, it can be understood as a phase of definition and a short explanation of the change. However, the roles of change initiator and change representative are often taken by the same person, so the two split phases can be tailored to work with common tasks, in a way that tasks that are defined to start only in the next phase can be started already in this phase. When it is raised the change request it should be arranged right away one meeting to discuss the goal and the steps to take with the change, including the initiator, the manager and if applicable the representative. In this case there would exist immediate feedback from the manager and the activities could start. In this phase it can be filled in the description, the products and materials affected, the expected costs, the participants list and the validation questions. In addition, the development of checklists or assessments and the timeline for the implementation can be started. After that, the manager proceeds to the confirmation of all the details and accepts the change request.

Planning Phase

In this phase, after the manager had accepted the request, the change representative has still some time to prepare the required documentation and information. As proposed, this task would have already been started previously, so the completion time is expected to be smaller. In this phase, a meeting should be arranged with the change main stakeholders to evaluate the timeline and possibly answer some questions. After all the data and documents are inserted, the ECR can move forward to the team approvals. Since the approvals can take a long time, it is suggested the creation of tasks within the platform that alerts the participants for the need of action. This can be used not only to trigger approvals but also to ask for documentation or to answer questions in earlier stages.

Processing and Validation Phase

At this phase, the responsibility shifts more to the change coordinator and to manufacturing, with the change proposal being tested and validated through a test production. However, there should be a regular exchange of information between the change coordinator and the representative to ensure that there is a clear understanding of all the developments and to ensure that the defined schedule is met. In this phase, the approval workflow should be set up in parallel with the validation, rather than waiting only for the end to do the final review. The reason behind this is that only manufacturing really needs the validation results. An alternative solution is also to split the current review in two, with the first containing all the approvals that can be done with the validation still ongoing and the second to the manufacturing approvers that need the results from validation to approve.

Implementation Phase

In this phase, it is crucial to ensure that deadlines are met. The implementation has to be done in the defined timeframe because the introduction of other changes is based on it. Therefore, the implementations must be synchronized, with constant communication between project manager and industrialization responsible. If there is not a good communication, a delay in a change can create a chain reaction of delays.

5.2 Development of ECR Status Tracking

The current status analysis and tracking of ECRs is very time-consuming and requires a lot of effort for the change representative to evaluate the status of each ECR and to perform the necessary activities, given its status. This creates the need to create a procedure or tool that can help and facilitate the work of the change representative. In this way, two different approaches will be presented so that change representatives can make a quick analysis and know immediately what action they should take regarding each ECR under their responsibility.

Preliminary Arrangements

The first required step is to sort out the different ECRs and group them in a way that is suitable for carrying out the analysis. For example, for a change representative, it would make sense to filter out only the ECRs for which he is responsible and those that are still in progress. In the company under study, this option was already available, but it was not used.

Approach 1

The first approach is to use the Microsoft Excel as a way to evaluate the deadlines for each ECR. After filtering all the data, the user can download an Excel file containing the information about the ECRs. With this data, the change representative or even the PM would be able to quickly perform an evaluation (Fig. 3) and see which ECRs need more attention. As shown in Fig. 3, an automatic algorithm can be used to assign a different color (red, yellow and green) according to deadline compliance (e.g. red identifies the ECRs that are overdue, yellow the ones that are less than one week from the deadline, and green the others). This ensures that every ECR is tracked, and none is forgotten or missed. In addition, this approach is relatively easy to apply, as there is already a good familiarity with the Excel tool.

ECR Number	ECR Title	Creation Date	ECR status	Process step	Deadline	Status	Comments
[illegible]	[illegible]	[illegible]	[illegible]	Customer Phase pending + Valid. P	[illegible]	■	Validation ongoing -> check activities w/ production
[illegible]	[illegible]	[illegible]	[illegible]	Change valid (with Customer Phase	[illegible]	■	implementation ongoing -> check activities w/ production
[illegible]	[illegible]	[illegible]	[illegible]	Formal Check pending + Customer	[illegible]	■	Formal check -> Trigger PDM
[illegible]	[illegible]	[illegible]	[illegible]	Formal Check pending + Customer	[illegible]	■	Formal check -> Trigger PDM
[illegible]	[illegible]	[illegible]	[illegible]	Planning Phase blocked for approva	[illegible]	▲	Needs approvals -> Push for approvals
[illegible]	[illegible]	[illegible]	[illegible]	Planning Phase blocked for approva	[illegible]	▲	Needs approvals -> Push for approvals
[illegible]	[illegible]	[illegible]	[illegible]	Pre-clarification Phase approved	[illegible]	●	ECR in a preparation stage -> Input information

Fig. 3. Excel table for the ongoing ECRs tracking

Approach 2

The second approach addresses the tracking of the ECRs using Microsoft Power BI. This tool would allow not only for a more graphical and interactive analysis (Fig. 4), where the user could easily search in detail, but it would also make it possible to capture an even greater volume of data and there would be direct integration with the company's ECM tool. While the primary objective is clear, the execution requires time to develop the necessary tools and structures capable of handling the substantial data volume while ensuring its validity. Initially, the user can still be responsible for exporting data from the ECM tool and importing it into Power BI.

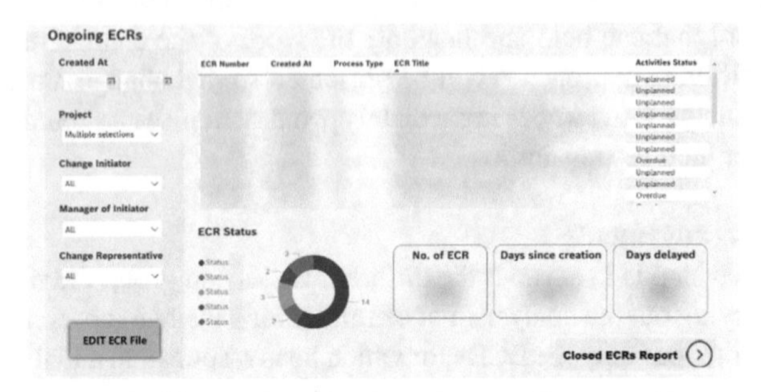

Fig. 4. Dashboard for ongoing ECRs tracking

This would already be a great improvement, since it would be possible to quickly have a report of the current status of open ECRs, including the list of delays, those that are almost reaching the deadline and the related information.

Although this may work as a preliminary solution, the real objective is to integrate the data and have a real time analysis in the Power BI platform. This would allow for this platform to work as a framework. Each person could easily see which were the pending tasks and the next steps. Even management could have a global report in terms of deadlines compliance and also in terms of total cost.

6 Discussion

This paper presents a review of the activities carried out along the different ECR phases, highlighting the different points to be initiated or completed in each phase. The importance of clearly defined activities is emphasized. By providing specific guidelines and recommendations, the clarity and structure of the overall process is enhanced. In addition, the proposed tools for tracking and managing multiple ECRs address a common challenge faced by project managers. These tools aim to improve the visibility and facilitate the analysis of a large number of ECRs, enabling project managers to efficiently monitor the status and progress of each ECR. This is in contrast to the existing literature, which often lacks detailed guidance on how to effectively manage a large number of ECRs.

However, it is important to acknowledge the limitations of this study. First, the research sample size was small because the model was developed based on the experience of a single company. Second, the specific culture of the organization may have influenced the implementation of the given suggestions, highlighting the need for validation across multiple companies. Future research should aim to validate the suggestions for other projects and organizations and assess their feasibility and appropriateness.

7 Conclusions

This paper has presented a review of the activities carried out along the different ECR phases, highlighting the different points to be initiated or completed in each one of the phases. In addition, the ECR's tracking approaches have been explored, given the current state of the ECM tool and the data available to be exported, but it is also pointed out the path that the company should follow to maximize the performance and benefits of the proposal. Both approaches would be an improvement and a way to ensure better compliance with the defined deadlines. Nevertheless, the second approach, if fully implemented with all the suggestions, can work as a framework for all the projects and all stakeholders to have a clear view of the status of the ECRs in real-time. It is expected that the application of the proposals will potentially reduce the time spent by the project manager on the activities, the overdue dates and ultimately the cost that the ECR process represents to the organization. The study contributed to the business process improvement literature and provided a new approach to address issues related to ECR tracking. The findings of this research can also be applied to other organizations facing similar change management and tracking challenges and may inspire future research in this area.

Acknowledgements. This work has been supported by FCT – Fundação para a Ciência e Tecnologia within the R&D Units Project Scope: UIDB/00319/2020.

References

1. Maceika, A., Toločka, E.: The motivation for engineering change in the industrial company. Bus. Theory Pract. **22**(1), 98–108 (2021). https://doi.org/10.3846/btp.2021.13042
2. Jarratt, T., Clarkson, J., Eckert, C.: Engineering change. In: Clarkson, J., Eckert, C. (eds.) Design Process Improvement: A Review of Current Practice, pp. 262–285. Springer, London (2010). https://doi.org/10.1007/978-1-84628-061-0_11
3. Perrotta, D., Faria, J., Araújo, M., Tereso, A., Fernandes, G.: Project Change Request: A Proposal for Managing Change in Industrialization Projects, pp. 1525–1529 (2017). https://doi.org/10.1109/IEEM.2017.8290148
4. Jarratt, T., Eckert, C.M., Caldwell, N.H., Clarkson, P.J.: Engineering change: an overview and perspective on the literature. Res. Eng. Design **22**, 103–124 (2011). https://doi.org/10.1007/s00163-010-0097-y
5. Wasmer, A., Staub, G., Vroom, R.W.: An industry approach to shared, cross-organisational engineering change handling - the road towards standards for product data processing. Computer-Aided Design **43**(5), 533–545 (2011). https://doi.org/10.1016/j.cad.2010.10.002.(Emerging Industry Needs for Frameworks and Technologies for Exchanging and Sharing Product Lifecycle Knowledge)
6. Wilberg, J., Elezi, F., Tommelein, I.D., Lindemann, U.: Using a systemic perspective to support engineering change management. Procedia Comput. Sci. **61**, 287–292 (2015). https://doi.org/10.1016/j.procs.2015.09.217. https://www.sciencedirect.com/science/article/pii/S1877050915030471 (Complex Adaptive Systems San Jose, CA November 2–4, 2015)
7. PMI. Managing Change in Organizations: A Practice Guide. Project Management Institute, Newtown Square (2013)
8. Eckert, C., Clarkson, P.J., Zanker, W.: Change and customisation in complex engineering domains. Res. Eng. Design **15**, 1–21 (2004). https://doi.org/10.1007/s00163-003-0031-7

9. Iakymenko, N., Romsdal, A., Semini, M., Strandhagen, J.O.: Engineering change management in the engineer-to-order production environment: insights from two case studies. In: Moon, I., Lee, G.M., Park, J., Kiritsis, D., von Cieminski, G. (eds.) APMS 2018. IAICT, vol. 535, pp. 131–138. Springer, Cham (2018). https://doi.org/10.1007/978-3-319-99704-9_17
10. PMI: A Guide to the Project Management Body of Knowledge (PMBOK®guide), 7th edn. Project Management Institute Inc., Newtown Square (2021)
11. Hamraz, B., Caldwell, N.H., Clarkson, P.J.: A holistic categorization framework for literature on engineering change management. Syst. Eng. **16**(4), 473–505 (2013). https://doi.org/10.1002/sys.21244
12. Iakymenko, N., Romsdal, A., Alfnes, E., Semini, M., Strandhagen, J.O.: Status of engineering change management in the engineer-to-order production environment: insights from a multiple case study. Int. J. Prod. Res. **58**(15), 4506–4528 (2020). https://doi.org/10.1080/00207543.2020.1759836
13. Hamraz, B., Clarkson, P.J.: Industrial evaluation of FBS linkage–a method to support engineering change management. J. Eng. Des. **26**(1–3), 24–47 (2015). https://doi.org/10.1080/09544828.2015.1015783
14. Habib, H., Menhas, R., McDermott, O.: Managing engineering change within the paradigm of product lifecycle management. Processes **10**(9), 1770 (2022). https://doi.org/10.3390/pr10091770
15. Saunders, M., Lewis, P., Thornhill, A.: Research Methods for Business Students, 6th edn. Pearson Education Limited (2012)
16. Yin, R.K.: Case Study Research and Applications: Design and Methods, 6th edn. SAGE Publications, Thousand Oaks (2018)
17. PMI: A Guide to the Project Management Body of Knowledge (PMBOK® guide), 6th edn. Project Management Institute Inc., Newtown Square (2017)
18. Cooper, R.G.: Perspective: the stage-gate® idea-to-launch process—update, what's new, and NexGen systems*. J. Prod. Innov. Manag. **25**(3), 213–232 (2008). https://doi.org/10.1111/j.1540-5885.2008.00296.x
19. Pan, Y., Stark, R.: An interpretable machine learning approach for engineering change management decision support in automotive industry. Comput. Ind. **138**, 103633 (2022)
20. Sabouni, C.E.: Improving Engineering Change Processes in a Local Forestry Machine Manufacturer. Savonia University of Applied Sciences, Finland (2023)

MITEC Model for Talent Management to Small and Medium Enterprises Develop with *Design Science Research*

Ricardo Perpétuo(✉) and Gabriel Pestana

Universidade Europeia, Campus da Quinta do Bom Nome, Estrada Correia 53, 1500-210 Lisboa, Portugal
{ricardo.perpetuo,gabriel.pestana}@universidadeeuropeia.pt

Abstract. This article seeks to explain how a model for talent management was built using the Design Science Research methodology to mitigate the risk of losing talent. The article tries to explain through the 6 stages of the methodology how the model was developed to monitor and analyse data on professional satisfaction, contributing data to organizations to identify patterns associated with risk situations (i.e., Talent Churn). An organization needs to have a mechanism to help professionals become aware of their status and monitor their progress, and at the same time it can help alert an organization to possible risk situations (e.g., loss of talent, lack of progress, signs of demotivation and dissatisfaction, recurrence of incidents, severity of alerts, lack of involvement such as professional commitment). An organization can act preventively to reduce possible losses of business or customers and proactively by using knowledge to create services or products. To do this, it is essential to monitor and mitigate the risk of losing Talent and adopt an interventionist approach to Human Capital. A limitation of the article is that the focus group was not carried out as the methodology recommends, and the future prospect is to carry out a pilot test of the model in a micro-enterprise or SME in Portugal in the services sector.

Keywords: MITEC Model · Talent Management · Design Science Research · Talent Retention · Small and Medium Enterprises

1 Introduction

According to the study Jobs of Tomorrow Mapping Opportunity in the New Economy [1] carried out by the World Economic Forum, the report on the future of work shows that Industry 4.0 needs Talent with skills, by 1) professionals with diverse competencies and with 2) a focus on lifelong learning. This transition implies the development of tools to monitor Talent.

Digital transformation implies an organizational effort to overhaul processes, services, and the traditional way of doing business. In addition to digitalization, it is necessary to review internal policies, assess needs and seek to address a higher level of customer satisfaction, especially in the online market in the services sector, it is important to create digital services and services [2].

Á. Rocha et al. (Eds.): WorldCIST 2024, LNNS 987, pp. 13–24, 2024.
https://doi.org/10.1007/978-3-031-60221-4_2

At the heart of all this is HC, and the question arises as to its impact on the business, the market, and the brand. Faced with competitiveness in the Small and Medium Enterprises (SME) sector and demands in the services market, an organization needs to create a strategic vision and have well-structured HC supported by Information and Communication Technology (ICT), which can become a differentiating factor in the Digital Transformation Era.

To draw up a strategic plan for SMEs in Portugal, [3] suggests that professionals get involved in the processes, in strategic decisions to innovate, apply knowledge and take advantage of Digital Transformation to create value, because they are fundamental in a Talent Value (TV) policy to increase satisfaction.

The challenge at the heart of this research is how to value talent - how can the organization implement policies to value talent by monitoring professional satisfaction to mitigate the risk of losing talent, which can be expressed through the following research question: "What Talent Management model can enhance and value the competitive advantage of an organization in Portugal?"

Taking into account the set of concerns reflected in the aforementioned studies, it is worth highlighting the need and relevance of: 1) have a talent value model that monitors HC satisfaction, 2) introduce new metrics to identify Talent, 3) have a tool that aggregates knowledge and generates alerts on the state of HC and 4) have a system to measure HC engagement, but also mitigate the risk of losing talent.

An innovative model was developed to support the valuation of talent centred on the professional, to help micro-enterprises and SMEs monitor the satisfaction of human capital, with the aim that the Monitoring Indicators of Talent Employees Career (MITEC) model can be adapted to any type of organization presented in the article [4]. The methodology used to build the model is Design Science Research (DSR).

2 Methodology: Design Science Research

For [5] Design Science is a method used to gather knowledge to conceptualize a solution or design a new artefact. [6] reinforces that Design Science involves science and design, using an approach to build an artifact by verifying its relationship through methods and techniques according to the context. Its purpose is to guide the conduct of scientific research for the construction of an artifact, helping to create new knowledge during this process [7].

The DSR is structured in 6 stages: 1) identification and motivation of the problem, 2) definition of the objectives for a solution, 3) design and development, 4) demonstration, 5) evaluation and 6) communication; with four possible entry points for a solution: a) problem, b) objective, c) design and development and d) customer or context [8].

According to [9], the following evaluation guidelines are used to conduct satisfactory research: 1) Design as an artifact, 2) Relevance of the problem, 3) Design evaluation, 4) Design contributions, 5) Research rigor, 6) Design as a research process and 7) Research communication.

For the evaluation of the artifact, the same authors suggest that an evaluation be carried out at each stage of the DSR to partially demonstrate the results, to verify that the research is achieving the proposed intentions [10], which are at the basis of the implementation of the DSR that we demonstrate below.

The authors [11] indicate five methods of evaluating the artifact: 1) Observational, 2) Analytical, 3) Experimental, 4) Test and 5) Descriptive through a case of a real study in the context of a Microenterprise and SME, however, another form of evaluation could be through Focus Group. The outcome of the research is intended to increase the effectiveness of the artifact and demonstrate reliability of the results.

3 Application Design Science Research for MITEC

Stage 1 - Identifying the Problem: This study seeks to identify and understand the context of the research problem and the construction of a model from the perspective of talent management (TM) for a Microenterprise and SME.

The first phase involved consulting and reading scientific articles, books, and dissertations on human resources (HR), where we surveyed the challenges in TM: challenges and costs for an organization, identifying Talent, the profile of a professional vs. Talent, tools for identifying a Professional or Talent and how to segment Talent.

We explored the concept of TM, which is made up of the phases Attract, Develop and Retain. In the Attract phase, we found that the Employer Branding concept is used by the scientific community as a strategy for organizations to develop actions in the market to attract talent to their organization, the Develop phase defines training actions and in the Retain phase what factors are used to retain the professional, concluding that there is no specific model for monitoring the valuation of talent.

In TM, we identified the bases of the concept, HR practices, their functions, and the skills profile of a HR professional and, finally, the HR Model for Talent Support, which showed an opportunity to explore the model into the concept of valuing talent, and the indicators of motivation, commitment, behaviour, and innovation.

To develop the model, the concepts used were: Employer Branding (EB), Gamification and Dynamic Capabilities (DC): 1) EB, to know what the Framework used in the TM cycle in the attraction Talent, 2) Gamification, which techniques or game mechanics have been tested in a real context and could be included in the application of the model in an organizational context to engage and motivate professionals and 3) DC this concept relates the terms: - knowledge, resources, capabilities, competencies, changing environments and culture, and since the loss of talent impacts the business and the loss of knowledge, we sought to innovate in HR with the addition of a term not explored in the literature, which enables HR to collect another type of information from professionals.

During this process, we found no articles linking the three concepts, which revealed an opportunity to explore them in terms of TM. We decided to carry out a systematic review of the concepts EB, Gamification and DC, to see what the latest suggestions were for developing the model.

In the second phase, we conducted a survey of international and national studies on the main indicators in Portugal, willingness to change companies, employment trends, reasons for dissatisfaction with the current job, the factors for a professional to stay or leave an organization, benefits by type of generation that professionals most prefer and levels of productivity in teleworking, concluding that the loss of talent has an impact on the business, through the loss of knowledge, value proposition, competitive advantage and customer relations.

We searched the web for interviews with HR directors to find out what their organization needs in terms of TM and what internal resources they have at their disposal to monitor and identify talent in their organization. We concluded that most of the microenterprise and SME organizations in Portugal are in a phase of transition to Digital Transformation and are more concerned with delivering value to their customers than with monitoring and evaluating the value of professionals.

We decided to interview 5 people in Lisbon (i.e. 3 people who work in a Microenterprise or SME and 2 from Large companies), in order to ask them about the current satisfaction assessment methods used in their organization, the data collection process, frequency and the difficulties experienced in HR, and we reconfirmed that there are differences in terms of the human, financial and technological resources that the two types use both internally (e.g., satisfaction monitoring) and externally (e.g., products, technology and services delivered to the customer).

This finding reinforced the decision as to which types of organization were the object of analysis in this study (i.e. Microenterprise and SME) and made it possible to define which professionals within these organizations we were going to collect data from. These interviews also served to present a small framework for the model to verify opinions and what opportunities for improvement could be implemented in it.

Stage 2 - Defining the Objectives: The general objective of the study was "To create a TM model that helps an organization monitor Talent satisfaction in order to mitigate the risk of losing Talent". We want to understand whether the Microenterprise and SME typologies in Lisbon's services sector identify which monitoring policies can have an impact on talent monitoring and satisfaction to mitigate the risk of talent loss (TL), and whether they recognize this challenge as an element that can impact strategy, knowledge, and competitive advantage in the market.

The starting question we sought to answer was presented in Chapter I "Which TM model can boost and enhance the competitive advantage of an organization in Portugal? " and specific objectives were defined for this question 1) To identify the interviewees' views on the subject, 2) What interest the organization has in implementing a monitoring model for the risk of losing Talent, 3) To pre-test the model in order to assess and validate the relevance of the model's dimensions and indicators and 4) Check what the relationship between the interviewees' views and the dimensions and indicators developed in the model, which seeks to help answer the starting question and the conceptual model, developed with the aim of testing the questions arising from the research, with the dimensions of the model and the interview questions.

Stage 3 - Design and Development: In design we define: 1) the concepts, 2) identify the participants who will contribute to the development of the model, 3) identify the procedures for construction and 4) verify the suggestions identified in the literature review for the construction of the model. In the development phase we consolidate the information used in the construction and gathering of requirements and metadata.

Explore the Concepts and Suggestions for the Model

We analysed the scientific concepts, EB, Gamification and CD - concepts, natures or techniques and the suggestions identified that will contribute to the realization of a conceptual model. EB aims to study in detail the structure that the model uses. EB involves promoting a clear vision of what makes the organization different and desirable as an employer for current and potential professionals [12, 13].

To involve Talent in the use of the model, we used Gamification, which, according to [14], makes it possible to influence and motivate human behaviour, make the work experience dynamic, and enhance the creation of participatory and committed communities in the organization. It is a human-centred solution, whose techniques and elements are based on games to promote professional involvement in the development of cooperation and knowledge-building activities and to help recognize patterns of demotivation.

DC can be defined as the routines or activities that operate together to grow, modify, or create capabilities [15]. We intend to identify the terms and concepts that can be added to HR to monitor Talent satisfaction to convey to the organization where it needs to strengthen its internal activities (e.g. communication, knowledge, innovation, and learning).

Participants in the Construction of the Model

The research was conducted by two main participants: the supervisor and the student. The advisor's role is to help build the study from a scientific point of view and the second is the one who carries it out. After constructing the fourth phase of the model, we interviewed 5 participants to give their views on the model and to make suggestions or improvements to the model.

Procedure for Building the Model

The model proposal is made within an interactive cycle of research in 5 stages: 1) search for information, 2) consolidate information and its relevance, 3) present the proposal and have it reviewed by the supervisor, 4) validate or redo it after the supervisor's intervention, 5) redo the previous stages until a proposal is consolidated. The aim is to make the process transparent and clear.

Each meeting between the mentor and the mentee is designed to organize what needs to be done during each phase of the model for HR managers, with intermediate cycles of information exchange and demonstration of the model's evolution. This cycle was repeated until the proposal was consolidated and presented to the interviewees to evaluate the final model.

The 7 Phases of Development the Model

Phase 1: We analysed the current state of the art in human resource management practices and the TM cycle, which is made up of attraction, development, and retention. The attraction phase uses EB, which uses the Employee Value Proposition to build a value proposition for professionals.

Phase 2: The HR Models of Collings & Mellahi and Engelman et al. reveal the opportunity to explore the concepts: Motivation, Commitment, Behaviour, and Innovation, both models did not present a descriptive method on the procedure to measure or evaluate the satisfaction of a Talent.

Phase 3: The first model proposal was based on knowledge from the literature review on HR, Talent, retention factors and TL. The model has five dimensions. At this stage of building the model, it was decided that HR practices would not be used because they are already explored, and the EB concept would only serve as a reference. The proposed model had 5 dimensions (i.e., Organization, Trust in Leadership, Health and Wellness, Meaning Work and Support Management).

Phase 4: We have finished reading the articles on the concepts mentioned in the previous phase, but we have changed the names of the dimensions to bring them more into line with the literature. We started reading the articles from the systematic literature review and began to include the suggestions referenced and added the indications from the Collings & Mellahi and Engelman model. At this stage we renamed the dimensions Purpose, Leadership, Health and Wellbeing, Digital Transformation and Support Management. The proposal presented has 5 dimensions (i.e., Purpose, Leadership, Health and Wellness, Digital Transformation and Support Management).

Phase 5: As we continued to read the articles in detail, it was decided that EB would contribute to the indicators, Gamification with the techniques to be used in the model, and to the indicators, while DCs would contribute to the dimensions and indicators. At this stage we identified that the literature refers to the importance of Engagement, so we decided to add new indicators in line with this concept. We restructured the model into a solution with 6 dimensions (i.e., Purpose, Digital Transformation, Health and Wellness, DC, Support Management and Engagement).

Phase 6: The proposed model at this stage is consolidated into 7 dimensions (i.e., Expectation, Organization, Agility, Digital Transformation, Learning, DC and Engagement). The first 6 dimensions made up of 7 indicators (i.e. we wanted to synchronize as many suggestions as possible with as many indicators as possible so that the monitoring process would gather the greatest level of information on the state of Talent through the various dimensions of analysis). The last dimension is made up of 5 indicators. The reason for having a smaller number of indicators is that when the term was identified as relevant it would be necessary to carry out a new literature review specifically on the term Engagement, to validate existing models, dimensions and indicators, limitations, and suggestions from the articles.

At this stage and after having a pre-consolidated version of the model, we carried out the 5 initial interviews to check the relevance of the model, to see if the nomenclatures were understood and what opportunities for improvement could be added to the model (i.e. dimensions and indicators). During this phase we developed the script for the interviews, the questions were framed with references in the literature, limitations, and the conclusions of the articles. We also started developing the questionnaire in parallel, because there were already going to be a few changes to the model, and we had to move on to the field research to assess the relevance of the study and the validation of the model.

Phase 7: The MITEC Model aims to position itself as a mechanism for monitoring the satisfaction and appreciation of Talent in an organization. The model has been consolidated into 1 strand, 7 dimensions and 47 indicators.

Model Metadata

In order to create the model, we began by trying to answer the following questions: what is the name of the aspect for assessing the level of satisfaction and appreciation of talent, what are the dimensions of analysis and indicators that contribute to collecting this information, what is the unit of measurement, what is the source of information (i.e., how to collect data from the professional), which category, which metric and how often to collect data from the talent, the weight (i.e. %) of the indicator for each dimension and the threshold (i.e. minimum, maximum and target value) to monitor the behaviour, satisfaction and appreciation of talent in the organization in order to mitigate the potential risk of loss.

Requirements and Rigor

We chose to develop the model because it allows us to aggregate the suggestions and concepts observed in the process of collecting and reviewing the literature and building the model. This construction sought to facilitate communication and establish a relationship between the target audience, the actor, and the language. This study can serve to guide new researchers in improving this model or adapting it to another context, such as the construction of DSR research.

Model Requirements

The requirements that the model must meet to be considered a satisfactory solution that has quality and can be useful to the target audience were collected.

During the construction process, we confirmed whether the model was fulfilling the defined requirements and what was missing for this solution to be considered satisfactory. The dimensions and indicators were readjusted according to the suggestions and opinions of the supervisor and the interviewees.

Cycle of Rigor and Relevance

A routine for reviewing information was established to make the process of systematizing the construction of the model easier. The model's criteria are shown in Table 2, indicating the type of result expected and its link to Table 1 to identify each criterion.

Stage 4 – Demonstration: The model was published in the article "Dynamics of Monitoring and Talent Management" at the WorldCIST 2022 conference [4], its organizational application for monitoring Human Capital and how to visualize the data presented through the dashboard, Self-awareness, self-regulation, and alert management.

Table 1. Requirements for building the model.

Requirement	The model prescribes
R01	Applied methodologies
R02	Participants in the conduct of research
R03	Vocabulary understandable for the user
R04	A procedure for constructing proposal
R05	Identification of concepts
R06	Collection of input from interviewees
R07	The model follows the identified suggestions as a guideline
R08	To the seven evaluation guidelines of the DSR method
R09	A solution that brings together the value of the work performed
R10	Real-time data
R11	Warning notifications and signs of demotivation
R12	Metadata Structure
R13	Model structure (i.e., aspect, dimensions, indicators)
R14	The actors who will use the model
R15	Application context description
R16	Organizational application (BPMN – data collection)
R17	Validation of model relevance

Stage 5 - Evaluation: The article Talent Analysis Model in Small and Medium Enterprises, which will be published at the ICITS 2024 conference, presents the results of the 40 interviews carried out with managers and directors to validate the model. In the process of evaluating the model, we were still experiencing the Covid19 pandemic, so we chose to use a mixed methods approach, a qualitative approach (i.e., interviews) and a quantitative approach (i.e., questionnaires). This approach uses a process of data collection, analysis, or a combination of quantitative and qualitative techniques in the same investigation and the interaction of the techniques provides better results.

Stage 6 – Communication: This study has already 2 publications, 1) Dynamics of Monitoring and Talent Management in 2022 [4], 2) MITEC Model Standardized Business Process (BPMN) for Talent Management in Small and Medium Enterprises in 2023 [16], 3) Talent Analysis Model in Small and Medium Enterprises, which will be published at the ICITS 2024, 4) this article itself, MITEC Model for Talent Management to Small and Medium Enterprises develop with DSR in 2024 and 5) The doctoral thesis in Universidade Eruropeia in Portugal.

Table 2. Rigor Cycle

Requirement	Rigor Cycle	Result
R01	DSR, SLR and Mixed Methods	Rigor + Contribution
R02	Participants related to the research	Requirement
R03	Easy-to-understand nomenclatures	Rigor
R04	Model building procedure	Requirement
R05	Concept covered by literature and interviewees	Rigor
R06	Seven DSR Criteria	Requirement + Rigor
R07	Participants' view	Contribution
R08	Definition of how the indicator is evaluated	Requirement
R08 + R12	Creation of indicator characterization metadata	Requirement
R08	What is the unit of measurement	Requirement
R08	What is the source of information for collecting data from professionals?	Requirement
R08	What is the category of the indicator?	Requirement
R08	What is the metric to evaluate the indicator	Requirement
R08	How frequently do you collect data on paper?	Requirement
R13	Model structure	Rigor
R13	Dimensions and Indicators	Rigor
R14	Model actors	Rigor
R15 + R10	Self-regulation techniques	Rigor
R15 + R10	Alert management for professionals	Rigor
R15 + R10	Gamification techniques	Rigor
R15 + R10	Self-awareness techniques	Rigor
R15 + R11	Hierarchical alert notifications	Rigor
R16	Organizational Application Process	Rigor
R17 + R09	Relevance of the model	Contribution + Requirement + Rigor

4 Design Science Research Conclusion

For the Criteria for evaluating the model, it is important to define the evaluation guidelines to verify that the solution presented was successful during its construction. It is not important to evaluate the efficiency of the model, nor the implementation rate or the acceptance rate, however this evaluation can be carried out in the future, now, the intention is to delimit the investigation and ensure that for the problem identified the model presents a satisfactory solution.

Design as a Model: The model developed is one that monitors the satisfaction and appreciation of Talent to mitigate the risk of loss. A survey of concepts was carried out to build the dimensions and indicators that will serve to evaluate the aspect under analysis. The model includes the vision or idea of each of the suggestions, which represents the work and value involved in the proposal, as well as the transparency of the process in identifying what is useful for the model.

Relevance of the Problem: 1) Attracting and retaining talent will be one of the biggest challenges an organization will face, 2) There was no specific model for monitoring talent development, 3) There are no studies linking talent development with CD, Gamification and EB, 4) There is a lack of studies on TL on TM and Talent Retention, 5) In the 5 initial interviews, 7 difficulties were detected and we present the 3 most relevant ones: - Organization with limited human and financial resources, - No Talent monitoring policy and - Lack of people and tools to motivate Talent and support HR Management, 6) The initial interviewees showed interest in implementing the model in their organization, 7) The loss of Talent has an impact on the business and for the organization through increased operating costs.

Evaluation of the Design: The model was built using the theoretical framework, with the interaction of the advisor and the opinion of the 5 interviewees. To evaluate the model, a study was carried out with 40 interviewees, answering a questionnaire and surveys to evaluate each dimension and indicator using a Likert scale [1…5].

Contribution of the Research: 1) Model built based on the literature review, 2) Dimensions and indicators to monitor and mitigate the risk of Talent Churn, 3) Gamification system to motivate and involve the user, 4) Definition of metadata for the functionality of the model, 5) BPMN for MITEC model in a organizational is structured into 3 macro-processes: 1) Employee Induction Process, 2) Employee Management Process and 3) TM Process [16], 6) Alert management, which can be further explored with the use of the model in a real context, 7) Creation of a list of automations for the model, 8) It serves as a basis for continuing to explore new aspects for monitoring Talent, 9) Model with the capacity to have data available in real time in order to mitigate the loss of Talent by being a support mechanism for TM and data collection (i. e., data collection). e., listening to the needs of professionals).

Rigor of the Research: The model is made up of 1 strand, 7 dimensions: Talent Expectation, Corporate Purpose, Organizational Agility, Digital Transformation, Organizational Learning, Dynamic Capability, Employee Engagement, adding 47 indicators to help assess Talent satisfaction. It was necessary to be more rigorous in building the model, describing the techniques to apply in an organization, i.e. gamification techniques, self-awareness, self-regulation and graphic representation, data collection technique, evaluation process, assigning a scale and drawing up a set of questions to evaluate the indicator.

Design as a Search Process: This model is not a continuation of a previous study, but rather consolidates the construction of new knowledge and encourages the actors to validate this model quantitatively and qualitatively. The design of the model required the use of scientific means (i.e. DSR, traditional research, systematic review, field study and mixed methods), the opinion of the advisor and the vision of the interviewees.

Communication of the Research: The doctoral thesis was submitted to the European University in Portugal.

5 Conclusion and Future Work

Conclusion

To respond to the challenge, this research generated MITEC. Made up of strands, dimensions of analysis and indicators that focus on monitoring events that contribute to investigating professional satisfaction to mitigate the risk of Talent Churn.

MITEC proposes certain criteria to demonstrate rigor, relevance, and contributions throughout the research. The construction of the model respected and complied with the requirements in the model evaluation criteria, presented in Table 2 to help the researcher review the results throughout the construction, inspecting and highlighting the rigor of the work carried out. A survey of the contextualization of the problem was carried out, in which the focus was on deepening the existing knowledge in the literature on Micro-enterprises and SMEs in Portugal, models, suggestions, problems identified in this type of organization and delimiting the research, extracting its relevance, because the creation of a Talent Enhancement model does not guarantee the reinforcement of its strategic position in relation to the competition, but the implementation of the model can reinforce satisfaction and recognize progress at work.

Limitation

The pandemic had an impact on the carrying out of the study, given that the limitation of people's movement had a direct influence on the Focus group, in order to assess the quality and validate the relevance of the model as indicated by the DSR, it would have been useful to carry out a brainstorming and listen to the opinion of 10 experts to verify points for improvement in the model, in terms of monitoring, dimensions and indicators.

Future Work

Study the implementation of new aspects, dimensions, and indicators, carry out the study in another geographical area in Portugal, carry out the study of professionals and other professional sectors in Portugal.

References

1. World Economic Forum. Jobs of Tomorrow Mapping Opportunity in the New Economy. Consultado em Fevereiro de 2023 (2020)
2. Mergel, I., Edelmann, N., Haug, N.: Defining digital transformation: results from expert interviews. Gov. Inf. Quart. **36** (2019)
3. Lopes, M.: Retenção de talentos: um estudo de caso em uma empresa de telecomunicações. UniCEUB - Centro Universitário de Brasília, pp. 1–27 (2017). Disponível em https://repositorio.uniceub.br/jspui/handle/235/11393

4. Perpétuo, R., Pestana, G.: Dynamics of monitoring and talent management. In: Rocha, A., Adeli, H., Dzemyda, G., Moreira, F. (eds.) Information Systems and Technologies. WorldCIST 2022. LNCS, vol. 469. Springer, Cham (2022). https://doi.org/10.1007/978-3-031-04819-7_1
5. Van Aken, J.E.: Management research based on the paradigm of the design sciences: the quest for field-tested and grounded technological rules. J. Manage. Stud. **41**(2), 219–246 (2004). https://doi.org/10.1111/j.1467-6486.2004.00430.x
6. Wieringa, R.J.: Design Science Methodology for Information Systems and Software Engineering. Springer, Heidelberg (2014). https://doi.org/10.1007/978-3-662-43839-8
7. Bax, M.P.: Design Science: filosofia da pesquisa em ciência da informação e tecnologia. In: XV ENANCIB (Encontro Nacional de Pesquisa em Ciência da Informação) "Além das Nuvens: Expandindo as Fronteiras da Ciência da Informação" (2014)
8. Peffers, K.E.N., Tuunanen, T., Rothenberger, M.A.M.A., Chatterjee, S.: A design science research methodology for information systems research. J. Manag. Inf. Syst. **24**(3), 45–77 (2007)
9. Lacerda, D.P., Dresch, A., Proença, A., Antunes, J.A.V., Jr.: Design Science Research: Método de Pesquisa Para a Engenharia de Produção. In: Gestão & Produção, vol. 20, pp. 741–761 (2013). http://www.scielo.br/pdf/gp/v20n4/aop_gp031412.pdf
10. Carcary, M.: Design science research: the case of the IT capability maturity framework (IT CMF). Electron. J. Bus. Res. Methods **9**(2), 109–118 (2011)
11. Hevner, A.R., Chatterjee, S.: Design research in information systems. Design Res. Inf. Syst. **28**, 75–105 (2004). http://desrist.org/design-research-in-information-systems
12. Backhaus, K.: Employer branding revisited. Organiz. Manag. J. (2016). https://doi.org/10.1080/15416518.2016.1245128
13. Backhaus, K., Tikoo, S.: Conceptualizing and researching employer branding. Career Dev. Int. **9**(5), 501–517 (2004)
14. Wortley, D.: Gamification and geospatial health management. IOP Conf. Ser.: Earth Environ. Sci. **20**(1), 012039 (2014). https://doi.org/10.1088/1755-1315/20/1/012039
15. Winter, S.G.: Understanding dynamic capabilities. Strateg. Manag. J. **24**(10), 991–995 (2003)
16. Perpétuo, R., Pestana, G.: MITEC Model Standardized Business Process (BPMN) for Talent Management in Small and Medium Enterprises. In: 2023 18th Iberian Conference on Information Systems and Technologies (CISTI), Aveiro, pp. 1–6 (2023). https://doi.org/10.23919/CISTI58278.2023.10211981

Proposal to Improve Risk Management in New Product Development Projects in an Automotive Company

Inês Laranjeiro[1,2], Anabela Tereso[3](✉), and João Faria[3]

[1] Master's in Industrial Engineering and Management, University of Minho, Guimarães, Portugal
Ines.Laranjeiro@pt.bosch.com
[2] Bosch Car Multimedia Portugal, S.A., Rua Max Grundig, Braga, Portugal
[3] ALGORITMI Research Centre/LASI, University of Minho, 4710-057 Guimarães, Portugal
anabelat@dps.uminho.pt, joao.faria@algoritmi.uminho.pt

Abstract. Risk Management is a crucial domain within Project Management for organizations, and it plays a fundamental role in the automotive sector, ensuring a competitive position in today's dynamic and constantly evolving environment. This research was carried out at Bosch Car Multimedia Portugal, with the main objective of analyzing and identifying areas for improvement in Risk Management practices in New Product Development projects of the company. As a result of the study, several measures were proposed to enhance and optimize the Risk Management process. The research strategy employed was a case study, incorporating various techniques for gathering the necessary information for both the theoretical foundation and the analysis of the current situation in the company, for the project under analysis. The research aimed to foster a culture of Risk Management within the project team, considering their needs, limitations, and expectations. It is important to promote Risk Management routines, ensure team involvement, and implement good practices that make the entire process effective, practical, and smooth. Following the comparative analysis between the observed reality and the theory, a model approach for Risk Management, tailored to the specific project, was proposed. Additionally, other suggestions were presented to facilitate and enhance the implementation of Risk Management practices within the company, as well as to broaden the overall knowledge of teams on the subject matter. The importance of cooperative and proactive work, the promotion of a risk-aware culture, and continuous investment in process improvement and team dynamics were highlighted.

Keywords: Project Management · Risk Management · New Product Development

1 Introduction

Globalization and increasing competitiveness are introducing new challenges that organizations must overcome by being innovative and introducing new ideas and projects. In the automotive industry, such as Bosch Group, the challenges of staying competitive

Á. Rocha et al. (Eds.): WorldCIST 2024, LNNS 987, pp. 25–34, 2024.
https://doi.org/10.1007/978-3-031-60221-4_3

are intensified by the fast evolution of Industry 4.0 technologies and the need to keep up with the demands. As such, it is critical for organizations to manage their projects and risks effectively to achieve their strategic goals and maintain a competitive edge [1, 2].

Risk Management (RM), as an integral part of PM, is crucial and plays a vital role in every project's success. For RM to be more effective, it should be ingrained in the organization's culture, philosophy, practices, and business processes rather than being treated as a separate task. When this is accomplished, all members of the organization actively participate in managing project risks. Taking a strategic approach to Risk Management also acknowledges the importance of implementing advanced practices across all stages of the decision-making process [3].

The study aimed to identify the challenges and opportunities of RM processes in the project under analysis and provide insights and recommendations that can assist in developing effective strategies for managing risks. By achieving these objectives, this study intended to contribute to the enhancement of RM practices in NPD projects within the context of an automotive company. The research approach utilized in this investigation was based on the case study methodology.

After this introduction, a theoretical background of the research topics under study is presented. Subsequently, the research methodology is outlined, followed by a chapter describing the current reality of Risk Management practices and routines i.e. the research environment. The succeeding chapter describes the structured framework and measures that were proposed to improve the RM process. Finally, the last chapter delineates the conclusions of the overall study and gives suggestions for further work.

2 Theoretical Background

2.1 Project Management

Project Management (PM) can be defined as a structured process by which an individual or group vision can be successfully converted into reality. PM is also commonly presented as the application of knowledge, methods, skills, tools, competencies, and techniques to achieve the project goals [4].

The PMBOK® Guide, "A Guide to the Project Management Body of Knowledge", published by the Project Management Institute (PMI), is an internationally recognized standard that contains guidelines and fundamentals of project management, that apply to a wide range of projects.

The PMI defines a project as an endeavor with a defined beginning and an end, that is meant to create a certain outcome [4]. Every project must be organized and managed in a certain way so that it is successfully performed and delivered within specific constraints.

Project Life Cycle (PLC)

The PMI has defined a standard PLC that includes the sequence of stages that a project goes through from its inception to its completion [5]. Each process group involves a specific set of activities that are performed by the project team, stakeholders, and other relevant parties to ensure that the project is successfully completed within the three key constraints: budget, time, and scope. The five stages of the PLC are the following: Initiating; Planning; Executing; Monitoring and Controlling; and Closing.

Knowledge Areas
The PMI has divided Project Management into ten knowledge areas, one of them the Risk Management. All the knowledge areas are intended to be integrated into most projects, most of the time (during all the PCL) [5, 6]. Each area is dedicated to a different subject and provides a structured framework for PM that can be adapted to various types of different projects, industries, and organizations. The project management knowledge areas are the following: Risk; Stakeholders; Procurement; Communication; Resources; Quality; Cost; Schedule; Scope and Integration [5].

Performance Domains
In a more recent version of the PMBOK® Guide, eight performance domains emerged as an alternative to the previously mentioned knowledge areas. These domains are based on twelve principles of PM, that serve as a guide for each domain's operations. The performance domains are the following: Stakeholder; Team; Development Approach and Life Cycle; Planning; Project Work; Delivery; Measurement and Uncertainty [4].

2.2 Risk Management

Risk Management (RM) is an essential component of a successful project, especially for New Product Development, as an NPD project involves a complex and challenging process that carries a high level of uncertainty [3, 7, 8].

It is important to highlight the distinction between positive risks, also called opportunities, and negative risks, referred to as threats. The main purpose of Risk Management is to minimize the negative impact of risks while maximizing the opportunities that may arise from them [4].

The 6th and 7th editions of the PMBOK® Guide provide a comprehensive and structured approach to RM, that can be applied to identify, analyze, and manage risks throughout the PLC. This approach has several phases; however, these sets of activities must be considered cyclical and interactive [5, 6].

The first RM process is Risk Management Planning, where the framework, procedures, routines, and objectives of the process are scoped and established. The Risk Management Plan (RMP) is the main output of this process, which outlines how the RM activities will be conducted throughout the PLC.

The Risk Identification process follows this initial planning phase and involves the identification of potential risks that could affect the project objectives. Following this, a thorough analysis of these risks is conducted in the Risk Analysis process, which can be further categorized into Qualitative or Quantitative approaches.

After this assessment, the next step should be the Risk Response Planning. In this process, the project team develops strategies and plans to respond to identified risks effectively. Once the response plans are in place, the Risk Response Implementation process follows, where the project team executes the planned risk responses.

Finally, there is the Risk Monitoring process, whose purpose is to continually monitor the identified risks and the effectiveness of the implemented risk responses [5].

As suggested in Fig. 1, this RM framework involves a continuous cycle of identifying risks, analyzing them, developing, and implementing response plans, and monitoring risks throughout the PLC. Plus, each iteration of the cycle involves improving the previous one [5].

Fig. 1. Risk Management processes. Adapted from [6]

3 Research Methodology

The underlying strategy for the research methodology employed in this project is the Case Study. This strategy enables a comprehensive understanding of the current situation, providing the researcher with an opportunity to suggest new practices. This research work is Cross-Sectional, as the topics addressed were studied and developed during a specific period, from February 2023 to July 2023.

In the development of the research, Mixed Methods were applied. Therefore, the researcher was able to gather both qualitative and quantitative data, which provided a more comprehensive understanding of the research problem and allowed for triangulation of the findings.

The Deductive approach was chosen to conduct the present research, which involves starting with a theory or set of theories and using them to generate predictions or hypotheses about what might be observed in the real world. This approach process involved analyzing the existing theories and literature related to Risk Management, relating it with the observations, and developing proposals for improvement when necessary. The adopted research philosophy was Interpretivism. This philosophy justifies the researcher's view of reality as subjective, subject to multiple interpretations and meanings. The researcher collected and analyzed both primary and secondary data. This combination of data sources allowed the researcher to gain a more comprehensive understanding of the research problem.

Internal documentation of the company was consulted, as well as external documentation, especially highlighting the PMBOK® Guide. In this case study, the researcher participated in the activities of the team under observation, being daily in touch with the Project Manager and the other team members of the project. Primary quantitative data were extracted, mainly in order to assess the overall risk profile of the project.

Furthermore, a questionnaire was developed to gather feedback on the status and relationship between the team and the project's Risk Management process. Finally, flexible interviews were conducted individually with experts in the field of PM and RM to better understand the standard processes within the company.

4 Case Study

The study being presented is centered on a project aimed at the development of a new product within the Automotive Electronics division of the company. Specifically, the project focuses on creating an advanced instrument cluster for motorcycles.

During the development of the study, the project being analyzed was in its initial stages, starting the production of its first samples. Moreover, the series production for the customer was expected within a two-year timeframe.

The product development lead team was structured into domains, all being coordinated by the overall Project Manager. Each domain has its own team and Sub-Project Manager, and it is important to note that these teams were geographically dispersed across different countries and continents. The core team for this project, as represented in Fig. 2, is composed of the following domains: Hardware; Software; Mechanics; Display; Plant; Purchase; Requirements; and Validation. The primary interface and source of information for the overall Project Manager are the Sub-project Managers of each domain.

This study focuses on the team responsible for delivering the Software domain managed by the Sub-project Manager for this domain that will be referred to as the "Project Manager".

Fig. 2. Project organization.

4.1 Current Risk Management Plan

During the initial stages of a project, characterized by numerous risks and uncertainties, it is crucial to address these challenges. Therefore, it was expected that the Risk Management Plan (RMP) would already be in progress to ensure a comprehensive RM approach. However, it was observed that Risk Management practices were still in a preliminary development state, with routines and activities still being defined, tested, and encouraged in the team's dynamics by the Project Manager. Given this lack of maturity in the RM process, the team displayed a relaxed approach and lacked a proactive attitude toward risk identification, analysis, and control.

The complete team (9 members) gathers once a week for a 1.5-h online meeting moderated by the Project Manager. During this weekly meeting, he is expected to conduct a follow-up on project risks with the team. However, due to the limited time available and the multitude of topics and issues brought up during the meeting, it was observed that the update and discussion of risks were seldom addressed. Moreover, it was usual for the Project Manager to document the identified risks during the meeting directly in the meeting minutes, which were screen-shared as a reference for discussion topics. Due to time constraints, the Risk List was not updated immediately during the meeting. This practice could potentially result in overlooked risks and loss of tracking over time, as they were not recorded in the appropriate location.

The company had an internal software designed for Risk Management.

The use of this tool starts once a risk is identified, meaning it needs to be documented in the project's Risk List. To input the risk in the platform, several fields can be used, such as the risk type, the causes, the rating for severity and impact, costs, and expected date of occurrence, among other information. Measures can be added to each risk. Each measure will be automatically added to the tasks open point list in the software, as for example, an activity to be completed by a certain responsible.

With these inputs, a Risk List is generated. Also, a Butterfly Risk Matrix is automatically mapped, giving a visual representation of the risks.

It is important to note that the ideal situation is for all team members to keep their own risks and measures updated, meet deadlines, and complete their risk-related activities. This proactivity will help to ensure project risks are being handled successfully and effectively.

When the current Risk List of the project was analyzed, several data were considered to evaluate the maturity of the current RM process.

First, it was noted that 100% of the identified risks were threats, indicating that the team is not giving that much importance when it comes to identifying opportunities. More than 42% of the risks were overdue, meaning that the expected risk event day had already passed, and no update was provided regarding the risk status. Approximately 75% of the identified risks had no created measures, inferring that the identified risks may not be receiving adequate attention. Observing most of the risk statements, the causal relationship between the risk and its potential impact was not explicitly stated, leading to an ambiguous risk description that may not be easily comprehensible to all members of the team.

Furthermore, it was noticed that some items on the list were not actually risks, but rather problems or situations that had already occurred.

With regard to risk analysis, it was recognized that no risk had been subject to quantitative analysis, however, some had information regarding probability and impact. This qualitative assessment was inconsistent since there were no established criteria for assessing the risks, which led each person to rely solely on their own perspective. This lack of a standardized approach results in varying levels of risk identification and prioritization among team members.

Finally, it was observed that over 80% of the risks were created by the Project Manager, and not by the risk responsible, which once again highlights the lack of proactivity of the team on this matter.

4.2 Team Maturity in Risk Management

To go further into detail about the team's perception and awareness regarding Risk Management, a questionnaire was distributed to the team. By gathering the obtained results from the questionnaire and considering the researcher's daily participation in project work and routines, further conclusions were drawn regarding the maturity of the RM process.

Many team members lacked a clear understanding of the concept of risk. Only 40% of the members recognized that opportunities can also pose risks, while approximately 30% mistakenly identified problems as risks. This demonstrates a significant misconception within the team. Furthermore, a concerning 70% of the team reported that they reviewed the Risk List monthly, which is a lengthy interval. This infrequent check raises concerns about the team's ability to effectively manage and address potential risks in a timely manner.

5 Proposal of Improvements

Regarding all the information that was acknowledged, the priority was to elaborate an efficient Risk Management Plan (RMP), to define a framework to ensure the team is successfully addressing risks in the project. This framework outlined several aspects: Methodology and Strategy; Roles and Responsibilities; Budgeting; Scheduling and Timing; Scoring; Categorization; Formats and Templates; and Tracing [9].

First, a workshop was designed and proposed to be performed with the team, to address the lack of knowledge on the RM domain. The session should provide a comprehensive overview of the key concepts within each step and highlight some of the best practices for effective RM that were proposed. The session also provided training on the RM software. This training was meant to equip team members with the necessary skills and knowledge to use the tool effectively and efficiently.

At the time, addressing risk-related topics was usually insufficient or forgotten, so an agenda for RM meetings was proposed to ensure risks were duly considered and followed up on. Initially, a dedicated meeting was proposed specifically to address risks. After the team is used to the RM process, extending the duration of the regular team meeting by twenty minutes is suggested as considered sufficient for RM.

When creating a risk on the RM tool, a standard guide of necessary information to input was established. Firstly, a standard format to describe and document risks was proposed to avoid ambiguity, with the "If…Then…" format, as shown in Fig. 3, which clearly mentions the trigger and the consequence of the risk. Furthermore, it has been determined that risks should never be identified without corresponding measures being developed. By implementing this practice, the team can ensure that risks are not merely acknowledged but are actively addressed with concrete measures in place to handle their potential impact.

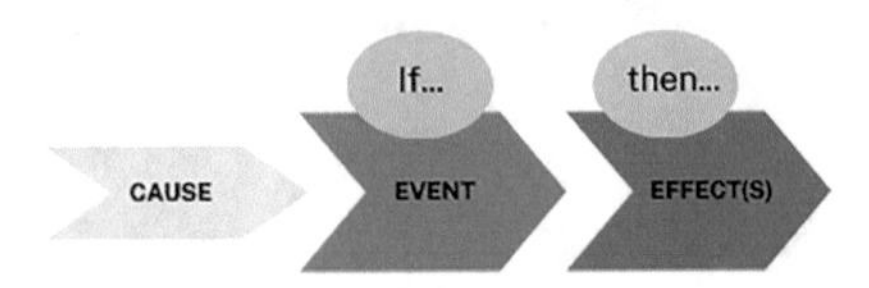

Fig. 3. Risk formulation representation.

The creation of *tags* was also suggested, to facilitate the filtering of risks and the risk list navigation in the software. A *tag* for each core team department was suggested, as well as an "ALERT" *tag* to add to risks rated as high or very high. Each risk can have more than one *tag*, and the purpose is to remark the main affected parties by associating its identification with the risk.

A standard practice for both qualitative and quantitative risk analysis was proposed to ensure these assessments were being carried out. Regarding qualitative analysis, a tailored guideline has been developed to rate the impact and probability of risks based on specific project criteria. This guideline improves assessment accuracy, minimizes ambiguity, and aligns different risk perspectives. Furthermore, it has been established that the risk assessment should involve a minimum of three stakeholders. By involving a diverse group of stakeholders, the assessment becomes more robust, comprehensive, and less reliant on individual biases or limited viewpoints.

To introduce quantitative assessment into the project, it has been proposed to schedule a meeting dedicated to this topic. The meeting will involve the participation of the Sales and Purchasing teams, along with other relevant stakeholders of the risks under analysis. The quantitative assessment was proposed to be mandatory for risks labeled with the "ALERT" *tag*. Due to the size of the company, it is often challenging or time-consuming to obtain costs associated with specific risks or potential response actions. To address the difficulty of this process, it is proposed to develop a user-friendly tool that enables quick access to cost-based risk analysis. This tool would enable stakeholders from various departments to easily input and access relevant cost information, supporting a comprehensive understanding of the financial implications associated with different risks.

A risk expert was suggested to be incorporated into the project to provide some support on RM activities, especially in this initial phase. Moreover, it was suggested to document lessons learned and problem-solving approaches to the faced risks to promote a culture of learning and collaboration within the organization.

6 Conclusions and Future Research

The primary objective of this case study was to provide recommendations for enhancing Risk Management in New Product Development projects within the specific context of an automotive company. Following an extensive study on the subject, challenges, and opportunities in the RM processes of the analyzed project were investigated and identified. Based on these findings, insights, and recommendations were proposed to improve the current project RM practices.

Through the conducted study, a significant maturity gap in the project under investigation was perceived, which was also evident in most NPD projects across the company. Referring to the project under analysis, several weaknesses were identified in the current RM process, as well as in the team dynamics and mindset. The process was unclear and not customized, leading the team to neglect the proper importance and attention to risks. Therefore, the proposed solutions aim to address these gaps and enhance the RM process, not only in this project but also potentially in other projects within the company.

Several opportunities for improvement were identified and structured with the intention of being implemented in the project.

This case study makes a contribution to the existing literature on Risk Management in New Product Development projects, particularly within the automotive industry. By conducting a thorough analysis of challenges, opportunities, and recommendations specific to an automotive company, this research not only provides practical implications for the industry but also expands the knowledge base in this field. The insights gained from this study offer valuable contributions to future research endeavors and the continuous improvement of Risk Management practices in project management.

The present study has addressed a highly relevant topic in the context of Project Management in the automotive industry, laying the groundwork for future investigations and opportunities for further studies. The findings and insights gained from this research can serve as a foundation for implementing standardized and customized Risk Management practices, not only in New Product Development projects in Bosch Braga but also across the global Bosch Group. By ensuring a consistent approach to RM across projects, industries can enhance their capabilities, improve decision-making processes, and ultimately achieve better project outcomes.

In summary, the proposals and achievements outlined in this case study present a road map for further enhancing Risk Management practices within automotive industries, emphasizing the importance of leadership support, fostering a risk-aware culture, and pursuing continuous improvements in the process. By building upon this research, the organization can continue to strengthen its RM framework and drive success in future projects.

Acknowledgment. This work has been supported by FCT – *Fundação para a Ciência e Tecnologia* within the R&D Units Project Scope: UIDB/00319/2020.

References

1. Kerzner, H.: Project Management Best Practices: Achieving Global Excellence. Wiley (2018) (9781119468851)

2. Lin, D., Lee, C.K., Lau, H., Yang, Y.: Strategic response to industry 4.0: an empirical investigation on the Chinese automotive industry. Indust. Manag. Data Syst. **118**(3), 589–605 (2018). https://doi.org/10.1108/IMDS-09-2017-0403
3. Rodrigues, B., Tereso, A., Araújo, M., Faria, J., Assis, P.: Improving risk management practice in industrialization projects: case study of an automotive company. In: 4th International Conference on Production Economics and Project Evaluation, pp. 205–211 (2018)
4. PMI. A Guide to the Project Management Body of Knowledge (PMBOK® guide), 7th edn. The Standard for Project Management (English). Project Management Institute (2021) (9781628256659)
5. PMI. A Guide to the Project Management Body of Knowledge (PMBOK® guide), 6th edn. Project Management Institute, Newtown Square (2017) (9781628251845)
6. Miguel, A.: Gestão Moderna de Projetos - Melhores Técnicas e Práticas, 8th edn. FCA (2019) (9789727228881)
7. Fernandes, I., de Sa-Soares, F., Tereso, A.: Modelling and analysing product development processes in the textile and clothing industry. Innov. Indust. Eng. **II**, 220–233 (2022). https://doi.org/10.1007/978-3-031-09360-9_19
8. Smolnik, T., Bergmann, T.: Structuring and managing the new product development process review on the evolution of the stage-gate® process. J. Bus. Chem. (2), 41–57 (2020). https://doi.org/10.17879/22139478907
9. Keshk, A.M., Maarouf, I., Annany, Y.: Special studies in management of construction project risks, risk concept, plan building, risk quantitative and qualitative analysis, risk response strategies. Alexandria Eng. J. **57**(4), 3179–3187 (2018). https://doi.org/10.1016/j.aej.2017.12.003

Limitations that Hinder the Successful Application of a Knowledge Management Model in Research Centers in a Higher Education Institution

Verónica Martínez-Lazcano[1](✉), Claudia Isabel Martínez-Alcalá[2], Iliana Castillo-Pérez[1], and Edgar Alexander Prieto-Barboza[3]

[1] Universidad Autónoma del Estado de Hidalgo, Pachuca, Hidalgo, México
{vlazcano,ilianac}@uaeh.edu.mx

[2] Consejo Nacional de Ciencia y Tecnología/Universidad Autónoma del Estado de Hidalgo, Pachuca de Soto, Hidalgo, México
cimartinezal@conacyt.mx

[3] Humboldt International University, Miami, FL, USA
prieto.barbosa@humboldtiu.com

Abstract. Knowledge management plays an important role in both organizations and higher education institutions, as they generate knowledge and through the implementation of a knowledge management model, it allows new knowledge to be generated. However, there are organizational and technological limitations or difficulties that may arise during the application of a knowledge management model in research centers of higher education institutions. The methodology of this research consisted of the application of a questionnaire with predefined questions using a Likert scale. The reliability analysis of the instrument was carried out using Cronbach's alpha coefficient applied to the identified limiting factors, obtaining a value of 0.93. In this study, seven organizational and technological limitations that arise during the application of a knowledge management model in higher education institutions were analyzed. The results identified areas of opportunity in five of the seven limitations to successfully implement a knowledge management model in research centers of higher education institutions.

Keywords: Knowledge Management Model · Organizational limitations · Technological limitations

1 Introduction

Knowledge is a key asset because it governs the way in which organizations function, and they understand the importance of its management, but it is not only important for organizations but also for educational institutions, so they consider it necessary to know the process of creating, storing, applying, disseminating, or transmitting and using knowledge [1, 2].

Á. Rocha et al. (Eds.): WorldCIST 2024, LNNS 987, pp. 35–44, 2024.
https://doi.org/10.1007/978-3-031-60221-4_4

The challenge established by UNESCO (United Nations Educational, Scientific and Cultural Organization) in which it refers to the goal of education for life in human beings, it is important to integrate knowledge management (KM), technology, science and above all, innovation in higher education, which is why Higher Education Institutions (HEIs) have been concerned with the creation of knowledge. Since they generate knowledge day by day, so their management allows them to renew their educational programs, training processes and teaching practices in accordance with institutional objectives, to be competitive and meet the needs of society in this changing and innovative world [1].

There are several concepts of knowledge management, so it can be mentioned that knowledge management is a process that allows finding, generating, collecting, storing, taking advantage of, and disseminating knowledge, as a production factor that allows organizations to achieve success, so the organization must achieve its effectiveness [3].

In this way, [4] they refer to knowledge management as an object that can be quantified, preserved, and transferred by technology, as well as knowledge as something that is in the brain and that can be shared through interaction, but it is also a capital whose objective is to improve performance and innovation, as it is considered as a competitive advantage.

Likewise, KM is a fundamental pillar in any institution, and its implementation in research centers in higher education institutions acquires crucial relevance. This research focuses on identifying the possible limitations that may arise in the application of a knowledge management model (KMM) in this specific context. In this sense, the main objective that guides this study is: What are the obstacles or limitations that arise when implementing a knowledge management model in research centers of a higher education institution?

After an analysis of organizational and technological limitations or difficulties that may arise during the application of a KMM in research centers of higher education institutions, seven were selected, which are: Absence of objectives, Lack of planning, Diffuse responsibility, Contextualization, Conceptual confusion, Lack of knowledge transfer culture and Lack of adequate technological infrastructure [5, 6]; several factors must be addressed in order to be successful in the application of the KMM.

2 Literature Review

For the development of this research, various sources of information [1–14] were consulted, mainly related to the issue of limitations that may arise in the application or implementation of a knowledge management model in higher education institutions. Among these, the research [7] entitled Mapping knowledge risks: towards a better understanding of knowledge management stands out, which classifies the drawbacks of knowledge at the organizational level into three categories, which are: human, technological and operational.

Likewise, in the research entitled Knowledge Management in Higher Education Institutions: Characterization from a theoretical reflection of [5], the authors point out the existence of some difficulties that limit the process of implementation of a KMM such as, the amount of information generated by HEIs remains as gray literature since it is not stored properly, other difficulties are that research centers lack the infrastructure and adequate human capital to generate new knowledge.

On the other hand, the research entitled Challenges in implementing organizational knowledge management: an integrative review [8] states that the exchange or transfer of knowledge between organizations can be a limitation when implementing a knowledge management model due to the nature of organizations, including time and cost constraints. Similarly, other limitations they mention [8] are the lack of commitment on the part of the staff, the lack of interaction, low motivation and the excessive number of technological tools used for knowledge management.

Likewise, the study called A review of knowledge management and its application in the contemporary business environment by [9] states that the process of capturing and exchanging information can be an important limitation if it is not carried out properly, since organizations must be able to capture, store and disseminate knowledge to improve their performance. In addition, the lack of use of technological tools can be another limitation for the application of KM. Similarly, the authors point out that there is complexity in the conversion of tacit to explicit knowledge, so this may be another limitation in the process of knowledge management.

Finally, [11] in their research entitled Information and Communication Technologies (ICT) in the processes of distribution and use of knowledge in HEIs, they state that the ICT tools used by the research centers of HEIs must be integrated into the corresponding activities, otherwise they become limitations in the implementation of the management. On the other hand, if it is used and disseminated through the intranet, repositories or virtual libraries, among other technological tools, it generates innovation, and the educational institution will be at the forefront of the latest trends so it will have competitive advantages.

As can be seen, the application of knowledge management models is very important in both organizations and higher education institutions, however, there are several aspects that can be important limitations when it comes to implementing a knowledge management model to improve performance, generate innovation and be competitive, among other advantages.

In this study, the organizational and technological limitations that arise during the application of a knowledge management model in higher education institutions were analyzed, which are: absence of objectives, lack of planning, ambiguous responsibility, contextualization, conceptual confusion, lack of knowledge transfer culture and lack of an adequate technological infrastructure to avoid them during the application of the Knowledge Management model [5, 6].

3 Material and Methods

A quantitative study was carried out, the design is non-experimental since the variable was not deliberately manipulated. The phenomena in their natural environment were analyzed. It is also cross-sectional because data were collected at a single point in time [15].

For the sample, 100% of the population involved was considered, corresponding to 684 professor-researchers who work in the research centers of a higher education institution in which the research was carried out. To calculate the sample size, a formula for finite and known populations was used [16], giving the sample value n = 247, it should be noted that a margin of error of 5% and a confidence level of 95% were considered. The sampling technique is non-probabilistic for convenience because it is considered that the selected population knows about the subject, so at the time the 247 records were obtained, the instrument was closed [17].

Regarding the instrument used to collect data, it was a self-created questionnaire with predefined questions, using a Likert scale with the values 1 (Never), 2 (Almost never), 3 (Sometimes), 4 (Almost always) and 5 (Always). The validation of the questionnaire was carried out by expert judgment approved by consensus and the method was individual aggregates [18].

Likewise, the reliability analysis of the instrument was carried out using Cronbach's alpha coefficient because this method allows us to know if the instrument measures accurately and if there is agreement between the scores issued by the participants' responses; obtaining a Cronbach's alpha of 0.93, which according to Herrera (1998) cited by [19] has an interpretation of excellent reliability.

In this way, this instrument was applied through Google Forms, which is a technological tool that facilitates the operation of the consultation and survey processes and because it is aimed at a controlled and specific audience to the research topic. It is important to mention that informed consent was included in this instrument.

The questionnaire included seven selected indicators corresponding to the dimension of limitations for the application of a knowledge management model, according to [6, 10, 11] which are: (1) Absence of objectives, (2) Lack of planning, (3) Diffuse responsibility, (4) Contextualization, (5) Conceptual confusion, (6) Lack of knowledge transfer culture and (7) Lack of adequate technological infrastructure.

As for the (1) Absence of objectives, this refers to the fact that the lack of clarity of the objectives does not allow the success of the project [6]. (2) Lack of planning, it is important that the educational institution carries out prior planning to improve the levels of performance and productivity of the staff, since a planned decision-making will allow the success of the project [10]. Regarding (3) Diffuse responsibility, it is desirable to create a KM team or department to oversee the entire process and various groups for the design, development, and evaluation of the KM process [6]. (4) Contextualization requires that KM projects be designed according to the characteristics of the HEI (values, objectives, and institutional functions) [6].

Likewise, (5) Conceptual confusion, it is important to define the entire KM process in the HEI, in order not to confuse KM with some other type of management [6]. With respect to the (6) Lack of knowledge transfer culture, it is important to provide physical and/or virtual spaces where all members can adequately capture knowledge for consultation and dissemination [11]. Finally, (7) Lack of adequate technological infrastructure, the HEI must have the adequate technological infrastructure that allows the members of the academic community to articulate, interact and socialize the knowledge generated in the institution [11].

In addition, Microsoft Excel was used since the results were obtained using Google Forms, a tool that provides the results in a spreadsheet. In addition, the descriptive statistical analysis technique was applied using SPSS software version 28.0.1.1 (15), so the frequency distribution table for each indicator was generated.

4 Results

The results were grouped by indicator for analysis. To this end, frequency distribution tables were prepared for each indicator, which include category, range, class mark, absolute frequency (*Af*), accumulated frequency (*af*), relative frequency (*Rf*); and percentage (%).

In this way, the data were grouped into categories according to their presence, which are: very strong, strong, weak, and very weak, to identify the strengths and weaknesses of the indicator.

It should be noted that the responses of the participants in each of the items of the instrument were grouped by category. Therefore, the responses of the participants who selected the options were always and almost always included in the very strong category; those who responded in the option of sometimes were grouped into the strong category; Answers with the option almost never in the weak category and finally, in the very weak category, responses with the option never.

Thus, the data that were integrated into the categories were calculated from their maximum and minimum value observed in each indicator. For the indicators, Absence of objectives, Diffuse responsibility, Contextualization, Conceptual confusion, Lack of knowledge transfer culture and Lack of an adequate technological infrastructure; the maximum value is 10 and the minimum value is 4, so the range (R) which is the maximum value minus the minimum, i.e., 10 – 4 = 6; R = 6. The number of intervals (K) is set to 4 (number of categories). The amplitude (A) is calculated, which is the range between the number of intervals (R/K), A = 6/4, A = 1.5; and is rounded to 2, so the breadth of each category for these indicators is 2.

For the indicator, Lack of planning; the maximum value is 12 and the minimum value is 4, so the range is 12 – 4 = 8; R = 8. The number of intervals (K) is set to be 4. The amplitude (A) is calculated, which is the range between the number of intervals (R/K), A = 8/4 = 2; so, the breadth of each category for these indicators is 2. Finally, these calculations can be seen in Table 1.

Similarly, Fig. 1 shows the percentages that indicate the presence of each of the indicators by category.

Table 1. Frequency distribution of Limitations for the application of a KMM in HEIs

Category	Rank	Class Mark	*Af*	*af*	*Rf*	%
Indicator: Absence of objectives						
Very Strong	[8 10]	9	169	169/247	0.68	**68**
Strong	[6 8)	7	48	217/247	0.19	19
Weak	[4 6)	5	23	240/247	0.09	9
Very weak	[2 4)	3	7	247/247	0.03	3
Total			247	1	1.00	100

(*continued*)

Table 1. (*continued*)

Category	Rank	Class Mark	*Af*	*af*	*Rf*	%
Indicator: Lack of planning						
Very Strong	[10 12]	11	129	129/247	0.52	**52**
Strong	[8 10)	9	99	228/247	0.40	40
Weak	[6 8)	7	15	243/247	0.06	6
Very weak	[4 6)	5	4	247/247	0.02	2
Total			247	1	1.00	100
Indicator: Diffuse responsibility						
Very Strong	[8 10]	9	135	135/247	0.55	**55**
Strong	[6 8)	7	53	188/247	0.21	21
Weak	[4 6)	5	41	229/247	0.17	17
Very weak	[2 4)	3	18	247/247	0.07	7
Total			247	1	1.00	100
Indicator: Contextualization						
Very Strong	[8 10]	9	89	89/247	0.36	**36**
Strong	[6 8)	7	63	152/247	0.26	26
Weak	[4 6)	5	54	206/247	0.22	22
Very weak	[2 4)	3	41	247/247	0.17	17
Total			247	1	1.00	100
Indicator: Conceptual confusion						
Very Strong	[8 10]	9	224	224/247	0.91	**91**
Strong	[6 8)	7	17	241/247	0.07	7
Weak	[4 6)	5	4	245/247	0.02	2
Very weak	[2 4)	3	2	247/247	0.01	1
Total			247	1	1.00	100
Indicator: Lack of knowledge transfer culture						
Very Strong	[8 10]	9	139	139/247	0.56	**56**
Strong	[6 8)	7	73	212/247	0.30	30
Weak	[4 6)	5	33	245/247	0.13	13
Very weak	[2 4)	3	2	247/247	0.01	1
Total			247	1	1.00	100
Indicator: Lack of adequate technological infrastructure						
Very Strong	[8 10]	9	134	134/247	0.54	**54**
Strong	[6 8)	7	71	205/247	0.29	29
Weak	[4 6)	5	29	234/247	0.12	12
Very weak	[2 4)	3	13	247/247	0.05	5
Total			247	1	1.00	100

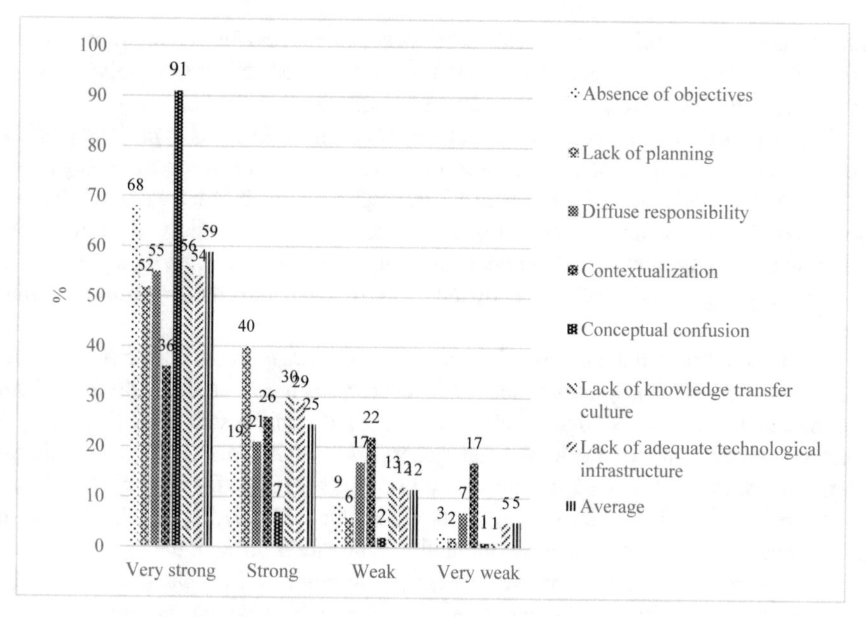

Fig. 1. Limitations for the application of a KMM in HEIs

5 Discussion and Analysis of Results

Answering the question of what limitations may arise in the application of a knowledge management model in research centers of a higher education institution. According to the analysis of the results, considering the indicators of the limitations referred to by [5, 6]. The analysis of the results showed that the predominant limitation of this study is Contextualization. This is because some of the participants report that they have never or almost never designed knowledge management projects based on institutional values, as well as have they not proposed knowledge management projects based on institutional objectives.

This coincides with [2, 6] who point out that knowledge management projects must be designed considering the characteristics of the educational institution as values and objectives, and this also requires the participation and integration in work teams of all the members who are involved with the knowledge management of the institution to achieve success in the project.

In this sense, another limitation that can be reinforced is the lack of planning, because some participating research professors mention that they have never or almost never carried out prior planning to improve their performance levels; nor do they usually plan their activities to improve their productivity levels.

In this regard, [7, 10] they report that KM in a HEIs requires prior planning by the members of the university community, which will reduce the lack or insufficiency of competencies and improve their levels of performance, productivity, and competitiveness of the educational institution's staff. This is because planned, evaluated and improvement-oriented decision making makes it possible to achieve success in the application of a knowledge management model in educational institutions.

In this way, this aspect confirms what [5, 8, 11] refers to in terms of providing infrastructure and making adequate use of information and communication technology tools that allow the application of a KMM in HEIs, since tools such as intranet, repositories, or virtual libraries, among others. They allow socializing, interacting, and articulating the knowledge generated by the members of the university community, for a greater flow of knowledge, generation of new knowledge and prevent information from becoming obsolete.

Finally, the limitation corresponding to Lack of culture of knowledge transfer is also important to consider improving the aspects corresponding to lack of culture to share, and disseminate the knowledge acquired, since the participants stated that they never or almost never disseminate the knowledge generated in their research, as well as do not adequately store the knowledge generated for consultation, and dissemination.

Therefore, in this last aspect [8, 9] they confirm that it is important to create in the university community a culture oriented to innovation through the appropriate use of technological tools that allow the transfer, assimilation and exchange of knowledge between educational institutions and even organizations with which the institutions converge, since it is important to overcome the barriers that allow the transfer of knowledge in multicultural contexts through the socialization, as well as the creation of suitable environments for the exchange of knowledge through the creation of structures, procedures and promotion of culture, all to facilitate the flow of knowledge.

6 Conclusions

Given the study carried out, it can be concluded that various aspects can be improved to avoid limitations or obstacles that may prevent the implementation of a KMM, in this way the results obtained in this research will allow higher education institutions to improve the identified aspects and achieve success in the application of a KMM.

In relation to the first limitation that was studied, it is important to mention that this limitation corresponds to the absence of objectives, which refers to the lack of clarity that those involved have in the objectives of the management project, which leads to limiting the success of the project, so it is important that the educational institution has institutional objectives that allow decisions to be made. Solve problems and overcome conflicts detected in the institution. In the same way, the lack of planning hinders the improvement of the levels of performance and productivity of the staff, so adequate planning allows correct decision-making, thus achieving the success of the project.

Another limitation is diffuse responsibility, which refers to the lack of a knowledge management team or department that is responsible for the entire process and various groups for the design, development, and evaluation of the knowledge management process. With respect to the limitation of contextualization, it is required that the projects be designed according to the characteristics of the higher education institution in which knowledge management is intended to be applied, so the values, objectives, structure, relational system, and institutional functions must be considered.

Similarly, another limitation identified is conceptual confusion, which refers to the importance of defining the entire process of KM in the HEI, in order not to confuse KM with any other type of management. Regarding the limitation of the Lack of knowledge transfer culture, it refers to the importance of properly storing information so that it does not become obsolete and can be used to generate new knowledge.

The last limitation identified is the lack of an adequate technological infrastructure, which refers to the fact that higher education institutions must have the adequate technological infrastructure that allows the members of the academic community to articulate, interact and socialize the knowledge generated in the institution.

Likewise, it is considered that the limitation in which greater attention must be paid to successfully apply a KMM in research centers of a higher education institution is Contextualization, this because it is important that the projects are aligned and directed towards the achievement of the objectives of the educational institution. They must also be designed with institutional values in mind. Therefore, it is recommended to socialize the mission, vision, objectives, and values of the educational institution so that KM projects are aligned based on them, and thus guarantee their success.

However, other limitations were also identified that were considered important to improve the knowledge management process in research centers of an HEI, such as lack of planning, lack of adequate technological infrastructure and lack of knowledge transfer culture. Therefore, being clear about the objectives, carrying out planning, identifying the assigned responsibility, knowing the KM process, and having the physical and virtual spaces, as well as the technological tools and the appropriate infrastructure to consult and disseminate the knowledge generated, will allow the success of the application of KM in the research centers of a Higher Education Institution.

References

1. Vázquez-González, G.C., Jiménez-Macías, I.U., Juárez, L.G.: Clasificación de Estrategias de Gestión del Conocimiento para impulsar la innovación educativa en Instituciones de Educación Superior. Rev. Int. Gestión Conocim. Tecnol. **10**(1), 18–35 (2022)
2. Mas, R., Meregildo, R., Torres, C., Cruz, R.: Gestión del conocimiento en la carrera de educación primaria en la Universidad Nacional del Santa, Perú. Telos **23**(2), 207–226 (2021). https://doi.org/10.36390/telos232.02
3. Santa Cruz, M.A., Córdoba, N.A., Cruz, J.J.: Gestión del conocimiento y efectividad organizacional en municipalidades de la provincia de San Martín, Perú, pp. 30–43. Instituto de Gobierno y de Gestión Pública (2021)
4. López, L., López, P., López, F.: Modelo de Gestión del Conocimiento para la Innovación. Rev. Administ. Organiz. **23**(45), 69–83 (2020)
5. Escorcia, J., Barros, D.: Gestión del conocimiento en Instituciones de Educación Superior: Caracterización desde una reflexión teórica. Rev. Ciencias Social. **XXVI**(3) (2020). https://www.redalyc.org/journal/280/28063519013/html/
6. Rodríguez, D.: Modelos para la creación y gestión del conocimiento: una aproximación teórica. Educar (37), 25–39 (2006). https://ddd.uab.cat/pub/educar/0211819Xn37/0211819Xn37p25.pdf
7. Durst, S., Zieba, M.: Mapping knowledge risks: towards a better under-standing of knowledge management. Knowl. Manag. Res. Pract. **17**(1), 1–13 (2019). https://doi.org/10.1080/14778238.2018.1538603

8. Ricardo, A., Boening, R., Jean, G., Dos Santos, N.: Challenges in implementing organizational knowledge management: an integrative review. Int. J. Develop. Res. **12**(8), 58343–58346 (2022). https://www.journalijdr.com/sites/default/files/issue-pdf/25136.pdf
9. Imhanzenobe, J., Adejumo, O., Ikpesu, I.: A review of knowledge management and its application in the contemporary business environment. Afr. J. Bus. Manag. **15**(10), 274–282 (2021). https://academicjournals.org/journal/AJBM/article-full-text-pdf/1E5947367874
10. Farfán, D.Y., Garzón, M.A.: La gestión del Conocimiento (Primera ed.). Editorial Universidad del Rosario, Bogotá (2006). https://repository.urosario.edu.co/bitstream/handle/10336/1207/BI%2029.pdf
11. Escorcia, J.H., Zuluaga-Ortiz, R.A., Barrios-Miranda, D.A., Delahoz-Dominguez, E.J.: Information and Communication Technologies (ICT) in the processes of distribution and use of knowledge in Higher Education Institutions (HEIs). Procedia Comput. Sci. 644–649 (2021). https://www.sciencedirect.com/science/article/pii/S1877050921025394/pdf?md5=20d0add32b172087e51ab376f25b17d6&pid=1-s2.0-S1877050921025394-main.pdf
12. Hughes, P., Hodgkinson, I.: Knowledge management activities and strategic planning capability development. Eur. Bus. Rev. 238–254 (2021). https://dora.dmu.ac.uk/bitstream/handle/2086/18625/PDF_Proof.PDF?sequence=1
13. Hamidreza, G., Farnaz, B.: Identifying the success factors of knowledge management tools in research projects (case study: a corporate university). Manag. Sci. Lett. **8**(8), 805–818 (2018). http://www.m.growingscience.com/msl/Vol8/msl_2018_65.pdf
14. Yousef, M., Collazos, C.A.: Collaborative strategies supporting knowledge management in organizations. Rev. Colomb. Comput. **21**(2), 6–12 (2020). https://revistas.unab.edu.co/index.php/rcc/article/download/4026/3337
15. Hernández, R., Fernández, C., Baptista, M.P.: Metodología de la Investigación. Qta. edn. McGraw Hill (2014). https://www.esup.edu.pe/descargas/dep_investigacion/Metodologia%20de%20la%20investigaci%C3%B3n%205ta%20Edici%C3%B3n.pdf
16. Otzen, T., Manterola, C.: Técnicas de muestreo sobre una población a estudio. Int. J. Morphol. **35**(1), 227–232 (2017). https://doi.org/10.4067/S0717-95022017000100037
17. Mucha-Hospinal, L.F., Chamorro-Mejía, R., Oseda-Lazo, M.E., Alania-Contreras, R.D.: Evaluación de procedimientos empleados para determinar la población y muestra en trabajos de investigación de posgrado. Rev. Cient. Cien. Soc. Humanid. Desafíos **12**(1), 50–57 (2021). https://doi.org/10.37711/desafios.2021.12.1.253
18. Universidad Adventista de Chile. Formato de validación por expertos. Guía para validar instrumentos de investigación (2017). https://www.unach.cl/wp-content/uploads/2018/06/INSTRUMENTOS_Validacion_expertos_cuestionario.docx
19. Nina-Cuchillo, J., Nina-Cuchillo, E.E.: Análisis de confiabilidad: Cálculo del Coeficiente Alfa de Cronbach usando el Software SPSS (2021). https://www.academia.edu/download/67404272/NINA_CUCHILLO_CONFIABILIDAD_CRONBACH_SPSS.pdf

Cyber Security Information Sharing During a Large Scale Real Life Cyber Security Exercise

Jari Hautamäki(✉), Tero Kokkonen, and Tuomo Sipola

Institute of Information Technology, Jamk University of Applied Sciences, Jyväskylä, Finland
{jari.hautamaki,tero.kokkonen,tuomo.sipola}@jamk.fi

Abstract. In the event of a cyber attack, the efficient production and utilisation of situational information is achieved by sharing information with other actors. In our research, we have discovered how information related to cyber security can be shared online as efficiently as possible between organisations. We used the constructive method to implement a cyber sercurity information sharing network using the Malware Information Sharing Project (MISP). The model was tested in a pilot exercise in fall 2021. The key findings in connection with the pilot showed that it is particularly important for the recipient of information security information how quickly and accurately the information security event is described. In order to help quick reaction, it would also be necessary to implement informal channels, through which security information can be shared easily without structured event descriptions.

Keywords: Cyber Security · Security Information Sharing · Situational Awareness · Threat Information Sharing · Indicator of Compromise

1 Introduction

The efficient production and utilisation of security information is intensified by sharing relevant information with other actors in the information distribution network quickly and reliably, without compromising the confidential information of actor's own organisation. It shall be recognized that, typically, organisations are cautious about sharing that information because it might include sensitive details of critical systems vulnerabilities. However, sharing and exchanging it is extremely valuable as it allows early warning of threats and new vulnerabilities for improved situational awareness initiating early mitigation processes. This action allows the recipient of the information to react in a timely manner to potential cyber security threats before they materialise in their own organisation [11,12,19].

The problem with cyber security information sharing is that information about data breaches and vulnerabilities usually cannot be passed to everyone or it comes too late to enable the prevention or mitigation of damage caused

Á. Rocha et al. (Eds.): WorldCIST 2024, LNNS 987, pp. 45–55, 2024.
https://doi.org/10.1007/978-3-031-60221-4_5

by attack or intrusion. On the other hand, the information provided might be too general to be used in defence procedures [10]. The exercise, where the implementation was tested, was executed in September 2021 in project the Healthcare Cyber Range HCCR Project [14].

The rest of the paper is constructed as follows. First the chosen research approach, the constructive and observational research methodology is introduced. It is followed by illustrations of the situational awareness and cyber security exercise concepts. After that, the technical implementation with the test case exercise scenarios are explained. Finally, the analysed results are presented and the whole study is concluded with found future research topics.

2 Research Approach

In this study, the constructive research methodology is used for finding the answer of the research question. By using the constructive research approach, state-of-the-art constructions are implemented as a resolution for domain-specific real world problems. Those implemented artefacts or constructions can for example be software components, tools, processes, or practices. When utilising the constructive research approach, both practical and theoretical aspects of the problem should be considered. Both quintessential research question and implemented practical solution should be bound to the theoretical basics of the phenomena [8,21].

The data acquisition methods used in the study were feedback discussion session immediately after the training ended, and a formal survey. The survey was administered to all participants about a week after the exercise. Feedback was collected in the so-called with a semi-structured method, where the organisers of the exercise prepare a frame for the issues and events to be handled in the feedback situation. The collection of feedback took place in the so-called through a protocol analysis, where the participants in the exercises brought out observations, their own interpretations and conclusions about the content and implementation of the exercise [22].

The research objective of this study is to research and develop a implementation of cyber security information sharing system and test that system during the national wide real life cyber security exercise. The research focuses answering for the following research question:

- How to develop a cyber security information sharing system that can be utilised for real life cyber security exercise of critical infrastructure actors? The main research question can be divided into the following sub-questions:
  - Can the implemented system be technically based on Malware Information Sharing Project (MISP)?
  - What is the proper information sharing architecture or hierarchy?
  - Is the implementation effective enough for real-life exercise usage?

3 Cyber Security Information Sharing

The sharing of cyber security information can be seen from two different perspectives. From an information sharing perspective, an event is a cyber security incident in an organisation. An incident will be reported in a way that information can be shared with other actors in the information-sharing community without a risk of sharing confidentional information from an organisation.

For the receiving organisation, the information appears as cyber security threat information. This is not necessarily an actual event in the recipient's organisation, but it is a potential threat. How potential it is depends the receiving organisation's technology environment, information and other relevant assets. In order to target the threat information as well as possible to organisations, it is advisable to share the information among actors in the same industry. For example, a few years ago, ransomware attacks were particularly targeted at hospital environments, where the sharing of threat information among industry players would have helped organisations respond in a timely manner to potential threats [23].

4 Cyber Security Exercise

A cyber security exercise is an event in which an organisation trains its preparedness for various cyber disruptions in the most appropriate way. The cyber security exercise is used to simulate or model cyber disruptions. This creates imaginary conditions in which the effects of the disorder and recovery can be tested.

There are different types of exercises depending on what kind of competences an organisation want to develop. Traditionally, technical exercises are supported by experts and technical maintenance expertise, while management and business exercises are provided to management. The co-operation exercise, on the other hand, applies to everyone and develops the skills of all staff and stakeholders. An exercise may also be a combination of different types of exercise, such as a technical-functional exercise [16–18].

In the cyber security exercise, participants are organised into different teams with their own roles. The teams are named by colours depending on what their role is. The Red Team (RT) represents the attackers and is responsible for planning and executing attacks according to training scenarios. Usually, the RT is made up of the technical experts of the training provider. The Blue Team (BT) is a defensive team made up of staff from the organisations participating in the exercise. The competence profile of the participants can vary a great deal depending on the focus of the exercise. In the wide-ranging exercises, the BT consists of a mix of technical experts, administrative staff, and other experts, such as communications experts. In large execrcise there can be several blue teams. The White Team (WT), also known as Exercise Control (EXCON), controls the entire exercise. It directs the cyber security exercise activity and monitors the progress of the exercise. Other color codes can also be used to describe the different actors in the exercise to clarify the tasks of the groups. The Purple Team (PT) can act as a combined offensive and defensive team. The Yellow Team (YT) is responsible for building the exercise environment. The Green Team (GT) is responsible for the technical support and maintenance of the training environment during the exercise [9,24].

5 Technical Implementation

The cyber security information sharing model was developed first for the HCCR project's and tested in pilot exercise in September 2021.

5.1 Implemented Construction

In the developed model, the cyber security information sharing architecture was formulated on three actors level: International level, national Information Sharing and Analysis Center (ISAC) level and enteprise/organisation level. The cyber security center located at the international level representing national CERT-FI (Computer Emergency Response Team - Finland). CERT-FI is responsible for cyber security information sharing between other national CERTs [1]. All national ISAC groups located on the ISAC level. Particularly the social welfare and health care ISAC was represented in the pilot exercise. Each ISAC group share security information from their responsible area of industry sector [2,3]. The lowest level actors represent companies and public organisations that utilise the threat information they receive in their operations and share threat information about cyber security breaches and verified events with other actors. The sharing of information takes place both horizontally, e.g., in the same industry and also vertically to the ISAC level and the international level (CERT).

In this information sharing architecture, the CERT shares and receives threat data that is expected to have a broader national impact. There are also vendors and providers on the ISAC level that receive and share threat information for their specific area. The sharing of cyber security-related information is based on verified events that are shared with others as threat information. Within the actor's own organisation, shared information is treated as cyber security incident information. Currently, the most popular cyber security information sharing solution is the MISP [20]. MISP can also be used as a threat intelligence platform to store and correlate targeted attacks, threat information, and vulnerability information. The European Union funds MISP development work.

5.2 Modeled Cyber Security Exercise a Real Life Test Event

The operating environment of the exercise included two hospitals and national services. In one of the hospitals modeled an intensive care unit (ICU) that included a patient simulator, a patient monitor, and a ventilator, which are commercial medical devices. Both hospitals had a modeled patient information system, referrals, prescriptions, imaging, and a laboratory. In addition, the exercise modeled national health-related services, such as DigiFinland's Omaolo service,

KELA's (Social Insurance Institution of Finland) OmaKanta service, prescription center and the patient information and imaging archive for hospital districts as well as THL's (Finnish Institute for Health and Welfare) health information services. The pharmacy function was also modeled, allowing prescriptions to be redeemed. The exercise was attended by three hospital districts and four authorities, as well as four companies providing digital services to healthcare providers, including the National Cyber Security Center [4,5,13–15].

The aim of the exercise was to test the planned healthcare training environment by producing disruption situations in the environment and to observe how different organisations react to them. Disruptions addressed to the training forces affected a single organisation or, on a large scale, multiple organisations. The exercise was a technical-functional exercise that lasted for three days, the first of which was set aside to familiarise the participants with the operating environment of the exercise. Cyber security status information was shared between organisations through the MISP application. The threat information identified during the exercise was routed to different actors through distribution groups configured in MISP.

Participating organisations were the following: (i) the three hospital districts participating in the pilot represented healthcare specific sector, (ii)other IAC groups and other actors represented security authorities, (iii) companies and service providers, (iv) social wellfare and healthcare ISAC represented the Finnish health and wellbeing services, (v) and Finnish Transport and Communications Agency - National Cyber Security Center (NCSC-FI) represented CERT-FI. In addition, one service provider participated in the exercise from outside the project partners. All the organisations had a representative in the exercise situation management (White Team, WT, 1 team) and a training force (Blue Team, BT, 6 teams) as shown in Fig. 1. Several organisations had sent an observer to follow up their own BT activities or to participate in the monitoring of the whole exercise. White team (WT) also represent European CERTs and International Service Providers. The total number of attendees to the exercise was 68 persons.

All these compositions had their own instance (MISP-server). In the MISP installation sharing of threat information was configured by "Sharing Group" method (Fig. 2). Sharing groups were following. SG: SHPT group was for sharing information between hospital districts, SG: Nat-Providers was for sharing between other ISAC's and national providers, SG: SOTE-ISAC was for information sharing between district hospitals and ISAC-SOTE actors, and SG: ISAC-Communities used to sharing information between different ISAC organisations. SG: INT-Providers and SG: INT-ISPs were groups for information sharing between national ISACs and international providers and ISPs. White teams represented those international actors. SG: EU-CERTs sharing group simulated information sharing between other international CERTs (in the exercise White Team) and national CERT-FI. All national cyber security threat information from hospital district level and ISAC level also shared with national CERT. All organsations also shared with method "Your organisation only" internally. Cyber security information sharing was classified ENISA [6] and TLP (Traffic

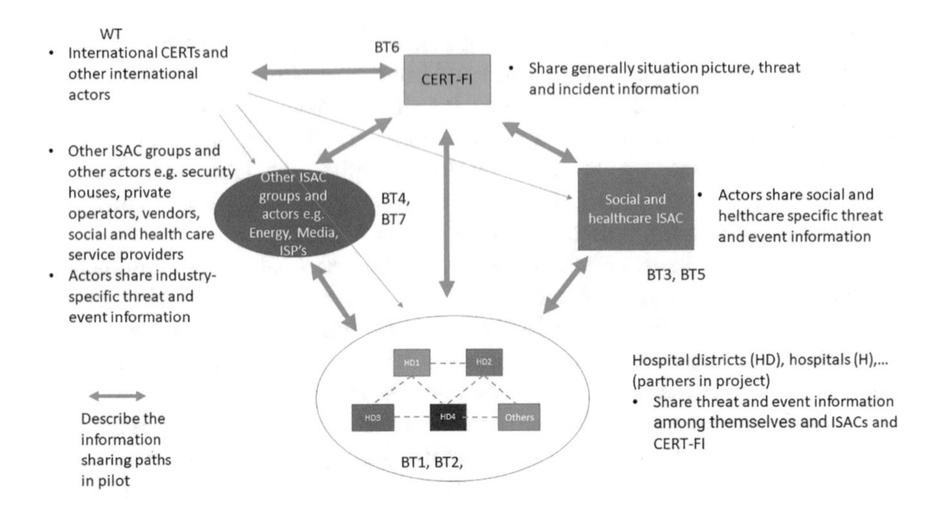

Fig. 1. Cyber security information sharing architecture

Light Protocol) taxonomy [7]. Information was also shared by non-formal text mode.

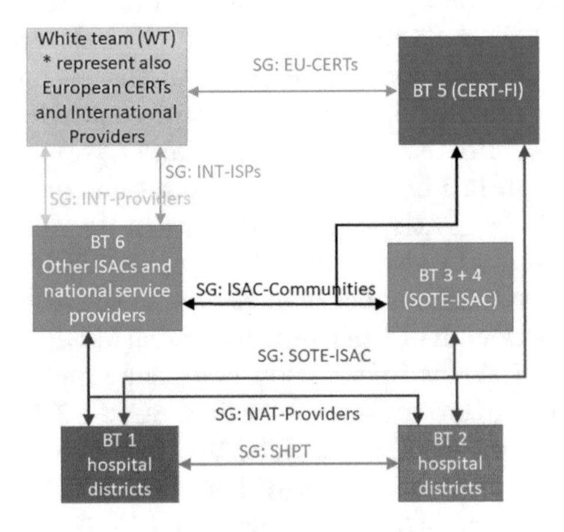

Fig. 2. Configuration of information sharing architecture in MISP platform

6 Results

The results from sharing cyber security information in the exercise were collected in two ways. In the end of the exercise instant feedback session was held, where all the attendees gave their firsthand feedback from the exercise. Feedback was also collected from the participants with a survey. A total of 67 people responded to the feedback (68 people participated in the exercise). The following are excerpts from the feedback and observations of the pilot exercise on the sharing of threat information.

6.1 Free-Form Feedback

A wider gathering of opinions and comments took place immediately after the exercise. This was carried out under the guidance of the leader of the exercise. A total of 31 answers and observations were written down. The results of these were classified into four categories; attitude towards the use of MISP, development proposals for the use of MISP, the usability of sharing information security information and development proposals for sharing information security information in general. Based on the results, a table (Table 1) was prepared according to the previously mentioned categories.

Table 1. Feedbacks at the end of exercise

Category	Positive	Negative	Neutral	Suggestions
Attitude for the using MISP	11	8	12	
Improvement suggestions for the use of MISP				13
Security information sharing tool is useful	11	6	14	
Improvement suggestions for the used sharing architecture				8

The results are fairly evenly distributed based on attitude to use MISP in exercise. About a third of respondents have a positive (11 responses) or neutral (12) attitude towards the use of MISP in sharing information security information. Accordingly, almost a third (8) did not find the use of MISP meaningful. "The use of the Data Sharing Application (MISP) would initially require a broader orientation before the exercise, e.g. where assembling MISP users to common training session". "I do not want to share unfinished data as is". A total of 31 proposals were made to develop the sharing of security information. In the development proposals (13) to the use of MISP emphasize e.g. the following: "There is a need for centralized maintenance and collection of situational awareness and communication to the management team." "It is important that the WT members leading the exercise are physically in the same space so that the information about the exercise spreads easily." "In the future, it would be advisable to authorize a special person to share information and use the MISP tool." and "Clear instructions are needed on which distribution group to use for data sharing." Accordingly, there were a total of 8 development proposals related to the general sharing of information security. Among these rose out e.g. a suggestion for a lighter tool to share first-hand threat information: "A lighter tool is needed, e.g. IRC channels, which could be used to share early observations even with rather imprecise data." and "Representatives from each area should be brought together in case of deviations."

6.2 Feedback Survey

The feedback survey was organised via electronic questionnaire. The main findings from the survey were: The majority (94%) of the respondents (N=66) agreed or totally agreed that the organised exercise developed the competence. Cooperation between organisations in knowledge and competence was reported to have developed (N=64) by the majority (84%). Cooperation between the knowledge and knowledge organisation and authorities was reported (N=64) by 80% (51) of the respondents. 89% (n=57) of respondents (N=64) report that they have identified development needs in their own competence. 85% (n=52) of the respondents (N=61) report that they have identified development needs in their own organisation's technologies or in its utilisation. In the organisation's code of conduct and processes, 91% (n=57) of respondents (N=63) identified areas for development. The majority (68%) of the respondents (N=59) either agreed or agreed with the following statement 'The environment used in the exercise (RGCE and the organisational environment) corresponded to the real world'. 80% of the respondents (N=58) either agreed or fully agreed with the statement "Data, systems and modelled processes in the training environment" from

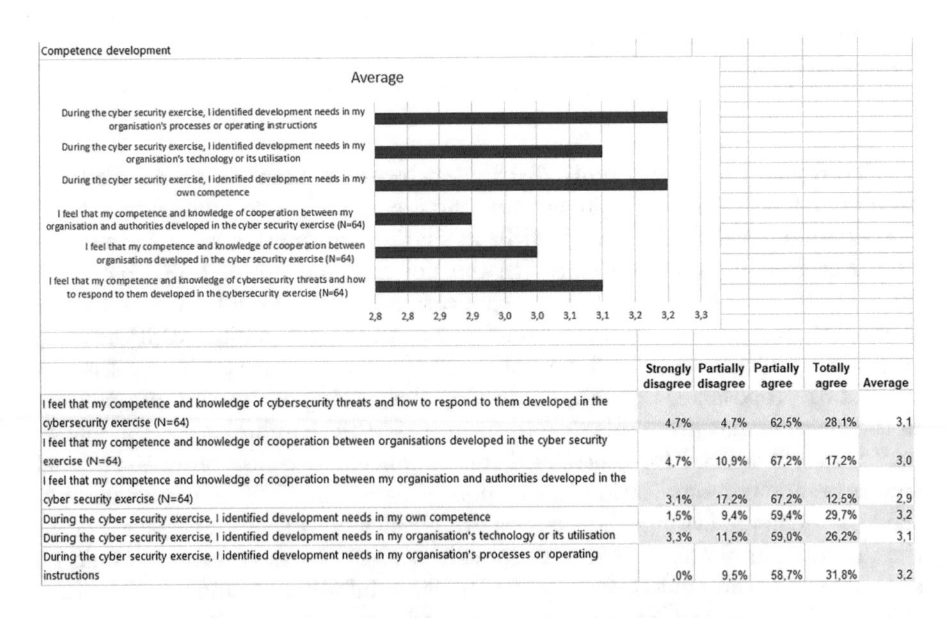

	Strongly disagree	Partially disagree	Partially agree	Totally agree	Average
I feel that my competence and knowledge of cybersecurity threats and how to respond to them developed in the cybersecurity exercise (N=64)	4,7%	4,7%	62,5%	28,1%	3,1
I feel that my competence and knowledge of cooperation between organisations developed in the cyber security exercise (N=64)	4,7%	10,9%	67,2%	17,2%	3,0
I feel that my competence and knowledge of cooperation between my organisation and authorities developed in the cyber security exercise (N=64)	3,1%	17,2%	67,2%	12,5%	2,9
During the cyber security exercise, I identified development needs in my own competence	1,5%	9,4%	59,4%	29,7%	3,2
During the cyber security exercise, I identified development needs in my organisation's technology or its utilisation	3,3%	11,5%	59,0%	26,2%	3,1
During the cyber security exercise, I identified development needs in my organisation's processes or operating instructions	,0%	9,5%	58,7%	31,8%	3,2

Fig. 3. Competence development in pilot exercise

a national point of view. The majority, i.e. 90% of the respondents (N=64), felt that their knowledge of cyber security threats and their reintegration developed during the pilot exercise. Results has been described in Fig. 3.

7 Conclusion

The data sharing architecture tested in the pilot exercise revealed a few bottlenecks in data sharing that can be summarised in the following observations. The transmission of sensitive information between operators must be confidential. The effectiveness of communication depends on how quickly and accurately a cyber security attack is described. Unnecessary information should be avoided in the sharing of information between actors in order to avoid a flood of information to detect relevant information. The pilot showed that researching and documenting cyber security attacks is time consuming and requires resources that may not be available at the same time. There is also a need for an informal channel such as chat channel for sharing threat information, where information can be shared in an informal form at a very early stage, when a possible attack is suspected. Such a communication channel allows the recipient of the information to tune in to a potential threat situation before transmitting more detailed threat information through a formal information sharing community.

The main research question was "How to develop a cyber security information sharing system that can be utilised for real life cyber security exercise of critical infrastructure actors?"

During the development of the cyber security information sharing system, three different levels of division were identified within the architecture: international, national industry specific and organisational levels. On this basis, a data sharing architecture was developed in which the sharing took place both vertically and horizontally using the sharing groups of the MISP platform. The configuration of the distribution groups was carried out according to plan. The implementation is in line with the Cyber Security Centre's plans for a national data sharing architecture.

The main research question was divided into the following sub-questions: "Can the implemented system be technically based on MISP?", "What is the proper information sharing architecture or hierarchy?" and "Is the implementation effective enough for real-life exercise usage?"

The technical implementation based on MISP platform. MISP platform is mostly used platform in cyber security information sharing and ENISA has encourage organisation in Europe to use it as common European level security information sharing by publishing its own cyber security information taxonomy. The cyber security information sharing architecture developed in the exercises is in line with how ISACs and CERT-FI in Finland have been organised. As the results show, the implemented cyber security information sharing architecture and system is a workable solution to implement the exercise. However, the system requires an in-depth study of the incident before the results of the investigation can be shared with others. This undermines the efficiency of data sharing and thus delays actions against potential threats. There is also a need for a lower-level information-sharing method that does not similarly require a formal description of incidental security incidents. The most suitable solution for this purpose would be a chat application for sharing pre-information.

Acknowledgements. This research was partially funded by the Regional Council of Central Finland/Council of Tampere Region and European Regional Development Fund as part of the Health Care Cyber Range (HCCR) project of JAMK University of Applied Sciences Institute of Information Technology (grant number A74537) and by the Resilience of Modern Value Chains in a Sustainable Energy System project, co-funded by the European Union and the Regional Council of Central Finland (grant number J10052). The authors would like to thank Ms. Tuula Kotikoski for proofreading the manuscript and Mr. Henri Tervakoski for installing all the MISP instances and configurations which were used in exercise.

References

1. CERT. https://www.kyberturvallisuuskeskus.fi/en/our-activities/cert
2. HAVARO Service | NCSC-FI. https://www.kyberturvallisuuskeskus.fi/en/havaro-service
3. ISAC Information Sharing Groups — NCSC-FI. https://www.kyberturvallisuuskeskus.fi/en/our-services/situation-awareness-and-network-management/isac-information-sharing-groups
4. My Kanta pages - Citizens. https://www.kanta.fi/en/my-kanta-pages
5. Omaolo Service. https://digifinland.fi/en/our-operations/omaolo-service/
6. Reference Incident Classification Taxonomy. https://www.enisa.europa.eu/publications/reference-incident-classification-taxonomy
7. CIRCL: Traffic Light Protocol (TLP) - Classification and Sharing of Sensitive Information. https://www.circl.lu/pub/traffic-light-protocol/
8. Crnkovic, G.D.: Constructive research and info-computational knowledge generation. In: Magnani, L., Carnielli, W., Pizzi, C. (eds.) Model-Based Reasoning in Science and Technology: Abduction, Logic, and Computational Discovery, pp. 359–380. Springer, Heidelberg (2010). https://doi.org/10.1007/978-3-642-15223-8_20
9. Diogenes, Y.: Cybersecurity - attack and defense strategies: infrastructure security with red team and blue team tactics (2018)
10. Goodwin, C., et al.: A framework for cybersecurity information sharing and risk reduction (2015). https://www.microsoft.com/en-us/download/confirmation.aspx?id=45516
11. He, M.: Perspectives on cybersecurity information sharing among multiple stakeholders using a decision-theoretic approach: cybersecurity information sharing. Risk Anal. **38**(2), 215–225 (2018). https://doi.org/10.1111/risa.12878
12. Imanimehr, F., Gharaee, H., Enayati, A.: An architecture for national information sharing and alerting system. In: 2020 10th International Symposium on Telecommunications (IST), pp. 217–221 (2020). https://doi.org/10.1109/IST50524.2020.9345861
13. JAMK University of Applied Sciences. Terveydenhuoltoalan kyberturvallisuus kehittyi yhdessä alan toimijoiden kanssa — Tech to the Future. https://blogit.jamk.fi/techtothefuture/2022/02/14/jamkissa-kehitettiin-terveydenhuoltoalan-kyberturvallisuutta-yhdessa-alan-toimijoiden-kanssa/

14. JAMK University of Applied Sciences. Real life medical equipment and simulated public health services in healthcare cyber security exercises (2021). https://jyvsectec.fi/2021/04/real-life-medical-equipment-and-simulated-public-health-services-in-healthcare-cyber-security-exercises/
15. JAMK University of Applied Sciences. Terveydenhuollon kyberharjoitusympäristön kehittäminen etenee (2021). https://blogit.jamk.fi/techtothefuture/2021/02/19/terveydenhuollon-kyberharjoitusympariston-kehittaminen-etenee/
16. JYVSECTEC: Kyberhäiriöiden hallinta - käsikirja terveydenhuollon toimijoille (2020). https://jyvsectec.fi/wp-content/uploads/2020/12/kyberhairioiden-hallinta-kasikirja-terveydenhuollon-toimijoille.pdf
17. Karjalainen, M., Kokkonen, T.: Review of pedagogical principles of cyber security exercises. Adv. Sci. Technol. Eng. Syst. J. **5**(5), 592–600 (2020). https://doi.org/10.25046/aj050572
18. Karjalainen, M., Kokkonen, T., Puuska, S.: Pedagogical aspects of cyber security exercises. In: 2019 IEEE European Symposium on Security and Privacy Workshops (EuroS PW), pp. 103–108 (2019). https://doi.org/10.1109/EuroSPW.2019.00018
19. Khajeddin, S.N., Madani, A., Gharaee, H., Abazari, F.: Towards a functional and trustful web-based information sharing center. In: 2019 5th International Conference on Web Research (ICWR), pp. 252–257 (2019). https://doi.org/10.1109/ICWR.2019.8765297
20. project, M.: Misp - open source threat intelligence platform & open standards for threat information sharing. https://www.misp-project.org/. Accessed 25 Jan 2022
21. Rautiainen, A., Sippola, K., Mättö, T.: Perspectives on relevance: the relevance test in the constructive research approach. Management Accounting Research **34**, 19–29 (2017). https://doi.org/10.1016/j.mar.2016.07.001
22. Steven, J.T., Robert, B., Marjorie, D.: Introduction to Qualitative Research Methods: A Guidebook and Resource, 4th edn. Wiley (2016). http://search.ebscohost.com.ezproxy.jamk.fi:2048/login.aspx?direct=true&db=nlebk&AN=1061324&site=ehost-live
23. Thamer, N., Alubady, R.: A survey of ransomware attacks for healthcare systems: risks, challenges, solutions and opportunity of research. In: 2021 1st Babylon International Conference on Information Technology and Science (BICITS), pp. 210–216 (2021). https://doi.org/10.1109/BICITS51482.2021.9509877
24. Traficom. Kyberharjoitusohje - traficomin julkaisuja 26/2019 - käsikirja harjoituksen järjestäjälle (2022)

Companies Amidst Evolving Digital Media Challenges in CSR Communication. Systematic Literature Review

Zoltán Rózsa[1](✉), Luboš Smrčka[2], Jan Kubálek[2], and Jiří Hermann[3]

[1] Faculty of Social and Economic Relations, Alexander Dubcek University of Trencin, Studentska 3, 911 50 Trencin, Slovakia
zoltan.rozsa@tnuni.sk

[2] Department of Strategy, Faculty of Business Administration, Prague University of Economics and Business, W. Churchill Sq. 1938/4, 130 67 Prague 3, Czech Republic
{lubos.smrcka,jan.kubalek}@vse.cz

[3] Dr. Hermann Asset Management, Ltd., A. Staska 1859/34, 140 00 Prague 4, Czech Republic
jiri@drhermann.cz

Abstract. Effective digital communication of corporate social responsibility (CSR) initiatives and efforts is critical to shaping stakeholder perceptions and opinions. Therefore, the aim of this article is to describe the current state of research on digital CSR communication and to formulate further directions in the form of a systematic literature review. Data collection of 46 relevant studies from 2015 – 2023 was conducted in the Web of Science research database. The PRISMA reporting guideline was used to enhance the transparency and quality of the systematic review. The results, through scientometric indicators and presented trends, identify two research streams: the ability of CSR to strengthen the brand-consumer relationship and the engagement of CSR stakeholders. A discrepancy was also found between the positive impact of digital CSR on E-WOM and the low engagement of digital news stakeholders. The presented results have some limitations, although they are beneficial for scholarship and practice as they provide a comprehensive overview of the current evidence on the issue addressed and recommendations for improving stakeholder engagement.

Keywords: CSR · Digital Communication · Social Media

1 Introduction

In an era of rapidly evolving technologies and constant progress, digital communication is the most crucial channel to establish genuine and lasting contact between the enterprise and its environment [1, 2]. Therefore, it is becoming a popular tool [2] which is also used in the field of corporate social responsibility [3] due to its flexibility [4], broad accessibility, and interactivity or the possibility to get immediate feedback [5]. Increasingly, therefore, it provides a unique opportunity for businesses seeking to capture stakeholder attention in order to enhance brand perception or increase their visibility and foster customer loyalty [6].

Á. Rocha et al. (Eds.): WorldCIST 2024, LNNS 987, pp. 56–65, 2024.
https://doi.org/10.1007/978-3-031-60221-4_6

However, the other side of the coin should be remembered. As the popularity of digital communication tools among companies is growing, so is the awareness of the possibilities of its use among stakeholders [7], as they are no longer pure recipients of messages but often seek out information themselves to confirm or refute the messages being disseminated [8]. Thus, businesses are no longer the sole custodians of their messages. Instead, they engage in a digital dialogue in which anyone can endorse, challenge, or even reject the published content, thus affecting the company's reputation and the trust of stakeholders [6].

Despite the widespread use of digital communication tools in business practice, the impacts of digital communication properties on the dissemination of CSR messages still need to be well documented because knowledge relies mainly on offline communication evidence [9]. Therefore, this paper builds on the abovementioned gap and aims to describe the current research on digital communication CSR and formulate further research directions using a systematic literature review.

In the theoretical part, the paper provides a foundation for understanding the research topic and thus explains concepts such as corporate social responsibility and digital communication of social responsibility. The methodological part explains the method of selecting the research studies analyzed. Afterward, it presents the results, focusing on formulating recommendations for further research. Furthermore, the conclusion also articulates the benefits for science and practice and explains the research limitation of the chosen approach.

2 Theoretical Background

The concept of Corporate Social Responsibility (CSR), which has gained considerable attention in recent years, is based on the idea that businesses must be ethical in all their activities, consider the impact of their actions on stakeholders, and at the same time contribute to global sustainability [10–13]. Most of the published works in the context of CSR, in addition to social responsibility [14], mention in particular some of the following key terms: stakeholders [15], economic responsibility [16], environmental aspects, the optimal use of environmental resources (Martínez et al., 2013), and also the willingness to support and implement activities that are not prescribed by law [14]. Above all, however, CSR offers the opportunity to minimize the risk of conflict with stakeholders and to help rebuild and gain the trust of a wider audience, as well as to promote a positive public image of the companies' activities [17].

Understanding stakeholders' expectations and perceptions of CSR communication is therefore very important, as it significantly influences business performance. Businesses that fail to meet stakeholder needs may face negative consequences, including a lack of employee support or loss of customers [18, 19]. However, when companies adopt CSR initiatives linked to stakeholder preferences and strategically allocate resources to them, the positive impact on their performance is enhanced [20]. It is also important to note that the dimensions of CSR vary because stakeholder perceptions are influenced by the company context [21].

Social media is a term that generally refers to online platforms where users can network and share information [22] and thus digitally interact with stakeholders. They shape the way we communicate and how we process information [23] and, therefore, affect different areas of our lives, such as public relations, advertising, marketing, and business [24]. Thus, it is possible to consider social media as dynamic and influential tool that changes communication [25].

However, it is necessary to realize that social media has simultaneously positive and negative effects [26]. It influences stakeholders' behavioral intentions by forming favorable attitudes toward the company [27] and improves non-financial performance by increasing stakeholders' supportive behavior [28, 29]. However, at the same time, social media can cause damage to the company's reputation or reduce stakeholder engagement [30–32]. Companies should, therefore, carefully manage their communication in the social media environment [33–35].

3 Methodology

The conducted research aims to describe the current state of research in digital CSR communication and to formulate further research directions in the field. In this context, we stated the following research questions:

RQ1: What are the leading scientometric indicators of the study area?

RQ2: What are the main research topics in the area of interest?

The collection of relevant research studies was conducted in the Web of Science research database, primarily used by academics and researchers to access scholarly articles and other research materials. That database was chosen because it provides selective and balanced coverage of the world's leading research and thus includes high-quality and influential research publications, ensuring the accuracy and reliability of the data [36]. We used the following keywords to identify relevant studies: "digital CSR communication". PRISMA reporting guideline was used to increase the transparency and quality of the systematic review guideline [37].

Of the total number of identified publications 76, we discarded 29. The resulting set consisted of 46 publications published between 2015 and 2023 (2015 – 1, 2016 – 2, 2017 – 1, 2018 – 3, 2019 – 2, 2020 – 8, 2021 – 14, 2022 – 8, 2023 – 7).

4 Results and Discussion

Research question 1 focused on the scientometric indicators of the area under study, namely where the debate is taking place and which articles (authors) are the most important regarding citation counts. We found that the discussion of the issue is spread over a relatively large number of journals with only one article (32). Among the journals with more than one published article (5 journals with 14 articles), we can include: Sustainability (6), Corporate Social Responsibility and Environmental Management (2), Journal Of Public Relations Research (2) Journal Of Theoretical And Applied Electronic Commerce Research (2), Revista De Comunicacion Peru (2).

The analyzed articles were cited 497 times without self-citations (H-index = 15). The average number of citations was 13.04, and the most frequent number (median) was only 6.5. Nine of the analyzed articles had no citations. The most cited article with 64 citations was “Exploring Digital Corporate Social Responsibility Communications on Twitter” [38] in the Journal of Business Research.

Regarding the second research question, we identified two research streams. The first stream focuses on the ability of CSR to strengthen the brand-consumer relationship in the digital era. The findings show that electronic word of mouth (E-WOM), in the form of any online expression by a consumer about a brand, product, or service, can significantly influence a customer's purchase decision. The second stream covers the studies dealing with stakeholder engagement, specifically perceptions of the complexity of social media use from the perspective of business owner-managers by specific communication strategies and tactics and the impact mentioned above on engagement, based primarily on content analysis of published messages and analysis of changes on corporate websites.

Within the first research stream, Cheng et al. (2021) results show that CSR activities are related to E-WOM and purchase intentions directly and indirectly through the brand attitude [3]. Zhang et al. 2022 also found that customer social responsibility activities conducted on social media have an impact on social media attitudes, customer E-WOM behaviors and purchase intentions; moreover, consumer feelings towards a brand can bridge the gap between customer-related CSR, E-WOM and purchase intentions [39]. Gupta et al. (2021) further confirmed that CSR communication influences customers through brand admiration and directly and indirectly through brand admiration, relationship with consumer loyalty, and purchase intentions [4].

Ma et al. (2021) investigated the impact of CSR social media communication on patients of four large hospitals. The SEM analysis findings confirmed that CSR positively influences E-WOM, and consumer-company identification (CCI) mediates this relationship [7]. Soudi et al. (2000) aimed to understand consumers' attitudes and perspectives on CSR commitments and initiatives of companies and their impact on reputation. The qualitative part of their study identified the main factors with influence, and the quantitative part confirmed a strong correlation between CSR and company reputation [40]. Finally, the research results on a sample of Romanian retail companies also indicate that the communication of CSR initiatives through social media significantly and positively influences company reputation, customer satisfaction, customer trust, and, ultimately, customer loyalty [1].

The above findings are also contained in other studies [5, 41–43] and thus overwhelmingly confirm the positive impact of digital communication, specifically concerning E-WOM, brand admiration, company reputation, consumer-company identification (CCI), customer trust, purchase intention customer satisfaction and customer loyalty. The only exception is the longitudinal study by Vogler et al. (2021), which did not confirm the impact of CSR communication on social media (Facebook) on reputation. However, it did highlight the importance of CSR reporting, confirming a positive impact, even in the case of negative publicity, in traditional media [6].

In the second research stream, Camillieri (2018), drawing on the Technology Acceptance Model [44], set out to explore the attitudes of hotel owners and managers toward digital media in order to promote their social responsibility and sustainability practices. The results showed that digital media tools could help promote their social and environmental behaviors and owner-managers perceived them as helpful in connecting with stakeholders [45]. These results were also confirmed in his second study, which revealed that young owner-managers from large SMEs are more likely to use digital media than their counterparts from smaller businesses [46].

The content of communication is addressed from several sides. Mentionable are the chosen communication strategies and perspectives [47–51] or the formation of strategic alliance partners [25]; visual content analysis [52], corporate podcasts [53], online reviews [54], hashtags [55–57] or targeting corporate discussions on social media [58]. A specific area then consists of the impact of legislation and global initiatives on CSR disclosures, mainly on company websites [59–66]. These themes are also elaborated in "A Review of Topics and Digital Communication Strategies' Success Factors" [67], so we do not address them in detail.

Unfortunately, research conducted mostly confirms low interest in published content in the form of generated interaction [56, 68–70], so it is evident that companies are not exploiting the full potential of digital marketing tools despite their relative availability [71]. This is probably also because companies do not exploit the potential of social media in the form of target groups' participation in content creation [38, 72] due to the fear of "hijacking" the company's communication initiatives [73] or little willingness to change established communication practices [62] and the resulting disconnection between the most published content and the content that generates interactions [74].

Future research should not primarily concern the justification in the contradiction between "soft" and "hard" data, namely the positive impact of CSR digital communication on E-WOM (brand admiration, company reputation, consumer and company identification (CCI), customer trust, purchase intention, customer pleasure and customer loyalty) found in stakeholder questionnaire surveys and low involvement of interested parties (in the form of generated interaction) found in the content analysis of digital messages. In this context, it is also necessary to examine the real interest of interested parties in the CSR issue because it is undoubtedly a fundamental factor that affects their involvement [27]. Research on the impact of transparency and authenticity and consistency of messages [75] on stakeholder engagement also seems sound, as it is transparent CSR communication that increases the credibility of CSR perception and consumers' attitude towards the company [1, 27]. Finally, longitudinal studies in various industries will also be welcome [39]. Specifically, it is also essential to focus on the change in the attitude of stakeholders over time [41].

5 Conclusion

The study aimed to describe the current state of research and further directions in CSR digital communication. The aim was inspired by the fact that, especially in recent years digital tools are increasingly used in social responsibility communication [28] to spread messages, involving stakeholders or build the trust or reputation of businesses [29, 76, 77].

Without any doubt, published research shows that spreading messages about corporate CSR activities in the digital environment positively affects E-WOM, brand admiration, company reputation, consumer-company identification (CCI), customer trust, purchase intention, customer satisfaction, and customer loyalty. Paradoxically, stakeholders do not show this positivity through increased interaction with the messages.

The presented results also have certain limitations, which are primarily based on the limited selection of keywords and the selected database, which, like the Scopus database, has been criticized for bias towards research produced in non-Western countries, research in a non-English language, and research in the field of arts, humanities, and social sciences [78].

Despite these limitations, this study is beneficial to science and practice, as it provides a comprehensive summary of the current evidence on the addressed issue in a better form than individual studies can provide, points out areas where further research is needed, and helps avoid unnecessary duplication of research efforts. And brings concrete recommendations for increasing the involvement of stakeholders in the dialogue on CSR in the digital environment.

Acknowledgments. The research was supported by the Scientific Grant Agency of the Ministry of Education, Science, Research, and Sport of the Slovak Republic and the Slovak Academy Sciences (VEGA), project No 1/0364/22: Research on the eco-innovation potential of SMEs in the context of sustainable development.

References

1. Topor, D.I., et al.: Using CSR communication through social media for developing long-term customer relationships. The Case of Romanian Consumers. Econ. Comput. Econ. Cybernet. Stud. Res. **56**(2), 255–272 (2022)
2. Oueslati, H., et al.: Importance and conditions of effectiveness of CSR communications in franchise networks. Int. J. Retail Distrib. Manag. 19 (2023)
3. Cheng, G.P., et al.: The relationship between CSR communication on social media, purchase intention, and E-WOM in the banking sector of an emerging economy. J. Theor. Appl. Electron. Commer. Res. **16**(4), 1025–1041 (2021)
4. Gupta, S., et al.: The relationship of CSR communication on social media with consumer purchase intention and brand admiration. J. Theor. Appl. Electron. Commer. Res. **16**(5), 1217–1230 (2021)
5. Ahmad, N., et al.: CSR Communication through social media: a litmus test for banking consumers' loyalty. Sustainability **13**(4), 15 (2021)
6. Vogler, D., Eisenegger, M.: CSR communication, corporate reputation, and the role of the news media as an agenda-setter in the digital age. Bus. Soc. **60**(8), 1957–1986 (2021)
7. Ma, R., et al.: The relationship of corporate social responsibility on digital platforms, electronic word-of-mouth, and consumer-company identification: An application of social identity theory. Sustainability **13**(9), 17 (2021)
8. Pérez, A., De Los Salmones, M.D.G.: CSR communication and media channel choice in the hospitality and tourism industry. Tourism Manag. Perspect. **45**, 16 (2023)
9. Gupta, S., et al.: Using social media as a medium for CSR communication, to induce consumer-brand relationship in the banking sector of a developing economy. Sustainability **13**(7), 16 (2021)
10. Sarkar, S., Searcy, C.: Zeitgeist or chameleon? A quantitative analysis of CSR definitions. J. Clean. Prod. **135**, 1423–1435 (2016)
11. Sharma, A., Choudhury, M., Agarwal, S., Sharma, R., Sharma, R.: Corporate social responsibility and roles of developers for sustainability in companies. In: Singh, P., Milshina, Y., Batalhão, A., Sharma, S., Hanafiah, M.M. (eds.) The Route Towards Global Sustainability: Challenges and Management Practices, pp. 313–332. Springer International Publishing, Cham (2023). https://doi.org/10.1007/978-3-031-10437-4_16

12. Sari, W.P., et al.: Corporate social responsibility (CSR): concept of the responsibility of the corporations. J. Critic. Rev. **7**(1), 241–245 (2020)
13. Rozsa, Z., et al.: Antecedents and barriers which drive SMEs in relation to corporate social responsibility? Literature review. Int. J. Entrep. Knowl. **10**(2), 107–122 (2022)
14. Dahlsrud, A.: How corporate social responsibility is defined: an analysis of 37 definitions. Corp. Soc. Responsib. Environ. Manag. **15**(1), 1–13 (2008)
15. Carroll, A.B.: The pyramid of corporate social responsibility: toward the moral management of organizational stakeholders. Bus. Horiz. **34**(4), 39–48 (1991)
16. Gupta, M. and Hodges, N.J., Corporate social responsibility in the global apparel industry: An Exploration of Indian Manufacturers' Perceptions. 2013
17. Warnaars, X.S.: Why be poor when we can be rich? Constructing responsible mining in El Pangui, Ecuador. Resources Policy **37**(2), 223–232 (2012)
18. Wood, D.J.: Corporate social performance revisited. Acad. Manag. Rev. **16**(4), 691–718 (1991)
19. Lee, M.-D.P.: A review of the theories of corporate social responsibility: its evolutionary path and the road ahead. Int. J. Manag. Rev. **10**(1), 53–73 (2008)
20. Boesso, G., Michelon, G.: The effects of stakeholder prioritization on corporate financial performance: an empirical investigation. Int. J. Manag. **27**, 470–496 (2010)
21. Madueño, J.H., et al.: Relationship between corporate social responsibility and competitive performance in Spanish SMES: empirical evidence from a stakeholders' perspective. BRQ Bus. Res. Q. **19**(1), 55–72 (2016)
22. Rani, P.U.: Impact of social media on youth. Int. J. Innov. Technol. Explor. Eng. **8**(11 Special Issue), 786–787 (2019)
23. Barrot, J.S., Acomular, D.R.: How university teachers navigate social networking sites in a fully online space: provisional views from a developing nation. Int. J. Educ. Technol. High. Educ. **19**(1), 19 (2022)
24. Moinuddin, S.: Mapping 'social' in social mediasphere in India. In: Moinuddin, S. (ed.) Digital Shutdowns and Social Media: Spatiality, Political Economy and Internet Shutdowns in India, pp. 35–56. Springer, Cham (2021). https://doi.org/10.1007/978-3-030-67888-3_2
25. Yang, A.M., Ji, Y.G.: The quest for legitimacy and the communication of strategic cross-sectoral partnership on Facebook: a big data study. Publ. Relat. Rev. **45**(5), 9 (2019)
26. Octarina, N.F., Bon, A.T.: Legalization of human rights violations. In: Proceedings of the International Conference on Industrial Engineering and Operations Management (2019)
27. Lee, A., Chung, T.L.D.: Transparency in corporate social responsibility communication on social media. Int. J. Retail Distrib. Manag. **51**(5), 590–610 (2023)
28. Thakur, S.: CSR Communication on social media as a driver of the non-financial performance of the firm: role of chief executive officers and senior executives in CSR communication: a viewpoint. In: Crowther, D., Seifi, S. (eds.) Corporate Social Responsibility in Difficult Times, pp. 3–17. Springer, Singapore (2023). https://doi.org/10.1007/978-981-99-2591-9_1
29. Zarzycka, E., et al.: Communication aimed at engendering trustworthiness: An analysis of CSR messages on Twitter. Bus. Ethics Environ. Respons. (2023)
30. Hu, M., et al.: It pays off to be authentic: an examination of direct versus indirect brand mentions on social media. J. Bus. Res. **117**, 19–28 (2020)
31. Pradeep Reddy, K., et al.: Organizational opportunities through digital and social media marketing. In: Strategic Management and International Business Policies for Maintaining Competitive Advantage, pp. 156–167 (2023)
32. Jeong, H.J., Kim, J.: Brand accommodation to informal communications on social media: with the mediation of communication appropriateness and the moderation of product involvement. Int. J. Internet Market. Advertis. **19**(1–2), 42–62 (2023)
33. Foroudi, P., et al.: Against the odds: consequences of social media in B2B and B2C. In: Beyond Multi-channel Marketing: Critical Issues in Dual Marketing, pp. 163–189 (2020)

34. Alanazi, T.M.: Impact of social media marketing on brand loyalty in Saudi Arabia. Int. J. Data Netw. Sci. **7**(1), 107–116 (2023)
35. Penttinen, V.: Hi, I'm taking over this account! Leveraging social media takeovers in fostering consumer-brand relationships. J. Bus. Res. **165**, 114030 (2023)
36. Birkle, C., et al.: Web of science as a data source for research on scientific and scholarly activity. Quant. Sci. Stud. **1**(1), 363–376 (2020)
37. Page, M.J., et al.: The PRISMA 2020 statement: an updated guideline for reporting systematic reviews. PLoS Med. **18**(3), e1003583 (2021)
38. Okazaki, S., et al.: Exploring digital corporate social responsibility communications on Twitter. J. Bus. Res. **117**, 675–682 (2020)
39. Zhang, L., et al.: Antecedents and consequences of banking customers' behavior towards social media: evidence from an emerging economy. Behav. Sci. **12**(12), 18 (2022)
40. Soudi, N., Mokhlis, C.E.: The impact of CSR communication on the company's reputation. Estud. Econ. Aplicada **38**(4), 12 (2020)
41. Puriwat, W., Tripopsakul, S.: Customer engagement with digital social responsibility in social media: a base study of COVID-19 situation in Thailand. J. Asian Finance Econ. Bus. **8**(2), 475–483 (2021)
42. Zhao, W., Cheng, Y., Lee, Y.I.: Exploring 360-degree virtual reality videos for CSR communication: an integrated model of perceived control, telepresence, and consumer behavioral intentions. Comput. Hum. Behav. **144**, 11 (2023)
43. Li, M., Liu, F., Abdullah, Z.: Analysis of online CSR message authenticity on consumer purchase intention in social media on Internet platform via PSO-1DCNN algorithm. Neural Comput. Appl. 36(5), 2289–2302 (2023)
44. Davis, F.D.: Perceived usefulness, perceived ease of use, and user acceptance of information technology. MIS Q. **13**, 319–340 (1989)
45. Camilleri, M.A.: The promotion of responsible tourism management through digital media. Tourism Plan. Developm. **15**(6), 653–671 (2018)
46. Camilleri, M.A.: The SMEs' technology acceptance of digital media for stakeholder engagement. J. Small Bus. Enterp. Dev. **26**(4), 504–521 (2019)
47. Fraustino, J.D., Connolly-Ahern, C.: Corporate associations written on the wall: publics' responses to fortune 500 ability and social responsibility Facebook posts. J. Publ. Relation. Res. **27**(5), 452–474 (2015)
48. Sun, Y., Kong, D.Y., Zhai, L.M.: To interact and to narrate: a categorical multidimensional analysis of Twitter use by US banks and energy corporations. IEEE Trans. Prof. Commun. **66**(2), 117–130 (2023)
49. Illia, L., et al.: Exploring corporations' dialogue about CSR in the digital era. J. Bus. Ethics **146**(1), 39–58 (2017)
50. Hesse, A., et al.: Consumer responses to brand communications involving COVID-19. J. Mark. Manag. **37**(17–18), 1783–1814 (2021)
51. Alonso, M.A., Esteban, P.P.: Health risk prevention awareness communication from the Corporate Social Responsibility programs, of the pharmaceutical company's set in Spain. The case of Novartis, Roche, Sanofi and Grifols. Revista Internacional De Relaciones Publicas **6**(11), 47–72 (2016)
52. Milanesi, M., Kyrdoda, Y., Runfola, A.: How do you depict sustainability? An analysis of images posted on Instagram by sustainable fashion companies. J. Glob. Fash. Market. **13**(2), 101–115 (2022)
53. Barrio-Fraile, E., et al.: Use of corporate podcasting as a SDGs communication tool in the main Spanish banks. Rev. Latina Commun. Soc. **81**, 97–122 (2023)
54. Bhattacharyya, S.S.: Exploration and explication of the nature of online reviews of organizational corporate social responsibility initiatives. Int. J. Organiz. Anal. **31**(6), 2280–2299 (2022)

55. Moyaert, H., Vangehuchten, L., Vallejo, A.M.F.: The CSR communication strategy of Iberdrola on Facebook and Twitter: a corpus-based linguistic and content analysis. Rev. Commun. Seeci **54**, 24 (2021)
56. Patuelli, A., et al.: Firms' challenges and social responsibilities during Covid-19: a Twitter analysis. PLoS ONE **16**(7), 30 (2021)
57. Confetto, M.G., Covucci, C.: A taxonomy of sustainability topics: a guide to set the corporate sustainability content on the web. TQM J. **33**(7), 106–130 (2021)
58. Kouloukoui, D., et al.: Mapping global conversations on twitter about environmental, social, and governance topics through natural language processing. J. Clean. Prod. **414**, 11 (2023)
59. Antonio, I., et al.: Communicating the stakeholder engagement process: a cross-country analysis in the tourism sector. Corp. Soc. Respons. Environ. Manag. **27**(4), 1642–1652 (2020)
60. Mason, A., et al.: Examining the prominence and congruence of organizational corporate social responsibility (CSR) communication in medical tourism provider websites. J. Hosp. Tourism Insights **6**(1), 1–17 (2023)
61. Siano, A., et al.: Communicating sustainability: an operational model for evaluating corporate websites. Sustainability **8**(9), 16 (2016)
62. Zeler, I., Oliveira, A., Morales, R.T.: Corporate social responsibility, and Covid-19 health crisis: communication of Spanish energy companies on Twitter. Rev. Commun.-Peru **21**(1), 451–468 (2022)
63. Zheng, E.Y., et al.: Institutionalizing corporate social responsibility disclosure: historical webpages of the Fortune global 500 companies, 1997–2009. Corp. Soc. Respons. Environ. Manag. **30**(2), 661–676 (2023)
64. Charumathi, B., Gaddam, P.: The impact of regulations and technology on corporate social responsibility disclosures - Evidence from Maharatna Central Public Sector Enterprises in India. Australas. Account. Bus. Financ. J. **12**(2), 5–28 (2018)
65. Kapoor, P.S., et al.: Effectiveness of travel social media influencers: a case of eco-friendly hotels. J. Travel Res. **61**(5), 1138–1155 (2022)
66. Vrontis, D., et al.: Stakeholder engagement in the hospitality industry: an analysis of communication in SMES and large hotels. J. Hosp. Tourism Res. **46**(5), 923–945 (2022)
67. Pilgrim, K., Bohnet-Joschko, S.: Corporate social responsibility on Twitter: a review of topics and digital communication strategies' success factors. Sustainability **14**(24), 24 (2022)
68. Capriotti, P., Zeler, I., Oliveira, A.: Assessing dialogic features of corporate pages on Facebook in Latin American companies. Corp. Commun. **26**(5), 16–30 (2021)
69. Conte, F., et al.: Designing a data visualization dashboard for managing the sustainability communication of healthcare organizations on Facebook. Sustainability **10**(12), 14 (2018)
70. Ozturan, Grinstein, A.: Impact of global brand chief marketing officers' corporate social responsibility and sociopolitical activism communication on Twitter. J. Int. Market. **30**(3), 72–82 (2022)
71. Zauskova, A., Reznickova, M.: SoLoMo marketing as a global tool for enhancing awareness of eco-innovations in Slovak business environment. Equilib.-Quart. J. Econ. Econ. Policy **15**(1), 133–150 (2020)
72. Paliwoda-Matiolanska, A., Smolak-Lozano, E., Nakayama, A.: Corporate image or social engagement: Twitter discourse on corporate social responsibility (CSR) in public relations strategies in the energy sector. Profesion. Inf. **29**(3), 16 (2020)
73. Vollero, A., et al.: From CSR to CSI analysing consumers' hostile responses to branding initiatives in social media-scape. Qualit. Market Res. **24**(2), 143–160 (2021)
74. Blanco-Sanchez, T., Moreno-Albarracin, B.: Instagram as a communication channel in the academic field. Comparison of the strategies of the best universities in the world. Rev. Commun.-Peru **22**(1), 35–51 (2023)

75. Lim, J.S., Jiang, H.: Linking authenticity in CSR communication to organization-public relationship outcomes: Integrating theories of impression management and relationship management. J. Publ. Relat. Res. **33**(6), 464–486 (2021)
76. Jiang, Y.N., Park, H.: Mapping networks in corporate social responsibility communication on social media: a new approach to exploring the influence of communication tactics on public responses. Publ. Relat. Rev. **48**(1) (2022)
77. Dhyani, S., Sharma, M.: Effect of communicating corporate social responsibility through social media on brand image. In: Digital Marketing Outreach: the Future of Marketing Practices, pp. 109–124 (2022)
78. Tennant, J.P.: Web of science and Scopus are not global databases of knowledge. Eur. Sci. Edit. **46**, 1–3 (2020)

Ga-starfish: A Gamified Retrospective

Luz Marcela Restrepo-Tamayo(✉) and Gloria Piedad Gasca-Hurtado

Universidad de Medellín, Medellín, Colombia
{lmrestrepo,gpgasca}@udemedellin.edu.co

Abstract. Technical and non-technical factors influence productivity in software development teams, with technical aspects being more commonly considered. However, non-technical ones hold significant sway in software development, especially within agile contexts. Achieving productivity within a software development team involves enhancing how tasks are executed. In agile work environments, improvement actions are carried out through retrospectives. However, effectively addressing retrospectives as a key event for software process improvement poses a challenge for teams, mainly when aiming to enhance productivity. This study provides a useful tool for conducting retrospectives, primarily focusing on analyzing non-technical factors that influence the productivity of software development teams. To achieve that goal, the proposal is to employ gamification elements by designing the tool based on an instrument design method and a list of non-technical factors. The most relevant outcome of this work is a technique called Ga-starfish, intended for conducting a retrospective based on gamification elements for agile software development environments. The proposed technique is a valuable alternative to improve the software development team's productivity, which is highly demanded given the challenges of the Fourth Industrial Revolution.

Keywords: Agile · Gamification · Retrospective · Social and human factors · Software development teams · Starfish

1 Introduction

Software engineering is one of the disciplines in demand in the framework of the fourth industrial revolution [1]. Software products are developed under projects, and successful delivery depends on team productivity [2]. However, team productivity depends not only on technical skills but also on non-technical factors [3]. Identifying and managing these factors is a fundamental aspect of maximizing the potential of teams, regardless of the software development model they use [4].

The productivity of software development teams is closely related to software process improvement. The retrospective is a relevant event in agile software development teams, in which the whole improvement process required for the team is carried out [5]. The retrospective is the meeting where teams reflect on their performance, identify areas for improvement, and define corrective actions [6]. However, effectively addressing these retrospectives represents a challenge for teams since the environment and how they are conducted is key to achieving success in the required improvement process [7].

Á. Rocha et al. (Eds.): WorldCIST 2024, LNNS 987, pp. 66–78, 2024.
https://doi.org/10.1007/978-3-031-60221-4_7

In this context, a tool that manages to properly conduct a retrospective, focusing the work team in a suitable environment with the continuous improvement process, would achieve one of the essential objectives of process improvement related to team productivity. Gamification constitutes a strategy that makes it possible to enrich retrospectives, taking advantage of game mechanics to foster motivation and commitment of team members [8] so that they actively participate when addressing non-technical factors, specifically social and human factors (SHF) that are perceived as influencing the productivity of software development teams.

Considering the above, this study focuses on designing a gamification strategy applied to retrospectives using the starfish technique. This new technique provides agile software development teams with a useful tool to improve aspects that influence their productivity as a key element for the continuous improvement of the team and the process it executes for product development. Therefore, the design of this technique considers the SHF that have been identified as influencing the productivity of software development teams [9]. The purpose is to propose a way to conduct a retrospective with gamification elements to motivate and engage the team, providing an environment conducive to the improvement of software processes and for the collective construction of an improvement plan in each cycle of an agile process. Game elements will mediate this environment to achieve the participants' motivation, engagement, well-being, and fun, as gamification principles are established and popularly used in these strategies [10].

The design of Ga-starfish is done using the Pedagogical Instrument Design (PID) method [11, 12] and the list of SHF that are perceived to influence the productivity of software development teams in order to deepen the understanding of such factors derived from previous research [9].

Since gamification is a strategy that promotes the transformation of the process being intervened, this premise transforms the reflection and analysis required in improvement processes, specifically in the retrospective when referring to agile software development teams. Ga-starfish is intended to promote a guided and open discussion regarding the actions related to critical SHF to improve the productivity of software development teams while contributing to conducting the retrospective by addressing the challenges related to non-technical aspects that directly influence the productivity of a work team.

Section 2 presents the context and provides relevant information about the retrospective, the starfish technique, and the importance of gamifying this process to analyze SHFs that are perceived to influence the productivity of software development teams. Section 4 presents the method used to gamify the starfish technique. Section 5 presents the proposed gamified strategy, and in Sect. 6, the results are discussed. Finally, in Sect. 7, conclusions are reported, and future work is related.

2 Background

2.1 Retrospective

Retrospective is a process closely related to continuous improvement in software development's framework of agile work approaches. It is consolidated through a periodic meeting where team members reflect on their past performance to identify opportunities for improvement and reinforce successful practices [13]. This technique facilitates the collective feedback of the work team and allows adjusting and defining actions that improve teamwork during a project [14].

The primary purpose of the retrospective is to promote continuous improvement. It is used to analyze the team's actions, identifying which actions should be replicated in the following cycles, which should be improved, and which should be eliminated. In addition, based on the team's reflection, the new actions to be started are defined [15]. Based on this classification, concrete strategies are defined to implement these changes.

2.2 Starfish Technique

There are specific recommendations, techniques, and dynamics that are useful for conducting a retrospective and implementing the recommendations of this meeting. Among the recommended dynamics are sailboat, futurespective, and story Oscars, to mention a few [7]. However, starfish is a widely used technique because of its ease of use, comprehension, and classification structure of the different categories of statements identified in a retrospective. It is a technique that provides a visual structure in the form of a star and allows to classify of the actions of the work team under five labels: 'Start doing', 'Stop doing', 'Doing more', 'Doing less', 'Keep doing', as shown in Fig. 1 [16].

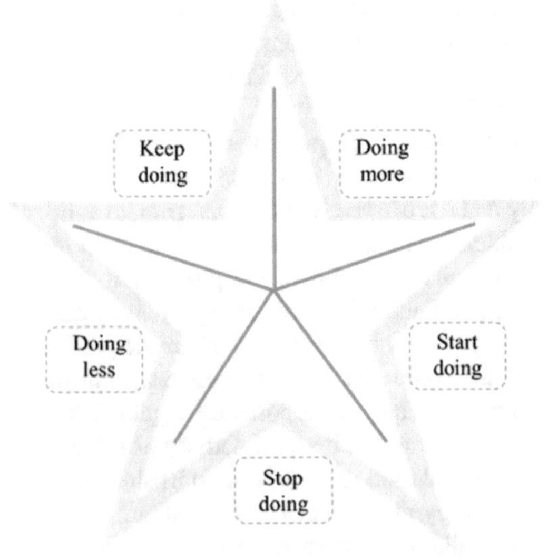

Fig. 1. Starfish structure (adapted from [16])

2.3 Gamification

Gamification can be defined as using game design elements in contexts where games are not usually played [17]. It aims to improve people's participation, motivation, and performance in activities that would not typically be considered games by incorporating elements that allow tasks to be more engaging [18]. In addition, it allows people to be involved in problem-solving [19]. The gamification elements that can be used are classified into dynamics, mechanics, and components [20, 21], as presented in Table 1.

Table 1. Gamification elements by category (adapted from [20, 21]).

Category	Description	Elements
Dynamics	General aspects of the gamification system must be considered without being explicitly in it	Relationships - Social interaction Constraints Restrictions Emotions Narrative Progression
Mechanics	Basic processes that drive action and player participation	Reward Continuous feedback Challenges Competition Cooperation Chance Transactions
Components	They are the tangible form of the elements called dynamics and mechanics. Each mechanic is related to one or more dynamics, and each component is associated with one or more mechanics	Mission or quest Leaderboard Achievement Content Unlocking Level Badges Points Avatar Team Gifts & Sharing Social Graphs Voting

Gamification offers an innovative perspective to address the inherent limitations of the starfish technique in agile software development team retrospectives. Including gamification elements during project execution can increase team member engagement, facilitate creative idea generation, and improve feedback [22]. This perspective overcomes the limitations of the starfish technique by providing a more dynamic, engaging, and participatory approach, thus driving a more complete and adequate evaluation of each project cycle.

2.4 Social and Human Factors

In software development teams, the starfish technique is mainly used to address non-technical aspects that affect team performance. Within that set of factors, particularly, 13 SHF perceived as influencing the productivity of software development teams have been identified, which have been classified into three categories [9]: 1) Factors associated with the person. 2) Factors associated with team interaction. 3) Capabilities and Experience (Fig. 2).

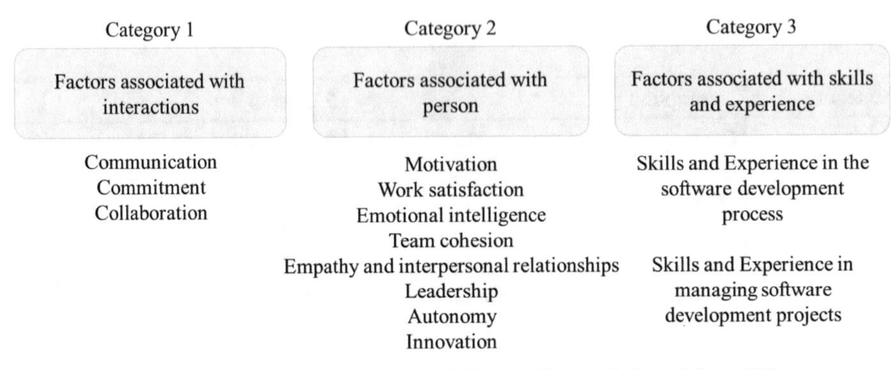

Fig. 2. Classification of Social and Human Factors (adapted from [9])

Finally, the starfish technique can be beneficial in boosting SHF. This proposal provides the opportunity to strengthen these factors to improve productivity and the work environment. Under this context, a gamification strategy can be designed that includes dynamic elements like Social interaction - Relationships, mechanic elements like competition, and component elements like leaderboard, points, and rewards [23] in such a way that it is possible to address these factors in a guided way encouraging the active and effective participation of the members of the work teams.

3 Related Work

Some research is interested in retrospective analysis, analyzing challenges, limitations, and positive aspects during these meetings. One of these studies creates a mapping between activities and potential problems that can be resolved to improve retrospectives in practice [7]. On the other hand, how people reflect in these spaces has been analyzed, even considering feelings [6].

Although the starfish technique facilitates the conduct of the retrospective and is widely used, some limitations may threaten the improvement process, mainly when it is intended to analyze SHF within the framework of team productivity. The lack of participation of team members and a conducive environment could limit the generation of ideas and solutions associated with the improvement plan expected as a result of the retrospective. Expressing opinions freely and spontaneously is one of the essential characteristics of achieving the right environment that tends to reduce bias and avoid unnecessary confrontations [7]. In this way, openness on the part of the team members is fundamental to make it possible to discuss and propose key actions for process improvement.

Another limitation of the different retrospective techniques is associated with the inability of agile teams to achieve the levels of reflection required for the retrospective to be successful. These levels are reporting, relating, reasoning, and reconstructing on continuous improvement [6].

In the retrospectives, limitations are associated with how team discussions are generated. These discussions may be affected by team members' biases and, in cases where they are not supported by hard evidence, may not reflect reality but rather a biased opinion. Therefore, a SHF roster that focuses the discussion and centers the meeting on a discussion geared toward improving team productivity can achieve control and decrease the complexity of running these meetings [24].

Concerning bias, a critical limitation can also be identified since a retrospective based mainly on the participants' experiences has the risk of memory bias, as people may remember events differently, leading to incorrect conclusions. Therefore, it is important to help software development teams recall accurate and joint visions of the projects, and there are different proposals, one of them being the one proposed in this work [25].

Therefore, software development teams must be offered a technique that reduces the mentioned limitations on the retrospective, focusing this meeting on the improvement process while waiting. Therefore, in this work, we propose the use of non-technical factors as an element that allows adequate levels of reflection of the work team to focus the analysis of the improvement actions on clear evidence within the framework of the team productivity and to diminish the bias, offering to the participants game elements that motivate their participation based on the reality of the events and remembering precise and joint vision.

4 Method

To gamify the starfish technique as a tool to define a concrete action plan related to SHF perceived to influence the productivity of software development teams, we adapted the method of designing pedagogical strategies based on gamification principles, called the Pedagogical Instrument Design (PID) method [11, 12]. In this case, the design method includes five phases covering all formal aspects of the strategy. The phases of this stage are detailed in Fig. 3.

Phase A: Planning	Phase B: Design	Phase C: Pilot test	Phase D: Scheduling	Phase E: Evaluation
Define population, purposes, subject matter and competencies to be developed	Define type of game, objectives of the game, roles, prizes, challenge, define material to be used, instructions, rules.	Design the test instrument, test the instrument and adjust the instrument	Establish date, time and physical space	Design instrument for participants to evaluate the activity

Fig. 3. Phases of the employed method (adapted from [12])

5 Results

5.1 Phase A: Planning

The gamified retrospective is oriented to agile software development teams. Its purpose is to evaluate the software development process when working with agile frameworks and to facilitate improving the software development process. Besides facilitating the analysis of the development process and establishing improvements from such analysis, using a gamified activity, it is possible to motivate the participation of the members differently, increasing the possibility of achieving accurate and joint visions of the most evident events in the development of the projects.

The theme addressed through this gamified strategy is the improvement of the software processes of agile teams, oriented to the identification of improvement opportunities as a team against non-technical factors, specifically SHF, that are perceived as influencing the productivity of software development teams (Fig. 2).

Taking into account the above and the respective definitions given by Werbach and Hunter [21], the gamification elements that are interesting to include in the gamified strategy are:

- Social interaction – Relationships: This element includes all the social dynamics that can occur inside and outside the gamification experience.
- Rewards: These are benefits obtained by reaching an achievement or action.
- Competition: Where the mechanic of competition refers to any form of win/lose struggle, combat is a concrete battle that is typically short-lived and part of the larger struggle.
- Leaderboard: Leaderboards are visual displays of player progression and achievement in rank order among some groups of players.
- Avatar: Visual or graphic representation of the player. Common in role-playing games. An avatar can be anything; the requirement is that it uniquely identifies the player. It provides identity to the player.
- Points: Points are numerical representations of game progression.

5.2 Phase B: Design

Each step involved in the design phase for gamifying the starfish technique is described below.

Game Objectives. The objective of the gamified strategy is oriented to collect data related to the process followed by the software development team, taking into account the following aspects:

- Analyze the progress made in each software product development cycle;
- Self-evaluate how the team has applied the established work methodology;
- Identify and describe specific actions to improve the process and apply them in future cycles.

Game Rules. The game's rules define the norms under which the activity is to be carried out, and each of these fulfills a purpose associated with the identified limitations of the retrospectives. In this case, they are the following:

- The gamified retrospective is carried out every time an agile development cycle ends. This rule aims to facilitate an adequate environment for the critical and constructive analysis required in an improvement process [7];
- A SHF is analyzed at a time, but as many factors can be analyzed as the team decides is appropriate. This rule is intended to contribute to the analysis of critical elements of the team's continuous improvement, such as non-technical factors that affect its productivity [9];
- All work team members must be present, but their participation in the writing of the actions should not be forced. This rule is intended to address the limitation related to the level of reflection required by members to achieve a successful challenge [6] and increase their levels of motivation and participation [26];
- Each team member should select a color of Post-it so that it is possible to identify to whom each action corresponds. This rule is defined with the purpose of including gamification elements that make the activities related to the retrospective more attractive [18];
- Each team member will write the actions he/she considers appropriate on the post-its of the selected color. This rule aims to reduce bias and avoid unnecessary confrontations that harm the environment required for a retrospective [7];
- Each team member should identify actions that can be assigned to each of the following labels: 'Start doing,' 'Stop doing,' 'Doing more,' 'Doing less,' 'Keep doing'. This rule reflects the conservation of the essential structure of Starfish [16];
- Focus only on the positive aspects that lead to continuous process improvement.

Game Instructions. The instructions to be followed to carry out the gamified starfish are as follows:

- Select a tool that facilitates the design of the retrospective based on the starfish technique;
- Tell the participants the purpose of the retrospective. The facilitator should address the three essential aspects: 1) Identify the team's accomplishments and highlight them to ensure continuity. 2) Provide key information to identify areas for improvement and take the necessary steps to move them forward. 3) Acknowledge those activities the team should abandon, examining the underlying reasons to justify their cessation.
- Agree with the team on the differentiated prize or gift given to each of the three people who have accumulated the most points.
- Select one of the SHF categories (Fig. 2);
- Select one of the SHFs from the selected category;
- Read the definition of the selected factor with the whole team;
- Ask participants to write down the actions that the team performs related to the selected factor;
- Ask participants to place the written actions on the corresponding label.
- Share all the actions of the 'Keep doing' label and ask the team members to rate each one of them in such a way that it is possible to assign them a priority order that allows the most important actions to boost the factor under analysis to be identified. The evaluation is done from 1 to 5, where 1 is the lowest priority level and 5 is the highest. For each action, average the ratings of all members.

- Actions Level 1: If the average is between 4 and 5, it is an action to be maintained. The Post-it color in this level will give the owner 10 points.
- Actions Level 2: If the average is between 3 and 4, it is an action that can be maintained at the team's discretion. The Post-it color at this level will give the owner 5 points.
- Actions Level 3: If the average is below 3, it is an action that is not recognized by the team and is therefore discarded. These actions do not assign points.

- Share all the actions of the 'Start doing' label and ask the team members to rate each one of them in such a way that it is possible to assign them a priority order that allows the most important actions for promoting the factor under analysis to be specified. The number of votes does the evaluation. The actions with the most votes, or those in case of a tie, will be implemented in the next cycle. The winning Post-it color will give the proponent 20 points.

Define Roles. In this case, the project's servant leader is in charge of coordinating the execution of the gamified starfish. On the other hand, the software development team members are involved in the project.

Set Points. Each participant who contributes to the actions under the 'Keep doing' label can earn 10 or 5 points depending on the average priority value assigned by the team. On the other hand, each participant contributing actions under the 'Start doing' label can earn 20 points if his or her proposed action receives the most votes.

Define Levels. Three levels are defined in descending order of the number of points each team member obtains. The person who obtains the most points is the 'Absolute Winner'. Whoever follows him/her is called an 'Innovator', and whoever follows him/her is called a 'Thinker'.

Establish Prizes or Gifts. Prizes are assigned to the three members with the most points. These prizes or gifts should be differentiated for each winner level and consistent with the organizational resources available. The prizes can be redeemable bonuses or even time off and should be known to the team members at the beginning of the gamified starfish.

Material to be Used. The following material is required to develop Ga-starfish in a face-to-face manner:

- The starfish structure can be drawn on a board or large sheet of paper, according to Fig. 3.
- Markers or pens will be used to label each section of the star and for each participant to write the actions.
- Sticky notes or post-it notes, where each participant will write his or her contributions related to each factor analyzed. As many colors of post-it notes as there are team members are required.
- Taxonomy with the definitions of the SHF.
- A list with the names of each team member, the respective avatar (post-it color), and the accumulated points.

5.3 Phase C: Pilot Test

All aspects defined in the design phase were socialized with three professional experts in agile methodologies and with knowledge in gamification, who made recommendations regarding drafting the game rules.

A pilot test was conducted with a team of 4 members in the last sprint of a software development project. The application of this pilot test highlighted the importance of each member having permanent access to the definition of the SHF being analyzed, so the corresponding taxonomy was included in the materials.

5.4 Phase D: Scheduling

Ga-starfish will be implemented during the first half of 2024 in software development teams of 4 members. A case study scheme with experimental and control groups will be used, using only post-tests [27].

5.5 Phase E: Evaluation

The evaluation of Ga-starfish will be carried out through the adaptation of a satisfaction survey used in another study [28] so that it is possible to obtain relevant information from each participant regarding their experience through the applied technique.

6 Discussion

Ga-starfish is a gamified technique that uses a leaderboard and points, two of the most used elements in software development [29]. However, it is complemented by Social interaction - Relationships, Reward, Competition, and Avatar, so it is possible to cover the previously presented limitations.

This novel proposal addresses some of the known limitations of the retrospectives by providing agile software development teams with a tool focused on analyzing their productivity. Therefore, it focuses such analysis in a precise way, reaching levels of reflection required by the team from using the SHF listing to relate, reason, and reconstruct the improvement of the software development process carried out in each cycle [6]. From the SHF listing, it will also be possible to analyze indicators and metrics that lead to obtaining sufficient and robust evidence of the process to reflect the reality of the process and reduce possible biases [24]. In addition, the design of the Ga-starfish technique includes a set of instructions and rules that facilitate its implementation. These rules promote teamwork and the construction of solid and integral visions of the work team [25].

The implementation of Ga-starfish may address the challenges related to the planning of the retrospective and the active participation of the software development team members, either as proponents or as voters. However, it is possible that during its implementation, other challenges related to the duration time or the clarity of the technique, aspects identified in other studies related to retrospectives in agile teams, may become evident [30].

Since Ga-starfish is a gamification-based technique, its implementation is expected to yield more positive results concerning increasing motivation and creativity than the traditional technique, as occurred with the adoption of other collaborative games [26].

Ga-starfish is a technique that can be easily implemented in educational contexts, as strategies related to knowledge transfer [28] or the identification of profiles of software development team members [31], to mention a few, have been introduced. This proposal addresses the activities that allow gamification to be evaluated in a software development context [32].

7 Conclusions and Future Work

This paper presented Ga-starfish, a valuable technique for conducting a gamification-based retrospective. This technique was designed following the Pedagogic Instrument Design (PID) method, focused on using gamification to select game elements as part of the design proposal. The gamified technique is based on the elements Social interaction - Relationships, Reward, Competition, Leaderboard, Avatar, and Points. The purpose of the gamified retrospective is to guide the identification of opportunities for improvement of the software development team against the SHF that influence their productivity, in addition to facilitating the openness of the team members against sharing those aspects that need to be intervened.

Evaluating the gamified retrospective and consolidating the results is part of the future work, considering the possibility of having evidence of its use and validation data that allow taking actions against possible improvements to consolidate the technique and the game rules. Likewise, future work includes the design of a software tool supported by machine learning that facilitates the use of gamified retrospectives even in distributed teams.

Validation using a quasi-experiment with a control group is another line of future work. It is expected to obtain relevant indicators that allow Ga-starfish to be strengthened by including additional gamification elements to the points and levels defined in this proposal.

References

1. Fitsilis, P., Tsoutsa, P., Gerogiannis, V.: Industry 4.0: required personnel competences. Int. Sci. J. "Industry 40" **3**, 130–133 (2018)
2. Pressman, R.: Ingeniería del Software: Un EnfoquePráctico, Séptima ed. Mc Graw Hill, New York (2010)
3. Sadowski, C., Zimmermann, T.: Rethinking Productivity in Software Engineering. Springer, New Yotk (2019). https://doi.org/10.1007/978-1-4842-4221-6
4. Pries-Heje, J., Johansen, J.: SPI Manifesto (2010)
5. Ciancarini, P., Missiroli, M., Zani, S.: Empirical Evaluation of Agile Teamwork. Springer, Cham (2021). https://doi.org/10.1007/978-3-030-85347-1_11
6. Andriyani, Y., Hoda, R., Amor, R.: Reflection in agile retrospectives. In: Baumeister, H., Lichter, H., Riebisch, M. (eds.) XP 2017. LNBIP, vol. 283, pp. 3–19. Springer, Cham (2017). https://doi.org/10.1007/978-3-319-57633-6_1

7. Matthies, C., Dobrigkeit, F., Ernst, A.: Counteracting agile retrospective problems with retrospective activities. Commun. Comput. Inf. Sci. **1060**, 532–545 (2019). https://doi.org/10.1007/978-3-030-28005-5_41
8. García, F., Pedreira, O., Piattini, M., et al.: A framework for gamification in software engineering. J. Syst. Softw. **132**, 21–40 (2017). https://doi.org/10.1016/j.jss.2017.06.021
9. Machuca-Villegas, L., Gasca-Hurtado, G.P., Puente, S.M., Tamayo, L.M.R.: Perceptions of the human and social factors that influence the productivity of software development teams in Colombia: a statistical analysis. J. Syst. Softw. **192**, 111408 (2022). https://doi.org/10.1016/j.jss.2022.111408
10. Oprescu, F., Jones, C., Katsikitis, M.: I PLAY AT WORK-ten principles for transforming work processes through gamification. Front. Psychol. **5**, 1–5 (2014). https://doi.org/10.3389/fpsyg.2014.00014
11. Gomez-Alvarez, M.C., Gasca-Hurtado, G.P., Manrique-Losada, B., Arias, D.M.: Method of pedagogic instruments design for software engineering. In: 11th Iberian Conference on Information Systems and Technologies (CISTI) (2016)
12. Gasca-Hurtado, G.P., Gómez-Álvarez, M.C., Manrique-Losada, B.: Using gamification in software engineering teaching: study case for software design. In: Rocha, Á., Adeli, H., Reis, L.P., Costanzo, S. (eds.) WorldCIST'19 2019. AISC, vol. 932, pp. 244–255. Springer, Cham (2019). https://doi.org/10.1007/978-3-030-16187-3_24
13. Derby, E., Larsen, D.: Agile Retrospectives Making Good Teams Great. The Pragmatic Bookshelf (2006)
14. Drury, M., Conboy, K., Power, K.: Obstacles to decision making in Agile software development teams. J. Syst. Softw. **85**, 1239–1254 (2012). https://doi.org/10.1016/j.jss.2012.01.058
15. Filipova, O., Vilão, R.: Software Development From A to Z (2018)
16. Mas, A., Poth, A., Sasabe, S.: SPI with retrospectives: a case study. Commun. Comput. Inf. Sci. **896**, 456–466 (2018). https://doi.org/10.1007/978-3-319-97925-0_39
17. Deterding, S., Dixon, D., Khaled, R., Nacke, L.: From game design elements to gamefulness. In: Proceedings of 15th International Acadamic MindTrek Conference on Envisioning Future Media Environments - MindTrek '11, p. 9 (2011). https://doi.org/10.1145/2181037.2181040
18. Pedreira, O., García, F., Brisaboa, N., Piattini, M.: Gamification in software engineering - a systematic mapping. Inf. Softw. Technol. **57**, 157–168 (2015). https://doi.org/10.1016/j.infsof.2014.08.007
19. Zichermann, G., Cunningham, C.: Gamification by Design Implementing Game Mechanics in Web and Mobile Apps, 1st edn. O'Reilly Media, Inc. (2011)
20. Werbach, K., Hunter, D.: For the Win: How Game Thinking can Revolutionize your Business. Wharton Digital Press, Philadelphia (2012)
21. Werbach, K., Hunter, D.: The Gamification Toolkit - Dynamics, Mechanics, and Components for the Win (2015)
22. Dorling, A., McCaffery, F.: The gamification of SPICE. In: 12th International SPICE Conference, pp. 295–301 (2012)
23. Machuca-Villegas, L., Gasca-Hurtado, G.P., Restrepo-Tamayo, L.M., Morillo Puente, S.: Gamification elements in software engineering context [Elementos de gamificación en el contexto de ingeniería de software]. RISTI - Rev Iber Sist e Tecnol Inf **2020**, 718–732 (2020)
24. Lehtinen, T.O.A., Itkonen, J., Lassenius, C.: Recurring opinions or productive improvements—what agile teams actually discuss in retrospectives. Empir. Softw. Eng. **22**, 2409–2452 (2017). https://doi.org/10.1007/s10664-016-9464-2
25. Bjarnason, E., Regnell, B.: Evidence-based timelines for agile project retrospectives – a method proposal. In: Wohlin, C. (ed.) XP 2012. LNBIP, vol. 111, pp. 177–184. Springer, Heidelberg (2012). https://doi.org/10.1007/978-3-642-30350-0_13

26. Przybylek, A., Kotecka, D.: Making agile retrospectives more awesome. In: Proceedings of the 2017 Federated Conference on Computer Science and Information Systems, FedCSIS 2017, pp. 1211–1216 (2017)
27. Sheskin, D.J.: Handbook of Parametric and Nonparametric Statistical Procedures, 2nd edn. Chapman & Hall/CRC, Boca Raton (2000)
28. Restrepo-Tamayo, L.M., Gasca-Hurtado, G.P., Galeano Ospino, S., Machuca-Villegas, L.: Knowledge transfer in software development teams using gamification: a quasi-experiment. Ingeniare **30**, 705–718 (2022). https://doi.org/10.4067/S0718-33052022000400705
29. Platonova, V., Bērziša, S.: Gamification in software development projects. Inf. Technol. Manag. Sci. **20**, 58–63 (2018). https://doi.org/10.1515/itms-2017-0010
30. Matthies, C., Dobrigkeit, F.: Towards empirically validated remedies for scrum retrospective headaches. In: Proceedings of the 53rd Hawaii International Conference on System Sciences, pp. 6227–6236 (2020)
31. Muñoz, M., Peña, A., Gasca-Hurtado, G.P., et al.: Gamification to identify software development team members' profiles. In: EuroSPI European Conference on Software Process Improvement, pp. 219–228 (2018)
32. Dubois, D.J., Tamburrelli, G.: Understanding gamification mechanisms. In: 9th Joint Meet Found Software and Engineering, pp. 659–662 (2013)

The Strategy Implementation of Account-Based Marketing (ABM)—The Stratio Case

Mandalena Abreu[1,2](✉) and Bárbara Jordão[1]

[1] Polytechnic Institute of Coimbra, Coimbra Business School, Quinta Agrícola - Bencanta, 3045-231 Coimbra, Portugal
mabreu@iscac.pt, iscac17293@alumni.iscac.pt
[2] CEOS.PP, ISCAP, Polytechnic of Porto, Porto, Portugal

Abstract. Account Based Marketing (ABM) is one of the terms that is currently attracting attention in the marketing field. Companies in Portugal are also trying to improve their offer with this strategy. The present study aims to contribute to the knowledge of the usefulness of the ABM strategy; moreover, the importance of this work lies in the fact that it is still a little studied topic. With this objective in mind, this paper examines a real situation. The methodology used is a case study. The company analysed is Stratio Automotive (STRA, S.A.), where the ABM strategy is being implemented. In this way, the steps for applying this strategy are highlighted, to make the company more efficient, more focused on its customers' needs and in a more advantageous position. Finally, it is expected that ABM will make the company closer to the market and therefore more successful, according to the future results of the application of this strategy.

Keywords: Marketing · B2B · ABM · Strategy

1 Introduction

The marketing environment is constantly evolving, driven by advances in technology and the need to differentiate in an increasingly competitive marketplace.

In this context, Account Based Marketing (ABM) has emerged as an effective strategy for companies looking to achieve more targeted and personalized results. ABM is a strategic approach that focuses on specific accounts, as opposed to the traditional approach of marketing to a broad audience. By identifying high-value accounts and segmenting them individually, ABM allows companies to tailor their messages and actions to meet the unique needs of each potential customer. In this way, ABM aims to build deeper and more meaningful relationships with these accounts, increasing the chances of conversion and return on investment.

This article examines the case of Stratio Automotive, a Portuguese company that has a solution that acts as a real-time and remote diagnostic tool, giving customers access to all sensor data and additional information on each fault, without the need to be physically close to the vehicle. Firstly, data is collected from the vehicles via the data receiver. The receiver, developed by Stratio, can easily adapt to the model/brand of any fleet, and

Á. Rocha et al. (Eds.): WorldCIST 2024, LNNS 987, pp. 79–88, 2024.
https://doi.org/10.1007/978-3-031-60221-4_8

collect huge amounts of data to feed the machine learning algorithms. The data is then processed on an ongoing basis in the cloud, in real time and remotely. In the third step, artificial intelligence (AI) and machine learning models are used to detect problems before they occur. Finally, the results are displayed on the Stratio platform or integrated into end-user software.

Stratio is implementing the ABM strategy throughout 2022 and 2023. It presents initial results and the benefits already achieved through ABM, including improved marketing efficiency, increased conversion rates, revenue growth and customer loyalty.

It explores the concept of account-based marketing and its application in the current marketing landscape. From identifying and segmenting strategic accounts to executing personalized campaigns, the key stages of ABM will be analyzed. It also discusses the tools and technologies available to support the implementation of ABM, as well as best practices to ensure the success of this approach.

ABM has many benefits. However, there are also challenges and limitations to consider. Issues such as the need for collaboration between marketing and sales teams, the complexity of large-scale customization, and the cost of implementation can be barriers for some organizations.

By the end of this article, you should have gained valuable insights into account-based marketing as an effective strategy for companies looking to increase the efficiency of their marketing campaigns, improve customer segmentation and drive business growth.

2 Literature Review

2.1 Marketing in the Business-to-Business Marketplace

Marketing can be divided into two main types: business-to-consumer (B2C) and business-to-business (B2B), according to [1], B2C marketing is a type of marketing that takes place directly between the company and the end consumer, and B2B marketing is used in transactions of products and services between companies for subsequent resale to third parties.

The main differences between B2B and B2C markets lie in the nature and complexity of industrial products and services, the diversity of industrial demand, the smaller number of customers, the higher volumes per customer and, finally, the closer and longer lasting relationships between supplier and customer [2]. However, although the B2B market has a smaller number of customers, the volume of transactions is usually much higher compared to the B2C market [2].

In some ways, the B2B market is similar to the B2C market in that both involve people taking on roles at the point of purchase and making decisions to satisfy needs. However, the business market involves much more money and products than the consumer market, and for this reason companies operating in it must try to know the market as well as the behaviour of their customers and, like the organisations selling to end buyers, act in ways that build profitable relationships with consumers and add value to them [2].

[2] have the perspective that a business-to-business market is characterized by the occurrence of an organizational purchase, which occurs when an organization formally establishes the need to acquire a product or service, initiating the decision process:

identification, evaluation, and selection among alternative suppliers. In this way, and in general, the B2B market is the place where organisations sell their products and services to other organisations to integrate them into their operations.

[3] states that before the concept of B2B marketing existed, it was known as industrial marketing, which focused mainly on the trade of raw materials such as wood, oil and iron ore for equipment and supplies used by other companies in their operations.

2.2 Relationship Marketing

The emergence of relationship marketing, as indicated by [4] and [5], signalled a paradigm shift from mass marketing to individualised or 'one-to-one' marketing, in which the customer and the supplier engage in a mutually beneficial co-production process [6].

The relational perspective is the result of a visible change since the beginning of the 1990s, resulting from various conjunctures such as the evolution of strategic thinking, greater competitiveness, the fragmentation of markets, consumers' access to information, their behaviour and allusive growth. The constant updating of products or services and their quality. This context has forced organisations to look for new market insertion strategies, as well as new ways to create value for the customer and generate competitive advantage [7].

According to [7], marketing challenges have been replicated, making him responsible for the development of various actions aimed at maintaining customer loyalty. When such actions have personalised characteristics, they are known as relationship marketing.

The value of customers is particularly relevant where efforts are made to maintain high levels of commitment to customers, especially those with higher value. As such, customer loyalty is seen as a key element of relationship marketing, as it is more expensive to acquire new customers than to retain existing ones, and there is a greater likelihood of repeat purchases over time (Antunes & Rita, 2008).

2.3 Account-Based Marketing

The history of account-based marketing dates back to the early 1990s. The first approaches to individualising strategies appeared in 1993 with the publication of The One-to-One Future by authors Don Peppers and Martha Rogers in 1996, which argued that the shift from the industrial to the information age meant the end of mass marketing and the beginning of a new paradigm focused on customisation and specialization [8].

The concept of account-based marketing only emerged in 2003, although the strategy had been adopted earlier. The concept was introduced by the American company ITSMA, and from there a new concept in business-to-business (B2B) marketing emerged [8]. Reflecting ITSMA's definition of "treating individual customers as markets in their own right", ABM is a structured process for developing and implementing highly customised marketing programmes for strategic customers, partners or prospects. It is a long-term programme that requires a commitment of resources. It can take more than a year to deliver significant returns. It is underpinned by careful analysis of the customer's key business issues [9].

According to [10], account-based marketing (ABM) is one of the most effective ways of attracting new customers for B2B companies, using campaigns tailored to specific accounts. The use of this type of campaign is an asset in attracting a specific company, as it works on an individual level and in the market itself. It can be said that ABM follows the following steps: identifying target companies, nurturing these companies with personalised content, and creating long-term relationships that convert into sales.

[9] go on to say that ABM makes such a measurable difference because this strategy is conceived through specific objectives aimed at a well-oriented public. Externally, it is an integrated and coordinated programme of activities that brings valuable suggestions and relevant ideas to customers. Internally, it fosters closeness and collaboration between marketing, account management and sales teams because it is only when everyone is involved with a customer that it is truly effective.

2.4 The Sales Funnel of Traditional Marketing and Account-Based Marketing

The implementation of account-based marketing strategies in the business-to-business market has increased significantly in recent years. However, there are still many companies that are not using ABM because they are unaware that it is one of the most effective strategies for acquiring new customers.

We are talking about a different technique that 'inverts' the traditional sales funnel - ABM focuses only on accounts with buying potential - and as a result most organisations still do not know what it is, how they can benefit from this strategy and where to start. The first thing to remember is that ABM does not replace traditional methods. It works as a complement to content or personalised marketing, directing your communications to defined accounts and pre-selected companies. In addition to facilitating the identification of target accounts by limiting the number of users to those that best match the ideal customer profile, it is also easier to monitor, since the set to be analysed in an ABM campaign is smaller than that of a traditional marketing campaign [11].

According to [12], ABM is a strategy based on specific accounts selected by the company. In other words, instead of creating campaigns, content and communications for a wide audience, marketing and sales teams define in advance which companies or leads they want to reach and personalise all their communications for these target accounts. The selected accounts to be targeted by this ABM strategy are evaluated. Based on this analysis, nurturing actions are developed.

2.5 Types of Account-Based Marketing

ABM focuses on the target companies it wants to reach, uses all available resources to attract them and nurtures them until they are ready to buy the product. Due to the growing interest in ABM strategies, companies have developed and implemented three different types of ABM [13].

According to [14], one-to-one ABM is the most personalised, meaning that companies build complete maps of accounts of interest through organisational charts, professional histories, etc. By building these maps, marketing and sales teams can come up with highly personalised actions. The limitation of one-to-one ABM is that it requires data to build the map, which takes time to verify and make credible. For this reason, this strategy is recommended for small accounts. Burgess & Munn (2021) add that this type of ABM is usually reserved for strategic accounts and is conducted on a one-to-one basis. One-to-one ABM is strategic ABM.

The one-to-few ABM strategy focuses on a larger group of strategic accounts or the next lower level of accounts, according to [13. Technology becomes more important here, helping to automate the process of account and stakeholder awareness, campaign execution and measurement. The advantage of this strategy is that it is less resource intensive - both in terms of people and budget - than one-to-one ABM. It works with clusters of accounts with similar characteristics, i.e. a group of similar companies is created, which allows the creation of campaigns and content that are slightly personalised [14].

Finally, one-to-many ABM works with many companies. This strategy allows the use of personalised content and campaigns, as the targeted companies are in the same or common industry, but the personalisation is not as high [14]. It is enabled by a new category of marketing technology that automates ABM-inspired tactics at scale across hundreds or even thousands of identified accounts. This is where most of the current buzz around ABM is coming from, driven in large part by the technology companies themselves (Burgess & Munn, 2021).

2.6 Benefits of the Account-Based Marketing Strategy

Account-based marketing strategies have their benefits. According to [15], the buzz around ABM strategies is growing as companies want to reach their customers more accurately.

Account-based marketing helps the business to work and communicate with high-value accounts as if they were individual markets. This, along with personalising content and tailoring all communications and campaigns to these specific accounts, will result in higher ROI and increased customer loyalty (Baker, 2020).

Marketing acts as a core member of the account, or as a member of the sales team, and adds value in several ways: (1) by researching the key issues/problems facing the account; (2) by mapping them to responsible individuals; (3) by developing tailored sales and marketing campaigns. This can deliver significant results for both the client and the sales and marketing teams.

2.7 Implementation of the Account-Based Marketing Strategy

As mentioned above, account-based marketing is a strategic approach that has grown in popularity in recent years. ABM focuses on identifying target accounts and creating personalised campaigns and messages to engage with them. By targeting specific accounts, ABM aims to improve lead quality, increase customer engagement and ultimately drive revenue growth.

The ABM strategy must be based on an ethos that encourages people to work together towards a common goal. Building collaborative cross-functional teams is increasingly recognised as a source of innovation and agility in successful organisations, and the ABM strategy plays all the relevant roles in determining how best to meet a client's particular business needs and objectives [9].

Account-based marketing is the common marketing practice of segmenting and selling to key accounts or organisations. The great advantage in developing an ABM strategy is to use the singularisation of events to target specific groups of people within an organisation to build a relationship. An ABM technique involves a change in data collection, so it's important to review the terms of data collection, synchronisation or integration that will help implement a plan for unique, personalised and singular celebrations. For example, one way to develop an ABM strategy and a personalisation strategy is to gather broader statistical data about an organisation, its customers, its accounts, and the variety of people relevant to that entity. Armed with a list of objectives and interests, experiences can be created that are tailored to the client's idea, expectations and preferences, producing unique events that aim to create successful sensations for their target audience, as well as communication campaigns centered around them (González, 2021).

While academics have long considered key account management issues from a sales perspective, marketing technology now offers companies the ability to develop and execute account-based marketing (Baker, 2020).

Building an account-based marketing strategy involves five stages: (1) identifying and defining accounts (customer companies), (2) creating the database, (3) designing the action plan and content to be offered, (4) creating the campaign, and (5) analysing the data [4].

For this strategy to be successful, the marketing and sales teams need to be well aligned. Therefore, it is necessary to decide what characteristics the companies must have in order to be considered as target companies and consequently to be mapped, characteristics that can range from the type of industry, company size, turnover, geographical location, etc. The next step is to outline the strategies to be implemented in mapping the companies, using the most appropriate tools. This will enable the project to be aligned with the company's values and objectives.

Once the marketing and sales teams have mapped the companies, it is necessary to outline the content strategy or offers for the campaign, focusing on maximum personalisation, considering all the needs of the decision-makers and the industry. This personalisation can be achieved by profiling the company, i.e., carrying out research to better understand its values, mission, and difficulties, so that the content produced is as personal as possible and the company feels identified with it. This content can be communicated in a variety of ways, such as targeted advertising, personalised emails, and so on. There are several ways to create an ABM campaign, such as importing the company list into paid media tools such as Facebook Ads, Google Ads and, most commonly, LinkedIn Ads. Using these tools makes it easier to connect with companies by creating specific content that they identify with.

Once the campaign has been launched, it is necessary to analyse the data and see which of these companies have now become a lead. This data analysis can be done by looking at the number of opportunities generated, the number of companies that have progressed through the different stages of the pipeline, how many of them have become customers, and other metrics that are suitable for analysing the success of these actions.

2.8 Measuring the Returns of the Account-Based Marketing Strategy

According to [9], ABM strategy delivers the highest ROI of any B2B marketing approach. This conclusion is based on ITSMA's extensive research on the subject over nearly two decades, including quantitative surveys, qualitative interviews, first-hand experience, the annual Marketing Excellence Awards, and in-depth work with members and clients.

[9] add that ABM strategy is not a short-term lead generation approach. It is a strategic initiative that requires sustained investment to achieve maximum results. Organisations that have been running their ABM programmes for three years or more are more likely to achieve higher returns from this strategy than from other marketing programmes and are the first to measure the return on their programme.

As mentioned earlier, measuring and evaluating the ABM strategy is something that happens in the long term, but since there isn't the time for this project work, a different approach is taken. To do this, it is necessary to lay the right foundations with the players in the teams involved and define a set of metrics that will work at each account level and expand as the ABM strategy grows.

The metrics of the ABM strategy should cover three categories: relationships, revenue and reputation. The metrics chosen to evaluate the success of the ABM strategy must be directly related to the goals set for the company as well as the individual goals set for each target account [9].

We can therefore conclude that the metrics also include several key indicators of the performance of the ABM strategy. These include factors such as the number of new executive relationships within target accounts, the number of executive meetings, the number of new relationships in new lines of business and the quality of the relationships. These metrics are indicators of future success, especially when the ABM strategy has not been in place long enough to produce concrete financial results.

Thus, it can be said that the main metric used to evaluate the implementation of the ABM strategy is the relationships achieved with target accounts, which can be measured in the following ways: (1) number of relationships the company has with target accounts, new executive contacts; (2) involvement of target accounts in attending events, responding to campaigns, engaging in social media channels, number of meetings to learn more about the product; (3) strength of existing relationships with customers (Burgess & Munn, 2021).

3 Methodology

This article follows the case study method. According to Yin (2001), the case study method can be used when researching and analysing events in their own context, in real time.

Furthermore, the methodology that will be used is action research for the implementation of the ABM strategy. According to [15], 'action research' can be described as a family of research methodologies that involve simultaneous action (or change) and inquiry (or understanding), using a cyclical or spiral process that alternates between action and critical reflection. In subsequent cycles, methods, data and interpretations are continually improved in the light of the experience (knowledge) gained in the previous cycle.

[16] add that action research offers the possibility of intervening in the entity being studied and analysing the results. It allows for an open-ended approach to the field of enquiry, making it possible to capture information that often cannot be predetermined. This strategy leads the researcher to be an active participant in any changes in a system.

4 Case Study–Stratio Automotive

4.1 Stratio Automotive in a Glance

Stratio Automotive (STRA, S.A.) is based at the Instituto Pedro Nunes (IPN) in Coimbra and has offices in Lisbon, London and Singapore. The company was founded in 2017 by Ricardo Margalho and Rui Sales, who currently hold the positions of CEO and President, respectively. Stratio is an automotive-focused company whose mission is to develop technological solutions that enable the early detection of failures and the verification of the status, condition and performance of vehicle systems (predictive maintenance). The Stratio platform is the world leader in predictive maintenance, with technology capable of accurately predicting vehicle failures in advance and in real time. The company's proprietary technology combines large-scale processing with the latest machine learning techniques. The result of this combination is invaluable to client companies, preventing hundreds of thousands of breakdowns every day. The practical impact on the lives of businesses and citizens in general is immense. In fact, millions of people are seeing their lives improved by reducing delays on public transport, virtually eliminating late deliveries *or the late arrival of essential goods, among other benefits.*

4.2 Implementing the Account-Based Marketing Strategy at Stratio Automotive

The Account Based Marketing strategy [17] was implemented at Stratio Automotive with the aim of increasing the effectiveness of the company's marketing strategy by focusing its efforts on a specific set of high value accounts (companies) and personalising the approach through collaboration between the marketing and sales teams.

The first stage of this strategy was to identify and define the accounts (companies). These companies were selected following a market study based on the following parameters: (1) location (region), (2) turnover, (3) number of vehicles (buses) each company owns. For example, the first ABM campaign was carried out in the DACH region (Germany, Austria, and Switzerland) and seven high-value companies with a fleet of more than 1,000 vehicles were selected.

When the companies were selected, the buyer personas that best fit each company were also selected. This selection was also made jointly by the marketing and sales team and was segmented into three parts: (1) the user (P1), this persona is the user of the solution and is the main beneficiary of using Stratio's software; (2) the manager/supervisor or head of the company (P2), who supervises the team using the solution and is the main decision maker; (3) the executive (P3), mainly concerned with ROI, this persona can become a blocker in acquiring Stratio's solution. For the seven companies selected from the DACH region, 172 personas were identified using LinkedIn and ZoomInfo, and the second stage of the ABM strategy, the creation of the database, was carried out.

The third stage of this strategy is to develop the action plan and the content to be offered. The action plan was a series of five emails with a fifteen-day interval between each email. Each email contains a personalized text for each of the companies and always offers content ranging from case studies, articles and videos of webinars held by Stratio, all of which are integrated into a CTA (call to action) in the body of the email. The entire email campaign was created using the company's CRM, Hubspot. For each email sent through Hubspot, we were able to access metrics that helped us understand whether the campaign had a positive impact on the seven selected companies. These metrics relate to the email sent: send rate, deliver rate, open rate, click-through rate, Unsubscribe rate and Unsent rate.

5 Results and Discussion

With this ABM email campaign, we expect to see a positive percentage of emails opened, which means that the companies have identified with it and at the end of the campaign they feel ready to have a meeting with the sales team to learn more about the Stratio solution and feel ready to buy the solution.

The steps for applying this strategy have been outlined up to the third stage.

The empirical results will be analyzed in the near future and it is expected that they will clearly show the way for the company to get closer to its customers and the general market.

6 Conclusion

This report presents a case study of Stratio Automotive and its implementation of an ABM strategy to increase customer loyalty.

With the implementation of this strategy, it is expected that many of the companies mapped will convert from leads to customers, accelerating revenue growth and consequently the growth of Stratio Automotive. It can also be concluded that there is a greater connection between the customer and the company. This closeness makes it easier to identify needs, difficulties, and strategies, and to act in a personalised and assertive way.

In a very practical sense, the results of this study will enable the company to identify the most effective ABM strategies, how to achieve well-mapped target accounts, what to do to achieve the most individualised strategies for each customer's needs, and to identify the ABM strategies that could bring the greatest return.

This study also aims to fill the gap in the literature on the subject. It is hoped that it will increase knowledge of ABM strategy, identify specific strategies and also increase knowledge of relationship marketing strategy in the B2B market, helping to improve the relationship between the company and its customers.

References

1. Oliveira, L.: Marketing Industrial: Uma Abordagem Das Suas Perspectivas No Crescimento Econômico Industrial (2016). www.simprod.ufs.br

2. Kotler, P., Keller, K.L.: Marketing Management, 15th ed., Pearson, UK (2016)
3. Lilien, G.L.: The B2B Knowledge Gap. Int. J. Res. Mark. **33**(3), 543–556 (2016). https://doi.org/10.1016/j.ijresmar.2016.01.003
4. Egan, J.: Relationship Marketing Exploring relational strategies in marketing (2011). www.pearsoned.co.uk/egan
5. Kotler, P., Armstrong, G.: Principles of Marketing – Global Edition, 15th ed, Pearson Education (2015)
6. AkrAkroush, M.N., Dahiyat, S.E., Gharaibeh, H.S., Abu-Lail, B.N.: Customer relationship management implementation: An investigation of a scale's generalizability and its relationship with business performance in a developing country context. Int. J. Commer. Manag. **21**(2), 158–191 (2011). https://doi.org/10.1108/10569211111144355
7. Ferreira, B., Marques, H., Caetano, J., Rasquilha, L., Rodrigues, M.: Fundamentos de Marketing (EDIÇÕES SÍLABO) (2021)
8. Ferreira, R.: Implementação Da Estratégia De Account-Based Marketing Em Startup De Tecnologia Da Informação (2021)
9. Burgess, B., Munn, D.: A Practitioner's Guide to Account-Based Marketing Praise for the Second Edition (2021)
10. Liminal. ABM - Account-Based Marketing: O que é? | Martech Magazine (2019). https://liminal.pt/martech-magazine/abm-account-based-marketing-o-que-e/
11. Trindade, R.: Account-Based Marketing: a estratégia de marketing e de vendas que veio para ficar. E agora? (2022). https://digitalks.pt/artigos/account-based-marketing-a-estrategia-de-marketing-e-de-vendas-que-veio-para-ficar-e-agora/
12. Salesforce, Account-Based Marketing (ABM): O que é? - Blog da Salesforce, February 15 (2021). https://www.salesforce.com/br/blog/2021/02/account-based-marketing-abm.html
13. Burgess, B.: Three types of account-based marketing (2017)
14. Cordeiro, M.: Account-Based marketing: O Marketing baseado em contas (2019). https://rockcontent.com/br/blog/account-based-marketing/
15. Santos, V., Amaral, L., Mamede, H.: Using the Action-Research Method in Information Systems Planning Creativity research (2013). https://www.researchgate.net/publication/261464482
16. Myers, M. D., Avison, D.: Qualitative Research in Information Systems (2002)
17. Baker, K.: The Ultimate Guide to Account-Based Marketing (ABM) (2020. https://blog.hubspot.com/marketing/account-based-marketing-guide

Approach to the Consumption of Fake News. The Case of Ecuador

Abel Suing(✉), Kruzkaya Ordóñez, and María Isabel Punín

Universidad Técnica Particular de Loja, San Cayetano, Calle Champagnat, Loja 11010, Ecuador
arsuing@utpl.edu.ec

Abstract. In the digital age, news consumption has undergone a radical transformation, providing instant access to an overwhelming amount of information. However, this phenomenon is not without its challenges, and one of the most pressing problems is the rampant spread of fake news. The phenomenon of fake news not only undermines the integrity of the information flow, but also poses serious threats to society, affecting decision-making, public trust and democratic stability. In this context, this research presents the importance and relevance of fake news consumption, analysing its profound implications and reflecting on the measures needed to address this contemporary challenge. At a time when information has become a vital resource, understanding the threat of fake news becomes essential to safeguard the integrity of an informed society. The methodology is quantitative through a survey conducted via a Google form, between 23 May 2022 and 31 January 2023. A total of 263 participations were received. It is concluded that the social actors that generate the most disinformation are political leaders and people who spread fake news. In the predominant practices of news consumption, a daily dedication to learn about the reality of the country stands out. The most used social networks are WhatsApp and Instagram. The sources of information that generate trust come from the media and state information. In the fight against information, it is important to contrast messages on social networks with the news published in the media.

Keywords: Disinformation · news · media consumption · media competition

1 Introduction

In recent years, the media, both traditional and digital, have lost the trust of citizens, possibly due to an increase in the levels of polarization and the clientelistic arrangements they have maintained with governments [1–3]. This has an impact on the dissemination of false information [4], which leads to the phenomenon of fake news or disinformation, which is increasing on a daily basis.

Disinformation, which always has authorship and intentionality, is information that is deliberately created and disseminated to cause harm, confuse and falsify [5]. It corresponds to "a cultural phenomenon historically linked to the dynamics of mass media intervened by state and commercial interest groups" [6]. Disinformation involves fake

Á. Rocha et al. (Eds.): WorldCIST 2024, LNNS 987, pp. 89–96, 2024.
https://doi.org/10.1007/978-3-031-60221-4_9

news broadcasts, misinformation, online propaganda, hyper-partisan information, infotainment, rumours and conspiracy theories, as well as diverse practices such as bots, trolls, fake news portals [7].

The presence and continuing rise of fake news questions the credibility of contemporary journalism [8]. Disinformation is also seen as news clutter that is produced to create doubts and false debates, and then to affect economic profitability or ideological gain [9].

Fake news spreads faster than real news through social media [10]. The highest consumption of fake news occurs through social media, in contrast to traditional print media, where measurements show low consumption of misinformation [11]. However, many falsehoods are transmitted through social media, and the risk of immediate dissemination is immediate [10], because it is individuals who mobilize false news.

The relationship with the official production of disinformation has been noted and polarizes citizens [12], but truthfulness in society is a shared responsibility of the media and institutions [13]. Paradoxically, fake news dismantles the monopoly of institutional disinformation coming from the mainstream media, a dangerous attempt to emancipate readers [14, 15].

To counter disinformation, local actions are implemented and articulated with governments in the so-called fight against disinformation, e.g. laws, media literacy, internet blackouts, arrests, proposals, working groups, reports, research [16]. Among the outstanding experiences are the EU Code of Best Practices against online disinformation, published in 2018, which is based on five areas: 1) advertising revenues, 2) transparency in political advertising, 3) fake accounts, 4) citizen empowerment and 5) encouraging the monitoring of disinformation [17].

Although it is complex to establish a homogeneous norm to regulate digital platforms due to their continuous fluidity, there are three main axes: 1) self-regulation of social media, 2) prevention of affectations in electoral campaigns, and 3) defence of freedom of information and expression.

In Latin America, efforts to regulate platforms focused on verification agencies, legislation to penalize fake news, dialogue with platforms, promotion of ethical standards and strengthening digital literacy [18]. The European Union defined as one of the main lines of work, in the face of disinformation, the raising of awareness through citizens' literacy to detect and counteract disinformation [19].

One of the effective ways to reduce disinformation is to promote Media and Information Literacy, which results in the participation of citizens in the media environment [20]. This is because protocols have yet to be proposed to measure the specific weight of disinformation generated from social networks, which is then presented in the conventional media, and influences citizens' decisions. For this same reason, the role of social media in the dissemination of fake news should be analysed, based on the fact that there are few limitations to social media in the management of any consumer's data.

Faced with sufficient material conditions to spread misinformation, there is an interest in studying people's behaviour in the face of fake news and their ability to identify it, as has already been done in other countries and communities [21]. The justification for the study is that verifying false information has become vital, necessary to critically interpret the data, and to create critical participations, based on ethical parameters.

Ecuador is considered because of its proximity to the subjects, but mainly because 76% of the population has access to the Internet. These users generate more than 16.3 million connections, showing access from more than one device per user. TikTok has established itself as one of the main social networks with around 12 million users. While Facebook and Instagram integrate 15.7 million accounts - users [22], i.e. there is a digital ecosystem prone to news consumption through social networks, with the foreseeable risk of receiving fake news.

The research questions are: 1) Which social actor generates more disinformation. 2) What are the predominant news consumption practices of Ecuadorians. 3) Which information source generates trust in Ecuadorians. 4) What are the predominant news consumption practices of Ecuadorians. 5) What are the predominant news consumption practices of Ecuadorians, and 6) What are the predominant news consumption practices of Ecuadorians. 4) What are the predominant anti-information practices of Ecuadorians?

2 Methodology

This report is descriptive of the phenomenon investigated. The methodology used is quantitative, using a questionnaire based on two studies.

1. A survey of digital users on information habits applied in Argentina by the organization "100 por ciento", made up of three organizations linked to journalism. This study was conducted by Cecilia Mosto, Francisco Corallini and Ariel Mosto, and sought to determine the level of disinformation among Argentines and how it impacts on their daily lives.
2. The research by Cerdá-Navarro et al. (2021) entitled "Fake or not fake, that is the question: recognition of disinformation among university students", which investigated the ability of university students to discern the veracity of the information they receive.

The questionnaire consisted of five blocks: 1) identification of fake news; 2) measuring the frequency of consulting news in the media and trust in them; 3) frequency of participation in forums and news comments; 4) perception of the prevalence of fake news and identification skills; and 5) socio-demographic questions.

Implementation was carried out through a Google form, in a snowball dynamic by accessing participants, between 23 May 2022 and 31 January 2023. 263 participations were received, corresponding to 112 men, 151 women. The average age is 26 years, by age group: 18–27: 74%; 28–37: 19%; 38–47: 5%; 48–57: 2%. The participants live in different cities of the country, and have different occupations.

3 Results

The results of the surveys are shown in the following graphs and tables, which show that there is greater trust in journalists than in politicians. Figure 1 shows that respondents indicate that many of the fake news would be generated by political leaders and citizens themselves.

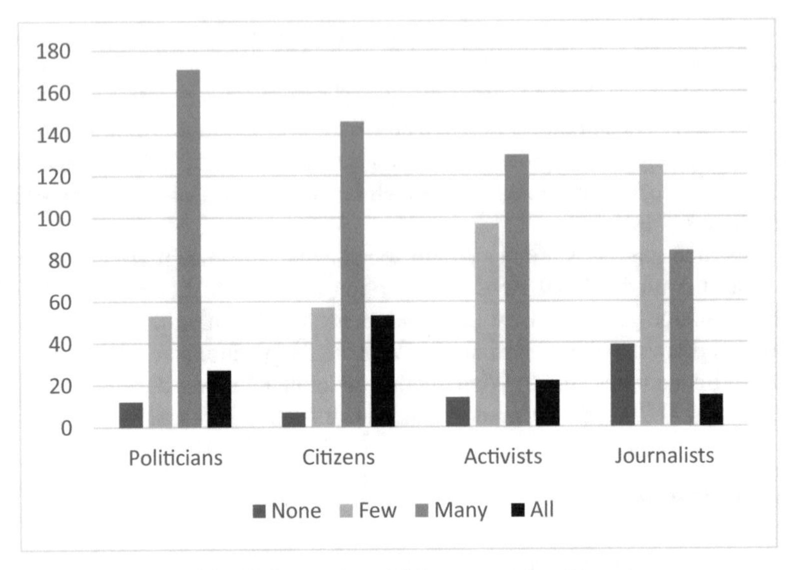

Fig. 1. Perception of fake news authorship.

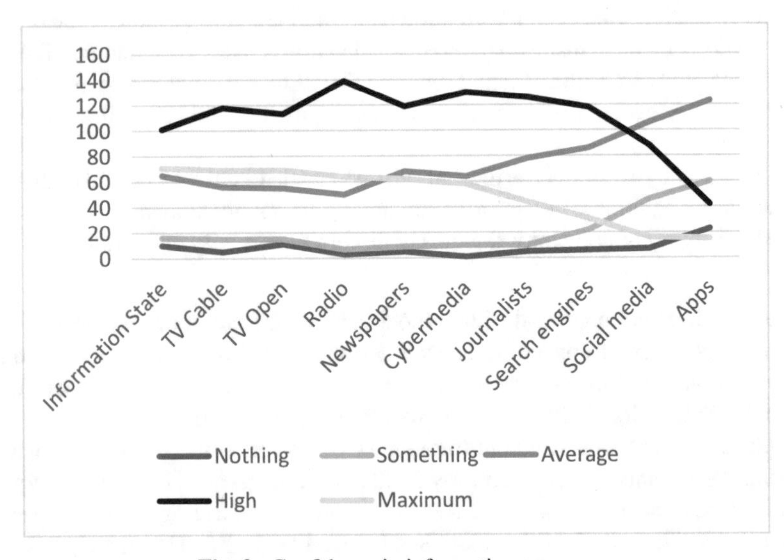

Fig. 2. Confidence in information sources.

The sources of information that citizens trust is radio, newspapers, television, and the state (Fig. 2).

In terms of news consumption, respondents indicate that they mostly follow the news on a daily basis, especially national news, so their self-perception is that they are highly or fully informed (Table 1). News consumption through social media occurs mostly on Instagram and WhatsApp, to which they spend between 1 and 3 h per day.

Table 1. News consumption practices.

Age group	Follow news				Consume news			Informed person			
	Rarely	Every month	Every week	Diary	Local	National	International	Few	Medium	High	Complete
18–27	6	13	80	95	35	111	48	6	80	82	26
28–37	3	4	13	30	10	32	8	1	16	25	8
38–47	0	1	4	9	2	11	1	0	4	6	4
48–57	0	0	0	5	2	3	0	0	2	3	0
Total	9	18	97	139	49	157	57	7	102	116	38
Age group	Preferred social media						Daily time spent on social media				
	Facebook	Twitter	YouTube	WhatsApp	Instagram	Other	Minus 1	1–3 h	3–5 h	More 5	
18–27	66	14	14	40	55	5	21	83	57	33	
28–37	15	6	2	10	12	5	7	30	5	8	
38–47	5	3	0	3	3	0	3	5	3	3	
48–57	1	3	0	0	0	1	0	3	0	2	
Total	87	26	16	53	70	11	31	121	65	46	

Table 2. Practices against disinformation.

Verify information	Ability to identify fake news					Verification strategy			
	None	Minima	Media	High	Absolute	In media	In Internet	View sources	Other
Never	1	2	3	3	1	5	1	0	4
Sometimes	0	13	59	45	13	70	32	14	14
Always	1	3	20	81	18	75	17	15	16
	2	18	82	129	32	150	50	29	34

Verify information	It is fake news because it is				Spreading fake news			Report fake news		
	Inconsistent	Social alarm	It comes from social media	Other	Never	Sometimes	Always	Never	Sometimes	Always
Never	6	2	0	2	9	0	1	9	1	0
Sometimes	70	22	26	12	113	15	2	61	44	25
Always	74	19	24	6	117	1	5	51	46	26
	150	43	50	20	239	16	8	121	91	51

In relation to the figures on trust in the media, the most common verification practice is to check against published news stories that are perceived to be false (Table 2). Citizens feel in a position to identify a false news item on the basis of its inconsistency, and once identified, they avoid re-broadcasting it and try to warn about published hoaxes.

4 Conclusions

Based on the results, the research questions are answered. 1) The social actors that generate the most disinformation are political leaders and people who spread false news. 2) The predominant news consumption practices of Ecuadorians include a daily dedication to learn about the reality of the country, which also involves the use of social networks, particularly between 1 and 3 h. The most used social networks are WhatsApp and Instagram. 3) The sources of information that generate trust come from the media and State information. 4) Among the fight against information is to contrast the messages of social networks with the news published in the media.

However, the credibility value of the media as a provider of reliable stories and data is restored. The evidence obtained reaffirms the perceived credibility of the media, which, in the face of fake news, is a key to solving the problem of the low quality of information, which must also be achieved through information literacy training.

It should be pointed out that social media consumption is based on images and short texts, which leaves little room for argumentation, leaving the interpretation of facts to individuals, based on their subjectivities, without the contexts provided by the media, leaving room for speculation.

Acknowledgement. This publication is a first advance of a research project on the dissemination of fake news in Andean countries and verification practices. Funded by a call for proposals sponsored by the Vice-Rectorate of Research of the Universidad Técnica Particular de Loja.

References

1. Gutiérrez-Renteria, M.: Mexico. In: Newman, N., Fletcher, R., Robertson, C., Eddy, K., Nielsen, R. (eds.) Reuters Institute Digital News Report 2022, pp. 153–184. University of Oxford. (2022)
2. Salazar, G.: Más allá de la violencia: alianzas y resistencias de la prensa local mexicana. CIDE (2022)
3. Newman, N.: Resumen ejecutivo y hallazgos clave del informe de 2019, Informe de noticias digitales. (2019)
4. Osmundsen, M., Bor, A., Vahlstrup, P.B., Bechmann, A., Petersen, M.B.: Partisan polarization is the primary psychological motivation behind political fake news sharing on Twitter. Am. Polit. Sci. Rev. **115**(3), 999–1015 (2021)
5. Wardle, C., Derakhshan, H.: Information Disorder. Toward an interdisciplinary framework for research and policymaking. Council of Europe (2017)
6. Echeverría, M., Cano, C.A.R.: ¿La alfabetización digital activa la incredulidad en noticias falsas? Eficacia de las actitudes y estrategias contra la desinformación en México. Revista de Comunicación (2023). https://doi.org/10.26441/RC22.2-2023-3246

7. Tucker, A., et al.: Social media, political polarization, and political disinformation: a review of the scientific literature. SSRN Electron. J. (2018). https://ssrn.com/abstract=3144139. Accessed 30 Nov 2023
8. Rodrigo-Alsina, M., Cerqueira, L.: Periodismo, ética y posverdad. Cuadernos. Info **44**, 225–239 (2019)
9. Del-Fresno-García, M.: Desórdenes informativos: Sobreexpuestos e infrainformados en la era de la posverdad. El Profesional de la Información **28**(3), 1–11 (2019)
10. Vosoughi, T., Deb, R., Aral, S.: The spread of true and false news online. Science **359**, 1146–1151 (2018)
11. Koc-Michalska, K., Bimber, B., Gomez, D., Jenkins, M., Boulianne, S.: Public beliefs about falsehoods in news. Int. J. Press Polit. **25**(3), 447–468 (2020)
12. Pemstein, D., et al.: The V-Dem measurement model: latent variable analysis for cross-national and cross-temporal expert-coded data. In: V-Dem Working Paper No. 21. 7th edition. University of Gothenburg: Varieties of Democracy Institute (2022)
13. Marcos, J., Sánchez, J., Olivera, M.: La enorme mentira y la gran verdad de la información en tiempos de la postverdad. Scire **23**(2), 13–23 (2017)
14. Tanz, J.: "El periodismo lucha por la supervivencia en la era posterior a la verdad", Wired (2017). https://www.wired.com/2017/02/journalism-fights-survival-post-truth-era/. Accessed 30 Nov 2023
15. Cytrynblum, A.: Periodismo social: Una nueva disciplina. La Crujía (2007)
16. Funke, D.: Global responses to misinformation and populism. In: Tumber, H., Waisbord, S. (eds.) The Routledge Companion to Media Disinformation and Populism, pp. 449–458. Routledge (2021)
17. Magallón, R. La (no) regulación de la desinformación en la Unión Europea. Una perspectiva comparada. *Revista de Derecho Político*, 1(106), 319–346 (2019)
18. Rauls, L.: How latin American governments are fighting fake news. Americas Quaterly (2021). Consultado en https://www.americasquarterly.org/article/how-latinamerican-governments-are-fighting-fake-news/. Accessed 30 Nov 2023
19. European Commission (Ed.) Action Plan against Disinformation. EEAS - European External Action Service - European Union (2018)
20. Wilson, C., Grizzle, A., Tauzon, R., Akyempong, K., Cheung, C.K.: Media and Information Literacy Curriculum for Teachers. United Nations Educational, Scientific and Cultural Organization (2011)
21. Cerdá-Navarro, A., et al.: Fake o no fake, esa es la cuestión: reconocimiento de la desinformación entre alumnado universitario. Revista Prisma Social **34**, 298–320 (2021)
22. Alcázar, P.: Ecuador Estado Digital, junio 2023. Mentinno Consultores (2023)

Syntax in Emoji Sequences on Social Media Posts

Alexandre Pereira[1](✉) and Gabriel Pestana[2]

[1] Centre for Research in Applied Communication, Culture and New Technologies (CICANT) – Universidade Lusófona de Humanidades e Tecnologias, Lisbon, Portugal
alexandre.pereira@ulusofona.pt

[2] School of Technology, Setúbal Polytechnic University, Setubal, Portugal
gabriel.pestana@estsetubal.ips.pt

Abstract. With the standardization of emojis, by the Unicode consortium, their use was quickly adopted in publications on social networks. Since then, several studies have been carried out on the functions of emojis in digital communication. Emojis were assigned meanings of emotion, sentiment, semantics, digital gestures, and syntax. We sought to understand the points of view of authors who defend the existence of emoji grammar or emoji syntax and those who, on the contrary, defend that emojis have a function of digital gestures.

The research for this review was conducted using the PRISMA model. In this study, we tried to understand whether social media users have a syntactic concern when using emoji sequences. We only considered studies that work with messages published on social networks, or equivalent data, such as messages exchanged via computer or mobile applications.

The literature on the use of emoji sequences on social media distinguishes the use of repeated emojis from the use of non-repeating emojis. In the latter, it is possible to find sequences with syntax features. Although these studies used small corpus sizes, we concluded that some users show syntactic concerns when using emoji sequences. The authors propose to use Emoji Semantic Extractor combined with EmojiNet to search for the existence of syntactic sequences of emojis in large blocks of messages.

Keywords: emoji grammar · emoji syntax · emoji data analytics · emoji language · digital gestures

1 Introduction

During the last decade, emojis have emerged as a new mediator of online communication, particularly on social media. The first set of emojis, created in 1998 by Shigetaka Kurita, contained 176 elements of 12 × 12 pixels to improve text communication via mobile phone [1]. In 2015, the Unicode consortium published the first standard version of emojis with 776 emojis. The current version comprises 3782 emojis (Unicode consortium, September 2023). The first set of emojis created by Kurita contained symbols other than emotional representations. Most emojis in the original set represented weather elements,

Á. Rocha et al. (Eds.): WorldCIST 2024, LNNS 987, pp. 97–107, 2024.
https://doi.org/10.1007/978-3-031-60221-4_10

moon phases, the zodiac, numbers on boxes, gadgets, tools, and more. Of the first emojis, only about twenty expressed specific emotions like smile and surprise. Although, in the beginning, the representation of feelings was not very expressive, the feeling became an essential element in the first version of Unicode that supported emojis. Five years later, a sentiment emoji (😂-face with tears of joy) became the Oxford Dictionary's Word of the Year. The increasing trend in using emojis on social media is clear and has served to express emotions on a broader range, refer to specific concepts more quickly, and express actions [2].

Sentiment analysis and the study of semantic nuances embedded in emojis have become a growing area of research, captivating numerous scholars who explore its implications from various angles. Wijeratne et al. [3] pioneered the creation of vector representations for emojis using a word embedding model derived from EmojiNet [4], effectively capturing their semantics. Miller et al. [5] delved into the interpretative variations of emojis, scrutinizing both semantic and sentiment aspects across different platforms. Their objective was to discern whether the semantics and sentiments attributed to emojis vary depending on the platform of consumption.

Berengueres & Castro's [6] study focuses on the emotional resonance of emojis, shedding light on the disparity in sentiment perception between those who employ emojis and those who interpret them. Eisner et al. [7] contributed by constructing vector representations for all Unicode-defined emojis, using Emoji2vec derived from concise emoji descriptions on the Unicode website. In evaluating their model, they conducted sentiment analysis tests on tweets, surpassing the performance of the skip-gram method by Barbieri et al. [8].

Barbieri et al. [9] argued that seasonal influences impact the usage of identical emojis, asserting that analysing emoji semantics is challenging due to their subjective nature, particularly concerning emotional descriptions. Na'aman et al. [10] posited that emojis serve semantic and affective purposes. They contend that emojis can convey textual content (semantics, prepositions, punctuation) or serve as affective markers, sometimes embodying both functions simultaneously.

Hu et al. [11] explored four types of emojis (positive, neutral, negative, and non-facial), encompassing 20 distinct emojis. Their findings indicated that emojis predominantly express emotions, intensify expression, and modulate tone. Notably, negative emojis are more frequently employed to convey emotions, while neutral emojis are preferred for expressing irony.

In 2017, Marcel Danesi proposed that emojis can function as an autonomous language, drawing insights from a study involving 323 individuals aged 18 to 22 who consider emojis integral to their digital interactions [12]. Grounded in semiotics, Danesi views emojis as symbols transcending literal representation, asserting their emotional and communicative expression capacity, encompassing intent, mood, and spirit. He contends that the Internet's influence has fostered diverse writing modes and argues that emojis, categorized as non-alphabetic text, constitute an artificial writing system with both pictographic and logographic functions, making them more universally applicable than traditional writing systems. Danesi also introduces two ways of constructing messages with emojis, both using implicit syntactic structures: one aligns with spoken language structure, while the other expresses interpretable ideas to create a conceptual framework. An example of the latter is a PETA campaign that Danesi identifies as having a syntactic structure representing a girl dreaming of beautification actions to become a princess. He concludes by regarding emojis as a universal symbolic language, comparable in certain aspects to Esperanto, noting their artificial and universal nature.

Several projects explore English-to-emoji translation. The "Emoji Translation Project" by Fred Benenson and Chris Mulligan focuses on translating texts, with Benenson having previously translated Herman Melville's Moby Dick into emoji sequences (Emoji Dick). Another project, the "Emoji-to-Text-Translator" utilizes a Word2Vec methodology to translate text into emojis. Joe Hale created pictorial representations of well-known books, such as Lewis Carroll's Alice's Adventures in Wonderland and Homer's Odyssey, using only emojis.

However, McCulloch & Gawne [13] present a contrasting view. Analysing sequences of emojis in a corpus of over a billion emojis from the SwiftKey smartphone keyboard, they searched for "emoji grammar". Their study of sequences of 2, 3, and 4 emojis compared to sequences of 2, 3, and 4 words revealed that over half of the 200 most used emoji sequences are repetitions, unlike word sequences where repetitions are rare. They argue that emojis should be regarded as digital gestures without a hierarchical structure or grammatical function.

Table 1. Top 10 emoji bigrams, trigrams, and quadrigrams in McCulloch & Gawne study

Top 10 bigrams	Top 10 trigrams	Top 10 quadrigrams
1. 😂😂	1. 😂😂😂	1. 😂😂😂😂
2. 😘😘	2. 😘😘😘	2. 😘😘😘😘
3. 😍😍	3. 😍😍😍	3. 😍😍😍😍
4. 😊😊	4. 😊😊😊	4. 😊😊😊😊
5. ❤❤	5. ❤❤❤	5. ❤❤❤❤
6. 😭😭	6. 😭😭😭	6. 💋💋💋💋
7. 😁😁	7. 👍👍👍	7. 👍👍👍👍
8. 👍👍	8. 😁😁😁	8. 😭😭😭😭
9. 😔😔	9. 💋💋💋	9. 😁😁😁😁
10. 😍😘	10. 😔😔😔	10. 😳😳😳😳

Despite having worked with a much smaller data set than the study above - they used only 269 posts methodically selected - Ge & Herring [14] and Herring & Ge [15] came to different conclusions than McCulloch & Gawne. In their studies, they discarded sequences of emojis with similar meanings (e.g., two similar smileys) and repeated emojis, and only considered sequences of different emojis. In conclusion, they assigned meaning to emoji sequences and observed that they can function like verbal utterances and stated that the emoji system is becoming a language.

According to Ge & Herring [14], few studies have focused on syntactic relationships within emoji sequences. On the other hand, Ferrari [16] adds that studies that address

Table 2. Emoji sequences and respective interpretation in Ge and Herring studies

Emoji sequence	Assigned meaning
	I'm just kidding; kiss.
	LOL I got a haircut.
	I love spicy food.
	I will run and stay away from beer and ice cream.
	Congratulations on receiving four awards.
	I love the gifts.
	You and your children, (we) cheer (you) on, become very happy.

linguistic issues in emojis, in general, work with corpora too small for the results obtained to be considered universal. Therefore, it is important to carry out a study that analyses the possible syntactic relationships in emoji sequences within a corpus big enough that allows drawing universal conclusions. Since the studies by McCulloch & Gawne and Ge & Herring, mentioned above, oppose emoji grammar to digital gestures, this literature review will focus precisely on these two themes and the opposition between them.

2 Research Methodology

We used the PRISMA model to elaborate the systematic literature review on emoji grammar [17]. In selecting the articles to analyse, we were only interested in those that worked on posts by users of social networks. Theoretical studies on the subject were not of interest to this study. We were interested in understanding whether there are users who show (implicit or explicitly) syntactic concerns when using sequences of emojis. The goal was to analyse if users write posts on social networks, expressing some level of concern with syntax when using emojis. As such, the research challenge intended to find scientific evidence to answer the following question:

> Is there a syntactic concern on the part of some users in sequences of non-repeated emojis, in posts on social networks?

We were interested in understanding whether there are users who have syntax concerns when writing on social networks. It didn't have to be all users; McCulloch & Gawne [13] showed that most of the observed emoji sequences are repetitions and digital gestures without associated syntax. But, users who try to write linguistically correctly, even if they experience difficulties, can try to be creative, as shown by Jing Ge [18]. However, according to the same author, emojis are not the best tool for structured writing.

2.1 Inclusion and Exclusion Criteria

The literature review selectively incorporates articles focusing on messages in social networks or equivalent platforms, specifically addressing emoji sequences, syntax, or digital gestures, to analyse the varied perspectives of authors and the supporting data regarding the role of emojis in social network communication.

Table 3. List of criteria considered for the SLR.

Inclusion criteria (topics covered)	Exclusion criteria (topic uncovered/omitted)
INC1 - emoji bigrams	EXC1 - does not address emoji sequences or syntax
INC2 - linguistic functions of emojis	EXC2 - addresses only sentiments or emotions
INC3 - the syntax of emojis	EXC3 - does not address any of the inclusion criteria
INC4 - sequences of emojis	EXC4 - no reference to emojis in the title or abstract
INC5 - the grammar of emojis	EXC5 - not written in one of the following languages: English, Portuguese, French, Spanish, Italian
INC6 - emojis as gestures	EXC6 - not a scientific paper or thesis
	EXC7 - does not use data from posts on social networks
	EXC8 - no emoji data analysis

2.2 Databases

We chose two databases to search for articles for our review. The first is b-on, a Portuguese database that searches ten other databases, including SCOPUS and EBSCO. The other is Google Scholar, through the Publish or Perish application. Both databases allow for advanced searches and export the complete list of articles as a CSV file and the bibliographic references in EndNote and BibTex formats.

Queries

We formulated two queries applied to both databases, generating four article sets; the first query required the presence of "emoji" while excluding "emoticon," incorporating terms like "semiotic," "linguistic," "grammar," "beat gestures," or "sequence," with the constraint that "emoji" must be in the title and articles published from January 2016 to July 2023 – the first Unicode standard for emojis was published in the summer of 2015. The first query became:

- Query text: (emoji AND (semiotic OR linguistic OR grammar OR beat gestures OR sequence)) NOT emoticon
- Title must contain: emoji
- Date range: from January 2016 to July 2023

The second query, simpler, focused on the term "bigram" in the title within the same date range, aiming to accumulate articles studying emoji sequences that specifically mention "bigram" instead of "sequence." The search was confined to peer-reviewed English articles via b-on, and due to the limitations of the Publish or Perish application, article quality and language were evaluated retrospectively. The second query became:

- Query text: bigram
- Title must contain: emoji
- Date range: from January 2016 to July 2023

3 Results Analysis

We ran both queries against both databases and got the results below. The first query on the b-on database returned 84 records. The same query applied in Google Scholar, through Publish or Perish application returned 144 records. The second query returned 4 records on b-on and 25 records on Publish or Perish. The total number of records returned by b-on was 88. The number of records returned by Publish or Perish was 169.

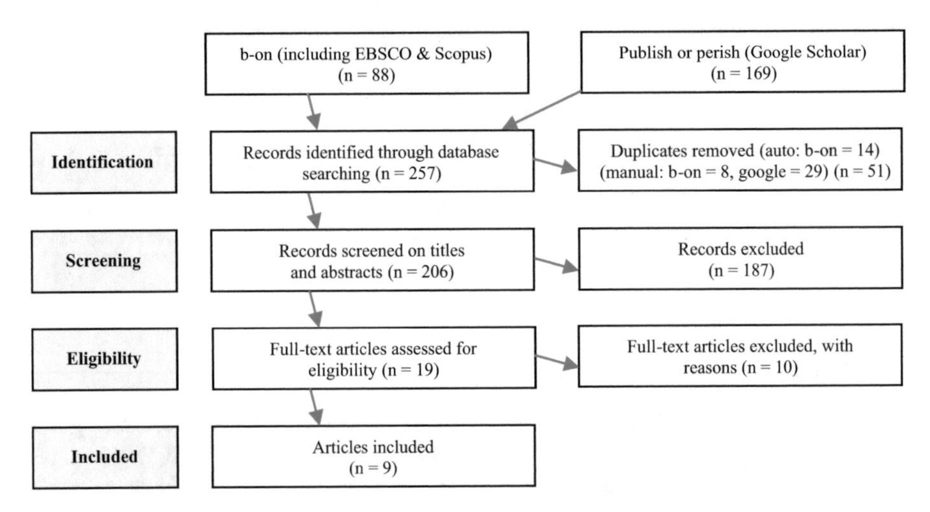

Of the 206 articles selected for titles and abstracts, 187 were excluded. The majority of excluded articles do not meet any of the inclusion criteria. Some articles are studies of emoji sentiments, some do not address emoji sequences or syntax, others still address the syntax of emojis, but do not use social media data. Included articles address emoji bigrams, discuss emoji sequences or syntax, or discuss emojis as gestures, which were the relevant topics to help answer the research question.

In the eligibility phase, 10 articles were eliminated (see Table 5). With the exclusion criterion EXC1, 4 articles were excluded. Pasternak & Tieu [19] compare emojis to gestures but don't study emoji syntax, sequences or bigrams. The articles by Arafah & Hasyim [20] and Saidi et al. [21] make a semiotic analysis of emojis, but does not address emoji sequences or syntax. Sun [22] mentions emojis as an Internet language but does not address emoji sequences or syntax.

The EXC3 criterion excluded 3 articles. In Stamatov study [23], participants communicated using emojis. Article analysed the order of emojis SVO, SOV, SV, SO, OS, OV, etc. After, they used t-tests, ANOVA, and correlations. The experience had 4 rounds, and the one that is relevant to us - mixed emoji-text communication, because it does not force the user to use syntax with emojis - is not reported in the document. Makhachashvili et al. [24] make a comparison between emojis and hieroglyphs. They do not analyse emoji sequences in messages created by social media users. Cohn & Foulsham [25] measured parameters related to understanding dyads involving emojis, but not sequences of emojis, and they do not use data produced on social networks.

Table 4. Articles excluded in the screening phase.

	b-on query1	b-on query2	PoP query1	PoP query2	TOTAL
INC1	0	0	0	1	**1**
INC2	3	0	4	0	**7**
INC3	0	0	1	0	**1**
INC4	1	0	2	1	**4**
INC5	2	0	1	2	**5**
INC6	1	0	0	0	**1**
TOTAL INC					**19**
EXC1	1	0	0	0	**1**
EXC2	9	2	14	10	**35**
EXC3	44	0	89	10	**143**
EXC4	1	0	0	0	**1**
EXC5	2	0	2	0	**4**
EXC6	0	0	2	0	**2**
EXC7	0	0	1	0	**1**
TOTAL EXC					**187**

The EXC7 criterion eliminated 2 articles. The study by Ferrari [16] is a literature review. Discusses emoji grammar, but does not show any data, tables, or results of emoji sequences data analysis. In the study by Thamsen & Engelen [26] participants didn't write using emojis. Instead, they answered questions about sentences written with emojis. The study names some difficulties that writing with emojis can have; namely, it cannot distinguish between active and passive voice.

The EXC8 criterion eliminated 1 article. Floc'h study [27] mentions an emoji syntactic data analysis but does not present the data on which he concludes. He adds that it is not possible to draw inferences from the grammatical scope in the use of emojis, due to the small size of the corpus (Tables 1 , 2, 3, 4, 6 and 7).

After applying the PRISMA model, nine articles were retained for analysis. In [15], Herring & Ge explored the order of emojis in sequences, investigating whether such sequences exhibit language-like properties, particularly syntax, utilizing a relatively small dataset of 300 emoji sequences. Ge & Herring [14] and Jing Ge [18] delved into the innovative and intricate ways users employ emojis, suggesting their evolution toward becoming a meaningful and effective form of language in computer-mediated communication. Khandekar [28] proposed an emoji-exclusive communication tool, hinting at the existence of an emoji grammar, while Khandekar et al. [29] assigned meaning to sequences of non-repeated emojis by social network users, emphasizing the intentional creation of meaning with these sequences. Ge-Stadnyk [30] challenged McCulloch's perspective that emojis primarily function as digital gestures, offering results from their study on emoji sequences. Cohn et al. [31] explored emoji grammar using experimental

Table 5. Articles excluded in the eligibility phase, with reasons

	b-on query1	b-on query2	PoP query1	PoP query2	TOTAL
INC3	0	0	1	0	**1**
INC4	2	0	1	1	**4**
INC5	1	0	0	3	**4**
TOTAL INC					**9**
EXC1	2	0	2	0	**4**
EXC3	0	0	3	0	**3**
EXC7	1	0	1	0	**2**
EXC8	1	0	0	0	**1**
TOTAL EXC					**10**

Table 6. Citations and h-index for the eligible papers.

Title	h-index	Citations
Communicative functions of emoji sequences in the context of self-presentation: A comparative study of Weibo and Twitter users	1	19
The grammar of emoji? Constraints on communicative pictorial sequencing	1	57
Communicative functions of emoji sequences on Sina Weibo	1	84
Emoji grammar as beat gestures	1	42
OPICO: A platform for collecting and analysing emoji usage	0	0
Emoji as digital gestures	3	126
Opico: a study of emoji-first communication in a mobile social app	1	22
Emoji sequence use in enacting personal identity	1	33
Do emoji sequences have a preferred word order	1	9

data, concluding that emojis lack grammatical structure. In the studies by McCulloch & Gawne [13] and Gawne & McCulloch (2019), the authors debated the existence of emoji grammar, contrasting it with beat gestures, ultimately concluding that emojis are predominantly used as gestures. The subsequent table presents the h-index and number of citations for each of the eligible documents.

The following table shows the h-index and number of citations for each of the authors of the previous documents.

There are few articles discussing the issue of whether users are using emojis in their posts with a syntax concern. The two more cited articles go in opposite directions. Ge & Herring [14] claim emoji is becoming a language, while Gawne & McCulloch [32] argue

Table 7. Citations and h-index for the authors.

Author	h-index	Citations
Jing Ge-Stadnyk	12	746
Neil Cohn	38	5351
Jan Engelen	22	2471
Joost Schilperoord	19	1681
Susan C. Herring	70	35695
Gretchen McCulloch	9	770
Lauren Gawne	16	1227
Sujay Khandekar	1	22
Yuanzhe Bian	2	288
Ranjitha Kumar	16	2052
Chae Won Ryu	1	22
Joseph Higg	2	22
Jerry O. Talton	9	854

that emojis are used as gestures. In fact, they are the most active authors discussing the possible existence of emoji grammar.

Nevertheless, the authors used different methods of analysis. Gawne & McCulloch analysed all the emoji sequences, not distinguishing between repeated emojis from non-repeated emojis. On the contrary, Ge & Herring analysed non-repeated emojis separately and obtained different results from Gawne & McCulloch.

4 Conclusions and Future Work

In response to Danesi's suggestion of a potential emoji grammar [12], various researchers have investigated syntactic relationships in emoji sequences within social network publications. The literature distinguishes between repeated and non-repeated emoji sequences, with some attributing linguistic functions, including syntax [14] and [28], while others contend that emojis are merely digital gestures [32]. This review focused on user behaviour, particularly in non-repeated emoji sequences on social networks, aiming to understand if users exhibit genuine syntactic concerns when using emojis. Notably, Jing Ge-Stadnyk and Susan C. Herring [14] argue for language-like properties and syntactic evidence in emojis. However, conflicting views from Cohn et al. [31] and Gawne & McCulloch [32] assert that emojis lack grammatical structure and are primarily used as digital gestures.

A large-scale examination of users employing emojis as a language, particularly in non-repeated sequences on social networks, is proposed to advance this study. Ge and Herring's [14] approach, which interprets the meaning of emoji sequences within messages, is considered but on a larger scale, utilizing the Emoji Semantic Extractor from Pereira and Pestana [33] combined with EmojiNet [4] for syntactic information. Similar to Gawne & McCulloch [32], the focus is on non-repeated sequences, with an additional proposal to analyse distinct message groups discussing various topics to identify characteristic emoji sequences. Addressing the critique of small corpus size [16], this proposed study aims to employ a substantial corpus, ensuring more meaningful results.

References

1. Negishi, M.: Meet Shigetaka Kurita, the father of emoji. Wall Street J. **26** (2014)
2. Ayvaz, S., Shiha, M.O.: The effects of emoji in sentiment analysis. Int. J. Comput. Electr. Eng. **9**, 360–369 (2017). https://doi.org/10.17706/IJCEE.2017.9.1.360-369
3. Wijeratne, S., Balasuriya, L., Sheth, A., Doran, D.: A semantics-based measure of emoji. Similarity **8** (2017). https://doi.org/10.1145/3106426.3106490
4. Wijeratne, S., Balasuriya, L., Sheth, A., Doran, D.: EmojiNet: building a machine readable sense inventory for Emoji. LNCS, vol. 10046, pp. 527–541 (2016). https://doi.org/10.1007/978-3-319-47880-7_33
5. Miller, H., Thebault-Spieker, J., Chang, S., Johnson, I., Terveen, L., Hecht, B.: "blissfully happy" or "ready to fight": Varying interpretations of emoji. In: Proceedings of 10th International Conference on Web and Social Media, ICWSM 2016, pp. 259–268 (2016)
6. Berengueres, J., Castro, D.: Differences in emoji sentiment perception between readers and writers. In: Proceedings of 2017 IEEE International Conference on Big Data 2017, pp. 4321–4328 (2018). https://doi.org/10.1109/BigData.2017.8258461
7. Eisner, B., Rocktäschel, T., Augenstein, I., Bosnjak, M., Riedel, S.: emoji2vec: Learning Emoji Representations from their Description, pp. 48–54, w16-6208 (2016)
8. Barbieri, F., Ronzano, F., Saggion, H.: What does this emoji mean? A vector space skip-gram model for twitter emojis. In: Proceedings of the 10th International Conference on Language Resources and Evaluation, LREC 2016, pp. 3967–3972. European Language Resources Association (ELRA) (2016)
9. Barbieri, F., Marujo, L. s., Karuturi, P., Brendel, W., Saggion, H.: Exploring emoji usage and prediction through a temporal variation lens. In: CEUR Workshop Proceedings 2130, 1–5 (2018)
10. Na'aman, N., Provenza, H., Montoya, O.: Mojisem: Varying linguistic purposes of emoji in (Twitter) context. In: ACL 2017 - 55th Annual Meeting of the Association for Computational Linguistics Proc. Student Res. Work, pp.136–141, P17–3022 (2017)
11. Hu, T., Guo, H., Sun, H., Nguyen, T.V.T., Luo, J.: Spice up your chat: The intentions and sentiment effects of using emojis. In: Proceedings of 11th International Conference Web Social Media, ICWSM 2017, pp. 102–111 (2017)
12. Danesi, M.: The Semiotics of emoji The rise of visual language in the age of the Internet. Bloomsbury Publishing Plc, London (2017)
13. McCulloch, G., Gawne, L.: Emoji grammar as beat gestures. In: CEUR Workshop Proceedings 2130, pp. 3–6 (2018)
14. Ge, J., Herring, S.C.: Communicative functions of emoji sequences on Sina Weibo. First Monday (2018)
15. Herring, S.C., Ge, J.: Do emoji sequences have a preferred word order? In: 14th International AAAI Conference on Web Social Media, pp. 1–5 (2020)
16. Ferrari, E.: The study of emoji linguistic behaviour: an examination of the theses raised (and not raised) in the academic literature. Commun. Soc. (Formerly Comun. y Soc. 36, (2023). https://doi.org/10.15581/003.36.2.115-128.

17. Page, M.J., et al.: Others: The PRISMA 2020 statement: an updated guideline for reporting systematic reviews. Int. J. Surg. **88**, 105906 (2021)
18. Ge, J.: Emoji sequence use in enacting personal identity. In: Companion World Wide Web Conference on WWW 2019, pp. 426–438 (2019). https://doi.org/10.1145/3308560.3316545
19. Pasternak, R., Tieu, L.: Co-linguistic content inferences: From gestures to sound effects and emoji. Q. J. Exp. Psychol. **75**, 1828–1843 (2022). https://doi.org/10.1177/17470218221080645
20. Arafah, B., Hasyim, M.: Linguistic functions of emoji in social media communication. Opcion **35**, 558–574 (2019)
21. Saidi, A.I., Puspitasari, D.G., Hermawan, F.F.: The Function Of Emoji In Digital Communication In Indonesia. Webology (2022)
22. Sun, Y.: Changes from the Internet Language to Emoji. In: 6th International Conference on Education, Language, Art and Inter-Cultural Communication (ICELAIC 2019), pp. 508–511 (2020)
23. Stamatov, E.G.: Do emoji use a grammar? Emergent structure in non-verbal digital communication, Arno (2017)
24. Makhachashvili, R., Bakhtina, A., Kovpik, S., Semenist, I.: Hieroglyphic Semiotics of Emoji Signs in Digital Communication. In: International Conference on New Trends Language, Literature Social Communication (ICNTLLSC 2021), pp. 182–192 (2021)
25. Cohn, N., Foulsham, T.: Meaning above (and in) the head: combinatorial visual morphology from comics and emoji. Springer (2022)
26. Thamsen, L., Engelen, J.A.A.: Impact of linearity on emoji sequences. Limitations of emoji, Arno (2019)
27. Floc'h, J.L.: Standardisation et différenciation des emplois des emoji sur Facebook : observations à partir d'un exercice pédagogique en DUT. Communiquer **28**, 35–51 (2020). communiquer.5397
28. Khandekar, S.: OPICO: A platform for collecting and analyzing emoji usage. Core (2018)
29. Khandekar, S., Ryu, C.W., Higg, J., Talton, J.O., Bian, Y., Kumar, R.: OPICO: a study of emoji-first communication in a mobile social app. In: Companion World Wide Web Conference 2019, vol. 18, pp. 450–458 (2019). https://doi.org/10.1145/3308560.3316547
30. Ge-Stadnyk, J.: Communicative functions of emoji sequences in the context of self-presentation: A comparative study of Weibo and Twitter users. Discourse Commun. **15**, 369–387 (2021). https://doi.org/10.1177/17504813211002038
31. Cohn, N., Engelen, J., Schilperoord, J.: The grammar of emoji? Constraints on communicative pictorial sequencing. Cogn. Res. **4**, 1–18 (2019). https://doi.org/10.1186/s41235-019-0177-0
32. Gawne, L., McCulloch, G.: Emoji as digital gestures. Language@Internet (2019)
33. Pereira, A., Pestana, G.: Is there meaning in the emoji sequences used on social media? the architecture of a model for emoji sequences analysis. World Conf. Inf. Syst. and Tech. Springer Int. Pub. (2022). https://doi.org/10.1007/978-3-031-04819-7_28

Trends and Collaborations in Information Systems and Technologies: A Bibliometric Analysis of WorldCIST Proceedings

Gonzalo Gabriel Méndez[1(✉)], Ronny Santana[2], and Oscar Moreno[1]

[1] Escuela Superior Politécnica del Litoral, Guayaquil, Ecuador
{gmendez,odmoreno}@espol.edu.ec
[2] Universidad de Guayaquil, Guayaquil, Ecuador
ronny.santanae@ug.edu.ec

Abstract. We present a bibliometric analysis of the WorldCIST proceedings from 2013, exploring key trends and collaborative dynamics within the conference. By examining WorldCIST's top contributors, thematic evolution, and co-authorship networks, we provide insights into the shifting landscape of information systems and technologies. Our findings highlight emerging research themes and the increasing international and interdisciplinary collaboration among participants. This condensed overview underscores WorldCIST's role as a forum for advancing knowledge in information systems and technology, reflecting its growing global influence and adaptability to new technological trends.

Keywords: Bibliometric analysis · Scientometrics · WorldCIST · Information Systems · Information Technologies

1 Introduction

The World Conference on Information Systems and Technologies (WorldCIST) has emerged as a prolific platform for scholars and practitioners. Since its inception in 2013, WorldCIST has been instrumental in shaping research directions and fostering collaborations across several disciplines related to information systems and technology. This paper presents a bibliometric analysis of WorldCIST proceedings, aiming to elucidate the evolving trends, collaboration networks, and the overall scholarly impact of the conference.

Bibliometric analyses in academic research serve as a tool for mapping the scientific landscape, understanding the dynamics of knowledge production, and identifying influential works and authors. In the context of WorldCIST, such an analysis is particularly valuable given the conference's role in highlighting cutting-edge research and technological advancements.

We focus on identifying the key research themes and topics that have dominated WorldCIST proceedings and uncover the patterns of collaboration and authorship, shedding light on the nature and extent of scholarly networks within the conference.

Á. Rocha et al. (Eds.): WorldCIST 2024, LNNS 987, pp. 108–117, 2024.
https://doi.org/10.1007/978-3-031-60221-4_11

This study contributes to a deeper understanding of the evolution and current state of research in information systems and technology. It also provides valuable insights for researchers, academics, and practitioners interested in the development and future directions of the field of Information Systems and Technologies.

2 Related Work

The field of bibliometric analysis has grown significantly, offering vital insights into the patterns and progressions in various academic domains. This section reviews relevant literature that underpins the methodology and context of our study, focusing on bibliometric analyses in information systems and technology, and previous assessments of academic conferences.

Bibliometric methods have been widely used to evaluate academic literature, offering insights into publication trends, citation impacts, and collaborative networks [1]. Tools like citation analysis, co-authorship mapping, and keyword frequency analysis have been particularly effective in revealing the intellectual structure and evolution of scientific fields [9].

Studies focusing on academic conferences have employed bibliometric techniques to understand the dynamics of knowledge sharing and collaboration. For instance, Moya-Anegón et al. demonstrated how conference proceedings could reflect the evolving trends and research networks within specific disciplines [7]. The analysis of co-authorship patterns in conferences has also been shown to offer insights into collaborative structures and trends over time [6].

The field of information systems and technology has been the subject of extensive bibliometric analyses. Studies like those by Leydesdorff & Zhou have explored the thematic and citation landscapes of information technology research, uncovering trends and emergent topics [5]. Similarly, analyses of specific journals or publication venues in this field have provided insights into the predominant research themes and influential works (e.g., [3]).

While there have been general bibliometric studies on information systems and technology, to our knowledge, there are not focused analyses on the scholarly production of WorldCIST. This gap underscores the novelty of our study, which provides a detailed bibliometric overview of the WorldCIST proceedings since 2013. The existing body of literature establishes the relevance and utility of bibliometric analyses in understanding academic landscapes. Our study builds upon these methodologies and insights, applying them to the context of WorldCIST to contribute to the understanding of trends, impacts, and collaborative dynamics in information systems and technology.

3 Data Collection and Processing

This bibliometric analysis focuses on the WorldCIST proceedings from 2013 to 2022. It does not include papers from 2023, as they are set to be published in early 2024, which is beyond the timeframe of this article's writing. We used the

APIs from OpenAlex[1] and Semantic Scholar[2] to collect the data. These APIs provided extensive bibliographic details including author information and titles. We also resorted to the dblp[3] computer science bibliography for verification and completeness purposes.

The resulting dataset was composed of a total of 1999 papers. The histogram of Fig. 1 depicts the number of WorldCIST papers published each year from 2013 to 2022. There's a fluctuating trend in publication volume over the years, with a notable peak in 2018, when 304 papers were published. The lowest publication count occurred in 2014, with 106 papers. Post-2018, there's a visible decline, with a slight rebound in 2021 and a decrease again in 2022. The analyses that follow are based on this data.

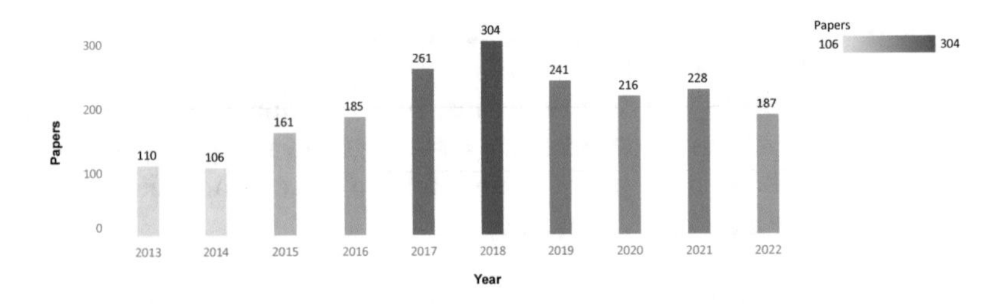

Fig. 1. Annual WorldCIST publication trend between 2013 and 2022.

4 Findings

In this section, we explore authorship trends, analyze institutional and geographical distribution, and conduct a thematic analysis to identify the dominant research themes within WorldCIST. Additionally, we investigate the collaboration patterns that emerge in the community to understand the dynamics of scholarly interactions.

4.1 Authorship Trends

Figure 2 provides a consolidated view of the scholarly contributions made by the 15 leading authors in WorldCIST. It outlines the annual publication patterns of contributors with 15 or more papers[4]. The year-by-year breakdown provides insight into the individual research activity levels, revealing both consistency and notable variations in output that could suggest changing research focus, project cycles, or collaborations.

[1] https://docs.openalex.org.

[2] https://www.semanticscholar.org/product/api.

[3] https://dblp.org.

[4] We have deliberately excluded paper counts from editorial roles to focus solely on research contributions.

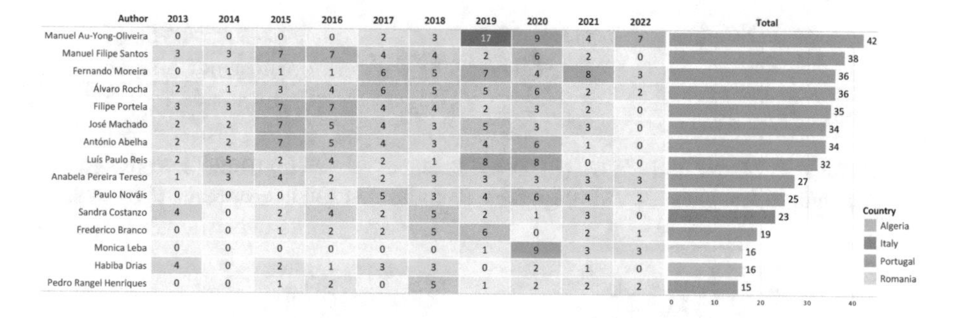

Fig. 2. Heatmap and bar chart representation of the publication frequency per author from 2013 to 2022, with a total publications count and a color-coded country association. The heatmap intensity indicates the number of publications per year, while the bar chart shows the cumulative publications for each author.

Manuel Au-Yong-Oliveira leads the group with a total of 42 publications, peaking sharply in 2019, with 17 contributions—serving as the first author in 3 of these. Close behind are Manuel Filipe Santos and Fernando Moreira, with 38 and 36 publications, respectively, indicating a consistent level of high output over the period under review. The figure also shows a significant representation of Portugal, with the top 10 authors in the ranking being affiliated with Portuguese institutions. Italy, Romania, and Algeria also make a significant appearance among the top contributors to WorldCIST.

4.2 Thematic Analysis

We performed a topic modeling analysis using the papers' abstracts[5] and applying the Mallet[6] toolkit. We used a latent Dirichlet allocation (LDA) algorithm, which is a probabilistic model that assumes documents are mixtures of topics and identifies these topics based on patterns of word frequencies and distributions across the document set. This makes LDA particularly suitable for uncovering hidden thematic structures in large text collections. To determine the optimal number of topics, we varied the number of topics and calculated coherence scores for each model. The coherence score is crucial as it evaluates the semantic relatedness of words within a topic, with higher scores indicating more meaningful connections. Our analysis revealed that a 14-topic model maximized the coherence score. Due to space constraints, we report here the five most salient topics from this model.

Topic 1 - Model and Technique Analysis (Fig. 3a): This topic seems to encompass terms related to modeling techniques and decision-making algorithms. It likely represents papers that focus on the development and application

[5] We did not have universal access to the papers' full texts.
[6] https://mimno.github.io/Mallet.

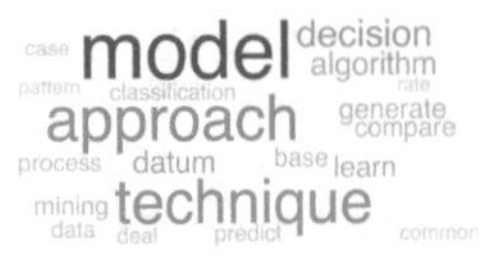

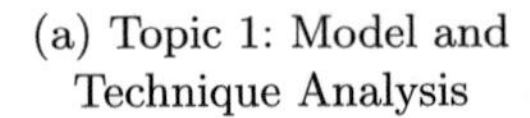
(a) Topic 1: Model and Technique Analysis

(b) Topic 2: Data Management Foundations

(c) Topic 3: Software Development Processes

(d) Topic 4: Systems Resource Management

(e) Topic 5: Organizational Information Systems

Fig. 3. Wordclouds of the top five topics (ordered by relevance) identified within World-CIST's scientific production.

of computational models and algorithms for data analysis and problem-solving within various domains.

Topic 2 - Data Management Foundations (Fig. 3b): Featuring words like 'datum', 'base', and 'service', this cluster pertains to foundational aspects of data management, such as database services, data quality improvement, and the underlying systems that support information management processes.

Topic 3 - Software Development Processes (Fig. 3c): The recurrence of terms such as 'software', 'development', and 'process' suggests a focus on the methodologies and life cycle processes associated with software engineering and application development.

Topic 4 - Systems Resource Management (Fig. 3d): With 'system' and 'resource' being prominent, this topic covers studies on systems engineering, resource allocation, and time-sensitive systems control—essential components in the optimization and management of technical and computational resources.

Topic 5 - Organizational Information Systems (Fig. 3e): This topic deals with the intersection of information systems and organizational contexts, including how businesses use information, social media, and digital resources to support operations, policy-making, and strategic objectives.

We also conducted a thematic evolution analysis using a WordStream [2]. The results show a dynamic shift in focus areas across paper abstracts from 2013 to 2022 (Fig. 4). Initially, in 2013, the dominant themes revolved around foundational software processes and information systems. As we progressed into 2014, there was a discernible pivot towards business-oriented themes, with terms like *business* and *area* becoming more prominent. The year 2015 shows an increased focus on data quality and organizational structures, which is indicative of a growing interest in the management and governance of information systems. By 2016, the attention had shifted towards social aspects and healthcare, with '*health*', '*content*', and '*social*' being prevalent, reflecting an expansion of information systems into more diverse and applied contexts. In the subsequent years, from 2017 to 2019, the emphasis seems to have moved toward performance measurement and the implementation of technologies within organizational settings. The year 2020 marks a significant transition to patient-centered themes, website service, and the experience of users, suggesting a response to the digital transformation and the user-centric approach that has become central to the field. This period likely corresponds with the increasing digitization efforts across sectors due to global events, such as the COVID-19 health crisis, which motivated a shift to online platforms and services. In the most recent years, 2021 and 2022, the WordStrean emphasizes terms like '*student*', '*education*', '*dataset*', and '*cancer*', suggesting a strong orientation towards educational technologies, data science, and health informatics. The emergence of 'cancer' as a key term reflects a notable intersection of information systems with specific health challenges and the application of data-driven approaches to address them.

These thematic transitions resonate with the findings from our topic modeling. The identified topics and their evolution depicted in the WordStream underline the shifting research interests over time and highlight the adaptability of WorldCIST to emerging research trends and societal needs.

Fig. 4. Wordstream visualization depicting the thematic evolution within the abstracts of WorldCIST from 2013 to 2022.

4.3 Collaboration Patterns

We constructed the WorldCIST co-authorship network, which encompasses 2394 authors and 4753 collaboration instances. Figure 5 shows the networks' four largest connected components, with authors represented as nodes and each collaborative effort as an edge between nodes. The size of each node encodes the number of connections of an author, reflecting their level of collaboration within the WorldCIST community. The node colors encode the geographical region to which each author belongs.

We used Cytoscape[7] [8] to identify the network's largest connected components, which represent the most densely interconnected clusters of researchers

[7] https://cytoscape.org.

within the WorldCIST community. These components are indicative of active collaboration and are often associated with strong research outputs and academic influence [4,10]. We also identified the central authors within each component based on their degree centrality. These individuals often serve as hubs of collaboration and are pivotal in the dissemination of ideas within the community.

The network is characterized by a considerable degree of fragmentation and its structure is dominated by two significantly large clusters. The most prominent component (Fig. 5 top left) groups 299 authors, while the second one (Fig. 5 top right) consolidates 199 authors. The vast majority of authors (n = 1896) are distributed across a myriad of smaller components beyond these two major clusters. The third (Fig. 5 bottom left) and fourth (Fig. 5 bottom right) largest groups contain 34 and 21 authors respectively. All other components comprise fewer than 20 authors each. This pattern of fragmentation suggests the presence of specialized research collaborations within the WorldCIST community.

The two largest connected components are heavily influenced by European, particularly Portuguese, authors, suggesting a strong regional network with deep academic ties. Both components are characterized by dense interconnections, denoting established collaborations likely built over successive conferences. This suggests the existence of a robust regional network characterized by deep academic ties within Europe, with Portugal playing a central role. Such dominance

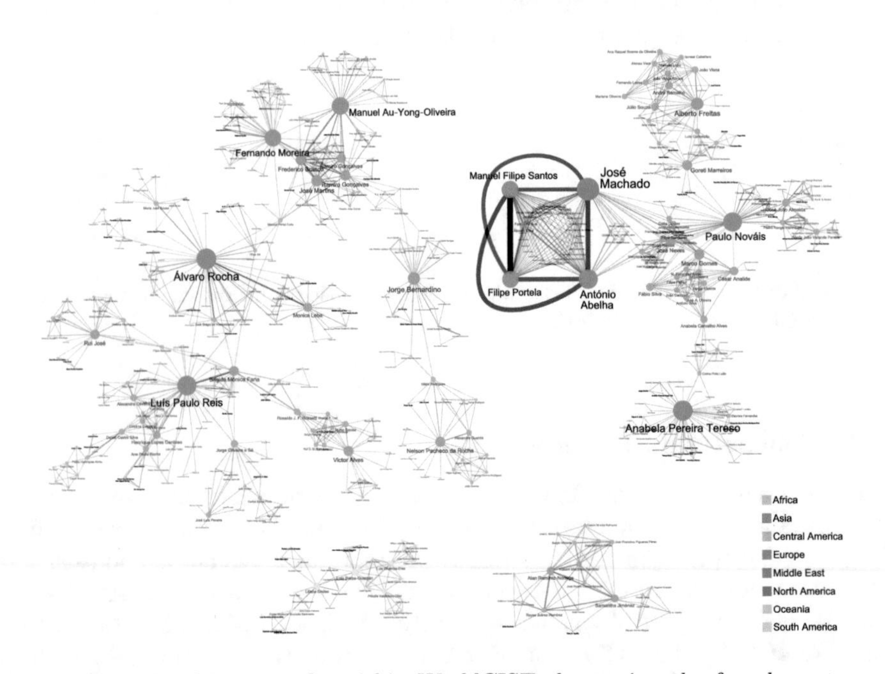

Fig. 5. Co-authorship networks within WorldCIST showcasing the four largest connected components. Each node represents an author, with node size reflecting the number of collaborations. Edges indicate co-authored works.

also points to established institutional partnerships or thematic research clusters that have evolved.

Álvaro Rocha holds the highest betweenness centrality in the network's largest connected component. This author has published under several Portuguese affiliations, including the University of Lisbon, the University of Coimbra, and the University of Porto. Fernando Moreira, Luís Paulo Reis, and Manuel Au-Yong-Oliveira, all affiliated with Portuguese institutions, are also distinguished by their high publication counts. Meanwhile, Manuel Pérez Cota, Ramiro Gonçalves, and Jorge Bernardino are notable for their roles as connectors bridging various sub-groups within the network. Notably, Manuel Pérez Cota, contributing from the University of Vigo in Spain, is the only one in this group not affiliated with a Portuguese institution.

The network's second-largest connected component prominently features Portuguese authors, notably affiliated with the University of Minho. This cluster, however, is not insular; it extends its academic reach internationally with notable collaborations involving authors from Nepal, the United Arab Emirates, the United States, Japan, Brazil, and Spain. Central figures like José Machado, António Abelha, Manuel Filipe Santos, and Filipe Portela emerge not just as the cluster's most prolific authors but as the most frequent collaborators within the entire conference. Also important in terms of their betweenness centrality are Anabela Pereira Tereso and Paulo Nováis.

The co-authorship network's third and fourth largest connected components are significantly smaller and less dense than the previous two. This could suggest emerging collaborations or interdisciplinary research efforts that are just beginning to coalesce. The third component showcases a considerable presence of South American scholars, predominantly from Ecuador. It includes key figures such as Luis Barba-Guamán, Liliana Enciso, and Luis Chamba-Eras, all affiliated with the Universidad Técnica Particular de Loja. The inclusion of European affiliations, like Barba-Guamán's connection to the Universidad Politécnica de Madrid (Spain), signifies active intercontinental collaboration, bridging South American research with European academic networks. This blend of regional focus and international outreach hints at the diverse, cross-border interactions that WorldCIST fosters, nurturing both localized research initiatives and broader, globally interconnected scholarly endeavors.

The fourth connected component of the WorldCIST co-authorship network displays a discernible concentration of academic collaboration within North America, specifically Mexico. Central to this component are Samantha Jiménez and Alan Ramírez-Noriega, affiliated with the Instituto Tecnológico de Tijuana and the Universidad Autónoma de Sinaloa, respectively. Reyes Juárez-Ramírez, also a prominent node, showcases the Autonomous University of Baja California's influence. The component's configuration, with these key individuals, underscores a cohesive network of Mexican scholars with connections extending to multiple institutions, indicative of a dynamic and collaborative research environment within the country's technological and university system.

5 Discussion and Future Work

The findings from our analysis of WorldCIST data present several noteworthy observations and implications for the field of information systems and technology research. Firstly, the authorship trends highlight prolific contributions from certain researchers, particularly from Portugal, illustrating the importance of individual scholarly efforts in shaping the conference's content. This raises questions about the diversity and inclusivity of voices in the conference, suggesting a potential need for broader representation from various global regions.

The institutional and geographical spread of contributions, while initially Eurocentric, has shown a promising expansion, indicating WorldCIST's growing global appeal. This shift towards a more diverse geographical representation could lead to richer, more varied perspectives in the field, fostering innovation and new approaches to tackling information systems and technology challenges.

Our thematic analysis revealed dominant research themes that reflect current and emerging interests in the field. The progression from foundational software processes to more applied themes—such as healthcare and education technology—mirrors the evolving societal needs and technological advancements. This trend underscores the responsiveness of the WorldCIST community to external factors, such as global health crises, and highlights the importance of adaptability in research.

Finally, the collaboration patterns observed in the co-authorship network reveal a robust structure of scholarly interactions, with notable regional clusters and indications of interdisciplinary collaborations. This network structure is vital for the dissemination of knowledge and the fostering of innovative research ideas. However, the observed fragmentation into many smaller groups suggests potential areas for fostering more extensive, cross-institutional collaborations.

Overall, our study provides valuable insights into the dynamics of the WorldCIST community, offering a lens through which we can understand broader trends in information systems and technology research. However, our study is limited in several aspects. First, our data primarily encompassed bibliographic details and abstracts, omitting full-text analyses, which could provide deeper thematic insights.

Future work could also focus on analyzing the conference's co-citation network to provide insights into the intellectual structure and evolution of WorldCIST's scientific dialogue. Such analysis could uncover underlying patterns of knowledge dissemination and thematic interconnections between research works. This approach would enable a more nuanced understanding of the conference's impact on the academic community, revealing key works and authors that have shaped its discourse. Such an analysis would complement our current findings, offering a more holistic view of the trends and influences within the information systems and technology research landscape.

Building on this study, future research can further explore the trends we have presented and their implications, thereby enriching our understanding of the field's development and trajectory.

6 Conclusion

Our analysis of the WorldCIST conference from 2013 to 2022 has revealed significant insights into authorship trends, thematic evolution, and collaboration patterns. The dominance of Portuguese authors underscores regional influences on academic contributions, while the thematic analysis reflects the conference's alignment with current technological and societal issues. The co-authorship networks highlight robust, often regionally focused, collaborative relationships. These findings enhance our understanding of WorldCIST's impact and evolution.

References

1. Aria, M., Cuccurullo, C.: Bibliometrix: an r-tool for comprehensive science mapping analysis. J. Informet. **11**(4), 959–975 (2017). https://doi.org/10.1016/j.joi.2017.08.007
2. Dang, T., Nguyen, H.N., Pham, V.: WordStream: interactive visualization for topic evolution. In: Johansson, J., Sadlo, F., Marai, G.E. (eds.) EuroVis 2019 - Short Papers. The Eurographics Association (2019). https://doi.org/10.2312/evs.20191178
3. Dwivedi, Y.K., et al.: Research on information systems failures and successes: status update and future directions. Inf. Syst. Front. **17**(1), 143–157 (2014). https://doi.org/10.1007/s10796-014-9500-y
4. Krumov, L., Fretter, C., Müller-Hannemann, M., Weihe, K., Hütt, M.T.: Motifs in co-authorship networks and their relation to the impact of scientific publications. Eur. Phys. J. B **84**(4), 535–540 (2011). https://doi.org/10.1140/epjb/e2011-10746-5
5. Leydesdorff, L., Zhou, P.: Nanotechnology as a field of science: its delineation in terms of journals and patents. Scientometrics **70**(3), 693–713 (2007). https://doi.org/10.1007/s11192-007-0308-0
6. Méndez, G.: Evolution of scientific collaboration communities in the AACE. In: EdMedia+ Innovate Learning. Association for the Advancement of Computing in Education (AACE), pp. 721–730 (2012)
7. Moya-Anegón, F., et al.: The impact factor of scientific journals: a bibliometric analysis. Scientometrics (2007)
8. Shannon, et al.: Cytoscape: a software environment for integrated models of biomolecular interaction networks. Genome Res. **13**(11), 2498–504 (2003). https://doi.org/10.1101/gr.1239303
9. Van Eck, N.J., Waltman, L.: Software survey: VOSviewer, a computer program for bibliometric mapping. Scientometrics (2010). https://doi.org/10.1007/s11192-009-0146-3
10. Vasilyeva, E., et al.: Multilayer representation of collaboration networks with higher-order interactions. Sci. Rep. **11**(1), 1–11 (2021). https://doi.org/10.1038/s41598-021-85133-5

A Comparative Assessment of Wrappers and Filters for Detecting Cyber Intrusions

Houssam Zouhri[1] and Ali Idri[1,2(✉)]

[1] Mohammed VI Polytechnic University, BenGuerir, Morocco
Houssam.Zouhri@um6p.ma, Ali.Idri@um5.ac.ma
[2] Software Project Management Research Team, ENSIAS, Mohammed V University, Rabat, Morocco

Abstract. The high number of features in network traffic data might overload intrusion detection systems (IDSs) and resulted in overfitting. Furthermore, duplicated, and unnecessary features may restrict an IDS's ability to learn and infer. Generally, feature selection methods can alleviate this issue. It is a data preprocessing step that can be applied before the classification phase and aims to improve classifier performance and interpretability by selecting only a few highly informative features. The present study aims at assessing the effects of four filters (Consistency-based subset selection, Pearson correlation, Double Input Symmetric Relevance and Chi2); and four wrappers (Boruta, BorutaShap, Recursive Feature Elimination, and Genetic Algorithm) on the classification effectiveness of four models: Random Forest, Multilayer Perceptron, eXtreme Gradient Boosting and Support Vector Machines using CICIDS2018 dataset. The findings suggested that combining ensemble models with RFE and CON techniques can effectively reduce the number of attributes without impacting the outcomes of classification.

Keywords: Intrusion detection systems · Feature selection · Filters · Wrappers

1 Introduction

With the expansion of internet use and networked systems, the confidentiality, integrity and security of digital resources are under constant threat. With such expansion in data traffic, network security becomes a major issue. Machine Learning-based IDSs (ML-IDS) are considered one of the most advanced solutions for securing sensitive data in networks [1]. However, the presence of unnecessary and irrelevant attributes in network traffic could decrease the efficacy of IDSs. Therefore, feature selection (FS) is required for intrusion detection technique that can be used to find and delete the useless features, boosting the accuracy and efficiency of IDSs.

According to a systematic literature review carried out in [2], researchers have mainly focused on FS as a pre-processing task to choose the optimum feature set and improve IDSs performances. FS has other benefits, including reducing the number of calculations, decreasing execution time and enhancing the transparency and interpretability of ML-IDS [3]. The researchers used different types of FS methods, covering filters,

Á. Rocha et al. (Eds.): WorldCIST 2024, LNNS 987, pp. 118–127, 2024.
https://doi.org/10.1007/978-3-031-60221-4_12

wrappers, embedded and hybrids. However, to the authors' knowledge, no studies have investigated and compared filter and wrapper techniques on IDS performance. Filter and wrapper techniques are the two basic types of FS methods. Filter techniques consist of evaluating features separately of any ML model, whereas wrapper approaches analyze the importance of features using a ML model [4]. In filter method, the approach evaluates the relevance and significance of individual features in a dataset without the involvement of a classifier algorithm, and can be classified into univariate and multivariate filters. While wrapper approach is used to calculate the importance or weights of features using a classification model to evaluate the performance of these inputs [5].

This study's primary goal is to evaluate and compare the effects of four wrappers FS techniques: Boruta (B) [6], BorutaShap (BSH) [7], Recursive Feature Elimination (RFE) [8] and Genetic Algorithm (GA) [9], and four filters: Pearson correlation (C) [10], Chi2 (K) [11], Double Input Symmetric Relevance (DISR) [12] and Consistency-based subset selection (CON) [13] on four classifiers' performance (Support Vector Machines (SVM), Random Forest (RF), XGBoost (XGB) and Multilayer Perceptron (MLP)) using CICIDS2018 datasets. The reason for choosing the above-mentioned filters lies on their proven robustness, demonstrating a balance between high accuracy and efficient computation times in our previous research [14], which involved a comprehensive evaluation of five univariate filters, and three multivariate filters including the CON and DISR. Furthermore, the selection of the aforementioned wrapper techniques is justified by the fact that they are the most widely used wrappers to select the most important feature subset in several IDSs research, such as [15]–[17]. Overall, this study uses an empirical evaluation of 32 combinations of FS-classifier: *32 = (4 classifiers * 1 datasets) * (4 filters + 4 wrappers*), and seeks at addressing the following research question (RQ): Does the performance of wrappers surpass that of filters, and are there any specific combinations of FS and classifiers that can improve the IDS accuracy?

This empirical study's contributions are summarized below:

- Proposing an innovative approach that combines the results of various filter and wrapper FS techniques to reveal the most critical features.
- Enhancing the selection of ML classifiers and FS methods through rigorous statistical analysis, including the Scott-Knott test and ranking methodologies such as the Borda Count.

The rest of this paper is structured as follows: Sect. 2 provides a summary of FS approaches. Section 3 includes some relevant work on the application of FS approaches, particularly in IDSs. The experimental design is described in Sect. 4. Results are presented and discussed in Sect. 5. Finally, Sect. 6 presents the conclusion and future work.

2 Background

This section gives a summary of FS approaches, focusing on wrapper-based FS techniques.

Wrapper FS strategies operate by iteratively including or omitting features from a subset and evaluating a ML model performance on that subset. The ultimate set of features chosen is the one which has the best performance.

a) **Boruta (B)** [6]**:** The B technique aims to identify and eliminate elements that are irrelevant to predict the target variable. It is a wrapper FS method that operates by comparing the importance of the original feature set with the importance of the shadow feature set. The shadow features are produced by randomly permuting the values of each column of the original features. This process is repeated until all unimportant features have been removed.
b) **Boruta-SHAP (BSH)** [7]: The BSH FS technique combines the B algorithm with Shapley values. This hybrid method employs the same working principle as B, but tt uses shaply values to calculate the relevance of each feature. The Shapley value is a concept from game theory that quantifies each feature's marginal contribution when combined with other features.
c) **Recursive feature elimination (RFE)** [8]: RFE is a wrapper FS method that removes attributes from the dataset iteratively and evaluates the detection rate of a ML model on the subset of remaining attributes. The features are deleted in declining order of relevance, until the model's performance begins to deteriorate.
d) **Genetic algorithm (GA)** [9]**:** GA are a kind of search algorithm that imitate the natural selection and genetics processes in biological systems. It operates by considering a population of individuals, each of them represents a subset of features. The best individuals are determined by how effectively a ML model performs on that subset of features. The best individuals are then combined and mutated to generate new feature subsets. The reproduction and mutation process are repeated until an end criterion is fulfilled, such as a maximum number of iterations or an achievement of a good performance.

3 Related Work

This section discusses some of the significant publications that dealt with FS approaches in the context of IDS, in particular filters and wrappers methods.

Zhiqiang et al. [17] presented a new wrapper FS approach using GA techniques and RF classifier. The IDS performance analysis is evaluated using the UNSW-NB15 and NSL-KDD datasets. The findings show that the RF algorithm achieved an accuracy of 96.12% and 92.06% using just 12 and 9 features for the NSL-KDD and UNSW-NB15 datasets, respectively. Sara Mohammadi et al.[18] developed a method based on FS and clustering approaches, employing a filter and wrapper called feature grouping-on linear correlation coefficient (FGLCC) technique and cuttlefish algorithm (CFA), respectively. The experiment was applied to the KDD Cup 99 dataset by using Decision tree (DT) algorithm. The results achieved a high accuracy of 95% using the proposed FGLCC-CFA algorithm. Halim et al.[15], presented work that focuses on improving network security and preserving critical data information with a minimal number of features using a GA-based FS method. The experiment was conducted on three network traffic benchmark datasets, including CIRA-CIC-DOHBrw-2020, Bot-IoT and UNSW-NB15, using three different ML models: SVM, K-NN and XGB. The results reached a maximum accuracy of 99.80%.

In addition, another recent work conducted by Awad et al.[16] proposed a novel approach employing RFE with cross-validation, while using a DT classifier (DT-RFECV) to select the relevant features from UNSW-NB15's dataset. Their DT-RFECV classifier achieved an accuracy of 95.30%. Megantara et al. [19] applied RFE and Gini importance to identify the important features for multi-classification task in the NSL-KDD dataset. The findings demonstrated that the used approach raises the accuracy when selecting the optimum subset and gives a better result than all features in the dataset. In [20], Yin et al. [20] presented a novel anomaly-based network IDS (IGRF-RFE) utilizing UNSW-NB15 dataset. The IGRF-FRE integrated Information Gain (IG) filter techniques with RF classifier to minimize the feature search space in its initial phase. Subsequently, it RFE wrapper technique with an MLP classifier. The outcomes demonstrate that through this approach, the accuracy of anomaly detection is improved by picking more relevant attributes and decreasing the feature space from 42 to 23. This improvement resulted in a heightened multi-classification accuracy for the MLP, increasing from 82.25% to 84.24%.

4 Experimental Design

This section discusses the intrusion detection dataset we used, as well as the processus employed to carry out the experiments.

4.1 Intrusion Detection Datasets

For this study, an intrusion detection dataset was used to facilitate our study. Table 1. gives a complete overview of this dataset after a rigorous pre-processing phase involving the elimination of infinite and missing values, the removal of irrelevant features and data balancing. In addition, valuable information such as the description, the number of features available, the total number of instances and the respective number of classes present in the dataset is presented.

Table 1. Data description.

Dataset	Description	#. of features/instance	#. of classes
CICIDS2018 [21]	This dataset stems from a collaborative effort involving both the Canadian Cybersecurity Institute and the Communications Security Establishment. It covers 7 attack scenarios including Heartbleed, Web attacks, Brute-force, Botnet, DoS, DDoS and Infiltration. The attacker's infrastructure is made up of 50 machines, whereas the victim organization is made up of 5 departments, totaling 420 machines and 30 servers	44/ 433,605	3

4.2 Proposed Method

Figure 1 outlines the entire process of evaluating and comparing the effects of filters and wrappers on the IDS performance using CICID2018 dataset. The performance of the four classifiers (SVM, MLP, RF and XGB) was evaluated using 10-fold cross-validation.

The methodology executed follows these steps:

- **Step 1**: Clean the dataset of duplicate rows, missing values, and unnecessary attributes before converting it to a useful and correct format. Next, extract the feature subsets from our dataset using the eight filters/wrappers. This will produce nine subsets including the original feature set.
- **Step2:** Build and assess the four models MLP, SVM, XGB, and RF by applying a 10-fold cross-validation to each subset of Step1. In total, we obtain 36 combinations (9 subsets * 4 classifiers).
- **Step3**: Cluster the classifiers using Scott-Knott (SK) test [22] according to their accuracy to evaluate the statistical significance of the classification performance. The SK test is employed to evaluate the importance of the results acquired during the model's evaluation and to compare the classifiers' performances.
- **Step4:** Use Borda Count (BC) [23] to rank the classifiers belong to the first best SK cluster on the basis of five metrics: Accuracy, Recall, Precision, AUC and F1 score.

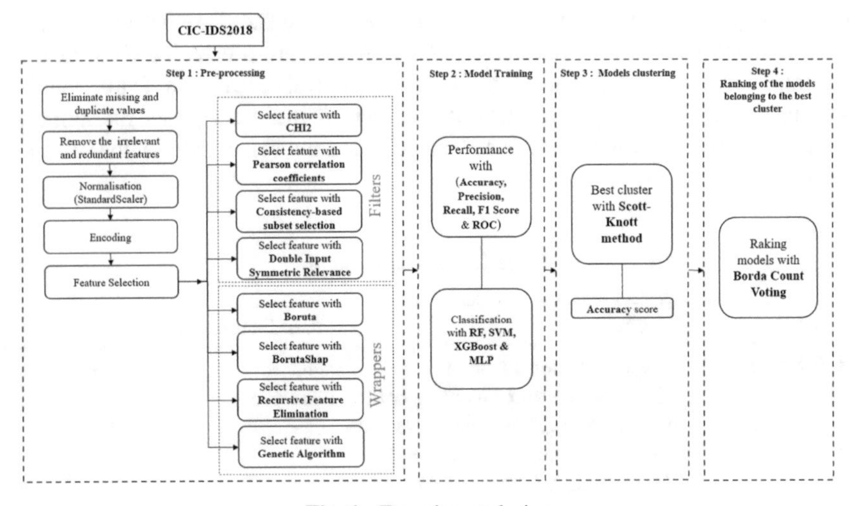

Fig. 1. Experiment design

To simplify, the created classifiers are named using the following abbreviations: C, K, CON, DSIR, B, BSH, RFE and GA denote Pearson correlation, Chi2, Consistency-based subset selection, Double Input Symmetric Relevance, Boruta, BorutaShap, Recursive Feature Elimination, and Genetic Algorithm. In addition, we used the number 100 for the classifiers built with the original feature set. For instance, RF100 refers to the RF formed with all features.

5 Empirical Results and Discussion

This section reports and discusses the empirical assessment's findings in relation to the RQs stated in Sec.1. FS and classification were performed using python libraries, and R Software was used to execute the Scott-Knott (SK) statistical test.

5.1 Cleaning and Transforming Data

Before moving on to the FS step, the CICIDS2018 dataset has been checked for missing values and irrelevant features. As a result, 94,296 instances of missing data were eliminated, and 38 attributes were eliminated due to duplication, constancy, and high correlation. The distribution of classes was then balanced using the under-sampling technique. After that, the Min-Max normalization technique was used to all features.

5.2 Comparison of Filters and Wrappers FS Techniques and the Best Combinations of FS-Classifier

This subsection shows and discusses the results of the step3 of the proposed method and compare the performance of filters and wrappers.

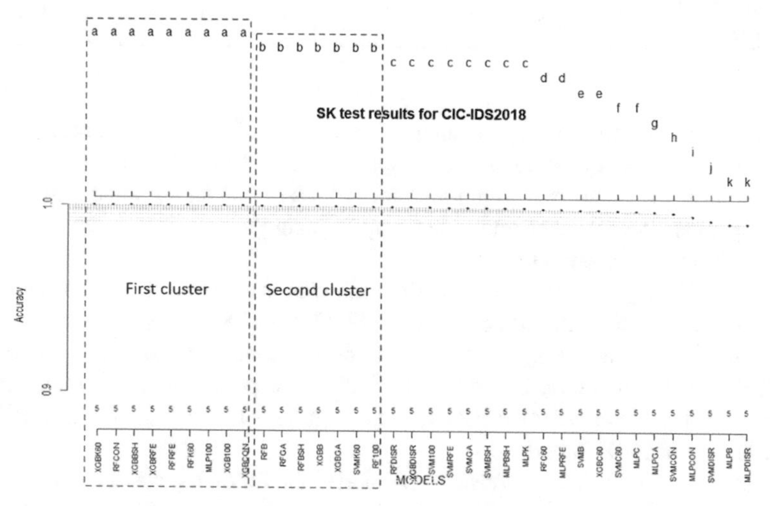

Fig. 2. SK test results for CIC-IDS2018 dataset by model. The x-axis represents the combinations generated by the FS-Classifier, while the y-axis represents the accuracy scores. The rankings of the best clusters are displayed from left to right.

Figure 2 shows the SK test results in terms of accuracy score, displayed through two axes. The x-axis illustrates the variations in classifiers, with best placements on the left, while the y-axis represents accuracy values. The results of fivefold cross-validation for each combination of FS-classifier are shown as a vertical line., and each dot at the center of these lines signifies the corresponding average accuracy values.

Eleven clusters were found by the SK test as shown Fig. 2. The best first cluster contains nine variants of classifiers, and the second best one contains seven. Among the 36 classifiers, the first and second SK clusters comprise nine and seven variants of classifiers, respectively.

In order to compare the performance of filters and wrappers, our statistical analysis is limited to the top two SK clusters. Table 2 illustrates the number of occurrences of filters and wrappers present in the top two clusters according to the SK test results. Considering the initial set of 16 combinations for both filters and wrappers, 50% of techniques present in the top two SK cluster are wrappers, while 31% are filters.

Table 2. The number and occurrence rate of FS-classifiers in the two best clusters based on FS techniques

Dataset	Percentage of classifiers based in filters (16)		#. Of occurrence		Percentage of classifiers based in wrappers (16)		#. Of occurrence	
	First cluster	Second cluster	First cluster	Second cluster	First cluster	Second cluster	First cluster	Second cluster
CSECICIDS2018	25%	6%	4	1	19%	31%	3	5

It can be concluded that the percentage of occurrence of wrappers techniques exceeds that of filters ones, wrappers techniques appear to outperform filters ones.

In addition, to find the best combinations of FS-classifier, the combinations that make up the top cluster were ranked and ordered by BC using five distinct. Metrics: accuracy, recall, precision, AUC and f1 score.

Table 3. displays the BC ranks of the nine classifiers from the top cluster in the CICIDS2018 dataset. With an accuracy of 99.99%, the XGBK classifier, which was built with K technique using 26 features, is ranked first. We observed that RFCON, XGBRFE and RFRFE took 2nd, 4th and 5th place, respectively, and formed with the smaller subset (6 features for RFECON, and 8 features for both XGBRFE and RFRFE), with accuracy of 99.99% both.

Considering these observations, we can derive the following conclusions:

- From the BC results, we demonstrate that several combinations performed well in different datasets.
- *For Classifiers*: The XGB and RF classifiers demonstrated robust and powerful performance, as they are present in the best cluster, while the SVM and MLP classifiers did not perform well, being absent from the best cluster.
- *For FS techniques*: The RF and XGB classifiers demonstrated strong performance when trained with most of the FS techniques, while the CON and RFE techniques performed well with subsets containing smaller number of features, giving the advantage of faster learning, a simpler model and easier interpretation.

Table 3. BC results, where the best combinations are written in bold.

FS-Classifier	Accuracy %	#. Of Features in the subset	Rank
XGBK6	99.99	26	1
RFCON	**99.99**	**6**	**2**
XGBBSH	99.99	23	3
XGBRFE	**99.99**	**8**	**4**
RFRFE	**99.99**	**8**	**5**
RFK6	99.99	26	6
MLP100	99.99	44	7
XGB100	99.98	44	8
XGBCON	99.98	6	9

6 Conclusion and Future Work

The aim of this study was to examine an empirical evaluation of the most widely used wrappers versus the best performing filters for IDSs using two public intrusion datasets. The subsets of features selected with wrappers and filters techniques were evaluated SVM, MLP, XGB and RF classifiers. It seems that wrappers perform well, as the best-performing classifiers are constructed on the basis of wrapper techniques. Nevertheless, CON filter techniques may still be beneficial for CICIDS2018 datasets.

This type of trustworthy comparison highlights the importance and necessity of using smaller feature sets in NIDS. It demonstrates the essential role played by the appropriate FS-classifier combination. This discovery could lead to the identification of influential features that impact flow classification in scenarios of attacks and network environments. We are optimistic that our results can be further enhanced by evaluating other combination methods like ensemble FS techniques. These aspects will be considered in future research. Moreover, these results will enable us to study the behavior of ML-IDS by investigating interpretability techniques.

References

1. Sinclair, C., Pierce, L., Matzner, S.: An application of machine learning to network intrusion detection. In: Proceedings - Annual Computer Security Applications Conference. ACSAC, vol. Part **F133431**, pp. 371–377 (1999). https://doi.org/10.1109/CSAC.1999.816048
2. Jović, K.B., Bogunović, N.: A review of feature selection methods with applications. In: 2015 38th International Convention on Information and Communication Technology, Electronics and Microelectronics, MIPRO 2015 - Proceedings, pp. 1200–1205 (2015), https://doi.org/10.1109/MIPRO.2015.7160458
3. Liu, H., Yu, L.: Toward integrating feature selection algorithms for classification and clustering. IEEE Trans. Knowl. Data Eng. **17**(4), 491–502 (2005). https://doi.org/10.1109/TKDE.2005.66

4. Bolón-Canedo, V., Sánchez-Maroño, N., Alonso-Betanzos, A.: A review of feature selection methods on synthetic data. Knowl. Inf. Syst. **34**(3), 483–519 (2013). https://doi.org/10.1007/S10115-012-0487-8/METRICS
5. Panthong, R., Srivihok, A.: Wrapper feature subset selection for dimension reduction based on ensemble learning algorithm. Proc. Comput. Sci. **72**, 162–169 (2015). https://doi.org/10.1016/J.PROCS.2015.12.117
6. Kursa, M.B., Rudnicki, W.R.: Feature selection with the boruta package. J. Stat. Softw. **36**(11), 1–13 (2010). https://doi.org/10.18637/JSS.V036.I11
7. BorutaShap: A wrapper feature selection method which combines the Boruta feature selection algorithm with Shapley values. https://doi.org/10.5281/ZENODO.4247618
8. Kuhn, M., Johnson, K.: Applied predictive modeling. Appli. Predictive Model. 1–600, (2013). https://doi.org/10.1007/978-1-4614-6849-3/COVER
9. Leardi, R., Boggia, R., Terrile, M.: Genetic algorithms as a strategy for feature selection. J. Chemom. **6**(5), 267–281 (1992). https://doi.org/10.1002/CEM.1180060506
10. Jany Shabu, S.,et al.: Research on Intrusion Detection Method Based on Pearson Correlation Coefficient Feature Selection Algorithm. J. Phys. Conf. Ser. **1757**(1), 012054, (2021). https://doi.org/10.1088/1742-6596/1757/1/012054
11. Liu, H., Setiono, R.: Chi2: feature selection and discretization of numeric attributes. In: Proceedings of the International Conference on Tools with Artificial Intelligence, pp. 388–391 (1995). https://doi.org/10.1109/TAI.1995.479783
12. Meyer, P.E., Schretter, C., Bontempi, G.: Information-theoretic feature selection in microarray data using variable complementarity. IEEE J. Sel. Top. Sign. Proces. **2**(3), 261–274 (2008). https://doi.org/10.1109/JSTSP.2008.923858
13. Dash, M., Liu, H.: Consistency-based search in feature selection. Artif. Intell. **151**(1–2), 155–176 (2003). https://doi.org/10.1016/S0004-3702(03)00079-1
14. Zouhri, H., Idri, A., Ratnani, A.: Evaluating the impact of filter-based feature selection in intrusion detection systems. Int. J. Inf .Secur., 1–27 (2023). https://doi.org/10.1007/S10207-023-00767-Y/TABLES/17
15. Halim, Z., et al.: An effective genetic algorithm-based feature selection method for intrusion detection systems. Comput. Secur. **110**, 102448 (2021). https://doi.org/10.1016/J.COSE.2021.102448
16. Awad, M., Fraihat, S.: Recursive feature elimination with cross-validation with decision tree: feature selection method for machine learning-based intrusion detection systems. J. Sensor Actuator Netw. 12(5), 67 (2023). https://doi.org/10.3390/JSAN12050067
17. Liu, Z., Shi, Y.: A Hybrid IDS using GA-based feature selection method and random forest. Int. J. Mach. Learn. Comput. **12**(2) (2022). https://doi.org/10.18178/IJMLC.2022.12.2.1077
18. Mohammadi, S., Mirvaziri, H., Ghazizadeh-Ahsaee, M., Karimipour, H.: Cyber intrusion detection by combined feature selection algorithm. J. Inform. Sec. Appli. **44**, 80–88 (2019). https://doi.org/10.1016/J.JISA.2018.11.007
19. Megantara, A.A., Ahmad, T.: Feature importance ranking for increasing performance of intrusion detection system. In: 2020 3rd International Conference on Computer and Informatics Engineering, IC2IE 2020, pp. 37–42 (2020). https://doi.org/10.1109/IC2IE50715.2020.9274570
20. Yin, Y., et al.: IGRF-RFE: a hybrid feature selection method for MLP-based network intrusion detection on UNSW-NB15 dataset. J Big Data **10**(1), 1–26 (2023). https://doi.org/10.1186/S40537-023-00694-8/TABLES/9
21. Sharafaldin, A.H.l., Ghorbani, A.A: .Toward generating a new intrusion detection dataset and intrusion traffic characterization. In: Proceedings of the 4th International Conference on Information Systems Security and Privacy, SCITEPRESS - Science and Technology Publications, pp. 108–116 (2018). https://doi.org/10.5220/0006639801080116

22. Scott, J., Knott, M.: A cluster analysis method for grouping means in the analysis of variance. Biometrics **30**(3), 507 (1974). https://doi.org/10.2307/2529204
23. Azzeh, M., Nassif, A.B., Minku, L.L.: An empirical evaluation of ensemble adjustment methods for analogy-based effort estimation. J. Syst. Softw. **103**, 36–52 (2015). https://doi.org/10.1016/J.JSS.2015.01.028

Multiclass Intrusion Detection in IoT Using Boosting and Feature Selection

Abderrahmane Hamdouchi[1] and Ali Idri[1,2](✉)

[1] Mohammed VI Polytechnic University, Benguerir, Morocco
Abderrahmane.Hamdouch@um6p.ma, Ali.Idri@um5.ac.ma
[2] Software Project Management Research Team, ENSIAS, Mohammed V University in Rabat, Rabat, Morocco

Abstract. The proliferation of Internet of Things (IoT) devices has resulted from the continuous evolution of interconnected computing devices and emergence of novel network technologies. Intrusion Detection Systems (IDSs) play a pivotal role in ensuring IoT network security. Nonetheless, the development of a robust IDS with high performance for detecting specific anomalies to prevent attacks remains a challenging task. This study systematically evaluated and compared two classifies XGBoost and LightGBM employing five filter-based feature selection methods (ANOVA, Kendall's test, Mutual Information, Maximum Relevance Minimum Redundancy, and Chi-square) with distinct selection thresholds. The evaluation was conducted over the NF-ToN-IoT-v2 dataset using the Scott-Knott test and Borda Count system. A total of 152 classifier variants were assessed to determine the most effective model. Results demonstrated that utilizing XGBoost in conjunction with 200 estimators, ANOVA and MI as the feature selection combination, and selecting the top 18 features, constituted an effective and powerful model compared to alternative methods.

Keywords: Intrusion detection system · ensemble learning · boosting · imbalanced data · IoT · NF · multiclass classification

1 Introduction

The Internet of Things (IoT) represents a plethora of devices that transmit and exchange data via wired connections. This transformative technology has the potential to improve several industries, including healthcare, energy, education, transportation, production, defense, and agriculture [1]. According to a report by IoT Analytics, 11.6 billion of the world's 21.2 billion connected devices qualify as IoT devices. Projections indicate that the number of IoT devices is likely to exceed 22 billion by 2025, assuming a 10% increase in the number of connected devices which increased the number of network vulnerabilities. Therefore, the significance of network security cannot be overemphasized. Consequently, identifying intrusions is essential for protecting networks against damaging actions and unusual occurrences [1].

Intrusion detection system (IDS) constitutes the first line of defense against cyber-attacks [2]. It refers to a device or program strategically placed at a particular network

Á. Rocha et al. (Eds.): WorldCIST 2024, LNNS 987, pp. 128–137, 2024.
https://doi.org/10.1007/978-3-031-60221-4_13

node to monitor all the traffic. IDS functions can involve reporting malicious behavior to a network administrator and identifying intruders [2]. IDS are classified into anomalies, signatures, or hybrid systems.

In recent years, machine learning (ML) has been extensively validated as a powerful technique for developing IDS models to secure IoT networks based on anomaly detection [3]. Despite the efficacy of single-learner ML techniques, they do not always perform well under all conditions because of problems such as generalization, high variance, and bias [4]. This has increased the focus of researchers on the use of ensemble learning techniques to enhance the overall performance, robustness, and generalizability of ML models [5].

Ensemble learning can prevent overfitting because using the average of multiple hypostases reduces the risk of selecting an incorrect hypothesis and improves the overall performance [4]. We can distinguish between a heterogeneous technique collected from several base learners and a homogenous technique composed of several variants of a single base learner [4]. Boosting is one of the most widely used homogeneous methods for this purpose.

The surveys [5] and [6] examined 170 and 124 primary papers on IDS, respectively. They highlighted preprocessing, feature engineering, model construction, and evaluation but observed insufficient experimental designs and a lack of comparative studies on FS techniques, ML algorithms, ensemble learning techniques, and statistical tests. These surveys [5] and [6] revealed significant limitations in existing research, motivating our study to address gaps: (1) no primary study has investigated FS techniques with boosting using various estimator numbers for multiclass classification using NetFlow standardization; (2) many studies used past datasets, and (3) prior research focused on classification accuracy, neglecting classifier robustness for IoT security. Our motivation was to empirically evaluate FS techniques, boosting algorithms using statistical tests to develop a powerful IDS model for IoT environments.

Therefore, the purpose of this research is to address these limitations by: (1) evaluating two boosting techniques: XGBoost (XGB) and LightGBM (LGBM), with four estimators (50, 100, 150,and 200); and (2) eliminating redundant and correlated features using different filters (Pearson's correlation [19], variance inflation factor (VIF), mutual information (MI), chi-squared (chi2), and minimum redundancy maximum relevance (mRMR) [7] for categorical features, and analysis of variance (ANOVA) and Kendall's tau (Kendall) methods for numerical features. The selection of these FS techniques is based on the fact that they use minimal computing resources and are the most popular filters for selecting the optimal subset of features in multiple domains; three thresholds were used to obtain the final feature subsets: 20%, 50%, and 80%, according to their power in previous studies [8, 9]. In addition, we selected the two boosting techniques since they are the most commonly used and powerful boosting methods [5, 9]; range of estimators is based in accordance with literature recommendations [10].

In this study, we evaluated the performance of 152 boosting ensembles ($156 = 2$ boosting techniques $\times$ 4 trees $\times$ (1 all features + (2 Numerical FS $\times$ 3 Categorical FS 2S) $\times$ 3 feature thresholds) for a multiclassification using a five-fold method over a recent IoT NetFlow IDS dataset named NF-ToN-IoT-v2. In addition, this study uses three performance criteria that are not sensitive to imbalanced data: Cohen's kappa (Kappa),

Matthew's correlation coefficient (MCC) [11], and Receiver Operating Characteristic Area Under the Curve (AUC), the Scott Knott (SK) statistical, and the Borda count (BC) ranking system to rank the highest-performing models.

This study attempts to answer the following research questions:

- **(RQ1)** What is the optimal number of estimators for each boosting technique for a multiclass classification in IoT contexts?
- **(RQ2).** Which combination of FS and boosting methods works best for each feature threshold and overall?

The main contributions of this study are as follows:

1. Determine the optimal number of estimators of boosting techniques for a multiclassification of intrusion detection in an IoT context.
2. Identify the optimal combination of FS methods, boosting techniques, and number of estimators for cyber-attack detection.

The remainder of this paper is organized as follows. Section 2 provides a summary of the relevant literature related to the subject of this study. Section 3 discusses the data used in the study, and explains the study's empirical methodology. In Sect. 4, the experimental results and discussions are presented. Section 5 presents remarks and recommendations for future research.

2 Related Works

Several significant works have been proposed for anomaly detection in IoT contexts using ML approaches without combination of categorical/numerical FS techniques, boosting techniques, SK test, and MCC and Kappa metrics. However, most of these studies focused on protecting either the device level or the network level and did not address both. A key contribution is the study of Sarhan et al. [12], who generated NF-ToN-IoT-V2 datasets in 2021, which were built on the foundational BoT-IoT dataset, highlighting the necessity of processing NetFlow features and employing an Extra Tree algorithm for classification. However, their feature processing has several limitations, such as using categorical variables as numerical (protocol features and port features), and deleting variables that may be useful instead of recategorizing them (IPs features). Awad et al. [13] focused on the NF-ToN-IoT-v2 dataset and developed a detailed preprocessing and classification schemes using models such as XGB and RF. Despite these promising findings, the inherent challenges in data preprocessing remain unresolved.

Research on anomaly detection [14] indicated that ensemble learning techniques can improve the efficiency of anomaly detection, often achieving superior predictive outcomes compared with any individual algorithm. Verma et al. [15] proposed an Ensemble Learning based Network Intrusion Detection System (ELNIDS) that incorporated several tree-based techniques and evaluated over the RPL-NIDDS17 dataset. Furthermore, this framework used all dataset features to construct complex ensemble models with a high computational overhead, which makes this system impractical.

3 Experimental Design

This section outlines the datasets, performance metrics, and the methodology used to carry out all the empirical evaluations.

3.1 Datasets Description

In this study, 43 NetFlow features based on standard version 9 were extracted and annotated from the original ToN-IoT dataset. The extraction process employed the nProbe tool developed by Ntop [16] The resulting dataset, denoted NF-ToN-IoT-V2, encompasses 16,940,469 instances distributed across nine classes with highly imbalanced distribution of the classes.

3.2 Performance Measure

Cohen's kappa (Kappa), Matthew's correlation coefficient (MCC), and receiver operating characteristic area under the curve (AUC) were used to assess the efficacy of the proposed classification models in multiclass classification. The macro-average approach was selected because it provides a comprehensive evaluation of the overall performance across all instances, which is particularly valuable when dealing with unbalanced datasets. These metrics were selected since they allow to identify the most effective and robust system and avoid overfitting when detecting the attack types [17].

3.3 Methodology

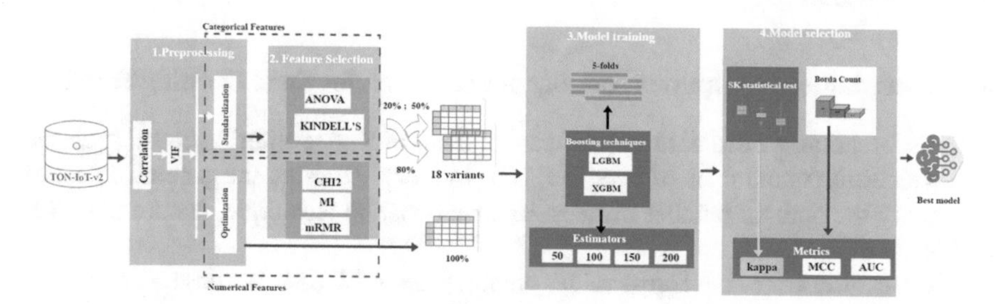

Fig. 1. Experimental process

Figure 1 illustrates the process followed to evaluate and compare the impacts of FS techniques and estimator numbers on the IDS classification performance. The performances of the two boosting techniques: LGBM and XGB with four estimators (50, 100, 150, and 200) were evaluated using a five-fold cross-validation approach, SK test, and BC voting system. The experimental process comprised the following steps:

- **Step 1**: Transform raw data into a usable and accurate format by removing missing values, duplicated rows, and irrelevant attributes, recategorizing categorical features, and applying the standardization transformations to numerical features.

- **Step2**: Five filters were used with three thresholds (20%, 50%, and 80%)[10]ANOVA and Kendall were used for numerical features, whereas MI, chi2, and mRMR were employed for categorical features, which produced 19 combinations = (3 thresholds × 3 Num FS × 2 Cat FS + 1 all features).
- **Step 3**: Construct and evaluate the performances of 152 models using the combination of two boosting techniques (LGBM and XGB) with four different numbers of estimators (50, 100, 150, and 200) on the 19 datasets of Step 2, in terms of kappa, MCC, and AUC using five-fold cross-validation. In addition, the SK test and BC system were utilized to rank the number of estimators for each boosting technique and the generated datasets of **Step 2**.
- **Step 4**: Compare each threshold using the SK test and BC system to determine the optimal combination of the FS method and boosting technique. To facilitate naming of the FS techniques in the Fig. 1. ANOVA, Kendall, chi2, MI, and mRMR were designated as A, K, C, M, and R, respectively. For example, A_M refers to the combination of ANOVA and MI.

4 Results and Discussions

This section examines the performances of LGBM and XGB when using different estimators (50, 100, 150, and 200), the five filters for a multiclass classification, and the four thresholds (20%,50%,80% and 100%) over the NF-ToN-IoT-v2 dataset in terms of Kappa, MCC, and AUC. A comparison of estimator performances within each generated model was conducted based on Kappa, as presented in the first subsection (RQ1). The second subsection compares and identifies the most efficient combinations of the FS techniques and boosting methods for each feature threshold using SK test and BC voting method (RQ2).

4.1 Determining the Optimal Number of Estimators for Each Boosting Technique

In this subsection, we evaluate and compare in terms of Kappa criterion boosting ensembles with different numbers of trees (50, 100, 150, or 200) using the generated datasets to identify the optimal number of trees for each boosting technique, feature threshold, and filter.

Figure 2 illustrates the kappa values obtained using LGBM. We observe that:

- For 20% of the features, optimal Kappa values were achieved with 50 estimators regardless the six FS combinations. The worst results were observed when: 200 estimators were used (five times), and 150 estimators were used once.
- For 50% of the features, optimal kappa values were obtained using 50 estimators across the six FS combinations. Conversely, the worst results were reported when employing 200 estimators (five times), and 150 estimators once.
- For 80% of the features, optimal kappa values were achieved using 50 estimators across the six FS combinations. In contrast, the worst results were observed when employing 200 estimators (five times), and 150 estimators once.
- Finally, using all the features, the highest kappa value was achieved using 50 estimators, and the worst kappa value was achieved using 150 estimators.

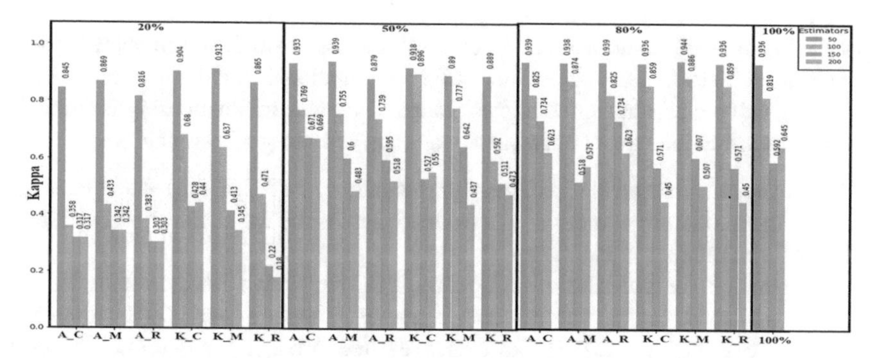

Fig. 2. Kappa values of the LGBM estimators.

To sum up, when using LGBM boosting technique, the best kappa values were achieved using a small number of estimators and the worst values were achieved using the largest number of estimators.

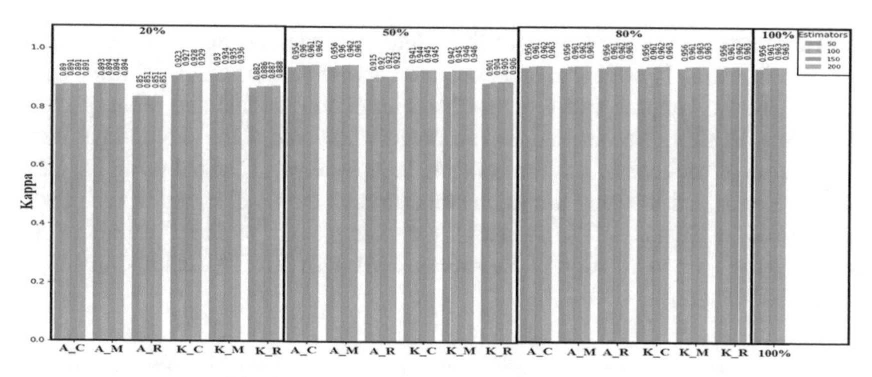

Fig. 3. Kappa values of the XGB estimators.

Figure 3 illustrates the kappa values obtained using XGB. We observe that:

- For 20% of the features, the best kappa values were simultaneously obtained using 200, 150, and 100 estimators (three times) in conjunction with ANOVA combined with chi2, MI, and mRMR. Additionally, XGB using 200 estimators yielded optimal results three times across the remaining FS combinations. Conversely, lower results were observed when employing 50 estimators across the six FS combinations.
- For 50% of the features, the best kappa values were simultaneously achieved twice using 200 and 150 estimators in combination with the pairing of Kendall with chi2 and MI. Furthermore, XGB utilizing 200 estimators yielded optimal results four times across the remaining FS combinations. In contrast, the worst results were observed when employing 50 estimators across the six FS combinations.
- For 80% of the features, optimal kappa values were simultaneously achieved once using 200 and 150 estimators in combination with the pairing of Kendall with MI. Furthermore, XGB utilizing 200 estimators yielded optimal results five times across

the remaining FS combinations. In contrast, the worst results were observed when XGB used 50 estimators across the six FS combinations.

- Finally, using all features, the highest kappa value was achieved using 200 and 150 estimators, and the worst kappa value was achieved using 50 estimators.

To sum up, when using the XGB boosting technique, the best Kappa values were realized with the maximum number of estimators, whereas the worst values were observed with a reduced number of estimators.

In conclusion, for multiclass classification, XGB has shown that increasing the number of trees increases kappa values because gradient-boosting-based DT learning algorithms grow trees depth wise [18]. However, a few trees produced superior LGBM results because they used a leaf-wise approach for splitting that designed unbalanced trees [19], in which a large number of trees involved overfitting and negatively affected the kappa values if the classes were highly imbalanced.

4.2 Optimal Combinations of FS and Boosting Methods Across Varied Feature Percentages

In this subsection, the combination of FS techniques and boosting methods, along with the best estimators, is evaluated and compared across various feature thresholds. The SK test, based on kappa, was employed to cluster the models and identify the best SK clusters, as illustrated in Fig. 4. The process involved using the optimal number of estimators of each variant of the boosting technique was selected using the SK test based on Kappa and utilizing the BC method to rank the models within each best SK cluster based on Kappa, MCC, and AUC. For example, the preferred variant of XGB, employing the FS combination of Kendall and MI with 20% of features, was the one with 200 estimators.

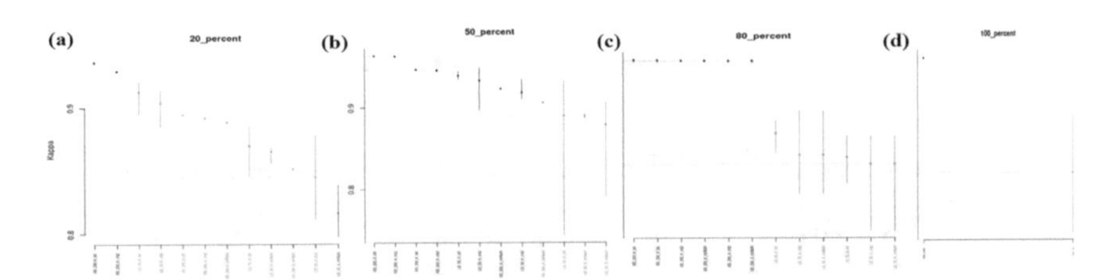

Fig. 4. Statistical comparison using SK for boosting techniques with the best estimators using (a) 20%; (b) 50%; and (c) 80% of features; in addition to (d) all features.

For the three thresholds (20%,50%,80%), a total of 12 classifiers (=2 boosting techniques $\times$ 3 categorical FS $\times$ 3 numerical FS) were evaluated for each threshold, and 2 classifiers were evaluated for all features. The SK test results in terms of Kappa scores are illustrated in Fig. 4.

The SK distribution for 20% of the features exhibits six clusters. Two classifiers based on XGB were assigned to the first cluster, and the remaining XGB variants were distributed as follows: three in the third cluster and one in the fifth cluster. In contrast, the distribution of LGBM variants is as follows: one in each of the second, third, fifth, and last clusters, and twice in the fourth cluster.

The SK test identifies two clusters for 50% of the whole features. The first cluster consisted of 8 classifiers, encompassing all XGB variants, except those based on the Kendall and mRMR combination. Additionally, three LGBM variants were part of this cluster, specifically those based on ANOVA with MI and chi2, and Kendall with chi2.The remaining LGBM variants are assigned to the second cluster.

The SK test provided two clusters when using 80% of features. The superior cluster comprises 6 classifiers that exclusively contain all the XGB variants. Conversely, all LGBMs were in the second cluster. In addition, the best cluster accommodates the XGB model by employing all the features, whereas the second cluster encompasses the LGBM variant.

It can be inferred that XGB consistently yielded optimal results across various feature percentages, as evidenced by the presence of 14 out of 17 models within the best SK clusters. The frequency of the occurrence of LGBM, totaling 3 variants, was considered acceptable.

Table 1. Best models for each feature threshold.

Percent	# Features	combination	Kappa	MCC	AUC
20%	7	K + MI + XGB_200	0.94	0.94	0.89
50%	**18**	**A + MI + XGB_200**	**0.96**	**0.96**	**0.91**
80%	25	K + MI + XGB_200	0.96	0.96	0.91
100%	35	XGB_200	0.96	0.95	0.91

The best model within each first cluster was determined using the BC system based on Kappa, MCC, and AUC metrics, as outlined in Table 1. The findings demonstrate the superiority of XGB classifiers across varying feature percentages. Performance stability was notably observed in cases involving 18 or more features. It can be inferred that the optimal model selection entails the utilization of a combination of ANOVA and MI with XGB, employing 200 estimators over the top 18 NetFlow features with Kappa = 0.96, MCC = 0.96, and AUC = 0.91.

5 Conclusion and Future Work

This study concluded with an examination of the results provided by an experimental comparison involving 152 boosting ensembles for a multiclass intrusion classification. LGBM and XGB were selected as boosting techniques using 50, 100, 150, and 200 estimators, along with three categorical feature-based FS techniques (chi2, MI, and mRMR) and two numerical feature-based techniques (ANOVA and Kendall). The evaluation of these models encompassed three performance metrics (Kappa, MCC, and AUC), the SK statistical test, and the BC system. The results indicated that XGB demonstrated improved kappa values with an increase in the number of trees, whereas LGBM showed superiority with a smaller number of estimators. XGB consistently delivered optimal results across various features. Notably, the combination of ANOVA and MI with XGB employing 200 estimators over the top 18 NetFlow features yielded the best overall performance.

These findings imply the importance of integrating boosting when designing resilient and adaptable IDS for IoT networks to detect the type of attack in multiclassification. Ongoing work focuses on leveraging deep ensemble learning to improve the model robustness and juxtapose it with boosting ensembles.

References

1. McKay, R., Pendleton, B., Britt, J., Nakhavanit, B.: Machine learning algorithms on botnet traffic: Ensemble and simple algorithms'. ACM Inter. Conf. Proc. Ser., 31–35 (2019). https://doi.org/10.1145/3314545.3314569
2. Eskandari, M., Janjua, Z.H., Vecchio, M., Antonelli, F.: Passban IDS: an intelligent anomaly-based intrusion detection system for iot edge devices. IEEE Internet Things J. **7**(8), 6882–6897 (2020). https://doi.org/10.1109/JIOT.2020.2970501
3. Chaabouni, N., Mosbah, M., Zemmari, A., Sauvignac, C., Faruki, P.: Network intrusion detection for iot security based on learning techniques. IEEE Commun. Surv. Tutorials **21**(3), 2671–2701 (2019). https://doi.org/10.1109/COMST.2019.2896380
4. Zhou, Z.-H.: Ensemble Learning. Encyclopedia of Biometrics, pp. 270–273 (2009). https://doi.org/10.1007/978-0-387-73003-5_293
5. Tama, B.A., Lim, S.: Ensemble learning for intrusion detection systems: a systematic mapping study and cross-benchmark evaluation. Comput. Sci. Rev. **39**, 100357 (2021). https://doi.org/10.1016/J.COSREV.2020.100357
6. Thakkar, A., Lohiya, R.: A survey on intrusion detection system: feature selection, model, performance measures, application perspective, challenges, and future research directions. Artif. Intell. Rev. **55**(1), 453–563 (2022). https://doi.org/10.1007/S10462-021-10037-9/METRICS
7. Peng, H., Long, F., Ding, C.: Feature selection based on mutual information: criteria of max-dependency, max-relevance, and min-redundancy. IEEE Trans. Pattern Anal. Mach. Intell. **27**(8), 1226–1238 (2005). https://doi.org/10.1109/TPAMI.2005.159
8. Nakashima, M., Kim, Y., Kim, J., Kim, J., Sim, A.: Automated feature selection for anomaly detection in. Network Traffic Data **1**(1), 27 (2018). https://doi.org/10.1145/1122445.1122456
9. Dhaliwal, S.S., Al Nahid, A., Abbas, R.: Effective intrusion detection system using XGBoost. Information **9**(7), 149 (2018), https://doi.org/10.3390/INFO9070149
10. Theodoridis, G., Tsadiras, A.: Using machine learning methods to predict subscriber churn of a web-based drug information platform. IFIP Adv. Inf. Commun. Technol. **627**, 581–593 (2021). https://doi.org/10.1007/978-3-030-79150-6_46/COVER
11. Baldi, P., Brunak, S., Chauvin, Y., Andersen, C.A.F., Nielsen, H.: Assessing the accuracy of prediction algorithms for classification: an overview. Bioinformatics **16**(5), 412–424 (2000). https://doi.org/10.1093/BIOINFORMATICS/16.5.412
12. Sarhan, M., Layeghy, S., Moustafa, N., Portmann, M.: NetFlow Datasets for Machine Learning-Based Network Intrusion Detection Systems. LNICST, vol. 371 pp. 117–135 (2021). https://doi.org/10.1007/978-3-030-72802-1_9/COVER
13. Awad, M., Fraihat, S., Salameh, K., Al Redhaei, A: .Examining the suitability of netflow features in detecting IoT network intrusions. Sensors **22**(16), 6164 (2022). https://doi.org/10.3390/S22166164

14. Sagi, O., Rokach, L.: Ensemble learning: a survey. Wiley Interdiscip Rev. Data. Min. Knowl. Discov. **8**(4), e1249 (2018). https://doi.org/10.1002/WIDM.1249
15. Verma, A., Ranga, V.: ELNIDS: ensemble learning based network intrusion detection system for RPL based Internet of Things. In: Proceedings - 2019 4th International Conference on Internet of Things: Smart Innovation and Usages, IoT-SIU 2019 (Apr 2019). https://doi.org/10.1109/IOT-SIU.2019.8777504
16. nProbe – ntop (2023). https://www.ntop.org/products/netflow/nprobe/# (Accessed 16 Feb 2023)
17. He, H., Ma, Y.: Imbalanced learning: Foundations, algorithms, and applications. Imbalanced Learning: Foundations, Algorithms, and Applications, pp. 1–210 (Jan 2013). https://doi.org/10.1002/9781118646106
18. Greedy Function Approximation: A Gradient Boosting Machine on JSTOR (2023). https://www.jstor.org/stable/2699986 (Accessed 15 Feb 2023)
19. Alzamzami, F., Hoda, M., El Saddik, A.: Light gradient boosting machine for general sentiment classification on short texts: a comparative evaluation. IEEE Access **8**, 101840–101858 (2020). https://doi.org/10.1109/ACCESS.2020.2997330

Is ChatGPT Able to Generate Texts that Are Easy to Understand and Read?

Andrea Sastre[1](✉), Ana Iglesias[1], Jorge Morato[1], and Sonia Sanchez-Cuadrado[2]

[1] Department of Computer Science Carlos III, University of Madrid, 28911 Madrid, Leganes, Spain
asastre@pa.uc3m.es, {aiglesia,jmorato}@inf.uc3m.es

[2] Department of Library and Information Science, Complutense University of Madrid, 28018 Madrid, Spain
sscuadrado@ucm.es

Abstract. The emergence of generative artificial intelligence applications in recent years has opened up the possibility of automatically generating text tailored to the needs of users. However, is the text generated by these tools easily readable and understandable? This study aims to analyze the ChatGPT's capacity to generate easy to understand texts in Spanish, while adhering to the recommended guidelines for Plain Language. The comprehensiveness of several websites belonging to the Public Administration in Spain has been analyzed. Moreover, ChatGPT was requested to enhance these texts in accordance with the guidelines of Plain Language in Spanish. The original texts and those generated by ChatGPT were then compared to analyze the primary linguistic indicators of different Plain Language guidelines. The study presents a quantitative analysis demonstrating that the new texts generated by ChatGPT do not adhere to Plain Language guidelines. Additionally, the study highlights the use of linguistic elements that should be avoided.

Keywords: Readability · ChatGPT · understanding of information · plain language · guidelines

1 Introduction

The Internet has become the primary channel for information, significantly altering the way citizens access and share information. It serves as a crucial communication medium between the administration and citizens. However, the use of technical jargon and complex administrative and legal concepts often results in content that is difficult for the general public to understand. Morato et al. identified issues with institutional websites designed for citizens, where issues of Language clarity and accessibility hinder the understanding of citizens and their ability to assert their rights [1]. Plain Language [2] simplifies communication and breaks down barriers, playing a transformative role in society. This paper presents a study that focuses on the use of new Artificial Intelligence (AI) tools, such as ChatGPT, to automatically transform text in order to facilitate communication in accordance with Plain Language guidelines.

Á. Rocha et al. (Eds.): WorldCIST 2024, LNNS 987, pp. 138–147, 2024.
https://doi.org/10.1007/978-3-031-60221-4_14

This article is structured as follows: Background section introduces an overview of the plain language in relation to ChatGPT. Next section proposes a methodology based on a comparative analysis. Finally, the main results and conclusion of readability indicators and guidelines are presented.

2 Background

Plain language is defined as "writing that is clear, concise, well-organized, and follows other best practices appropriate to the subject or field and intended audience" [2]. Various global laws and guidelines [1–4] have been implemented in different countries regarding the use of clear and understandable language, particularly in public information and services. These include the European Accessibility Act and specific laws in countries such as Spanish Law 34/2002 and Spanish Law 19/2013. Additionally, the Web Accessibility Initiative (WAI) of the World Wide Web Consortium (W3C) provides recommendations for making content more accessible to users, including those with disabilities. The European Commission has made efforts to improve the readability of governmental texts across Europe through initiatives such as the Clear Writing for Europe Conference [4] and the Commission Style Guide [5]. These guidelines emphasize readability features such as using shorter sentences and everyday language. They highlight the importance of adapting e-government information to the intended audience while maintaining a clear and logical structure.

There are different methods for obtaining plain language texts: use of user surveys; application of recommendations, automatically or manually; and use of AI (Fig. 1).

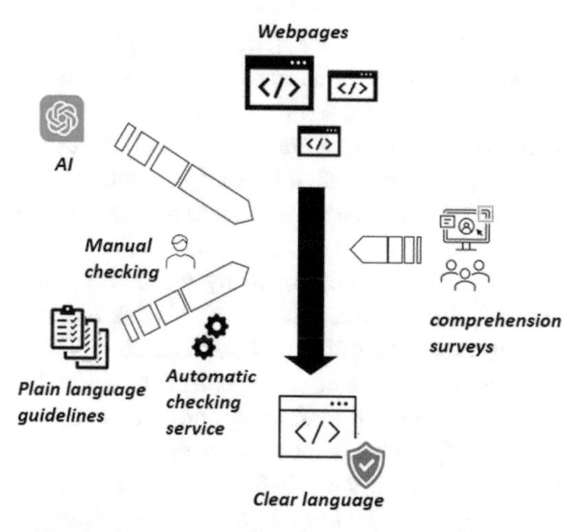

Fig. 1. Methods for obtaining plain language texts.

Generative Artificial Intelligence (GAI) is a broad concept that encompasses any automated process that uses algorithms to create, modify, or synthesize data, usually in the form of images or human-readable text. [6].Scientific literature emphasizes ChatGPT's versatile role [7], particularly in improving health communication and enhancing clarity for non-native English speakers [8–12]. Most of these studies found that the texts were more readable, even for patients with low levels of education. However, they also encountered a major problem with information loss and errors in the generated texts. Hence, a completely automated conversion is not advisable as it may oversimplify or eliminate crucial elements [9, 11, 12, 15]. In the educational field, ChatGPT was positively evaluated for its clear language, which was found to enhance students' learning [13]. In the organizational domain, it successfully provided information in a clear and understandable manner [14]. In legal research, ChatGPT has been successfully used as an artificial lawyer [15, 15], highlighting the benefits of communicating legal information in a clear, understandable, and user-friendly language across different dimensions.

Most scientific literature focuses on the English language, particularly in the medical and legal fields. However, no previous studies have been conducted on language simplification in Spanish. Moreover, there is a lack of research on the ability of generative AI tools to produce plain language, as the published studies on the topic conclude. This is highlighted by the limited analysis of linguistic features in relation to plain language guidelines, which could identify potential issues.

The novelty of this study, therefore, is to analyze the ability of ChatGPT, one of the generative AI tools, to adapt texts extracted from Spanish e-Government websites according to Plain Language guidelines. Additionally, linguistic features will be analyzed in depth to further evaluate the tool's effectiveness and its adherence to the recommended guidelines.

3 Methodology

An analytical-comparative study was conducted in several phases to check the adherence of text generation to Plain Language guidelines. In the first phase, the Plain Language guidelines were scrutinized to pinpoint key linguistic indicators (described in Sect. 4.1). Subsequently, a corpus was created (see Sect. 4) to analyze texts in Plain Language, sourced from original content on Spanish e-Government web pages, alongside aligned texts generated by ChatGPT (v 3.5). The results from the analysis were examined in the comparison phase.

The software tool Midolec was developed by the authors of this paper to automatically analyze the linguistic characteristics of systems based on Plain Language guidelines. Midolec is an evolution of Comp4Text Checker [17], a Natural Language Processing (NLP) tool that identifies morphological, syntactic, and semantic indicators. Additionally, the module calculates the readability and complexity index of the text using the Flesch-Kincaid index [18]. This index considers seven levels of readability, ranging from very-easy (90–100) to very-difficult (0–30), according to the Flesh-Kincaid interval. The Flesch-Kincaid formula (1) takes into account linguistic features such as word length and sentence length to determine readability, where L is the readability, P is the average number of syllables per word, Fm is the average number of words per sentence, and L corresponds to the readability of the text.

$$L = 206.84 - 0.60P - 1,02F \tag{1}$$

4 Design of the study

This section outlines the linguistic indicators that were considered in the creation of the corpus in accordance with the Plain Language Guidelines.

4.1 Text Comprehensibility Indicators

The study examined several Plain Language guidelines, including a) the European Commission Guideline, b) the Plain Language guideline for the Public Administration of the Autonomous Community of the Region of Murcia, c) the Clear Communication Practical Guide Community of Madrid, d) Artext and e) Pragmalinguistics. Table 1. summarizes the 20 common linguistic criteria identified from these guidelines. In summary, the guidelines agree that using a simplified linguistic structure improves comprehension. It is recommended to avoid using the subjunctive, passive voice, or other complex structures. Additionally, the guidelines advise against using certain forms of adjectives, adverbs, and long words, among others.

Table 1. Plain language guidelines.

Recommendation	Guidelines
Avoid infinitive	c, e
Avoid subjunctive	e
Avoid subordinate sentence	e
Avoid past tenses	e
Avoid passive voice	a, b, c, d, e
Avoid problematic structures	e
Avoid gerund	b, c, d, e
Personal verbs	a, b. e
Avoid adverbs ending in '-mente'	e
Avoid long words	d, e
Avoid participles	b, c., d, e
Avoid superlatives	d

4.2 Corpus

The corpus was compiled using public web information from the Spanish Administration. The pages did not have specific plain language labelling. The web pages were selected randomly from a set of 600 representative pages proposed by the heads of various Spanish government offices. Please see Table 2. for the ten selected pages. The informative text was extracted from each web page as of November 2023. ChatGPT was used to process these pages, with the express instruction to modify them to follow clear language recommendations.

5 Results

This section presents the primary findings of a thorough analysis of the linguistic indicators in the corpus. Table 3 (Appendix 1) displays the frequency of each type of linguistic indicator in each text of the corpus, which will be explained in the following sections.

5.1 Word Count Analysis

Table 2 summarizes the word count for each document. ChatGPT generally shortens texts by reducing the number of words, except in one specific case (Text 4). As mentioned in the background section, previous research in literature has focused on the information that ChatGPT omits when summarizing. However, this study will focus on ChatGPT's adherence to Plain Language guidelines. The extent of shortening in the text generated by ChatGPT becomes more pronounced as the number of words in the original text increases. This is attributed to the reduction in overall word count. If the number of elements to be avoided is similar to those in the original text, the percentage will be higher, resulting in a more prominent. This can make the analysis more complex. The next section will check if the reduction is indeed beneficial from a Plain Language perspective. ChatGPT sometimes loses information in the process of simplifying texts, especially if they are long. It is therefore necessary to proceed carefully to identify missing elements.

5.2 Readability Index

The text presents the results of the readability analysis of the corpus using the Flesh-Kincaid index for the Spanish language [18]. Figure 2 displays the readability scores (Flesch-Kincaid index) of the original texts from Spanish e-Government websites (light blue) and ChatGPT texts (dark blue). The difference between the two sets of texts is not significant. In more than half of the texts generated by ChatGPT, the readability according to the Flesch-Kincaid index is worse than the original ones. Notably, in some cases, even the readability category changes. For example, in Text 5, the original text has a score of 60,99 (normal difficulty), whereas the text generated by ChatGPT has a score of 56,67 (fairly difficult).

5.3 Simple Linguistic Structures

In some instances, the guidelines are adhered to the Plain Text guidelines recommend using simple sentence structures, which means avoiding subjunctive, passive voice or other problematic constructions (such as those with "se"). However, it has been observed that these recommendations are often ignored (see Appendix I). The original texts (Text 7 and Text 8) use the subjunctive mood, but ChatGPT's conversion omits it. In some cases, ChatGPT has made the linguistic structures more complex. For instance, in Text 9 or 10, it has used problematic structures or the passive voice.

Table 2. Words occurrences in original and ChatGPT version text

Text	Document	Original	ChatGPT	Urls
Text 1	Electronic Public Administration	439	270	https://www.hacienda.gob.es/es-ES/Areas%20Tematicas/Administracion%20Electronica/Paginas/default.aspx
Text 2	Digital Certificate	323	269	https://www.sede.fnmt.gob.es/certificados/persona-fisica
Text 3	DGT Appointment	2924	326	https://sede.dgt.gob.es/es/otros-tramites/cita-previa-jefaturas/
Text 4	Commitment activity	68	81	https://www.sepe.es/HomeSepe/Personas/distributiva-prestaciones/compromiso-actividad.html
Text 5	Business days	1885	721	https://administracion.gob.es/pag_Home/atencionCiudadana/calendarios/diasInhabiles.html#.XkCarzFKiUk
Text 6	Single payment	1922	441	https://www.sepe.es/HomeSepe/autonomos/prestaciones-para-emprendedores-y-autonomos/capitaliza-tu-prestacion.html
Text 7	Autocalculation of Unemployment Benefits	191	176	https://www.sepe.es/HomeSepe/que-es-el-sepe/comunicacion-institucional/noticias/historico-de-noticias/2019/detalle-noticia.html?folder=/2019/Septiembre/&detail=calculo-prestacion
Text 8	Unemployed Family Help	1228	329	https://www.sepe.es/HomeSepe/Personas/distributiva-prestaciones/he-dejado-de-cobrar-el-paro/cargas-familiares.html
Text 9	Autonomous Accredit Termination	85	83	https://www.sepe.es/HomeSepe/autonomos/cese-actividad.html
Text 10	Assistance for Unemployed 45 years old	956	317	https://www.sepe.es/HomeSepe/Personas/distributiva-prestaciones/he-dejado-de-cobrar-el-paro/mas-45-anos-no-cargas-familiares.html

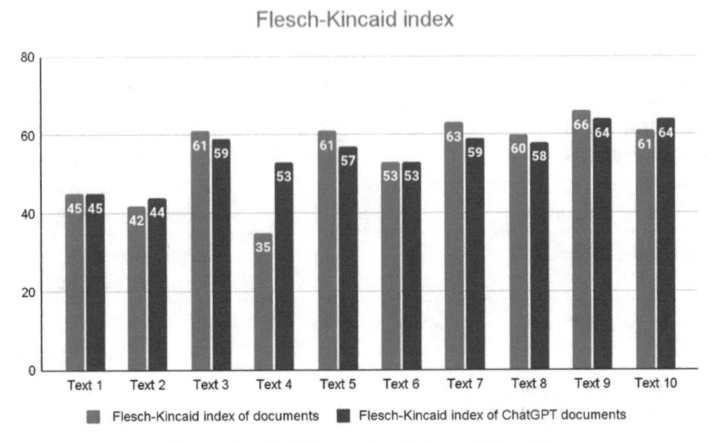

Fig. 2. Readability index (Flesch-Kincaid).

Types of Verbs

Firstly, it is important to note that seven out of the nine types of verbs that are recommended to be avoided were present in the text generated. Not only they occur more frequently, but they also appear for the first time in the texts generated by ChatGPT (Appendix 1). The recommendation includes using objective language, avoiding the use of subjunctives and personal verbs. In the texts generated by ChatGPT, the use of subjunctive was largely absent and reduced in the remaining texts. The use of personal verbs adheres to the recommended guidelines in both instances. However, it has been observed that in cases where the regulations are not met for personal verbs, there is a significant difference between the original text and the text generated by ChatGPT (Fig. 3).

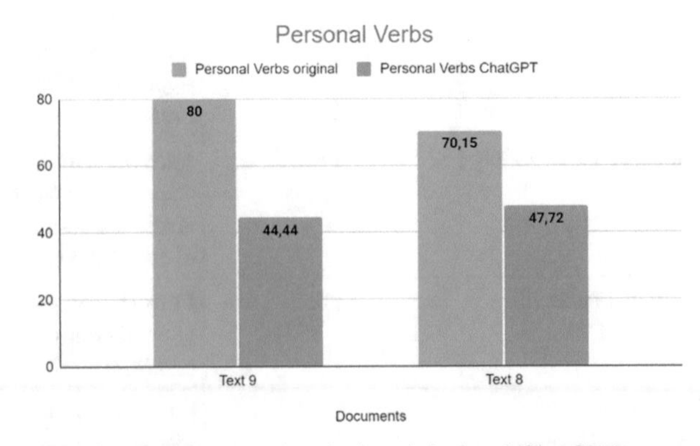

Fig. 3. Use of personal verbs in original and ChatGPT texts.

Adverbs

To enhance writing style, it is recommended to use shorter terms instead of adverbs ending in '-mente'. Out of the ten original texts, six contained such adverbs, but only two of the texts generated by ChatGPT included them. However, one ChatGPT-generated text did include this element, despite its absence in the original text.

Long Words.
Previous studies [19] have identified that the most comprehensible words are those containing between 6 and 11 letters. For the purposes of this study, a word is considered notoriously difficult to understand if it contains 14 letters or more.

Long words appear in all but one of the corpus texts (original and ChatGPT-generated). Despite the recommendations, the incidence of long words is higher in 7 of the 9 texts generated by ChatGPT. However, the variation between the frequencies of the original text and the one produced by ChatGPT is not relevant.

6 Conclusion and Discussion

This paper presents a study on ChatGPT's ability to generate easily understandable text in Spanish according to Plain Language guidelines. The corpus consisted of 10 web pages from the Spanish government, which ChatGPT automatically translated. The study included an in-depth analysis of the 20 texts in the corpus to identify various linguistic and readability features. To the best of our knowledge, no previous studies have been conducted on language simplification in Spanish or on language simplification for e-Government purposes using ChatGPT.

The study's main outcome reveals that ChatGPT has limitations in translating texts into Plain Text. Unfortunately, the new texts generated by the Artificial Intelligence tool do not adhere to most of the recommendations outlined in the Spanish language guidelines. Of all the elements analyzed (see Table 1. - 12 linguistic indicators), only two recommendations have been met: the avoidance of superlatives and the subjunctive. The most significant challenges are posed by verbs or verbal structures, and they are not simplified by using more straightforward constructions. The use of long words or adverbs ending in '-mente' should be avoided. Additionally, the Flesch index indicates that the texts are more difficult to understand than those found on websites. In ChatGPT practice, simplifying text can result in a loss of information, especially for longer texts. However, for shorter texts, it can be effective in fulfilling the given request.

This study had some limitations. Firstly, it is important to note that ChatGPT was not specifically trained for this research as it only evaluates the system's initial response. Furthermore, limitations were found in evaluating certain linguistic attributes, some of which cannot be automatically verified even with tools like Midolec. In future research, we will train ChatGPT using the provided guides and evaluate its performance. Additionally, we will assess the performance of other AI text generation applications.

Acknowledgements. This research has been partially funded by INSIGHT research project (PID2022-137344OB-C32) and the grant for research contracts PEJ-2021-AI/SAL-21563. We also thank Santiago Moreno Fernandez, Victor Muñoz-Velasco and Changao Wu for their support in the PLN analyzer.

Appendix I. Linguistic Elements Analyzed

Table 3 shows the comparative data in original text (O) and generated text (G) using Chat-GPT system for the 10 texts extracted from the Spanish Government web pages. **Boxed bold text** is used to highlight when the generated text is worse than the recommendation.

Table 3. Recommendations and number of occurrences in each original and generated text.

Recommendation	Text 1		Text 2		Text 3		Text 4		Text 5		Text 6		Text 7		Text 8		Text 9		Text 10	
	O	G	O	G	O	G	O	G	O	G	O	G	O	G	O	G	O	G	O	G
Avoid infinitive	8	**10**	17	11	195	15	1	**2**	26	10	65	26	6	6	21	19	-	**3**	13	13
Avoid Subjunctive	2	**4**	3	-	50	-	-	-	33	5	37	4	1	-	23	-	-	-	19	1
Avoid subordinate sentence	6	**11**	9	1	125	2	2	2	65	19	78	17	14	5	41	5	1	1	33	8
Avoid past tenses	3	3	-	-	3	-	-	-	1	-	13	1	2	2	11	1	-	-	8	3
Avoid passive voice	-	**1**	-	**1**	11	1	-	-	3	1	1	-	-	-	-	-	1	1	-	**3**
Avoid problematic structures	2	**3**	1	-	7	2	1	1	39	24	24	2	4	1	10	-	-	**1**	9	3
Avoid gerund	3	2	-	**1**	10	-	1	1	8	3	10	5	-	**1**	1	1	-	-	1	-
Personal verbs	39	**38**	20	**19**	265	**25**	4	4	139	**49**	110	**32**	24	**19**	94	**21**	4	4	69	**25**
Avoid adverbs ending in '-mente'	2	1	-	**1**	24	5	-	-	2	3	5	1	-	-	4	-	-	-	2	-
Avoid long words	15	15	4	3	23	8	-	**1**	38	16	21	5	-	-	10	5	1	-	6	6
Avoid Participles	8	3	6	5	29	1	2	-	40	13	20	1	-	**1**	18	3	1	2	20	2
Avoid Superlatives	-	-	-	-	-	-	1	-	-	-	1	-	-	-	1	-	-	-	-	-

References

1. Morato, J., Iglesias, A., Campillo, A., Sanchez-Cuadrado, S.: Automated readability assessment for Spanish e-Government information. J. Inform. Syst. Eng. Manag. **5**(1) (2021). e-ISSN: 2468–4376
2. Public Law 111–274-oct. 13, 2010. 124. Stat. 2861. 111th Congress. Plain Writing Act of (2010). https://www.govinfo.gov/content/pkg/PLAW-111publ274/pdf/PLAW-111publ274.pdf
3. Campillo, A, Sánchez-Cuadrado, S, Iglesias, A, Morato, J.: Influence of term familiarity in readability of spanish e-Government web information. In: Future Technologies Conference, FTC 2020, 5–6 November (2020)
4. European Commission. 2019 Clear writing for Europe Conference (2019a). https://ec.europa.eu/info/sites/info/files/clear_writing_conference_notes_for_website.pdf. (Accessed 30 Apr 2020)
5. European Commission Commission Style Guide (2019). https://ec.europa.eu/information_society/newsroom/. (Accessed 30 Apr 2020)
6. Fruhlinger, J.:. ¿Qué es y cómo funciona la IA generativa? Computerworld Spain, NA (2023). https://link.gale.com/apps/doc/A758619517/IFME?u=anon~7f439a8d&sid=googleScholar&xid=660a6e76
7. Olite, F.M.D., Suárez, I.D.R.M., Ledo, M.J.V.: Chat GPT: origen, evolución, retos e impactos en la educación. Educación Médica Superior **37**(2) (2023)
8. Teixeira da Silva, J.A.: Can ChatGPT rescue or assist with language barriers in healthcare communication?. Patient Educ. Counseling **115**, 107940 (2023). https://doi.org/10.1016/j.pec.2023.107940

9. Ali, S.R., Dobbs, T.D., Hutchings, H.A., Whitaker, I.S.: Using ChatGPT to write patient clinic letters. Lancet Digital Health **5**(4), e179-e81 (2023) . https://www.thelancet.com/journals/landig/article/PIIS2589-7500(23)00048-1
10. Moons, P., Van Bulck, L.: Using ChatGPT and Google Bard to improve the readability of written patient information: a proof of concept. Euro. J. Cardiovascular Nursing zvad087 (2023). https://doi.org/10.1093/eurjcn/zvad087
11. Ayre, J., Mac, O., McCaffery, K., et al.: New Frontiers in health literacy: using ChatGPT to simplify health information for people in the community. J. Gen. Intern. Med. (2023). https://doi.org/10.1007/s11606-023-08469-w
12. Lyu, Q., Tan, J., Zapadka, M.E., Ponnatapura, J., Niu, C., Myers, K.J., et al.: Translating radiology reports into plain language using ChatGPT and GPT-4 with prompt learning: results, limitations, and potential. Visual Comput. Indust. Biomed. Art **6**(1), 9 (2023)
13. Akiba, D., Fraboni, M.C.: AI-supported academic advising: exploring chatgpt's current state and future potential toward student empowerment. Educ. Sci. **13**(9), 885 (2023). https://doi.org/10.3390/educsci13090885
14. Ayinde, L., Wibowo, M.P., Ravuri, B., Emdad, F.B.: ChatGPT as an important tool in organizational management: a review of the literature. Bus. Inf. Rev. **40**(3), 137–149 (2023). https://doi.org/10.1177/02663821231187991
15. Tan, J., Westermann, H., and Benyekhlef, K.: ChatGPT as an Artificial Lawyer?. Artificial Intelligence for Access to Justice (AI4AJ 2023) (2023). https://ceur-ws.org/Vol-3435/short2.pdf
16. Westermann, H, and Benyekhlef, K. (2023). Justicebot: A methodology for building augmented intelligence tools for laypeople to increase access to justice. Proceedings of the Nineteenth International Conference on Artificial Intelligence and Law
17. Iglesias, A., Cobián, I., Campillo, A., Morato, J., Sánchez-Cuadrado, S.: Comp4Text Checker: an automatic and visual evaluation tool to check the readability of spanish web pages. In: Miesenberger, K., Manduchi, R., Covarrubias Rodriguez, M., Peňá, P. (eds.) ICCHP 2020. LNCS, vol. 12376, pp. 258–265. Springer, Cham (2020). https://doi.org/10.1007/978-3-030-58796-3_31
18. Fernández Huerta, J.: Medidas sencillas de lecturabilidad. Consigna **214**, 29–32 (1959)
19. Rello, L., Baeza-Yates, R., Dempere-Marco, L., Saggion, H.: Frequent words improve readability and short words improve understandability for people with dyslexia. In: Kotzé, P., Marsden, G., Lindgaard, G., Wesson, J., Winckler, M. (eds.) INTERACT 2013. LNCS, vol. 8120, pp. 203–219. Springer, Heidelberg (2013). https://doi.org/10.1007/978-3-642-40498-6_15

Does a Lack of Transparency Lead to the Dark Side of Social Media? a Study in South Western Europe According to Gender and Age

André Filipe Silva[1], Daniela Durães[1], Inês Azevedo[1], João Aparício[2], and Manuel Au-Yong-Oliveira[2,3,4(✉)]

[1] Department of Languages and Cultures, University of Aveiro, 3810-193 Aveiro, Portugal
{andrefsilva,dduraes,inesfazevedo}@ua.pt

[2] Department of Economics, Management, Industrial Engineering and Tourism, University of Aveiro, 3810-193 Aveiro, Portugal
{j.aparicio,mao}@ua.pt

[3] INESC TEC, Porto, Portugal

[4] Research Unit on Governance, Competitiveness and Public Policies (GOVCOPP), 3810-193 Aveiro, Portugal

Abstract. Social media's globalization has had an impact on young people by exposing them to unachievable beauty standards and an abundance of meticulously edited images and seemingly perfect bodies. Does a lack of transparency lead to the dark side of social media? The purpose of this study is to shed light on the ways in which social media has impacted young people and the harmful tactics that these platforms use to manipulate people's body image. To obtain data for this study, a survey with 151 responses was conducted. The data was then analyzed using the chi-square test of independence as well as Fisher's Exact Test (as the prerequisites of the chi-square test were not satisfied) by using the IBM SPSS Statistics software package. We can conclude that age and gender are not independent of experiencing different levels of pressure (women are more affected, as are younger people up to 25 years old) to maintain a perfect body through interacting on social media. More transparency is hence needed so that our youths are not led down the wrong path.

Keywords: Body image · strategy · social media · chi-square · beauty standards

1 Introduction

We live in a time marked by great technological advances and where it is increasingly essential to have access to the internet. We are constantly connected and in virtual communication with the rest of the world and, as a result, social networks have emerged as a predominant force that shapes human interactions and that, over the last few years, has grown exponentially.

The virtual world is saturated with meticulously edited images, filtered selfies and seemingly perfect bodies, which creates an unrealistic standard of beauty. Young people

Á. Rocha et al. (Eds.): WorldCIST 2024, LNNS 987, pp. 148–159, 2024.
https://doi.org/10.1007/978-3-031-60221-4_15

are particularly vulnerable to these visual narratives that often promote unattainable ideals. By constantly being exposed to a stream of images that are usually doctored, they end up participating in a digital culture that prioritizes perfection over authenticity, which has negative consequences for self-esteem and body image. In addition, as beauty standards are constantly changing, a scenario is created in which people are constantly chasing "perfection", which creates an opportunity for companies in the beauty and wellness sectors to create strategies to capitalize on these insecurities, which are often perpetuated by them and by social networks.

This research aims to analyze how social media perpetuates unhealthy standards of beauty, leading to a general feeling of inadequacy among young users from Portugal (South Western Europe). By examining the strategies used by social media platforms, we aim to explore the impact they have on young people's mental and emotional well-being.

2 Literature Review

2.1 Beauty Standards

Since the dawn of time, the definition of beauty has been an intricate and dynamic topic, the subject of constant debate and reflection [1]. The inherent complexity of understanding beauty lies in its ever-evolving nature, which manifests itself in different ways over time and in different cultures, making it challenging to establish a universally accepted definition. However, despite this diversity, it is fascinating to observe the almost innate ability of human beings to identify characteristics that are considered universally attractive.

Recent studies, such as those conducted by [1] and [2], highlight the presence of physical characteristics that seem to transcend cultural boundaries in the assessment of beauty. Facial uniformity, symmetry, skin homogeneity and sexual dysmorphism emerge as elements that are recognized as indicators of attractiveness. These characteristics, although they may vary in intensity and application, play a significant role in the perception of beauty in different social and historical contexts.

2.2 Beauty Standards Over Time

According to [2], the concept of beauty has undergone dynamic transformations over time, with distinct ideals in various civilizations.

In Ancient Egypt, women were depicted with high, thin waists, narrow hips, dark black hair, and golden skin, often accentuated with black eyeliner [3]. In contrast, Ancient Greek women favored a lack of eyebrows and often bleached their hair, emphasizing symmetry and beauty in men.

Medieval Japan appreciated beauty in long hair, fair skin, round, rosy cheeks, and small, curved lips. The Renaissance era in Italy introduced criteria such as a high forehead and blond hair, associating beauty with virtue, while a rounded figure was esteemed during this period. In eighteenth-century France, beauty standards included an oval face with a double chin and rosy cheeks, and the Edwardian era celebrated the hourglass figure, characterized by a tiny waist, full chest, and equally rounded hips.

[4] highlights the historical significance of small feet in China, leading to the painful practice of "foot binding" until its prohibition in 1912, which highlights the lengths that some individuals are willing to go through to perpetuate a specific beauty standard.

Despite the process of globalization, regional, racial, and culturally distinct beauty trends have persisted not only over the centuries, but to this very day.

2.3 Globalization of Beauty Standards

Social media and traditional media—such as television and fashion magazines—have been defining beauty standards for years, and this has had a significant impact on how people view themselves [5]. Globalization of media has spread the idea that youth, whiteness, and thinness are essential components of beauty, as exemplified by the Western ideal of a slim body, blue eyes, and fair complexion [6; 7]. This supports the ideal of beauty that is based on privilege and conforms to the idealized portrayal of whiteness in the media.

Although there are certain aspects of beauty that are considered universal, opinions about beauty differ greatly among cultures, ethnicities, and geographical areas, highlighting how subjective and dynamic international beauty standards are [8]. The diverse historical, cultural, and social influences that shape Western, Asian, and African cultures result in differing ideals of beauty.

Beauty standards in Western cultures are heavily influenced by historical preferences and media representation. These emphasize youthful appearance, a slim body, and Eurocentric features, which can negatively impact women's self-esteem and lead to body dissatisfaction [9]. The Asian beauty frenzy, fueled by social media and societal pressures, emphasizes features like clear skin, almond-shaped eyes, and a V-shaped face [10]. This pursuit of an ideal leads to increased demand for skin-lightening products and a rise in body image issues and plastic surgery requests [11, 12]. African beauty standards, diverse and rooted in history, traditionally celebrate dark skin, natural hair, and unique facial features. However, Westernized ideals have been introduced by the globalization of beauty, challenging these conventions and leading to practices like hair straightening and skin bleaching [10].

3 Social Media and Young People's Body Image

Nowadays, social networks have seamlessly integrated into the lives of young people, becoming pivotal in shaping their worldview. In 2023, out of a population of 8.01 billion people, 4.76 will have active social media accounts [13], which corresponds to 60% of the world's population. According to [13], the average person spends 2 h and 31 min a day on social media, which is a new record. A global survey carried out in the first quarter of 2023 revealed that 61 per cent of teenagers use social networks for various reasons, which indicates the widespread use of social networks among teenagers around the world [14].

According to several studies analyzed by [15], the greater the user's exposure to their body image, the greater the number of comparisons with their body image with other people. Therefore, research indicates that the more people share and view content related to physical appearance on social media, the worse their perception of their body becomes [16]. These elements are intrinsically related to problems with self-esteem, which in the case of adolescents, affects their levels of anxiety and body perception [15] and which can, in some cases, lead to changing one's appearance [17].

Although there are common usage patterns among platforms, there are differences in the effects on users' body image [18]. The most negative effects on body image are linked to appearance-focused content on photo-based platforms like Instagram [19]. The impact is increased by the prominence of "likes" beneath uploaded images. According to [20], there is a correlation between the importance people place on the quantity of likes and an increase in physical comparisons and body dissatisfaction. According to qualitative research [21], likes reinforce conventional standards of physical beauty by acting as markers of users' status and popularity [16].

A study conducted by [22] found a correlation between excessive use of social networks and behaviours related to eating disorders. Appearance-related social media use is associated with poor eating and a worse body image, which increases the need for interventions and additional attention [23]. Although there are various causes, exposure to social media is considered one of the factors that most contribute to eating concerns [24].

3.1 Negative Strategies Adopted by Social Media to Exploit Personal Insecurities

Social networks, seemingly benign platforms for social interaction, play a crucial role in perpetuating and amplifying personal insecurities, especially in terms of body image. This analysis takes a hard look at the negative strategies employed by these platforms, aiming to capitalize on users' psychological vulnerabilities.

Based on the research presented in the article by [25], it was found that advertising on the TikTok social media platform and the image of beauty products have a significant impact on consumer purchase decisions. More specifically, product image was found to have the strongest influence on consumer purchasing decisions [25], especially in the case of lipstick products endorsed by celebrities [25]. It is therefore possible to infer that TikTok has the potential to positively influence consumers' purchasing decisions for beauty products, particularly when the product image is attractive and endorsed by popular figures [25].

These strategies profit from people's insecurities and can create an emotional connection with consumers, capitalizing on their aspirations and desires by imposing idealized beauty standards. In conclusion, companies exploit appearance-related insecurities to drive consumer purchasing decisions, creating a demand based on the quest for social acceptance and self-improvement.

A study conducted by [26] thoroughly examined the relationship between social media use and the consumption of dietary supplements, anabolic steroids and SARMs among male gym users. The results revealed a significant association between image-centered use on social media and a negative body image, highlighting a direct correlation with the search for potentially unhealthy substances. This analysis explores the implications of this complex phenomenon.

Use of Algorithms. As seen above, the use of social networks brings numerous insecurities to its users. However, one of the main strategies that social networks use is to set up algorithms to keep the user active on the platform, by recommending images and videos that the user spends more time looking at or that are their favorites. For example, if the user likes to watch videos of ideal or significantly defined bodies, the algorithm recognizes this and continues to create other images/videos of ideal bodies to keep them hooked on the network. In this sense, the user forms the idea that it will be necessary to buy the products that these influencers promote so that they can achieve this "ideal" body [27].

According to [27], the owners of social networks are aware of the danger caused by the way they present their content. Some revelations have shown that the company Meta (which owns Facebook, Instagram and WhatsApp) is aware of the mental health effects of its products on young people, with a Facebook executive revealing internal documents that prove exactly this on young people's body image [28].

In the case of TikTok, *The Wall Street Journal* carried out a study that demonstrated how the company's algorithm manipulates what its users see in an extreme way, with the aim of retaining users for longer through content that arouses their emotions [29]. TikTok even tried to remove these types of videos, but those that were not in English ended up evading the filters of the company's policies and rules, meaning that those who spent more time on the platform tended to be more exposed to unmonitored content [30].

Young people tend to have more addictive behaviors when it comes to using social networks, it being difficult to break free from them when the algorithms are specifically designed to keep users hooked on the content that disturbs them the most [29].

The Help of AI. One strategy that social networks have been adopting for some time now is AI-controlled algorithms. These contain various techniques for personalizing the content that each user sees, the most widely used being those based on collaborative filtering recommendation systems. "Content-based recommendation algorithms analyze the content of the data uploaded by users, while collaborative filtering makes recommendations based on similarities with other users" [31]. The use of these algorithms can restrict the content to which the user is exposed and the possibility of creating a personalized bubble for each user, better known as "filter bubbles" [31].

With filter bubbles, users "receive content that corresponds to their tastes and beliefs and are often unaware that there is more diverse content and opinions" [31], due to the fact that the information produced on social networks is very significant and this leads to its simplification by the filter bubbles, making it "delimit the dimension of reality and have effects on users such as the polarization of ideas" [31].

In this sense, these filters lead to real relationships being affected due to a single view of reality on the part of the user, not considering numerous interpretations of it on a given subject, coupled with the fact that the user is often unaware of the existence of filter bubbles and the need to act against them. In this sense, according to [32], being exposed to a certain type of body from the filter bubble causes individuals to change their perceptions of a healthy body over time, often unrealistically.

4 Methodology

This work stems from our curiosity to understand the fact that social networks use strategies in order to survive / make more profit and consequently gain a larger market share through practices that jeopardize the well-being of their own customers.

We can also see that the aforementioned strategies have a more significant impact on young people, which could pose serious problems for the future of society.

Does a lack of transparency lead to the dark side of social media?

To carry out this study, we used a set of scientific articles from Scopus, ScienceDirect, Springer and the Wiley Online Library, as well as analyzing news from reliable websites. The main topic of this study focuses on the impact of young people's use of social networks on their perception of their body image in relation to other ages. Young people are considered to be all individuals under the age of 25.

In the initial stage, we collected a total of 124 studies exclusively from databases. After carefully screening the information, 61 studies were deemed relevant and retained for further examination. Subsequently, we thoroughly reviewed these 61 studies, eliminating an additional 9 due to insufficient relevant information. This resulted in a final selection of 52 studies, with 46 of them being used to support our research.

A 12-question questionnaire survey was carried out, comprising a sample of 151 individuals, with an average age of 25.1 years. The survey was launched online (via Google Forms) on November 3, 2023 and closed on November 18 of the same year. All responses were considered valid. The questions in the survey were structured as follows: single choice answers, yes/no questions, short answers and long answers, which are described in the appendix.

In order to conclude the study, we analyzed the most relevant issues based on our literature review, using the Chi-Square test for this purpose.

5 Results and Discussion

For our survey, we obtained a sample of 151 individuals, all of whose responses were considered valid. To analyze the responses, we used the chi-square test of independence, as well as Fisher's Exact Test, with the help of IBM SPSS Statistics software to provide the results.

As the survey was carried out by Portuguese students and shared with Portuguese people, all the responses were from people of Portuguese nationality (100%) (Portugal is in South Western Europe).

According to the survey, the social network most used by respondents is Instagram, with more than half choosing this network (52.98%), the least used being Reddit (0.66%).

Regarding the question "How often do you use social media?", a significant majority of respondents answered that they used it daily (97.35%), followed by "2 to 3 times a day" and "Monthly", both with 2 respondents (1.32% each).

In the survey, respondents were asked if they felt pressured to have a "perfect body" based on what they saw in images and videos on social networks, to which all respondents replied (100%), with the majority answering "Yes" (67.55%) and the remainder "No" (32.45%).

However, in response to the question "Do you believe that social media has contributed to you feeling more concerned about your body's appearance?", the "Yes" answer increased significantly (to 80.79%), with the "No" answer decreasing (to 19.21%). In this sense, we can see that, although there is pressure to obtain perfect bodies through social networks, we cannot rule out the belief on the part of those who do not feel this pressure that social networks influence the way their body looks.

Next, respondents were asked if they had ever had negative thoughts about their body through content viewed on social media, to which 52.98% answered "Yes", 17.88% answered "Maybe" and the remainder answered "No". For the purposes of this study, negative thoughts mean "an individual who is unhappy with the way they look" (Spann, 2022).

Regarding the question "Have you ever had cosmetic operations or gone on extreme diets because of the influence of social media on your body image?", 90% of respondents answered "No" and 10% answered "Yes".

Regarding the question "Have you ever sought professional help (e.g. therapist, psychologist) to deal with body image problems related to social media?", a significant proportion of respondents answered "No" (91.39%) and the rest "Yes" (8.61%).

6 Chi-Square Tests

The chi-square test of independence was carried out using the IBM SPSS Statistics software (v.26) (Tables 1–4).

Null hypothesis H0: Visualizing images/videos on social networks does not lead to feeling pressure to have a perfect body according to one's age.

Alternative hypothesis H1: Visualizing images/videos on social networks does lead to feeling pressure to have a perfect body according to one's age.

As the prerequisites of the chi-square test are not fulfilled (37.5% of cells have expected count less than 5; the value cannot be higher than 20%) (Table 2) we then resorted to Fisher's Exact Test. As the p-value is 0.051 (only marginally higher than 0.05) (table 2) we conclude that the null hypothesis may be rejected at the 5% level of significance. The variables are apparently not independent and one's age is apparently associated to feeling pressure to have a perfect body. Younger people (18–25 years of age) appear to be more affected.

Null hypothesis H0: Visualizing images / videos on social networks does not lead to feeling pressure to have a perfect body according to one's gender.

Alternative hypothesis H1: Visualizing images / videos on social networks does lead to feeling pressure to have a perfect body according to one's gender.

Looking at the Continuity Correction p-value (due to there being only one degree of freedom) of 0.000 (Table 4) we conclude that the null hypothesis may be rejected at the 1% level of significance). The variables are apparently not independent and one's gender is apparently associated to feeling pressure to have a perfect body. Female people appear to be more affected than male people.

Table 1. Fisher's Exact Test.

			3. Já se sentiu pressionado a ter um "corpo perfeito" tendo em conta aquilo que visualiza por via de imagens/vídeos nas redes sociais?		
			Não	Sim	Total
Idade	18 - 21	Count	20	57	77
		Expected Count	25,0	52,0	77,0
	22-25	Count	19	38	57
		Expected Count	18,5	38,5	57,0
	26-29	Count	5	5	10
		Expected Count	3,2	6,8	10,0
	30+	Count	5	2	7
		Expected Count	2,3	4,7	7,0
Total		Count	49	102	151
		Expected Count	49,0	102,0	151,0

Table 2. Fisher's Exact Test.

	Value	df	Asymptotic Significance (2-sided)	Exact Sig. (2-sided)
Pearson Chi-Square	7,750[a]	3	,051	,046
Likelihood Ratio	7,315	3	,062	,085
Fisher's Exact Test	7,416			,051
N of Valid Cases	151			

a. 3 cells (37,5%) have expected count less than 5. The minimum expected count is 2,27.

Table 3. Chi-Square Tests of Independence

			3. Já se sentiu pressionado a ter um "corpo perfeito" tendo em conta aquilo que visualiza por via de imagens/vídeos nas redes sociais?		
			Não	Sim	Total
Sexo	Feminino	Count	16	79	95
		Expected Count	30,8	64,2	95,0
	Masculino	Count	33	23	56
		Expected Count	18,2	37,8	56,0
Total		Count	49	102	151
		Expected Count	49,0	102,0	151,0

Table 4. Chi-Square Tests of Independence

Chi-Square Tests

	Value	df	Asymptotic Significance (2-sided)	Exact Sig. (2-sided)	Exact Sig. (1-sided)
Pearson Chi-Square	28,469[a]	1	,000	,000	,000
Continuity Correction[b]	26,582	1	,000		
Likelihood Ratio	28,348	1	,000	,000	,000
Fisher's Exact Test				,000	,000
N of Valid Cases	151				

a. 0 cells (0,0%) have expected count less than 5. The minimum expected count is 18,17.

b. Computed only for a 2x2 table

6.1 Conclusions Regarding Fisher's Exact Test and the Chi-Square Test

After analyzing the results of Fisher's Exact Test, we can see that young people, when exposed to content on social networks, are the ones who suffer most from insecurities about their bodies due to the beauty standards that are transmitted through videos and images published on social networks. These situations lead to the formation of negative thoughts about their bodies due to the pressure of these standards, leading to the need for professional help and, in more extreme cases, drastic diets and operations.

Additionally, we can see that the pressure exerted by social networks depends on gender, i.e., women tend to suffer more from negative thoughts about obtaining a perfect body than men, given that the p-value of the chi square test is lower than the significance level of 0.01 (0.000).

Considering our study's findings, we can conclude that this study confirms the studies carried out by [34].

7 Conclusion and Future Research

"Beauty is no longer in the eye of the beholder - now it's in the hands of a digital designer" [35].

In conclusion, we can safely say that social media's widespread influence has had negative effects on body image. Being constantly exposed to curated images and unrealistic beauty standards on these platforms has led to increased body dissatisfaction and self-esteem issues. The pressure to conform to idealized physical ideals, often perpetuated by influencers, creates feelings of inadequacy. Moreover, the prevalence of photo-editing tools blurs the line between reality and fiction, and creates unattainable standards that impact individuals' perceptions of their bodies. It is important to acknowledge and address these detrimental consequences and put in place measures to counteract them.

Promoting body-positive content is one of the tactics used to celebrate diversity and subvert conventional notions of beauty [36]. Platforms can also start educational campaigns to improve media literacy, giving users the tools, they need to identify and combat the negative effects of unrealistic body representations on their sense of self [37].

According to research, hiding Instagram likes may improve mental health by lowering the negative and lonely feelings connected to receiving social validation. It is hypothesized that hidden like counts could reduce the stress associated with social competition and promote a more positive online atmosphere [38].

Platforms use algorithms to identify potentially harmful content and flag malicious comments, encouraging users to reevaluate their actions. One recent example is Instagram's "Comment Warning" feature, which flags offensive comments automatically and suggests that users edit them. [37].

However, platforms ought to acknowledge their role in promoting offensive speech in addition to providing technical fixes. Addressing the root causes is essential, such as the psychological effects of upward social comparison. If social media companies want to lower online negativity, they should emphasize comprehending and addressing these underlying causes [38].

Acknowledgments. This work was financially supported by the Research Unit on Governance, Competitiveness and Public Policies (UIDB/04058/2020) + (UIDP/04058/2020), funded by national funds through FCT - Fundação para a Ciência e a Tecnologia.

References

1. Sisti, A., Aryan, N., Sadeghi, P.: What is Beauty? Aesthetic Plast. Surg. **45**(5), 2163–2176 (2021)
2. Dimitrov, D., Mayra B.C. Maymone, Kroumpouzos, G.: Beauty perception: A historical and contemporary review. Clinics in Dermatology, 41(1), 33–40 (2023)
3. Aly, A., Tolazzi, A., Soliman, S., Cram, A.: Quantitative Analysis of Aesthetic Results: Introducing a New Paradigm, Aesthetic Surgery Journal, 32(1), January, 120–124 (2012)
4. Brown, M.J.: Footbinding, industrialization, and evolutionary explanation. Hum. Nat. **27**(4), 501–532 (2016). https://doi.org/10.1007/s12110-016-9268-5
5. Calado, M., Lameiras, M., Sepulveda, A.R., Rodriguez, Y., Carrera, M.V.: The association between exposure to mass media and body dissatisfaction among spanish adolescents. Womens Health Issues. Sep-Oct **21**(5), 390–399 (2011)
6. Chang, F.C., Lee, C.M., Chen, P.H., Chiu, C.H., Pan, Y.C., Huang, T.F.: Association of thin-ideal media exposure, body dissatisfaction and disordered eating behaviors among adolescents in Taiwan. Eat. Behav. **14**(3), 382–385 (2013)
7. Kozee, L.: Unequal Beauty: Exploring Classism in the Western Beauty. Thesis Georgia State University (2016)
8. Dimitrov, D., Mayra B.C. Maymone, Kroumpouzos, G.: Beauty perception: A historical and contemporary review. Clinics in Dermatology, 41(1), 33–40 (2023)
9. The Effects of the Homogenization of the Western Beauty Standard on JSTOR. (2019). Jstor.org
10. Fakhro, A., Hyung Woo, Y., Yong Kyu, K., Nguyen, A.H.: The evolution of looks and expectations of asian eyelid and eye appearance. Semin. Plast. Surg. **29**(03), 135–144 (2015)
11. Rodrigo-Caldeira, D.: White Skin as a Social and Cultural Capital in Asia and Its Economic Markets. Social Science Research Network (2016)
12. Zhang, L., Huang, Q., Fu, H.: To be thin but not healthy - The body-image dilemma may affect health among female university students in China. PLOS ONE, 13(10) (2018)

13. Digital 2023: Global Overview Report — DataReportal – Global Digital Insights. January 26. (2023)
14. Statista.: Global teens reasons for using social media 2023. Statista (2023)
15. Çimke, S., Yıldırım Gürkan, D.: Factors affecting body image perception, social media addiction, and social media consciousness regarding physical appearance in adolescents. Journal of Pediatric Nursing (2023)
16. Chua, T. H. H., Chang, L.: Follow me and like my beautiful selfies: Singapore teenage girls' engagement in self-presentation and peer comparison on social media. Computers in Human Behavior, 55(A), 190–197 (2016)
17. Pan, W., Mu, Z., Tang, Z.: Social Media Influencer Viewing and Intentions to Change Appearance: A Large-Scale Cross-Sectional Survey on Female Social Media Users in China. Frontiers in Psychology, 13 (2022)
18. Vandenbosch, L., Fardouly, J., Tiggemann, M.: Social media and body image: Recent trends and future directions. Curr. Opin. Psychol. **45**, 101289 (2022)
19. Engeln-Maddox, R., Loach, R., Imundo, M.N., Zola, A.: Compared to Facebook, Instagram use causes more appearance comparison and lower body satisfaction in college women. Body Image **34**, 38–45 (2020)
20. Tiggemann, M., Hayden, S., Brown, Z., Veldhuis, J.: The effect of instagram "likes" on women's social comparison and body dissatisfaction. Body Image **26**(1), 90–97 (2018)
21. Dumas, T.M., Maxwell-Smith, M., Davis, J.P., Giulietti, P.A.: Lying or longing for likes? Narcissism, peer belonging, loneliness and normative versus deceptive like-seeking on instagram in emerging adulthood. Comput. Hum. Behav. **71**, 1–10 (2017)
22. Zhang, J., Wang, Y., Li, Q., & Wu, C.: The Relationship Between SNS Usage and Disordered Eating Behaviors: A Meta-Analysis. Frontiers in Psychology, 12 (2021)
23. De Valle, M.K., Wade, T.D.: Targeting the link between social media and eating disorder risk: A randomized controlled pilot study. Int. J. Eat. Disord. **55**(8), 1066–1078 (2022)
24. Sidani, J.E., Shensa, A., Hoffman, B.L., Hanmer, J., Primack, B.A.: The association between social media use and eating concerns among us young adults. J. Acad. Nutr. Diet. **116**(9), 1465–1472 (2016)
25. Darmatama, M., Erdiansyah, R.: The Influence of Advertising in Tiktok Social Media and Beauty Product Image on Consumer Purchase Decisions. Proceedings of the International Conference on Economics, Business, Social, and Humanities (ICEBSH 2021), 888–892 (2021)
26. Hilkens, L., Cruyff, M., Woertman, L., Benjamins, J., Evers, C.: Social media, body image and resistance training: Creating the perfect "me" with dietary supplements, anabolic steroids and sarm's. Sports Medicine - Open, 7(1) (2021)
27. Harriger, J.A., Evans, J.A., Thompson, J.K., Tylka, T.L.: The dangers of the rabbit hole: reflections on social media as a portal into a distorted world of edited bodies and eating disorder risk and the role of algorithms. Body Image **41**(1), 292–297 (2022)
28. Wells, G., Horwitz, J., Seetharaman, D.: Facebook Knows Instagram Is Toxic for Teen Girls, Company Documents Show. Wall Street Journal, September 14 (2021)
29. Smith, B.: How TikTok Reads Your Mind. The New York Times, December 5 (2021)
30. Kastrenakes, J.: TikTok will start automating video removals for nudity and more in the US. The Verge (2021)
31. Reina, R., Theophilou, E., Hernández-Leo, D., Medina-Bravo, P.: The power of beauty or the tyranny of algorithms. How do teens understand body image on instagram? In Prosumidores emergentes: Redes Sociales, Alfabetizácion Y Creación De Contenidos (pp. 429–451). Dykinson S.L (2021)
32. Tiggemann, M., Slater, A.: Facebook and body image concern in adolescent girls: A prospective study. Int. J. Eat. Disord. **50**(1), 80–83 (2017)

33. Guimarães, A. M.: Estatística: Teste exato de Fisher e teste de Qui-Quadrado usando R. Medium (2019)
34. Hargreaves, D.A., Tiggemann, M.: Idealized media images and adolescent body image: "Comparing" boys and girls. Body Image **1**(4), 351–361 (2004)
35. King, T.: Looking good, feeling… not so good: The impact of advertising and social media on body image. Cranfield School of Management; Cranfield (2021)
36. Rodgers, R.F., Wertheim, E.H., Paxton, S.J., Tylka, T.L., Harriger, J.A.: #Bopo: enhancing body image through body positive social media- evidence to date and research directions. Body Image **41**, 367–374 (2022)
37. Park, J., Kim, B., Park, S.: Understanding the behavioral consequences of upward social comparison on social networking sites: the mediating role of emotions. Sustainability **13**(11), 5781 (2021)
38. Lee, J.K.: The effects of social comparison orientation on psychological well-being in social networking sites: Serial mediation of perceived social support and self-esteem. Curr. Psychol., 1–13 (2020). https://doi.org/10.1007/s12144-020-01114-3

Analysis of the Impact of Orthogonality in the Readability of the OBO Foundry Ontologies

Francisco Javier Redondo-Aniorte, Francisco Abad-Navarro, and Jesualdo Tomás Fernández-Breis(✉)

Departamento de Informática y Sistemas, Universidad de Murcia, CEIR Campus Mare Nostrum, IMIB Pascual Parrilla, 30100 Murcia, Spain
jfernand@um.es

Abstract. There is an increasing use of ontologies in the biomedical domain, so ensuring their high quality is required. Reuse of content is a best practice in ontology engineering, whose non-optimal implementation may lead to lead to loss of information and, therefore, of readability of the ontology for both humans and machines. We present twelve metrics for analyzing lack of human readability in ontologies. We applied these metrics to the OBO Foundry repository, which contains an orthogonal collection of biological and biomedical ontologies, and has content reuse as a design principle. For this study, we have also generated an orthogonal view of the repository, which can be seen as a single ontology representing the merged repository. The comparison between both perspectives permitted to identify that the readability for some types of entities and annotations decrease due to how reuse is implemented.

Keywords: biomedical ontologies · quality assurance · readability · OBO Foundry

1 Introduction

The biomedical community has now developed a significant number of ontologies which are used for different purposes such as data interoperability and integration [1,2], domain knowledge standardization [3], or natural language processing [4]. In November 2023, BioPortal [5] hosted more than 1,000 ontologies and the Open Biological and Biomedical Ontology (OBO) Foundry [6] had more than 250 ontologies. The number of ontologies in these repositories is continuously increasing, which demonstrates their relevance and impact. The relevance of ontologies is likely to increase due to the advent of knowledge graphs, since recent proposals for standardized schemas and languages are based on ontologies [7].

Therefore, the quality assurance of the content of biomedical ontologies is becoming of paramount importance. Briefly speaking, ontologies describe a domain using terms/classes, properties and instances that are implemented using a formal language. Ontology entities have natural language annotations that make them understandable by humans. At the same time, such meaning is also

Á. Rocha et al. (Eds.): WorldCIST 2024, LNNS 987, pp. 160–169, 2024.
https://doi.org/10.1007/978-3-031-60221-4_16

provided to the machines in the form of logical axioms. In practical terms, the content of ontologies must be understandable by both humans and machines.

The ontology community has traditionally paid attention to the quality assurance of the ontologies from the axiomatic, logical perspective, that is, the content for machines [8]. In some cases, the human-oriented content, in natural language, has been used for evaluating the alignment between the content for humans and for machines [9] or for improving the axiomatization of the ontology [10,11]. In most cases, the analysis is performed at the level of single ontologies, without exploiting the fact that ontologies are usually stored in repositories and are reusing content from other ontologies from the repository.

The OBO Foundry promotes the development of an orthogonal collection of ontologies that share a set of design principles, what requires the reuse of terms for building that orthogonal set of ontologies. Orthogonality could be used when terms can be jointly applied to describe complementary but distinguishable perspectives on the same biological or medical entity. The reuse in ontologies has been studied by different authors [12–14]. More concretely, our previous work revealed that roughly 2,000 terms and 20,000 hidden axioms, on average, could be automatically reused [14]. By hidden axioms, we mean axioms not reused when reusing the terms. This means that if the ontologies were exploited in the context of the complete OBO Foundry repository, more axioms would be available to the machine.

Again, those studies focused on analyzing and promoting term reuse and the reuse of axioms but ignored the human-oriented content expressed in natural language. In addition to this, it should be noted that the content in natural language is not only used by humans but more and more by machines given the advent of machine learning approaches based on embeddings (see, for instance [15]). In this work we focus on the human readability of biomedical ontologies, also paying attention to its reuse. Recently, we have released the HURON framework [16], which provides different metrics regarding the human readability of ontologies. In this work we define and apply new metrics related to readability but with a focus on how readability related content is reused across a corpus of ontologies.

An innovative aspect of our work is that we will consider the orthogonal feature of the OBO Foundry repository to study its readability. Given that orthogonality implies reuse, the method developed for our readability study will take into account how the natural language content is reused by the OBO Foundry ontologies. For this purpose, we will define new concepts to capture non-readable reuses of ontology entities. Our study will compare the readability of the OBO Foundry repository when considered as a set of individual ontologies and as an orthogonal collection. This study will generate knowledge about the readability of the OBO Foundry repository and complementary information to our previous study on the reuse of axioms. Both studies contribute to generate knowledge about the effects of orthogonality in the use and reuse of biomedical ontologies.

2 Materials and Methods

2.1 Human Readability Properties in OWL Ontologies

In this study we assume OWL2 ontologies, which consist of classes, properties, individuals and constraints. Classes are defined by datatype properties and object properties, and described by annotation properties. Annotation properties are the ones used for providing human readability, so these are the relevant ones for this work. An excerpt of the OWL2 model is shown in Fig. 1. There, we can see that every entity has an IRI. In the lower part of the figure we can see examples of the annotation properties used by different ontologies for providing human-oriented content in natural language: rdfs:label, IAO_0000115, hasExactSynonym. These are just a subset of the annotation properties used in biomedical ontologies for human readability.

Despite there are many types of annotation properties that can be used in ontologies, in this work we focus on those related to names, synonyms and descriptions. Table 1 provides the complete set, which extends the one used in our previous work [17], that we use in this work. The table is organized by name, synonym, and description.

Table 1. Identified annotation properties for encoding labels, synonyms and descriptions.

Name	http://www.w3.org/2004/02/skos/core#prefLabel
	http://www.w3.org/2000/01/rdf-schema#label
	http://schema.org/name
	http://ncicb.nci.nih.gov/xml/owl/EVS/Thesaurus.owl#P108 (Preferred Name)
	http://purl.obolibrary.org/obo/IAO_0000589 (OBO Foundry unique label)
	http://purl.obolibrary.org/obo/IAO_0000111 (editor preferred term)
	http://xmlns.com/foaf/0.1/name
Synonym	http://www.w3.org/2004/02/skos/core#altLabel
	http://www.geneontology.org/formats/oboInOwl#hasExactSynonym
	http://www.geneontology.org/formats/oboInOwl#hasRelatedSynonym
	http://www.geneontology.org/formats/oboInOwl#hasBroadSynonym
	http://www.geneontology.org/formats/oboInOwl#hasNarrowSynonym
	http://ncicb.nci.nih.gov/xml/owl/EVS/Thesaurus.owl#P90 (fully qualified synonym)
	http://purl.obolibrary.org/obo/IAO_0000118 (alternative label)
	http://purl.obolibrary.org/obo/OBI_9991119 (FGED alternative term)
	http://purl.obolibrary.org/obo/OBI_9991118 (IEDB alternative term)
	http://purl.obolibrary.org/obo/OBI_0001847 (ISA alternative term)
	http://purl.obolibrary.org/obo/OBI_0001886 (NIAID GSCID-BRC alternative term)
Description	http://purl.obolibrary.org/obo/IAO_0000115 (definition)
	http://www.w3.org/2004/02/skos/core#definition
	http://www.w3.org/2000/01/rdf-schema#comment
	http://purl.org/dc/elements/1.1/description
	http://ncicb.nci.nih.gov/xml/owl/EVS/Thesaurus.owl#P97 (definition)

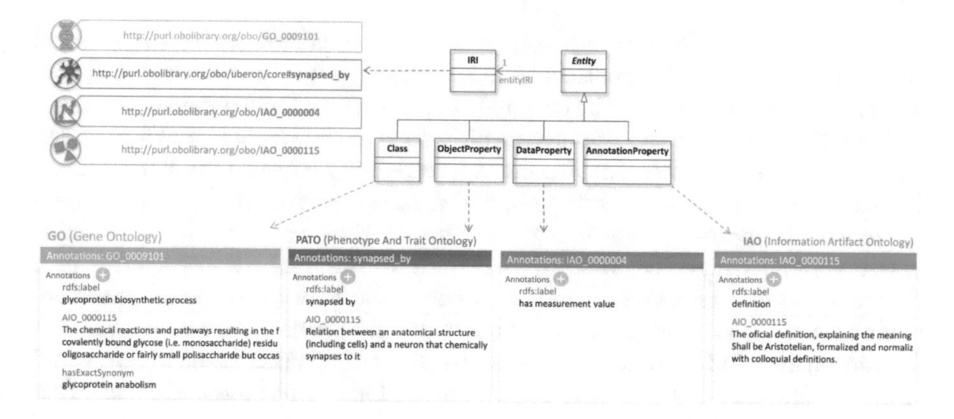

Fig. 1. Examples of entities and annotations in OWL2.

2.2 The Orthogonal View of the OBO Foundry Repository

In this work, we will study the corpus of ontologies included in the OBO Foundry repository. The OBO Foundry repository contains a set of biological and biomedical ontologies which has been conceived as an orthogonal collection of ontologies. Consequently, the content of this repository can be approached as a set of independent ontologies (non-orthogonal view), or as an orthogonal collection, in which, the entities of all the ontologies are merged into one ontology (orthologonal view).

In the orthogonal view, the classes and properties associated with the same IRI in different ontologies are merged, avoiding redundancy. For example, if the Gene Ontology term http://purl.obolibrary.org/obo/GO_0030641, whose label is 'regulation of cellular pH', is reused by other ontologies, the label being also reused, the orthogonal view would only include one occurrence of the label.

2.3 The Reuse-Based Readability Metrics

There two main ways for reusing entities from an ontology:

- Reuse of entity: OWL provides an *import*[1] mechanism through which we can reuse in our ontology all the content from the imported ontology.
- Reuse of URI: In an ontology, we can use the URI of an entity of an external ontology without importing all the content associated with that entity or with its source ontology. In this case, if we want to obtain the label, description or other properties of that entity, we will have to access the source ontology or to reuse explicitly those properties, which would not happen automatically. This is the type of reuse that can provoke readability loss, and it is very common according to our previous results [14].

[1] https://www.w3.org/TR/owl2-syntax/#Imports.

Let us suppose the case of a corpus of three ontologies, where the external class A is reused in all the 3 ontologies in the corpus, but its name is included only in 2 of those ontologies, and the external class B is reused in 2 ontologies without including its name. We call **non-readable reuse** to the loss of readability due to not reusing the properties carrying out the readability information. For example, if class A has a value X for *rdfs:label*, and an ontology reusing A does not have an annotation property *rdfs:label* with value X, that would be a non-readable reuse. We can define the **lack of reuse readability** (LRR) of a term in a corpus for a given annotation property, as the ratio between the number of non-readable reuses, and the number of reuses. In the previous example, such lack for the *rdfs:label* of A would be 1/3 and 2/2 for the *rdfs:label* of B. We could then average the results to say that the mean for the *rdfs:label* of A and B is 0.67, what stands for the degree in which the classes are reused in the corpus of ontologies without reusing their names. If we know build the orthogonal view for that corpus of three ontologies, there would be one appearance of class A (with one *rdfs:label*) and one appearance of B (with no *rdfs:label*). In this case the lack of reuse readability would be 0.5 (0 for A and 1 for B), since the *rdfs:label* of B would not be included in the orthogonal view.

We have defined 12 metrics by applying the concept of **non-readable reuse** to ontologies, which are the result of considering three aspects represented as annotation properties (names, synonyms and descriptions) in four ontology types of entities (classes, object properties, datatype properties, annotation properties):

- LRR-NC: Mean of the LRR for the names of each class.
- LRR-SC: Mean of the LRR for the synonyms of each class.
- LRR-DC: Mean of the LRR for the descriptions of each class.
- LRR-NO: Mean of the LRR for the names of each object property.
- LRR-SO: Mean of the LRR for the synonyms of each object property.
- LRR-DO: Mean of the LRR for the descriptions of each object property.
- LRR-ND: Mean of the LRR for the names of each datatype property.
- LRR-SD: Mean of the LRR for the synonyms of each datatype property.
- LRR-DD: Mean of the LRR for the descriptions of each datatype property.
- LRR-NA: Mean of the LRR for the names of each annotation property.
- LRR-SA: Mean of the LRR for the synonyms of each annotation property.
- LRR-DA: Mean of the LRR for the descriptions of each annotation property.

3 Results and Discussion

3.1 Experimental Setup

We used a set of 182 ontologies, downloaded from OBO Foundry that we generated and used in previous studies[2]. The OWL API [18] was used for parsing and extracting data from the ontologies, and R software was used for further

[2] https://doi.org/10.5281/zenodo.6363060.

analysis. From the initial set of ontologies, there were 38 ontologies that could not be analyzed due to parsing errors mainly caused by conflicts with the OWL API version used. Thus, we computed the 12 LRR metrics for 144 ontologies. The complete results are available in a GitHub repository[3].

3.2 Comparison by Number of Entities

Table 2 shows the size of the repository snapshot by counting entities, spread over the corpus of 144 ontologies. We can appreciate that the largest difference in relative terms between the orthogonal and the non-orthogonal views is for object properties and annotation properties, whereas the difference is smaller for classes and datatype properties. The number of entities in the orthogonal view represents unique occurrences in the corpus, and the *factor* column estimates the mean number of appearances of every entity in the non-orthogonal view. This would mean that object properties and annotation properties are more reused than classes and datatype properties.

Table 2. Number of entities in the two views based on the OBO Foundry snapshot.

OWL2 entity	Orthogonal view	Non-orthogonal view	Factor
Class	4,605,558	4,853,996	1.05
Object Property	1,983	12,212	6.15
Data Property	401	467	1.16
Annotation Property	1,868	10,901	5.83

3.3 Analysis of Human Readability

The LRR metrics described in Sect. 2.3 were calculated for the 144 OBO Foundry ontologies by considering both the non-orthogonal and the orthogonal vision of the repository, as explained in Sect. 2.2. Table 3 includes the results obtained for the orthogonal and the non-orthogonal vision of the OBO Foundry corpus for each of the LRR metrics. Additionally, Fig. 2 depicts this information with a line plot for comparing the orthogonal and the non-orthogonal view of the repository in terms of their resulting metrics values. As our LRR metrics quantify the lack of readability, lower values for a metric mean higher readability.

Both the orthogonal and the non-orthogonal views have a similar trend, although the orthogonal vision presents slightly lower values in general (see numeric values in Table 3). The value of metrics LRR-NC LRR-NO and LRR-ND is near 0 for both views, indicating that the names of the classes, the object properties and the datatype properties are being reused.

In Sect. 3.2 we have mentioned that object properties and annotation properties are the most reused properties, in contrast with classes and datatype

[3] https://github.com/fjredondo/lexical-analysis-obo-foundry.

Table 3. LRR metrics values obtained for the non-orthogonal and the orthogonal views

Metric	Non-orthogonal	Orthogonal	Ratio
LRR-NC	0.00014	0.00012	1.17
LRR-ND	0.01068	0.00998	1.07
LRR-NO	0.02523	0.01765	1.43
LRR-DD	0.17249	0.17207	1.00
LRR-DO	0.40083	0.37771	1.06
LRR-NA	0.47410	0.41381	1.15
LRR-DA	0.64680	0.56638	1.14
LRR-SC	0.76892	0.76823	1.00
LRR-DC	0.77103	0.77061	1.00
LRR-SO	0.86584	0.84367	1.03
LRR-SA	0.92687	0.92077	1.01
LRR-SD	0.97506	0.97506	1.00

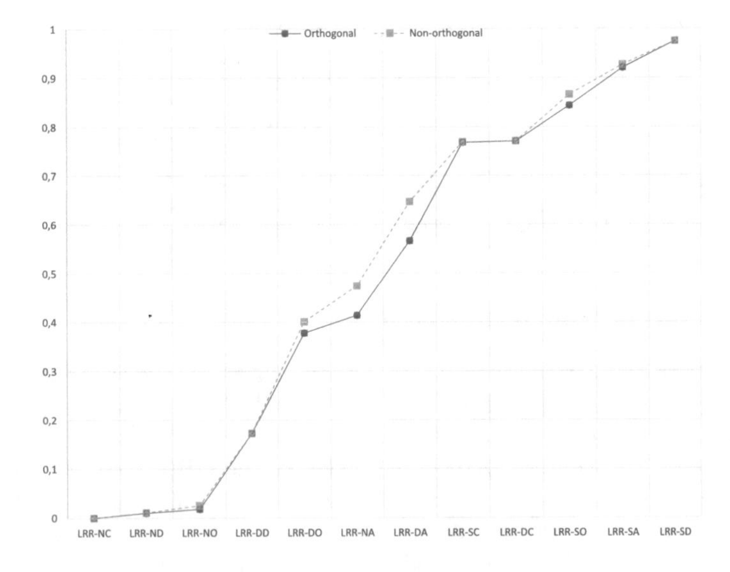

Fig. 2. Readability metrics in repository views, orthogonal (blue) and non-orthogonal (orange). The metrics are sorted by increasing value of the metric for the non-orthogonal view.

properties. The results show that the lack of reuse seems to be more related to the type of readability property than to the ontology entity, if we analyze the organization of the values of the metrics shown in Fig. 2. The lowest values are generally obtained for names, then for descriptions and, finally, the largest values are obtained for synonyms. The existence of a pattern based on the type of entity is not that evident.

It should be noted that the largest difference in absolute value between both views happens for annotation properties for names (LRR-NA) and descriptions (LRR-DA), and that the largest one using the ratios happens for the names of object properties (LRR-NO). These results are relevant since object properties and annotation properties are highly reused (6.15, and 5.83 in average, respectively). The difference between both views indicates that some object properties are being reused without including the corresponding names and, analogously, there are annotation properties being reused without names and descriptions.

It should be noted that a lot of biomedical ontologies apply the MIREOT approach [19] when reusing content. MIREOT provides the minimum information to reference to an external term, but sets the reuse of the human readable content as recommended but not compulsory. Our results show that the names are commonly reused since the best values are obtained for LRR-NC, LRR-NO and LRR-ND (< 0.025 in all the cases). The names of annotations are reused less (LRR-NA $= 0.47$), but this is also due to their lack of names since the value of LRR-NA for the orthogonal view is 0.41. The descriptions come next according to the values of the metrics, whereas the synonyms are the types of annotations with higher scores, for both non-orthogonal and orthogonal views. The values of the metrics in the non-orthogonal version are limited by the values of the orthogonal ones. 0.56638 for the descriptions of datatype properties in the orthogonal version means that 56% of the datatype properties have no descriptions. Consequently, the difference between 0.56638 and 0.64680, which is the score for the non-orthogonal version, measures to what extend the reuses of the 44% of the descriptions of the datatype properties are not reused. The ratio shown in the last column of Table 3 captures the difference with the standard for the different types of entities and types of annotations. Despite the value of the metric is small, the largest ratio is for the names of object properties, whereas the ratio is 1 for the descriptions (classes and datatype properties), synonyms (classes and datatype properties).

It is remarkable that there are metrics whose values are near 1 for both orthogonal and non-orthogonal vision of the corpus. For example, LRR-SO, LRR-SA, and LRR-SD indicate a lack of synonyms for the properties in the entire corpus. Additionally, the LRR-DC metric indicates that there is a high number of classes without any description along all ontologies of the corpus.

In summary, the results show that there is some loss of human readable information because how reuse is implemented when considering the OBO Foundry ontologies in the non-orthogonal view. A non-orthogonal view was adopted in our study of reuse of axioms [14]. If we bring both results together, there is a loss in both human and machine readable content. This can be avoided by acting in two directions: (1) ensuring that the human-readable information is reused: (2) exploit the OBO Foundry ontologies as a whole and not as individual ontologies, as it is done by the query interface of Ontobee [20].

4 Conclusions

In this article we have investigated the human readability of the OBO Foundry repository by taking advantage of a particular feature of this repository, their orthogonal conception and principle of ontology design. The results show that the metrics proposed permit to identify which are the most frequent cases of suboptimal reuse in this repository of ontologies and that there is a lack of human readability due to how reuse is implemented. In this work, we have focused on the readability for humans of the content of ontology entities. This framework can be combined with some FAIR principles [21] to include other ontology metadata that could contribute to the readability of ontologies for humans. As further work, we will also include these results in our HURON framework[16].

Acknowledgements. This research has been funded by MCIN/AEI/10.13039/501100011033 [grant number PID2020-113723RB-C22].

References

1. Legaz-García, M.C., Miñarro-Giménez, J.A., Menárguez-Tortosa, M., Fernández-Breis, J.T.: Generation of open biomedical datasets through ontology-driven transformation and integration processes. J. Biomed. Semant. **7**(1), 1–17 (2016). https://doi.org/10.1186/s13326-016-0075-z
2. Reimer, A.P., Milinovich, A.: Using UMLS for electronic health data standardization and database design. J. Am. Med. Inform. Assoc. **27**(10), 1520–1528 (2020)
3. He, Y., et al.: CIDO, a community-based ontology for coronavirus disease knowledge and data integration, sharing, and analysis. Sci. Data **7**(1), 181 (2020)
4. Gaudet-Blavignac, C., Foufi, V., Bjelogrlic, M., Lovis, C.: Use of the systematized nomenclature of medicine clinical terms (SNOMED CT) for processing free text in health care: systematic scoping review. J. Med. Internet Res. **23**(1), e24594 (2021)
5. Whetzel, P.L., et al.: BioPortal: enhanced functionality via new web services from the national center for biomedical ontology to access and use ontologies in software applications. Nucleic Acids Res. **39**(suppl_2), W541–W545 (2011)
6. Smith, B., et al.: The OBO foundry: coordinated evolution of ontologies to support biomedical data integration. Nat. Biotechnol. **25**(11), 1251–1255 (2007)
7. Lobentanzer, S., et al.: Democratizing knowledge representation with BioCypher. Nat. Biotechnol. **41**, 1–4 (2023)
8. Zheng, F., Abeysinghe, R., Cui, L.: Identification of missing concepts in biomedical terminologies using sequence-based formal concept analysis. BMC Med. Inform. Decis. Mak. **21**, 1–14 (2021)
9. Rector, A., Iannone, L.: Lexically suggest, logically define: quality assurance of the use of qualifiers and expected results of post-coordination in SNOMED CT. J. Biomed. Inform. **45**(2), 199–209 (2012)
10. van Damme, P., Quesada-Martínez, M., Cornet, R., Fernández-Breis, J.T.: From lexical regularities to axiomatic patterns for the quality assurance of biomedical terminologies and ontologies. J. Biomed. Inform. **84**, 59–74 (2018)
11. Quesada-Martínez, M., Mikroyannidi, E., Fernández-Breis, J.T., Stevens, R.: Approaching the axiomatic enrichment of the Gene Ontology from a lexical perspective. Artif. Intell. Med. **65**(1), 35–48 (2015)

12. Halper, M., et al.: Guidelines for the reuse of ontology content. Appl. Ontol. **18**(1), 5–29 (2023)
13. Kamdar, M.R., Tudorache, T., Musen, M.A.: A systematic analysis of term reuse and term overlap across biomedical ontologies. Semant. Web **8**(6), 853–871 (2017)
14. Quesada-Martínez, M., Fernández-Breis, J.T.: Studying the reuse of content in biomedical ontologies: an axiom-based approach. In: ten Teije, A., Popow, C., Holmes, J.H., Sacchi, L. (eds.) AIME 2017. LNCS (LNAI), vol. 10259, pp. 3–13. Springer, Cham (2017). https://doi.org/10.1007/978-3-319-59758-4_1
15. Chen, J., et al.: OWL2Vec*: embedding of OWL ontologies. Mach. Learn. **110**(7), 1813–1845 (2021). https://doi.org/10.1007/s10994-021-05997-6
16. Abad-Navarro, F., Martínez-Costa, C., Fernández-Breis, J.T.: HURON: a quantitative framework for assessing human readability in ontologies. IEEE Access **11**, 101833–101851 (2023)
17. Abad-Navarro, F., Quesada-Martínez, M., Duque-Ramos, A., Fernández-Breis, J.T.: Analysis of readability and structural accuracy in SNOMED CT. BMC Med. Inform. Decis. Mak. **20**(10), 1–21 (2020)
18. Horridge, M., Bechhofer, S.: The OWL API: a java API for OWL ontologies. Semant. Web **2**(1), 11–21 (2011)
19. Courtot, M., et al.: MIREOT: the minimum information to reference an external ontology term. Appl. Ontol. **6**(1), 23–33 (2011)
20. Ong, E., et al.: Ontobee: a linked ontology data server to support ontology term dereferencing, linkage, query and integration. Nucleic Acids Res. **45**(D1), D347–D352 (2017)
21. Wilkinson, M.D., et al.: The FAIR guiding principles for scientific data management and stewardship. Sci. Data **3**(1), 1–9 (2016)

Do Habit and Privacy Matter in a Post Pandemic-Era? Mobile Apps Acceptance of the Private Healthcare Sector in Portugal

Diana Gouveia[1] and Filipa Jorge[2](✉)

[1] University of Trás-os-Montes and Alto Douro, Quinta dos Prados, 5000-801 Vila Real, Portugal
[2] Research Unit on Governance, Competitiveness and Public Policies (GOVCOPP), University de Aveiro, DEGEIT, University of Aveiro, Campus Universitario de Santiago, 3810-193 Aveiro, Portugal
filipa.eira.jorge@ua.pt

Abstract. During the pandemic, services in general have been carried out through digital channels whenever possible. Healthcare services were no exception, all the services that could be digitalized have taken place through digital means. One of the most common forms of digitalization used by healthcare organizations is the use of mobile apps, which allow users to carry out healthcare tasks remotely, access medical information, or even contact tracing with COVID-19 infected people. After the pandemic period, in which people have developed a stronger habit of using applications for health purposes, it is important to understand whether this is a driver of the acceptance of this technology. Currently, this is also a context in which privacy concerns are barriers to the utilization of technologies that involve the sharing of health-related data, so it becomes relevant to understand whether this could be an inhibitor to the acceptance of this technology. This study aims to identify the antecedents of users' acceptance of mobile applications to access private healthcare services in Portugal in a post-pandemic context. To accomplish this purpose, this study collects data through a survey and analyses these data using Structural Equation Modelling. The sample is composed of 401 individuals who live in Portugal and who use mobile apps to access private healthcare services. The results evidence perceived usefulness and habit as antecedents of intention to use mobile applications to use healthcare services and this intention is an antecedent of the real use of these applications.

Keywords: Mobile Apps · TAM · Healthcare Sector · Digital Transformation

1 Introduction

The importance of mobile applications during the pandemic period, particularly in terms of tracking contacts with infected people, may have changed the way people interact with a type of mobile applications and the fact that they use it more or less [1]. During this period, several individuals developed the habit of using mobile applications for health purposes, so it is important to analyze if this habit remains relevant to explain the intention to use mobile applications for healthcare tasks.

Á. Rocha et al. (Eds.): WorldCIST 2024, LNNS 987, pp. 170–180, 2024.
https://doi.org/10.1007/978-3-031-60221-4_17

Additionally, privacy remains one of the biggest concerns of people when deciding to use a technology [2]. People are afraid to share their private data, particularly health-related data, with entities that may use these data for other purposes. Therefore, privacy concerns may be one of the most relevant antecedents of technology in the healthcare sector.

The study of the acceptance of this technology in the post-pandemic period is still scarce. More research is necessary to analyze whether the classic theoretical frameworks of the acceptance study remain valid and which determinants are most relevant in explaining usage behavior.

This study aims to analyze, in the post-pandemic period, the determinants of the use of mobile applications to enjoy private health services in Portugal. Companies must be able to update and adapt to the context of a post-pandemic world. Investigating how users accept applications aimed at the health sector is extremely important, to identify which attributes and characteristics should be incorporated into these applications. By following this process, when the applications are made available to users, there will be a greater chance of success.

The structure of this work is organized into different parts, including an introduction, a theoretical framework that addresses the main model of user acceptance of technology, a justification of the hypotheses and conceptual model, a research methodology that describes the methods used for data collection and analysis, conclusions, and bibliographical references.

2 Conceptual Model

In his 1989 article, [3] proposed the Technology Acceptance Model (TAM), which is the original, and at the same time the most popular, theoretical model for explaining consumer acceptance [4]. In his proposal, [3] suggested that user motivation can be explained by three main factors: perceived usefulness, perceived ease of use, and attitude toward using the system. This author states that the user's attitude towards technology is one of the main determinants of whether the user uses or rejects a technology. Finally, [3] hypothesizes that both perceptions, perceived usefulness, and perceived ease of use, are directly influenced by the characteristics of the technology. TAM also assumes that intention to use a technology always precedes its actual use.

Perceived usefulness is the degree to which a person believes that using a given technology would increase their performance [5]. According to [3], perceived ease of use influences perceived usefulness. If users find a particular technology easy to use, they will tend to perceive this technology as useful. Additionally, perceived ease of use is related to an individual's belief about how easy or effortless the task of learning to use a technology is [3]. It is assumed that the greater the perceived usefulness, the greater the intention to adopt a technology [6]. On the other hand, the greater the perceived ease of use, the greater the intention to use a particular technology [3]. Given the above, this research aims to propose the influence of ease of use on perceived usefulness as well as on the intention to use mobile applications related to the private health sector, according to the following hypotheses:

- H1a: Perceived Ease of Use influences the intention to use mobile applications for access to healthcare services.
- H1b: Perceived Ease of Use influences the perceived usefulness.

The perceived usefulness construct is related to the extent to which an individual believes that a given technology can improve their productivity or performance in some task of their work, i.e. the extent to which they will benefit in some way when performing a task [3]. Several previous studies have shown that the greater the perceived usefulness of a technology, the greater the intention to use it, for instance, the study of [7]. Given the above, this research aims to propose the influence of perceived usefulness on the intention to use technology, according to the following hypotheses:

- H2: Perceived Usefulness influences the intention to use mobile applications for access to healthcare services.

Several pioneering studies on privacy have attempted to conceptualize and operationalize privacy concerns in a more detailed way [8]. Among the various privacy-related constructs examined in the literature, the privacy concerns construct is one of the most widely used variables in information systems research and has consistently proven to be one of the strongest predictors of privacy-related behavior [8, 9]. Furthermore, in the context of the health sector, privacy concerns are an important determinant of healthcare patients' acceptance to use mobile applications [10]. With this information, it is suggested to analyze the influence of privacy concerns on the intention to use technology, so the following hypothesis was formulated:

- H3: Privacy concerns negatively influence the intention to use mobile applications for access to healthcare services.

According to [11], habit is defined as the extent to which people tend to perform behaviors automatically due to learning. This concept has been operationalized in two different ways, on the one hand, it is seen as a previous behavior [12] and, on the other hand, habit is measured as the degree to which an individual believes that the behavior becomes automatic [11]. As already evidenced in the study by [11], habit influences the intention to use a technology. Based on this previous empirical evidence, it is suggested to analyze the influence of habit on the intention to use technology, and formulate the following hypothesis:

- H4: Habit positively influences the intention to use mobile applications for access to healthcare services.

Usage is a construct that measures the frequency with which applications are used in their various functionalities [13]. Previous studies, such as [13] and [3], found that the actual use of a technology is influenced by the behavioural intention to use it. Given above, this research aims to propose the influence of the intention to use technology on its actual use, according to the following hypothesis:

- H5: The intention to use mobile applications for access to healthcare services positively influences the actual use of it.

Considering all the hypotheses presented above, the figure below shows the conceptual model proposed to explain the use of mobile applications for access to healthcare services, which is made up of the hypotheses previously presented (Fig. 1).

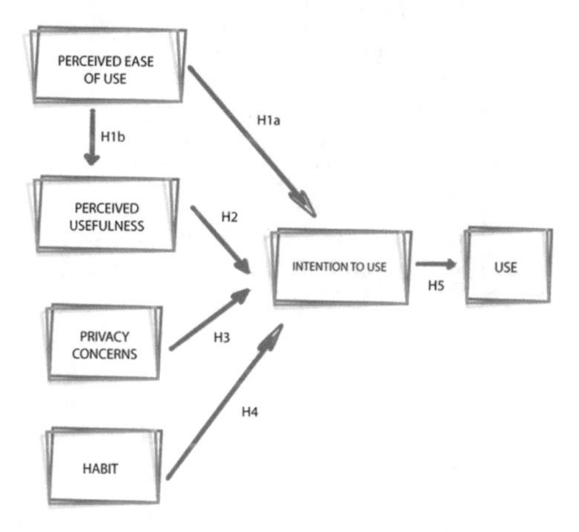

Fig. 1. Conceptual model

3 Methodology

This study uses a quantitative approach to examine the determinants of mobile apps use behaviour. Data was collected through a questionnaire spread online.

The questionnaire has three main sections. The first part has only one question whether the individual uses mobile applications. If they answer no, the questionnaire ends immediately, if they answer yes, the questionnaire continues. The second part of the questionnaire collects data from the sociodemographic characteristics of the individuals. The third part collects individuals' perceptions about mobile apps for health purposes. To develop the third part of the questionnaire, previous literature was used to select the measurement instruments to assess each construct presented in the conceptual model. The following table presents the authors of the scales used to measure each construct (Table 1).

After the data collection process, data analysis was performed using Structural Equation Modelling through the Partial Least Squares (PLS) technique to evaluate the empirical significance of the conceptual model hypotheses using SMARTPLS 4.0.9.5 [15]. The PLS was used because the data collected did not follow a normal distribution, according to the Kolmogorov-Smirnov test.

This research sample is composed of 401 individuals who use mobile applications in use private health care sector in Portugal. The mean age of the sample is 40.82 years old. The sample is formed almost equally by men (52%) and women (48%). The majority of respondents are single (52%), have completed secondary education (36%) or a degree (34%), are working, and have a monthly income of €1,000 or less (72%). Regarding their health profile, the majority of the sample do not have a diagnosed chronic disease (70%) but do have private health insurance (73%).

Table 1. Measurement Instruments

Construct	Scale adapted from
Perceived Ease of Use	Davis (1989) [3]
Perceived Usefulness	Davis (1989) [3]
Privacy Concerns	Xu et al. (2008) [14]
Habit	Venkatesh et al. (2012) [13]
Intention to use	Venkatesh et al. (2012) [13]
Use	Adapted from the functionalities that private health service applications in Portugal offer their users

4 Results

4.1 Measurement Model Evaluation

To assess the empirical significance of the hypotheses presented in the conceptual model, data will be analyzed using Structural Equation Modelling and, more particularly, through the PLS technique. This technique was selected since data does not follow a normal distribution, as Kolmogorov-Smirnov's test evidence ($p < 0.01$). To conduct this data analysis, SMARTPLS 4.0.9.6 was used [15].

The first step is measurement model evaluation, in particular, the assessment of construct reliability, item reliability, convergent validity, discriminant validity, and the absence of multicollinearity issues.

Item reliability criterion is validated if the loading value is higher than 0.7, according to [16]. To accomplish this criterion, one item was eliminated, in particular, HAB_2. All the items that remain in the model accomplish this criterion. Construct reliability is assessed with Composite Reliability (CR), whose value should be higher than 0.7, following the indications of [17]. As the table below evidence, all constructs have reliability. Internal consistency criterion is evaluated through Cronbach's Alpha indicator which should be higher than 0.7, according to [18]. As Table 2 expresses, all constructs accomplish this criterion. Convergent validity was assessed using Average Variance Extracted (AVE), which must be higher than 0.5 to accomplish this criterion, as suggested by [19]. As exposed in Table 2, all constructs have convergent validity. Discriminant validity was assessed using two criteria. First, this study uses the Fornell Lacker criterion and to verify this criterion, the square root of AVE (diagonal values in bold in Table 3) should be higher than the correlation between the remaining constructs [19]. Second, this study also uses the HTMT criterion, which is verified if these ratio values are smaller than one, according to [20]. Both criteria are verified, which indicates that our constructs have discriminant validity. Finally, the absence of multicollinearity issues is verified if VIF values are lower than 5 [18]. To accomplish this criterion one item was eliminated. All the remaining items accomplish this criterion (Table 4).

Table 2. Measurement model indicators

Construct	Item	Cronbach's Alpha	CR	AVE	Loading	VIF
Perceived Ease of Use (PEOU)	PEOU_1	0.915	0.934	0.701	0,810	2,934
	PEOU_2				0,839	3,443
	PEOU_3				0,840	2,539
	PEOU_4				0,867	2,833
	PEOU_5				0,833	2,623
	PEOU_6				0,831	2,379
Perceived Usefulness (PU)	PU_1	0.929	0.946	0.779	0,859	2,492
	PU_2				0,865	2,884
	PU_3				0,918	4,329
	PU_4				0,891	3,463
	PU_5				0,879	3,019
Habit (HAB)	HAB_1	0.863	0.917	0.786	0,870	2,277
	HAB_3				0,861	2,047
	HAB_4				0,928	3,087
Privacy Concerns (PC)	PC_1	0.923	0.942	0.764	0,808	2,449
	PC_2				0,891	3,440
	PC_3				0,868	3,259
	PC_4				0,909	4,631
	PC_5				0,892	3,572
Intention to Use (INT)	INT_1	0.931	0.945	0.743	0,849	2,972
	INT_2				0,910	4,945
	INT_3				0,864	3,310
	INT_4				0,802	2,303
	INT_5				0,859	3,396
	INT_6				0,885	3,734
Use (USE)	USE_1	0.969	0.972	0.656	0,777	3,442
	USE_2				0,813	4,148
	USE_3				0,796	2,954

(*continued*)

Table 2. (*continued*)

Construct	Item	Cronbach's Alpha	CR	AVE	Loading	VIF
	USE_4				0,780	2,981
	USE_5				0,839	4,492
	USE_6				0,845	3,993
	USE_8				0,832	3,673
	USE_9				0,795	3,065
	USE_10				0,853	3,839
	USE_11				0,825	3,527
	USE_12				0,792	3,276
	USE_13				0,818	3,683
	USE_14				0,804	3,272
	USE_15				0,856	4,876
	USE_16				0,839	4,203
	USE_17				0,783	3,055
	USE_18				0,780	2,810
	USE_19				0,737	2,794

Table 3. Fornell-Lacker criteria

Construct	PEOU	PU	HAB	PC	INT	USE
PEOU	**0,837**					
PU	0,800	**0,883**				
HAB	0,548	0,614	**0,887**			
PC	0,266	0,205	0,342	**0,874**		
INT	0,623	0,671	0,797	0,329	**0,862**	
USE	0,160	0,164	0,479	0,256	0,416	**0,810**

4.2 Structural Model

Having validated the measurement model, this study estimated the structural model using 5000 bootstraps. The table below presents the results of the structural model (Table 5).

The model presented explains 17.3% of the variance in the behaviour of using mobile applications for health purposes. Furthermore, this model also explains 69.8% of the behavioural intention to use these applications. According to the results, two hypotheses presented in the conceptual model are not empirically supported, namely H1a and H3. On the one hand, perceived ease of use does not directly and significantly impact the intention

Table 4. HTMT ratio

Construct	PEOU	PU	HAB	PC	INT	USE
PEOU						
PU	0,858					
HAB	0,611	0,686				
PC	0,287	0,217	0,382			
INT	0,670	0,720	0,884	0,355		
USE	0,174	0,168	0,512	0,282	0,420	

Table 5. Structural Model Results (direct effects)

H	Independent Variable	Dependent Variable	Coefficient	t-statistics	f-square	p-value
H1a	Perceived Ease of Use	Intention to use	0.134	1.775	0.021	0.076
H1b	Perceived Ease of Use	Perceived Usefulness	0.800	31.489	1.773	0.000
H2	Perceived Usefulness	Intention to Use	0.191	2.419	0.038	0.016
H3	Privacy Concerns	Intention to Use	0.053	1.246	0.008	0.213
H4	Habit	Intention to Use	0.587	10.790	0.651	0.000
H5	Intention to Use	Use	0.416	7.708	0.209	0.000

to use applications for health purposes. Similar results have already been reported by [21]. On the other hand, concerns about the privacy of the data provided in these applications also have no impact on the intention to use them. Despite that previous literature has reported over the years the influence of privacy concerns on the intention to use, the studies [22] and [23] also evidence that this influence is not significant. Bearing in mind that these studies are recent, it seems to us that this is evidence of a change in the significance of privacy concerns on the acceptance of this technology for health purposes. Furthermore, the results also evidence that the perceived ease of use of mobile apps for health purposes has a positive and significant influence on the perceived usefulness of these apps, being supported by H1b. This finding is consistent with the theoretical proposal of TAM [3] and with other studies of mobile apps acceptance. Additionally, the intention to use these mobile apps is also positively influenced by the habit of using them, providing empirical support to H4. This evidence is consistent with the theoretical proposal of UTAUT2 [13] and with other acceptance studies of mobile apps used in the health context [24]. The intention to use the mobile apps is also influenced by the perceived usefulness of these apps, confirming H3. This result also subscribes to the theoretical proposal of TAM as described by [3] and has been also reported by [25].

Finally, the results also evidence that the real use of mobile apps for health purposes is influenced by the intention to use this technology, being supported by H5. Similar results have also been reported by [26].

5 Conclusion

This research analyses the antecedents of mobile applications' acceptance of use private health care services in the post-pandemic period in Portugal. According to the results, perceived usefulness and habit are relevant antecedents of intention to use these mobile apps, and this intention is an antecedent of the real use of these mobile apps. The findings also evidence that privacy concerns and perceived ease of use have no relevant impact on the intention to use mobile apps for health purposes.

The present study intends to contribute to the scientific literature by identifying the antecedents of mobile apps acceptance after the pandemic. Since the pandemic has increased the use of digital technologies for simple tasks of the quotidian, such as access to private healthcare services, it is relevant to analyse if habit becomes a relevant antecedent of the intention to use the mobile apps to access to it. Besides, privacy is also a concern that has been discussed in the are literature as a relevant barrier to the intention to use several technologies. Therefore, in the context of the present study, privacy concerns are not a barrier to the use of this technology. Additionally, this research also contributes by evidencing that TAM, as a theoretical framework, remains resilient and effective, since the variables continue to be confirmed in the post-pandemic context despite the challenges imposed by the pandemic.

This study may be relevant to healthcare professionals who are related to mobile apps creation and development, since, from a pragmatic perspective, the results generated by this research allow these companies to realise the ability to define the areas where they should improve their applications to substantially increase the likelihood of them being used. According to the results, mobile applications must accomplish tasks that users value to they perceive these applications as useful and, consequently, to increase their intention to use these mobile applications. Additionally, the habit of using these mobile applications in the quotidian is very important to develop the intention to use it. So, it is suggested that healthcare professionals introduce some rewards in mobile apps for users to increase the likelihood of developing the habit of using these applications. Finally, users must have the intention to use the mobile applications to increase the likelihood to occur the real use of these applications.

This study has some limitations. First, this study uses a quantitative approach that allows us to identify the antecedents of mobile applications but does not allow us to understand why privacy concerns and perceived ease of use are not relevant antecedents. Therefore, future studies may use qualitative approaches to understand how the use of mobile apps acceptance can be facilitated and how the main barriers to the use of these apps. Second, this study only collects data in one moment in time, which does not provide an understanding of the evolution of the topic. Future studies should use longitudinal data collection to have a deep understanding of the evolution of this topic.

Acknowledgements. This work was financially supported by the research unit on Governance, Competitiveness and Public Policy (UIDB/04058/2020)+(UIDP/04058/2020).

References

1. Dzandu, M.D.: Antecedent, behaviour, and consequence (a-b-c) of deploying the contact tracing app in response to COVID-19: evidence from Europe. Technol. Forecast. Soc. Chang. **187**, 122217 (2023)
2. Malhotra, D.: Trust and reciprocity decisions: the differing perspectives of trustors and trusted parties. Organ. Behav. Hum. Decis. Process. **94**(2), 61–73 (2004)
3. Davis, F.: Perceived usefulness, perceived ease of use, and user acceptance of information technology. MIS Q. **13**, 319–340 (1989)
4. Parreira, P., Proença, S., Sousa, L.B., Mónico, L.: Technology Assessment Model (TAM): Modelos percursores e modelos evolutivos. Empreendedoras no Ensino Superior Politécnico: Motivos, influências, serviços de apoio e educação (2018)
5. Davis, F.: A technology acceptance model for empirically testing new end-user information systems: theory and results (Ph.D. dissertation, MIT) (1985)
6. Kurtz, R., Macedo-Soares, T., Ferreira, J.B., Freitas, A.S.D., Silva, J.F.D.: Impact factors on attitude and intention to use m-learning: an empirical test. REAd. Revista Eletrônica de Administração (Porto Alegre) **21**, 27–56 (2015)
7. Xia, M., Zhang, Y., Zhang, C.: A TAM-based approach to explore the effect of online experience on destination image: a smartphone user's perspective. J. Destin. Mark. Manag. **8**, 259–270 (2018)
8. Weber, J., Malhotra, D., Murnighan, J.K.: Normal acts of irrational trust: motivated attributions and the trust development process. Res. Organ. Behav. **26**, 75–101 (2004)
9. Stewart, K., Segars, A.: An empirical examination of the concern for information privacy instrument. Inf. Syst. Res. **13**(1), 36–49 (2002)
10. Tomczyk, S., Barth, S., Schmidt, S., Muehlan, H.: Utilizing health behavior change and technology acceptance models to predict the adoption of COVID-19 contact tracing apps: cross-sectional survey study. J. Med. Internet Res. **23**(5), e25447 (2021)
11. Limayem, M., Hirt, S.G., Cheung, C.M.K.: How habit limits the predictive power of intention: the case of information systems continuance. MIS Q. **31**(4), 705 (2007)
12. Kim, S., Malhotra, N.: A longitudinal model of continued IS use: an integrative view of four mechanisms underlying postadoption phenomena. Manage. Sci. **51**(5), 741–755 (2005)
13. Venkatesh, V., Thong, J., Xu, X.: Consumer acceptance and use of information technology: extending the unified theory of acceptance and use of technology. MIS Q. **36**, 157–178 (2012)
14. Xu, D., Liao, S., Li, Q.: Combining empirical experimentation and modeling techniques: a design research approach for personalized mobile advertising applications. Decis. Support Syst. **44**(3), 710–724 (2008)
15. Ringle, C.M., Wende, S., Becker, J.-M.: SmartPLS 4 (2022). www.smartpls.com
16. Churchill Jr., G.A.: A paradigm for developing better measures of marketing constructs. J. Mark. Res. **16**(1), 64–73 (1979)
17. Straub, D.W.: Validating instruments in MIS research. MIS Q. **13**(2), 147 (1989)
18. Hair, J., Hult, G., Ringle, C., Sarstedt, M., Danks, N., Ray, S.: Partial Least Squares Structural Equation Modeling (PLS-SEM). Springer, Cham (2017). https://doi.org/10.1007/978-3-030-80519-7
19. Fornell, C., Lacker, D.: Evaluating structural equation models with unobservable variables and measurement error. J. Mark. Res. **18**, 39–50 (1981)
20. Henseler, J., Ringle, C., Sarstedt, M.: A new criterion for assessing discriminant validity in variance-based structural equation modeling. J. Acad. Mark. Sci. **43**(1), 115–135 (2015)
21. Dhagarra, D., Goswami, M., Kumar, G.: Impact of trust and privacy concerns on technology acceptance in healthcare: an Indian perspective. Int. J. Med. Informatics **141**, 104164 (2020)

22. Fox, G., Clohessy, T., van der Werff, L., Rosati, P., Lynn, T.: Exploring the competing influences of privacy concerns and positive beliefs on citizen acceptance of contact tracing mobile applications. Comput. Hum. Behav. **121**, 106806 (2021)
23. Kamal, S., Shafiq, M., Kakria, P.: Investigating acceptance of telemedicine services through an extended technology acceptance model. Technol. Soc. **60**, 101212 (2020)
24. Cho, H., Chi, C., Chiu, W.: Understanding sustained usage of health and fitness apps: incorporating the technology acceptance model with the investment model. Technol. Soc. **63**, 101429 (2020)
25. Cramer, R.J., et al.: Alternative sexuality, sexual orientation and mobile technology: findings from the National Coalition for Sexual Freedom technology and health enhancement feasibility study. Psychol. Sex. **13**(2), 344–359 (2022)
26. Bhattacharjya, S., Cavuoto, L.A., Reilly, B., Xu, W., Subryan, H., Langan, J.: Usability, usefulness, and acceptance of a novel, portable rehabilitation system (mRehab) using smartphone and 3D printing technology: mixed methods study. JMIR Hum. Factors **8**(1), e21312 (2021)

The Role of Online Product Information in Enabling Electronic Retail/E-tailing

Abdallah Houcheimi and József Mezei(✉)

Faculty of Social Sciences, Business and Economics, and Law, Åbo Akademi University, Turku, Finland
Jozsef.Mezei@abo.fi

Abstract. This paper examines the impact of online product information on consumer purchasing decisions in the context of e-commerce. With the rapid growth of electronic retail (e-tailing), understanding how online information influences consumer behavior has become crucial. The study investigates the key aspects of online product information, including availability, searchability, and presentation, and their effects on customers' intentions to purchase more online. The research employs a configurational approach, analyzing survey data from users of e-tailing services. A key focus is on how product descriptions and information availability contribute to online shopping decisions. Additionally, the study considers the effects of information search processes as motivators for consumer behavior. The findings indicate that product information significantly enhances online shopping experiences and positively influences purchasing decisions. This highlights the critical role of product information in e-commerce and its ability to boost sales. It also suggests that different combinations of product information conditions can lead to higher levels of online shopping intentions. The paper contributes to academic research in e-tailing by providing insights into the importance of product information in electronic retail and identifying key conditions that lead to higher online purchases.

Keywords: Electronic retailing · FsQCA · Product Information · Information Search

1 Introduction

The rapid growth of e-commerce is a universal trend based on the unique characteristics of this business environment. E-commerce represents the fastest-growing sector within the technology industry resulting in a surge in the number of online customers as they increasingly shift from physical to electronic retail [19]. Electronic retail is a business-to-consumer e-commerce type, as it involves online transactions between buyers and sellers. Online customers are highly motivated to engage in online purchasing due to the many features of e-commerce, such as timesaving, information availability, round-the-clock accessibility, and variety of products [1].

The electronic commerce offers businesses with new channels of revenue, for example, Amazon's online stores offer millions of items for purchase, enabling customers to

Á. Rocha et al. (Eds.): WorldCIST 2024, LNNS 987, pp. 181–190, 2024.
https://doi.org/10.1007/978-3-031-60221-4_18

search for, evaluate, and buy desired products entirely online. The availability of product information and intense online products and services are also fundamental aspects of e-commerce markets [10]. Furthermore, online product information and online shopping experience significantly impact the customer's purchasing decision favorably [13]. For example, Ishtari online store provides its customers with crucial product information, such as price, color, availability, and delivery details which allow them to search for and evaluate specific products. However, customers have varying preferences in product features, leading to different needs for product information [5]. Thus, online businesses are highly interested in understanding the different ways in which online product information influences online customers [11].

Therefore, the primary objective of this research paper is to assess the relationship and impact of online product information on online customers' intentions to make more online purchases. To achieve this goal, the following research questions will be addressed:

- RQ1: Why online product information variables are important for online shopping?
- RQ2: How online product information influence online shopping?

The rest of the paper is structured as follows: literature review, methodology, data and analysis, and conclusion.

2 Literature Review

The growth of online retail is outpacing that of traditional retail, with younger customers increasingly completing their purchases online. In response, internet retailers are investing in web technologies to enhance the convenience of online shopping by providing detailed product-oriented information. The provision of online product information is a critical factor in e-commerce website readiness to satisfy the needs of online customers by providing them with timely product and service information. According to information processing theory, people collect and use information to make informed decisions [16, 17]. Several studies found that customers' decisions are influenced by information [9]. According to several studies online shopping is influenced by many aspects of product information such as online product information [2], information availability [15, 22] information search [8, 12]. In modern times, firms use internet and web technologies both before and after sales to provide customers with product and service information. Walmart for example, provides its online customers with visual and textual details of all its products offered online. Online product information also enables customers to determine in real-time whether the product they are seeking is available or not [7].

2.1 Product Information and Information Availability

Product information can be defined as the necessary and supplementary details and descriptions about several product's aspects and characteristics such as name, color, price, and delivery details. Another definition, product information includes product descriptions and technical specifications in addition to other information that can be provided by the seller such as celebrity endorsements and marketing information [11].

Online retailers are offering a vast range of products, each accompanied by detailed descriptions and necessary information. Online product information can be conveyed through visual and textual means, and this information has a significant impact on customers' knowledge about the product [2]. According to [15], visual and textual online product presentations enable customers to obtain product information which positively influences their shopping performance. In Song's study [22], it was found that customers' behavior and purchasing decisions vary based on the characteristics of the products offered. The availability of online product information has significant impacts on consumers' purchasing decisions [24]. These impacts can be either positive or negative, depending on several factors related to the characteristics of the information, such as the size of the choice set, the popularity information, and its presentation [24].

The importance of product information in general and the availability of specific information that are necessary for online customers while making their purchasing decisions have encouraged online customers to use variety of information sources such as online reviews. Online reviews became a vital source of information for online customers as their products choices are significantly influenced by the reviews of other customers [21]. In the absence of online reviews, product attributes still have substantial impact on customers' choices [23]. However, online customers' reviews play an essential role in customers' decision-making processes [4].

2.2 Information Search

Information search is the process of finding enough information about a specific product [3]. While online advertisements can capture customers' attention and interest in a particular product, further information is often required before a purchase decision is made. Using the results of information search, customers can decide whether to buy the product online or from a physical store [7]. Online product information aids customers in evaluating products to make informed purchase decisions [14]. Thus, online search for product information enables customers to explore product attributes [8], and information search is a crucial factor of the customer's buying decision process [3]. Moreover, information search and ease of use (convenience) are important motivators for online shopping, but information search is a higher motivator than convenience [12].

2.3 Conceptual Framework

Based on the discussed literature the following conceptual framework as presented in Fig. 1 suggests the relationship between the online product information components (independent variables) including: product information, information availability, and information search and their impacts on the online purchases (dependent/outcome variable). The constructs are defined based on the literature review as follows. In this research, product information: refers to the e-commerce website readiness with the necessary online product information that attracts customers to visit that website and browse for products in it. Information availability refers to the availability of the product's information that are considered necessary by the customer to evaluate the product and make his/her decision. Finally, information search refers to the online customers actions while looking for the product information and exploring its features. The following are the main research propositions which will be investigated in the data findings and analysis:

- Proposition 1: different combinations of online product information components will result in higher online purchases.
- Proposition 2: online product information components by themselves are necessary to result in higher online purchases.
- Proposition 3: online product information components by themselves are sufficient to result in higher online purchases.

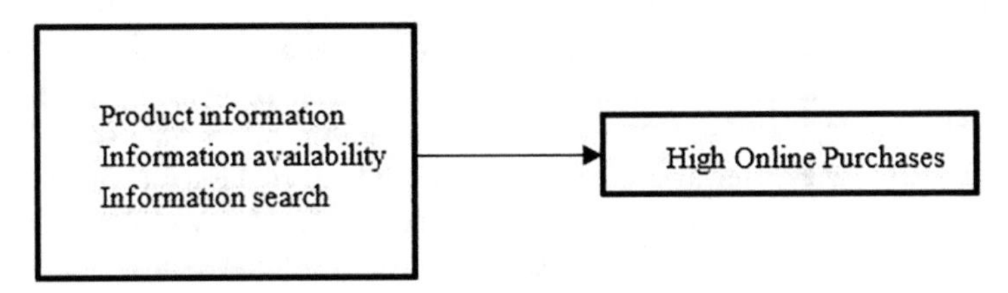

Fig. 1. Product Information and Online Purchases Conceptual Framework

3 Methodology

Configurational theories, such as the use of FsQCA (fuzzy-set qualitative comparative analysis), are increasingly being applied in the field of customer behavior research. Unlike traditional statistical analysis-based methods, configurational methods can capture complex relationships between variables and provide more nuanced causal explanations [20]. Using FsQCA involves several steps, including data calibration and truth table analysis, which will be described in the following.

FsQCA, and other configurational approaches in general have some distinct advantages over traditional correlation-based statistical models that make them appropriate in analyzing highly complex problems with a set of interconnected variables. First, FsQCA can help in capturing non-linear relationships, which is particularly relevant in the context of online retail platforms where there may be complex interactions between different factors that influence customer behavior, especially when product information-related variables are considered. Second, FsQCA is developed with the focus of handling complex causal configurations of condition variables. Third, the results of the analysis, i.e., a set of causal configurations of antecedent variables resulting in an outcome of interest, provide a more nuanced understanding of causality by identifying the combinations of conditions that are necessary or sufficient for a given outcome. Finally, FsQCA analysis can provide practical implications for businesses and marketers by identifying specific combinations of factors that are necessary or sufficient for a given outcome. This can be particularly relevant in the context of online retail platforms where there may be multiple factors that influence customer behavior in different ways, and businesses may want to target specific customer segments or behaviors. For these reasons, we have identified

FsQCA as the choice of methodology to understand the relationship between factors related to online product information and purchase behavior on online retail platforms. In performing FsQCA analysis, we have followed the guidelines of [18]. All the computations and analysis were performed using R statistical programming language, with FsQCA analysis conducted using the tools in the QCA library [6].

4 Data and Analysis

The research survey was comprised of several questions that were answered on a scale of 1 to 5, where 1 indicates strong disagreement and 5 indicates strong agreement. The survey link was distributed electronically to participants in Lebanon who had prior online shopping experience. A total of 448 participants were invited to take part in the survey, and 306 responses were received, which was deemed an adequate sample size for representing the target population. As we will discuss later in more details, 75% of all the possible configurations of our variables are present in the data from these 306 respondents. The survey participants included individuals of different genders, age groups, and career levels, such as students, employees, and business owners. Of the total participants, 57.2% (175) were female and 42.8% (131) were male. Furthermore, 78.5% of the participants fell into the middle age group of 20 to 40 years old. Importantly, there were no invalid responses or missing values in the collected survey records.

The constructs included in the research include Product Information, Information Availability, Information Search, and Online Purchases. The outcome of interest, Online Purchases, was assessed with the statement 'The online information about the products and services have significantly increased my online purchases.' In fsQCA analysis, an essential step is to convert the original data into fuzzy-sets, which is called data calibration. In the analysis, we have utilized direct calibration method, and used statistical characteristic measures to transform the original values into fuzzy membership values within the range of 0 to 1. To achieve this, three threshold values must be specified: non-membership (transformed value of 0), cross-over point (transformed value of 0.5), and full membership (transformed value of 1). In this study, the 5%, 50%, and 95% quantiles of the variables were used as threshold values, and intermediate points were transformed using a logistic function. This family of functions allows for a smooth and S-shaped membership function with control over inflection point and shape. Furthermore, it

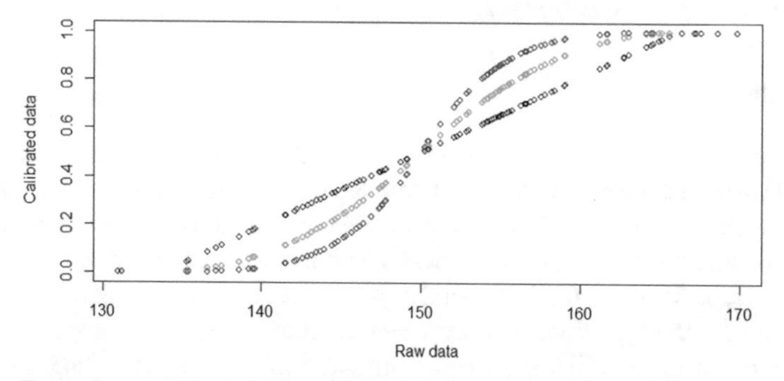

Fig. 2. Examples of calibrated data with logistic function and three different parameters

effectively handles extreme values in the data. Figure 2 illustrates different logistic-type membership functions (including a linear) [6].

For instance, when considering the variable Information Search, if a respondent indicates high importance (i.e., above the mean median value of the data), a transformed value between 0.5 and 1 will be assigned; the higher the original value is, the closer the transformed value is to 1. The calibration thresholds for the variables (with additional descriptive statistical measures) can be found in Table 1. The binary variable gender was coded by assigning 0 to female respondents and 1 to male respondents.

Table 1. Calibration Threshold Values and Descriptive Statistics

	Mean	Standard Deviation	5%	50%	95%
Product Information	4.07	0.81	2.50	4.00	5.00
Information Search	4.19	0.75	2.75	4.33	5.00
Information Availability	4.17	0.81	2.63	4.50	5.00
Online Purchases	3.81	1.03	2.00	4.00	5.00

After the calibration, the next step is necessity analysis, that aims to find information related variables that are always associated with high level of online purchases. The consistency and coverage values for the presence and absence of these attributes are shown in Table 2. The necessity relation was tested for the presence/high level of Online Purchases with both the presence and absence of the information related variables. A consistency value of over 0.9 indicates a necessary condition, while coverage reflects the importance of the relationship. As we can observe from the results, there is no single antecedent variable that is necessary for a customer to have increased online purchases as a result of various factors related to the information in online retail platforms. The highest value can be observed for Product Information, implying that it is very important to have sufficient information to consider it as a relevant factor in impacting the final purchase decision, however, this is not necessarily the case for all the customers. Similarly, the information availability and quality of information search have an important role in the overall impact of information on online purchase decision, but we cannot claim that it holds for (almost) all the customers.

4.1 Results

In the following we present the main results of the FsQCA analysis. Based on the calibrated variables, the truth table is constructed. In our analysis, we have four condition variables, which implies that 2^4 = 16 possible combinations need to be tested. In our data of 306 samples, we found 12 of these 16 combinations present. As the sample of customers is representative for the general population of online retail platform customers, the observed 12 configurations also represent the most important configurations. After the truth table is created, the sufficient configurations that result in high level of the outcome variable are identified. We have set the threshold of consistency as 0.8, the

Table 2. The Results of the Necessity Analysis

Condition	Consistency	Coverage
Gender	0.42	0.73
Not Gender	0.58	0.61
Product Information	0.82	0.72
Not Product Information	0.47	0.57
Information Search	0.78	0.73
Not Information Search	0.51	0.56
Information Availability	0.74	0.75
Not Information Availability	0.54	0.56

generally recommended value [20], and the frequency cut-off as 6, since we have a moderately large sample. The results of the analysis are presented in Table 3. In the solutions, there is no case of a low value of a condition contributing to a sufficient configuration, consequently, we simply marked with (•) when a variable is present in the solution configuration. The overall consistency of the solution set is 0.82, and the coverage is 0.88.

Table 3. The Results of the Sufficiency Analysis

	Solution 1	Solution 2	Solution 3
Gender			•
Product Information		•	•
Information Search		•	
Information availability	•		
Consistency	**0.75**	**0.79**	**0.73**
Coverage	**0.74**	**0.71**	**0.33**

According to the found solutions, we can notice that proposition 1 holds true as different combinations of online product information conditions can result in higher online purchases. For proposition 2, there was no single condition of the product information conditions appeared in all solutions and hence there is no necessary condition found which means that proposition 2 didn't hold true. Finally, proposition 3 hold true only in solution 1 where the information availability appeared to be sufficient for online customers to purchase online. While in the other two solutions there were no single condition appeared to be sufficient to result in higher online purchases.

4.2 Discussion

The results in Table 3 highlight three sufficient configurations. As we can observe from the three solutions, each of the conditions included in the study appears in at least one of the configurations as sufficient in realizing the positive impact of information on the online purchase decision, According to Solution 1, if the customers perceive that there is information available on the platform, it is a sufficient motivator for them to have a positive impact on the final purchase decision. This is in line with existing research, and this simple rule states that this condition alone is sufficient for a large number of customers. Solution 2 highlights the interrelated role of information search and product information: if a customer perceives both of them positively, then it is sufficient to increase online purchases on the platform. Solution 3 presents a configuration for the case of male customers. Specifically, the solution suggests that males who perceive that there is a lot of product information available on an online platform are more likely to make purchases online. According to this rule for males, online search is not required, but product information alone is sufficient, while according to Solution 2, the combination of these two variables is sufficient for both males and females. This finding may suggest that providing more detailed and accessible product information on an online platform could be an effective way to increase online sales among male customers.

4.3 Main Contributions

This paper has a significant contribution to academic research in the realm of e-tailing. First, the results of the study underscore the crucial role that detailed online product information plays an important role in influencing consumer purchasing decisions. We show how comprehensive product descriptions and availability of information can significantly enhance the online shopping experience resulting in increased purchases. Second, the study proved that different combinations of product information conditions lead to higher levels of online shopping intentions. Third, and to the best of our knowledge, this is the first study that investigates the impacts of various product information aspects (product information, information availability, and information search) on the shopping intentions of online customers. By utilizing a configurational approach, we show how a more sophisticated analysis of complex relationships between variables can offer insights into non-linear interactions and causal configurations that influence online consumer behavior.

The results of this study can help e-tailers to attract and retain customers and improve their sales through providing their customers with enough and accurate product information on their e-commerce websites and platforms. The paper results provide an effective guidance to e-tailers about providing their customers with the necessary information about their products and services. This might entail a strategy of action that help e-tailers to ensure that the necessary online product information are always accessible, available, and searchable.

5 Conclusion

In this paper, we discussed the role of product information in customer's purchase decision on online retail platforms, and presented the preliminary results from a survey conducted with respondents who are users of online retail platforms. Regarding RQ1, the results in Table 3 demonstrate that different combinations of online product information conditions can result in higher online purchases. Furthermore, each of the product information conditions included in the study appeared in at least one of the solution configurations as sufficient in realizing the positive impact of information on the online purchase decision. Therefore, it can be concluded that online product information is a crucial factor that can positively influence the purchasing decisions of online customers and lead to higher online purchases. As for RQ2, the finding implies that providing more detailed and accessible product information on an online platform could be an effective way to increase online sales among male customers. Overall, online product information is an essential factor that plays a significant role in influencing online shopping and online purchasing decisions. The study contributes to the existing literature by providing insights into the importance of product information in electronic retail and identifying the key conditions that lead to higher online purchases. It also provides valuable recommendations for online retailers to improve their product information offerings to enhance the online shopping experience of their customers.

References

1. Al Karim, R.: Customer satisfaction in online shopping: a study into the reasons for motivations and inhibitions. IOSR J. Bus. Manag. **11**(6), 13–20 (2013)
2. Blanco, C.F., Sarasa, R.G., Sanclemente, C.O.: Effects of visual and textual information in online product presentations: looking for the best combination in website design. Eur. J. Inf. Syst. **19**(6), 668–686 (2010)
3. Comegys, C., Hannula, M., Väisänen, J.: Longitudinal comparison of Finnish and US online shopping behaviour among university students: the five-stage buying decision process. J. Target. Meas. Anal. Mark. **14**, 336–356 (2006)
4. Constantinides, E., Holleschovsky, N.I.: Impact of online product reviews on purchasing decisions. In: 12th International Conference on Web Information Systems and Technologies, pp. 271–278. SCITEPRESS, Rome (2016)
5. Dash, A., Zhang, D., Zhou, L.: Personalized ranking of online reviews based on consumer preferences in product features. Int. J. Electron. Commer. **25**(1), 29–50 (2021)
6. Duşa, A.: QCA with R: A Comprehensive Resource. Springer, Cham (2018). https://doi.org/10.1007/978-3-319-75668-4
7. Gao, F., Su, X.: Online and offline information for omnichannel retailing. Manuf. Serv. Oper. Manag. **19**(1), 84–98 (2017)
8. Häubl, G., Trifts, V.: Consumer decision making in online shopping environments: the effects of interactive decision aids. Mark. Sci. **19**(1), 4–21 (2000)
9. Huber, O., Seiser, G.: Accounting and convincing: the effect of two types of justification on the decision process. J. Behav. Decis. Mak. **14**(1), 69–85 (2001)
10. Joseph, P.T.: E-commerce: An Indian Perspective. PHI Learning Pvt. Ltd. (2019)
11. Kang, T.C., Hung, S.Y., Huang, A.H.: The adoption of online product information: cognitive and affective evaluations. J. Internet Commer. **19**(4), 373–403 (2020)

12. Kumar, A., Thakur, Y.S.: Beyond buying to shoppers: motivation towards online shopping. BVIMSR's J. Manage. Res. **8**(1), 31–36 (2016)
13. Lazaris, C., Sarantopoulos, P., Vrechopoulos, A., Doukidis, G.: Effects of increased omnichannel integration on customer satisfaction and loyalty intentions. Int. J. Electron. Commer. **25**(4), 440–468 (2021)
14. Lee, Y., Lin, C.A.: Exploring the serial position effects of online consumer reviews on heuristic vs. Systematic information processing and consumer decision-making. J. Internet Commer. **21**(3), 297–319 (2022)
15. Li, M., Wei, K.K., Tayi, G.K., Tan, C.H.: The moderating role of information load on online product presentation. Inf. Manage. **53**(4), 467–480 (2016)
16. Newell, A., Shaw, J.C., Simon, H.A.: Elements of a theory of human problem solving. Psychol. Rev. **65**(3), 151 (1958)
17. Norman, D.A.: Toward a theory of memory and attention. Psychol. Rev. **75**(6), 522 (1968)
18. Pappas, I.O., Woodside, A.G.: Fuzzy-set Qualitative Comparative Analysis (fsQCA): guidelines for research practice in Information Systems and marketing. Int. J. Inf. Manage. **58**, 1–23 (2021)
19. Ramadan, Z.B., Farah, M.F., Daouk, S.: The effect of e-retailers' innovations on shoppers' impulsiveness and addiction in web-based communities: the case of Amazon's Prime Now. Int. J. Web Based Communities **15**(4), 327–343 (2019)
20. Rihoux, B., Ragin, C.C.: Configurational Comparative Methods: Qualitative Comparative Analysis (QCA) and Related Techniques. Sage Publications, New York (2008)
21. Singh, J.P., Irani, S., Rana, N.P., Dwivedi, Y.K., Saumya, S., Roy, P.K.: Predicting the "helpfulness" of online consumer reviews. J. Bus. Res. **70**, 346–355 (2017)
22. Song, J.D.: A study on online shopping cart abandonment: a product category perspective. J. Internet Commer. **18**(4), 337–368 (2019)
23. Von Helversen, B., Abramczuk, K., Kopeć, W., Nielek, R.: Influence of consumer reviews on online purchasing decisions in older and younger adults. Decis. Support Syst. **113**, 1–10 (2018)
24. Yu, Y., Liu, B.Q., Hao, J.-X., Wang, C.: Complicating or simplifying? Investigating the mixed impacts of online product information on consumers' purchase decisions. Internet Res. **30**(1), 263–287 (2020)

Habitat Suitability Assessment of Three Passerine Birds Using Ensemble Learning with Diverse Models

Omar El Alaoui[1] and Ali Idri[1,2](✉)

[1] ENSIAS, Mohammed V University in Rabat, Rabat, Morocco
ali.idri@um5.ac.ma
[2] Mohammed VI Polytechnic University Benguerir, Ben Guerir, Morocco

Abstract. The application of machine learning (ML) for predicting species habitat suitability has become increasingly popular. However, using a single ML algorithm may not provide optimal predictions for a given dataset, making it challenging to achieve high accuracy. Therefore, this study proposes a novel approach to assess habitat suitability of three redstarts species (P. Moussieri, P. Ochruros, and P. Phoenicurus) based on ensemble learning techniques. Initially, eight ML models namely MLP, SVM, KNN, Decision Trees (DT), Gradient Boosting Classifier (GB), Random Forest (RF), AdaBoost (AB), and Quadratic Discriminant Analysis (QDA) were trained individually. Then, based on the diversity of these base-learners, seven heterogeneous ensembles of two up to eight models were constructed for each species dataset. This study presents a thorough modeling framework for implementing heterogeneous ensembles and enhancing the overall comprehension of their performance in comparison to single models. The performance of this experiment was evaluated using: (1) six performance measures (AUC, sensitivity, specificity, accuracy, Kappa, and TSS), (2) Borda Count ranking method, and (3) Scott Knott statistical test. Results showed the potential of heterogeneous ensembles for predicting habitat suitability of the three redstarts birds. The proposed ensembles consistently outperformed single models across all datasets.

Keywords: heterogenous ensemble learning · species distribution modeling · ecological niche models · habitat suitability · machine learning · redstart birds

1 Introduction

Biodiversity conservation plays a crucial role in protecting and preserving the variety of species, their habitats, and ecosystems [1]. All species on Earth are interdependent, the extinction of one species can pose a significant threat to the survival of other species, underscoring the delicate balance within our planet's ecosystems [2]. To address this challenge, ecologists have implemented various strategies to protect species and conserve their habitats, including species distribution modeling [3]. SDMs, also known as Habitat Suitability Models, have gained popularity as an effective tool in ecology for evaluating

Á. Rocha et al. (Eds.): WorldCIST 2024, LNNS 987, pp. 191–201, 2024.
https://doi.org/10.1007/978-3-031-60221-4_19

species suitability and identifying potential habitats [4]. SDM models provide ecologists with valuable insights by analyzing data of species occurrences and their correlation with environmental conditions, making them a favored method for studying and conserving species habitats [3, 5].

Machine learning (ML) has gained popularity in recent years for SDM due to its ability to handle complex and non-linear relationships between environmental variables and species distribution [3, 4]. Numerous studies have demonstrated that ML models can provide more accurate predictions than classical statistical methods. However, the use of a single model may not always provide the best results due to the potential drawbacks and limitations of each model, which can be highly dependent on the specific domain of interest. To address this issue, researchers have investigated ensemble learning techniques [6, 7], that can assimilate and merge multiple sources of information by combining multiple models, each with their unique strengths and weaknesses [8]. Ensembles can be homogeneous (combining instances of the same model) or heterogeneous (combining different models). Heterogenous ensembles typically tend to have higher variance but can reduce the bias of the model [8].

In the context of species distribution modelling several studies investigated the use of heterogenous ensembles to enhance models' performance. The study [9] proposed a heterogenous ensemble using eight algorithms as base learners combined using weighted voting to predict the distribution of some medicinal plants located in Egypt. In [10], authors firstly trained 4 ML models (RF, SVC, DT, LR), and then constructed 11 heterogeneous ensembles combined using soft voting to predict the potential distribution of mosquito species in Germany. Studies [9–11] showed that ensembles generally outperform single models in terms of performance. However, these studies have revealed certain limitations: (1) the studies have not covered all the necessary pre-processing steps. Inadequate pre-processing of data can lead to biased results due to overfitting. (2) The evaluation process used in comparing ensembles and single models was insufficient due to the lack of appropriate statistical tests and insufficient assessment of the dispersion of performance metrics on various data subsets. (3) the experimental design of these studies was unclear and did not provide a comprehensive modelling framework for using heterogenous ensembles. Therefore, this study aims to address these limitations by presenting a comprehensive modelling framework for using heterogeneous ensembles and offers better insight into their performance compared to single models.

This paper aims to model and predict the suitable habitats of the three redstarts birds (P. Moussieri, P. Ochruros, and P. Phoenicurus) located in Morocco using single machine learning algorithms and heterogeneous ensembles. Initially, eight ML algorithms (SVM, KNN, MLP, GB, DT, RF, AB, and QDA) were trained as singles. Then, based on the diversity of these base-learners, seven heterogeneous ensembles of two up to eight models were constructed for each species dataset. The performance of the current experiment was evaluated using six performance measures, namely AUC, sensitivity, specificity, accuracy, Kappa, and TSS. The BC ranking method was then used to rank the models based on these six metrics, allowing for a comparison of their performance across multiple criteria. Additionally, the SK test was applied to assess whether there was a significant difference in AUC values between the proposed models. This study aims to address and examine the following research questions:

- **RQ1:** How well are the eight ML techniques in modeling and predicting the distribution of the three redstart birds?
- **RQ2:** Does the performance of the heterogeneous ensembles constructed based on diversity differ significantly from their respective single models?

This study presents several key contributions, including:

- A comprehensive evaluation of the performance of eight ML algorithms (SVM, KNN, MLP, GB, DT, RF, AB, and QDA) in modeling habitat suitability of three redstarts species in Morocco.
- Constructing heterogeneous ensembles based on base-learners diversity.
- An evaluation of the performance of the heterogeneous ensembles in comparison to their singles to determine whether ensemble modeling improves predictive accuracy.

The rest of this paper is divided into different sections. Section 2 presents the material and methods used in this study. Section 3 presents and discusses the results obtained. Lastly, Sect. 4 outlines the conclusion and future works.

2 Material and Methods

2.1 Birds' Dataset

This study focused on the occurrence data of three bird species belonging to the Phoenicurus genus group. Phoenicurus is a genus of passerine birds classified within the Muscicapidae family, which comprises eleven species commonly known as Redstart. Table 1 provides details of the occurrence data for the three bird species used in this study. The data was collected from the Global Biodiversity Information Facility (gbif.org) [12] and comprises a total of 9,925 observation records (Download link: https://doi.org/10.15468/dl.m8efvg). Note that the scientific names of these birds are all valid and met the standards set by ITIS (Integrated Taxonomic Information System), and each species is associated with a distinct and permanent TSN (Taxonomic Serial Number), as detailed in Table 1.

Table 1. Description of the three redstart birds

Image	Scientific Name	Common Name	#Observations	#TSN
	Phoenicurus Moussieri	Moussier's Restart	5148	562035
	Phoenicurus Ochruros	Black Redstart	3316	562036
	Phoenicurus Phoenicurus	Common Redstart	1461	562037

2.2 Environmental Data

The present study used three main types of environmental data: climatic, topographic, and vegetation variables. The climatic data comprised of 19 bioclimatic variables obtained from the Worldclim global database [13], which were used to capture the current climate conditions across the study area. These variables were used at a resolution of 2.5 arcminutes, which corresponds to an approximate spatial resolution of 5 km across the entire region of Morocco. The bioclimatic variables provide information on temperature, precipitation, and other aspects of climate that influence species distributions. The topographic variable consisted of elevation data with a resolution of 2.5 arcminutes, which was obtained from Digital Elevation Database (SRTM) [14]. Finally, the vegetation variable consisted of a land cover variable obtained from NASA using the MODIS Land Cover Type Product (MCD12Q1 v6) (https://lpdaac.usgs.gov/products/mcd12q1v006/). The MODIS data provides a global view of land cover at a spatial resolution of 500 m. The LC_type4 dataset, which provides an 8-class land cover classification scheme including categories such as grass vegetation, broadleaf vegetation, and urban/built-up lands, was used for this study.

2.3 Experiment Workflow

After collecting data on species occurrences, it is important to minimize sampling bias and autocorrelation by applying spatial filtering techniques [15]. This study applied a minimum inter-point distance of 2 km to thin the data using spThin R package. Next, to balance the presence and absence classes for building SDM models, we generated pseudo-absence data. The random method for generating pseudo-absence data is commonly used, but it may not be the best choice. Therefore, we generated pseudo-absence data by excluding environmentally similar locations using a buffer of 250 km around each presence point.

The subsequent step in this experiment involves pre-processing the data, which starts with identifying and handling outliers. In this study, the Interquartile Range (IQR) method was used to detect extreme outliers. These outliers accounted for less than 8% of the total data, thus we opted to remove them to improve the performance of the models. Moreover, a correlation-based approach, using Pearson's correlation coefficient and mutual information (MI) [16], was employed to detect and eliminate irrelevant features. Variables with high correlation coefficients ($|r| \geq 0.8$) were evaluated using MI, and the less informative variable was removed. The final stage of this phase consists of normalizing the data to ensure that all numerical variables are scaled uniformly. To achieve this, the z-score normalization technique was applied to normalize all the numerical predictors to a unified scale.

After pre-processing the data on species, we start building the distribution models. Initially, eight machine learning algorithms (SVM, KNN, RF, MLP, GB, DT, AB, and QDA) were trained as singles to model the distribution of the three redstarts species (P. Moussieri, P. Ochruros, and P. Phoenicurus). Thereafter, to select the base learners of the heterogenous ensembles from the eight single models, we calculated the diversity of base-learners, we used the Q-statistic pairwise measure [17] to evaluate the diversity of singles. Q varies between -1 and 1, and the closer Q is to 0 the more independent and diverse the classifiers are. Seven combinations were selected with this strategy: ensembles of two, three, four, five, six, seven, and eight singles. The process followed in this selection strategy illustrated in the following three steps:

- **Step (1):** Calculate the Q-statistic diversity for all combination pairs and choose the pair with the best (lower) Q value.
- **Step (2):** Calculate the Q-statistic of the best pair from step (1) with each remaining singles (one by one) and add the one that yields the best Q value to the pair.
- **Step (3):** Repeat step (2) by incrementally increasing the ensemble size until reaching the ensemble of size eight.

After selecting the base learners for the heterogeneous ensembles, we combined them using the soft voting method, assigning equal weights (w = 1) to the base learners. All eight single models in the heterogeneous ensembles were trained on 10 folds using the K-fold cross-validation technique without parameter tuning, using the Scikit-learn library in Python (Fig. 1).

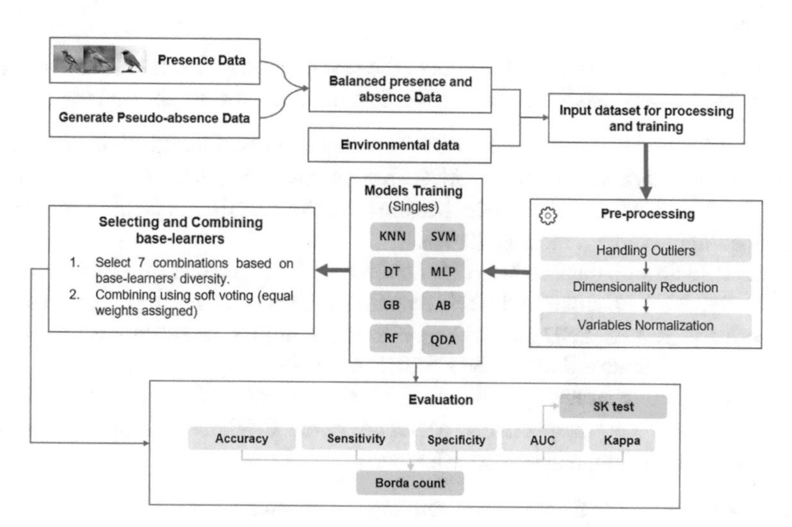

Fig. 1. Experiment workflow of the proposed approach.

3 Results and Discussions

This section presents and discusses the performance of eight singles models and seven heterogenous ensembles on the three species datasets.

3.1 Evaluation of the Eight Single Machine Learning Techniques

The performance of the eight single models in predicting the distribution of the three redstart species was assessed using six evaluation metrics: Accuracy, Sensitivity, Specificity, AUC, Kappa, and TSS. Their average values were computed on 10-folds CV and summarized in Table 2. The results indicated that the RF model achieved the best performance for most metrics in all three species datasets. Specifically, for the P. Moussieri dataset, RF had the highest values for Accuracy, AUC, Kappa, and TSS, which were 81.9%, 0.82, 0.64, and 0.64, respectively. KNN had the best specificity of 81.9%, while QDA had the highest sensitivity reached 87.1%. Similarly, for the P. Ochruros species dataset, RF had the highest values for Accuracy, Specificity, AUC, Kappa, and TSS, which were 80%, 77.2%, 0.80, 0.60, and 0.60, respectively, while QDA had the highest sensitivity of 90.7%. For the P. Phoenicurus species dataset, RF again had the highest values for Accuracy, Specificity, AUC, Kappa, and TSS, which were 74.7%, 72.8%, 0.74, 0.49, and 0.48, respectively, and AB had the highest sensitivity reached 87%. The results of BC ranking method indicated that RF outperformed all other models, ranked first in all three species datasets based on the six-evaluation metrics used, while QDA algorithm achieved the worst results, ranking last in all the three datasets.

3.2 Evaluation of the Heterogeneous Ensembles Compared to Their Single Models

Table 3 presents the performance results of the seven heterogeneous ensembles over the three species datasets using the six metrics. The obtained results showed that:

- For the P. Moussieri dataset, the E5-SBD ensemble outperformed all other ensembles in terms of accuracy, sensitivity, AUC, Kappa, and TSS, achieving 82.8%, 86.5%, 0.83, 0.66, and 0.66, respectively. Meanwhile, the E8-SBD ensemble achieved the best specificity value, reaching 79.9%.
- For the P. Ochruros dataset, E5-SBD achieved the best results in terms of accuracy, specificity, AUC, Kappa, and TSS, reaching 80.7%, 75.4%, 0.81, 0.60, and 0.61, respectively. The E7-SBD ensemble achieved the highest sensitivity value (88.8%).
- For the P. Phoenicurus dataset, the E5-SBD ensemble outperformed all other ensembles in terms of accuracy, specificity, AUC, Kappa, and TSS, reaching 74.8%, 69%, 0.75, 0.49, and 0.50, respectively. Meanwhile, the E2-SBD ensemble achieved the best sensitivity value, reached 86.8%.

Table 4 displays the Borda Count rankings of seven heterogeneous ensembles and eight single models over the three distinct species datasets based on the six metrics, (accuracy, sensitivity, specificity, AUC, Kappa, and TSS). The results revealed that:

- For the P. Moussieri species dataset, 4 ensembles outranked all singles. E5-SBD ranked the first in this dataset, followed by E8-SBD, E6-SBD, and E7-SBD. In contrast, RF outranked the remaining three ensembles: E4-SBD, E2-SBD and E4-SBD.
- For P. Ochruros species dataset, three ensembles outranked all single models. E5-SBD ranked the first in this dataset, E7-SBD the second, and E8-SBD the third. However, RF ranked the fourth, means that is ranked over four ensembles: E6-SBD, E3-SBD, E4-SBD, and E2-SBD.
- For P. Phoenicurus species dataset, three heterogeneous ensembles (E5-SBD, E6-SBD, E8-SBD) outranked all the single models. Specifically, E5-SBD ranked the first, E6-SBD the second, whereas E8-SBD ranked the third. Notably, RF outranked the four remaining ensembles: E4-SBD, E7-SBD, E3-SBD, and E2-SBD.

Table 2. Results in terms of accuracy, sensitivity, specificity, AUC, Kappa, and TSS of the eight ML models over the three species datasets.

Species	Models	Accuracy (%)	Sensitivity (%)	Specificity (%)	AUC	Kappa	TSS	BC
P. Moussieri	RF	81.9	84.5	79.4	0.82	0.64	0.64	1
	GB	80.8	82.9	78.8	0.81	0.62	0.62	2
	KNN	81.1	80.3	81.9	0.81	0.62	0.61	3
	MLP	80.8	81.2	80.4	0.81	0.62	0.61	4
	SVM	79.1	82.0	76.4	0.79	0.58	0.58	5
	AB	79.2	81.5	76.9	0.79	0.58	0.57	6
	DT	79.0	79.1	78.8	0.79	0.58	0.56	7
	QDA	73.9	87.1	61.1	0.74	0.48	0.49	8
P. Ochruros	RF	80.0	82.7	77.2	0.80	0.60	0.60	1
	GB	78.7	85.1	72.0	0.79	0.57	0.58	2
	KNN	77.8	82.5	72.9	0.78	0.56	0.56	3
	AB	77.6	84.2	70.7	0.77	0.55	0.54	4
	DT	77.1	78.6	75.5	0.77	0.54	0.53	5
	MLP	74.1	79.2	68.9	0.74	0.48	0.50	6
	SVM	73.8	82.1	65.3	0.74	0.48	0.48	7
	QDA	72.4	90.7	53.4	0.72	0.44	0.44	8
P. Phoenicurus	RF	74.4	76.1	72.8	0.74	0.49	0.48	1
	KNN	72.8	77.2	68.3	0.73	0.46	0.40	2
	DT	71.5	72.4	70.6	0.72	0.43	0.45	3
	MLP	71.0	72.4	69.7	0.71	0.42	0.44	4
	GB	70.6	74.4	66.8	0.71	0.41	0.44	5
	AB	69.5	87.0	51.1	0.70	0.38	0.42	6
	SVM	69.4	80.6	57.4	0.70	0.38	0.39	7
	QDA	69.2	71.8	67.2	0.69	0.38	0.36	8

Figure 2 displays the results of the test carried out on AUC values for three different redstart species to determine whether there was a statistically significant difference in the AUC values between the heterogenous ensembles and singles. It is observed that:

- In the P. Moussieri dataset, the Scott-Knott test resulted in three clusters. The first cluster contains 5 heterogeneous ensembles (E4-SBD, E5-SBD, E6-SBD, E7-SBD, and E8-SBD) and 4 single models (RF, GB, KNN, and MLP). The second cluster contained two ensembles (E3-SBD, and E2-SBD) and three singles (AB, SVM, and DT), while QDA presented in the last cluster.

Table 3. Performance assessment of seven heterogenous ensembles for three redstart bird species using accuracy, sensitivity, specificity, AUC, and Kappa metrics

Species	Ensembles	Accuracy (%)	Sensitivity (%)	Specificity (%)	AUC (%)	Kappa	TSS
P. Moussieri	E2-SBD	78.8	79.7	78.2	0.79	0.58	0.58
	E3-SBD	79.2	80.4	78.0	0.79	0.58	0.60
	E4-SBD	81.5	84.6	78.5	0.82	0.63	0.65
	E5-SBD	82.8	86.5	79.2	0.83	0.66	0.66
	E6-SBD	82.5	85.6	79.4	0.82	0.65	0.66
	E7-SBD	82.4	85.2	79.7	0.82	0.65	0.64
	E8-SBD	82.6	85.4	79.9	0.83	0.65	0.65
P. Ochruros	E2-SBD	77.9	70.4	75.3	0.78	0.56	0.55
	E3-SBD	78.5	84.8	71.9	0.79	0.57	0.58
	E4-SBD	77.9	84.9	70.7	0.79	0.56	0.57
	E5-SBD	80.7	85.1	75.4	0.81	0.60	0.61
	E6-SBD	79.8	88.1	71.2	0.80	0.59	0.61
	E7-SBD	80.2	88.8	71.4	0.80	0.60	0.61
	E8-SBD	80.1	87.4	72.6	0.80	0.60	0.61
P. Phoenicurus	E2-SBD	69.4	86.8	51.4	0.69	0.39	0.41
	E3-SBD	70.9	71.3	70.6	0.71	0.42	0.41
	E4-SBD	74.1	79.3	68.8	0.74	0.48	0.47
	E5-SBD	74.8	80.0	69.0	0.75	0.49	0.50
	E6-SBD	74.7	80.8	68.5	0.75	0.49	0.49
	E7-SBD	73.9	79.7	67.8	0.74	0.48	0.46
	E8-SBD	74.5	80.8	68.5	0.74	0.49	0.47

- In the P. Ochruros dataset, the Scott-Knott test resulted in 2 clusters. The first cluster contained all 7 ensembles and 5 single models (RF, GB, AB, KNN, and DT), indicating that these models have similar performance in this dataset. The second cluster contained 3 single models (MLP, SVM, and QDA), which implies that these models exhibit different performance in comparison to the models in the first cluster.
- In the P. Phoenicurus dataset, the Scott-Knott test resulted in 2 clusters. The first cluster contained 5 ensembles (E4-SBD, E5-SBD, E6-SBD, E7-SBD, E8-SBD) and 2 singles (RF and KNN), indicating that these models have similar performance in this dataset. The second cluster contained 2 ensembles (E2-SBD, and E3-SBD) and the remaining 6 singles (DT, MLP, GB, AB, SVM, QDA), which implies that these models exhibit different performance in comparison to the models in the first cluster.

Table 4. Borda Count ranks of the heterogeneous ensembles and their single models over the three redstart species.

BC ranking	P. Moussieri	P. Ochruros	P. Phoenicurus
1	E5-SBD	E5-SBD	E5-SBD
2	E8-SBD	E7-SBD	E6-SBD
3	E6-SBD	E8-SBD	E8-SBD
4	E7-SBD	RF	RF
5	RF	E6-SBD	E4-SBD
6	E4-SBD	GB	E7-SBD
7	GB	E3-SBD	KNN
8	KNN	E4-SBD	DT
9	MLP	E2-SBD	MLP
10	E3-SBD	KNN	E3-SBD
11	SVM	AB	GB
12	AB	DT	AB
13	DT	MLP	E2-SBD
14	E2-SBD	SVM	SVM
15	QDA	QDA	QDA

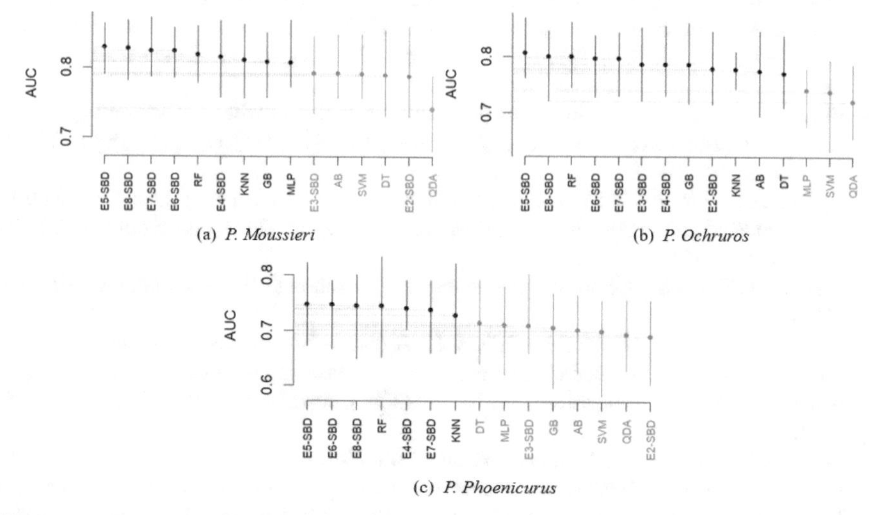

Fig. 2. SK test results of the seven heterogenous ensembles with their singles across the three redstart birds

4 Conclusion and Future Works

To sum up, the heterogenous ensembles, showed better performance than single models, they ranked over single models in all three species datasets when taking into account the five-evaluation metrics using Borda Count method. Moreover, in terms of consistency and stability over the 10 data splits (using 10-fold CV), ensembles varied less and showed less dispersion than single models. However, when statisti-cally comparing the differences between accuracy values of ensembles and singles using SK test, it is found that there is no significant difference in accuracy values. In future work, we aim to extend our research by testing the proposed ensemble models on diverse datasets to generate a comprehensive understanding of their ap-plicability across varying ecological contexts.

References

1. IUCN/SSC, Strategic Planning for Species Conservation: A Handbook. Version 1.0 (2008)
2. Koh, L.P., Dunn, R.R., Sodhi, N.S., Colwell, R.K., Proctor, H.C., Smith, V.S.: Species coextinctions and the biodiversity crisis. Science **305**(5690), 1632–1634 (2004). https://doi.org/10.1126/science.1101101
3. Lawler, J.J., Wiersma, Y.F., Huettmann, F.: Using species distribution models for conservation planning and ecological forecasting. In: Predictive Species and Habitat Modeling in Landscape Ecology: Concepts and Applications (2011). https://doi.org/10.1007/978-1-4419-7390-0_14
4. Padonou, E.A., Teka, O., Bachmann, Y., Schmidt, M., Lykke, A.M., Sinsin, B.: Using species distribution models to select species resistant to climate change for ecological restoration of bowé in West Africa. Afr. J. Ecol. **53**(1), 83–92 (2015). https://doi.org/10.1111/aje.12205
5. Guisan, A., Thuiller, W.: Predicting species distribution: offering more than simple habitat models. Ecol. Lett. **8**(9), 993–1009 (2005). https://doi.org/10.1111/j.1461-0248.2005.00792.x
6. El Alaoui, O., Idri, A.: Predicting the potential distribution of wheatear birds using stacked generalization-based ensembles. Ecol. Inform. **75**, 102084 (2023). https://doi.org/10.1016/j.ecoinf.2023.102084
7. El Alaoui, O., Idri, A.: Heterogeneous ensemble learning for modelling species distribution: a case study of redstarts habitat suitability, pp. 105–114, July 2023. https://doi.org/10.5220/0012118100003541
8. Zhou, Z.H.: Ensemble Methods: Foundations and Algorithms (2012). https://doi.org/10.1201/b12207
9. Kaky, E., Nolan, V., Alatawi, A., Gilbert, F.: A comparison between Ensemble and MaxEnt species distribution modelling approaches for conservation: a case study with Egyptian medicinal plants. Ecol. Inform. **60**, 101150 (2020). https://doi.org/10.1016/j.ecoinf.2020.101150
10. Früh, L., Kampen, H., Kerkow, A., Schaub, G.A., Walther, D., Wieland, R.: Modelling the potential distribution of an invasive mosquito species: comparative evaluation of four machine learning methods and their combinations. Ecol. Model. **388**, 136–144 (2018). https://doi.org/10.1016/j.ecolmodel.2018.08.011
11. Dong, J.Y., Hu, C., Zhang, X., Sun, X., Zhang, P., Li, W.T.: Selection of aquaculture sites by using an ensemble model method: a case study of Ruditapes philippinarums in Moon Lake. Aquaculture **519**, 734897 (2020). https://doi.org/10.1016/j.aquaculture.2019.734897
12. GBIF.Org User, "Occurrence Download." The Global Biodiversity Information Facility (2023). https://doi.org/10.15468/DL.M8EFVG

13. Fick, S.E., Hijmans, R.J.: WorldClim 2: new 1-km spatial resolution climate surfaces for global land areas. Int. J. Climatol. **37**(12), 4302–4315 (2017). https://doi.org/10.1002/joc.5086
14. CGIAR-CSI, SRTM 90 m Digital Elevation Database v4.1, Consortium for Spatial Information
15. Guélat, J., Kéry, M.: Effects of spatial autocorrelation and imperfect detection on species distribution models. Methods Ecol. Evol. **9**(6), 1614–1625 (2018). https://doi.org/10.1111/2041-210X.12983
16. Kraskov, A., Stögbauer, H., Grassberger, P.: Estimating mutual information. Phys. Rev. E Stat. Phys. Plasmas Fluids Relat. Interdiscip. Topics **69**(6) (2004). https://doi.org/10.1103/PhysRevE.69.066138
17. Kuncheva, L.I., Whitaker, C.J.: Ten measures of diversity in classifier ensembles: limits for two classifiers. IEE Colloquium (Digest) **50**, 73–82 (2001). https://doi.org/10.1049/ic:20010105

Budgeted Recommendation with Delayed Feedback

Kweiguu Liu[1(✉)], Setareh Maghsudi[2], and Makoto Yokoo[1]

[1] Faculty of Information Science and Electrical Engineering, Kyushu University, Fukuoka 819-0395, Japan
liu@agent.inf.kyushu-u.ac.jp, yokoo@inf.kyushu-u.ac.jp

[2] Faculty of Electrical Engineering and Information Technology, Ruhr-University Bochum, 44801 Bochum, Germany
Setareh.Maghsudi@ruhr-uni-bochum.de

Abstract. In a conventional contextual multi-armed bandit problem, the feedback (or reward) is immediately observable after an action. Nevertheless, delayed feedback arises in numerous real-life situations and is particularly crucial in time-sensitive applications. The exploration-exploitation dilemma becomes particularly challenging under such conditions, as it couples with the interplay between delays and limited resources. Besides, a limited budget often aggravates the problem by restricting the exploration potential. A motivating example is the distribution of medical supplies at the early stage of COVID-19. The delayed feedback of testing results, thus insufficient information for learning, degraded the efficiency of resource allocation. Motivated by such applications, we study the effect of delayed feedback on constrained contextual bandits. We develop a decision-making policy, delay-oriented resource allocation with learning (DORAL), to optimize the resource expenditure in a contextual multi-armed bandit problem with arm-dependent delayed feedback.

Keywords: Budget Constraints · Delayed Feedback · Online Learning · Resource Allocation

1 Introduction

The contextual bandit problem is a well-known variant of the seminal multi-armed bandit problem: A decision-maker (agent, hereafter) observes some random context, i.e., features, at each round of decision-making. It then pulls one of the available arms and immediately receives a reward generated by the random reward process of the selected arm. Given the context, the agent maximizes each round's reward while effectively exploring the potential alternatives. The state-of-the-art applications of the contextual bandit problem include online advertising and personalized recommendation; nevertheless, the current research neglects those applications with additional constraints on resources [2] that cause the following challenges: (i) The exploitation becomes limited by the resource

Á. Rocha et al. (Eds.): WorldCIST 2024, LNNS 987, pp. 202–213, 2024.
https://doi.org/10.1007/978-3-031-60221-4_20

available for exploration; (ii) The agent can no longer seek to maximize the instantaneous rewards as the arm with the highest reward can be an expensive one. Consequently, the work opted for maximizing the accumulated rewards. One main assumption in budgeted learning is immediately observable feedback, but feedback is usually late in real-world applications. Delayed feedback exacerbates the difficulty of exploration because the information about suboptimal arms procrastinates. Consequently, delayed feedback makes resource allocation inefficient for exploration and exploitation. A motivating example is the distribution of medical supplies at the early stage of COVID-19. During the outbreak, medical supplies, e.g., protective kits and ventilators, do not support urgent needs. That renders an optimal allocation of scarce resources imperative; nevertheless, delayed feedback about testing results hinders health administrators from accomplishing that goal, as inaccurate estimates aggravate the difficulty of making decisions for vital resources. To address this real-world challenge, the Greek government collaborated with the schools from the United States; [3] designed and deployed a national scale learning system named Eva to save lives. Eva's goal is to efficiently allocate scarce testing resources to identify as many infected passengers as possible while striving for more accurate estimates of COVID-19 prevalence from passengers. The challenge Eva faced was the delayed feedback of COVID test results because delayed feedback brings the following adverse effects in budgeted learning:

1. **Over-exploration:** Delayed feedback yields an inaccurate estimation of unknown parameters. Thus, the agent over-explores to retrieve the information that it could have simply obtained in a non-delayed scenario.
2. **Inefficient allocation:** Over-exploration results in ineffective budget expenditure because over-exploration depletes the resources that could have been used to explore arms with higher rewards and a longer response time.
3. **Ineffective allocation:** The delayed feedback in Eva was guaranteed to return in a shorter period. However, feedback in other real-world applications is usually not guaranteed to return, and such unresponsive feedback makes resource allocation challenging.

To tackle the challenges above, one shall incorporate delays in budget planning, i.e., one needs to consider expected delayed rewards of arms because resource allocation is also affected by delayed feedback. Hence, we disentangle the adverse effects of delayed feedback by gradually filtering out less-responsive arms, i.e., arms with excessively long or unbounded delays. Next, we formulate a delay-oriented linear program to handle online resource allocation. Specifically, our proposed algorithm *delay-oriented resource allocation with learning* (DORAL) consists of two stages: (i) using a fraction of the budget, the first stage identifies a set of top responsive arms that are likely to return feedback within a time window; (ii) with the remaining budget, the second stage uses the obtained set of top responsive arms to form delay-oriented linear programming to optimize resource allocation.

Contributions. Our proposed two-stage algorithm ensures an efficient online resource allocation in a general setting, where arm-dependent delays can be excessively long or unbounded. The previous works in constraint learning circumvent the issues caused by delayed rewards via posterior sampling or patiently waiting for feedback. Such remedies are applicable given prior information about delays or when the waiting time is relatively short. Our alternative solution handles delayed feedback directly through the joint allocation of resources and learning time. Also, we propose a delayed version of the robust top arms identification method.

2 Related Work

Recent research on learning with delayed feedback includes diverse applications such as [7,16] on personalized recommendation, [8,10] on edge computing,[1] on the military, and [6] on communication networks. The cutting-edge research categorizes delayed feedback into two classes, namely, bounded- and unbounded delays. For example, a medical result that arrives within 48 hours is a bounded delay, whereas a customer's feedback that usually never returns is unbounded. The state-of-the-art methods combine existing learning algorithms with the following concepts to handle delayed feedback: *waiting* or *cut-off*. The cut-off is a predetermined waiting window discarding any feedback outside its boundaries. Waiting for feedback to update the estimators is a popular method; nonetheless, it is appropriate when delays are bounded [14,19]. In cases of significant delays, cut-off is the best fit, because waiting indefinitely without updating estimators can worsen bias and increase storage overhead. Reference [17] applies the concept to delayed linear bandits. Similar applications appear in [15].

Resource allocation with delayed feedback has gained attention due to various challenges in real-world applications. One solution to allocate resources with delays is via posterior sampling to estimate delayed feedback. The Greek government and [3] designed the system Eva for the urgent allocation of limited medical resources during the COVID outbreak, and Eva circumvents delayed feedback, i.e., testing results, by applying the empirical Bayes procedure to estimate the prevalence of passengers from different countries. Similarly, Northern American ride-sharing company Lyft allocates resources to ad campaigns across periods to attract more drivers, but prospective drivers cannot hit the road until finishing the mandatory requirements. Thus, [12] applies Thompson Sampling to predict potential drivers, i.e., delayed rewards for resource allocation. Compared to the previous work, our work considers possible non-returned feedback that makes allocated resources ineffective, so we need to identify which arms are less responsive to minimize ineffective resource allocation.

3 Problem Formulation

We consider an environment with finite classes of contexts $\mathcal{X} = \{1, \dots, J\}$. Let $\pi_1, \dots, \pi_J$ denote a known distribution over contexts. Each type of context is characterized by an unknown parameter vector $\theta_j \in \mathbb{R}^k$, where k denotes the number of features. A finite set of arms (actions) is $\mathcal{A}$, and $|\mathcal{A}| = A$. At each time t, a subset of $\mathcal{A}$, denoted by $\mathcal{A}_t$, is available. Each arm $a_t \in \mathcal{A}_t$ has a feature vector $\mathbf{f}_{a_t} \in \mathbb{R}^d$ and a corresponding fixed cost c_{a_t}; $\mathbf{f}^{(j)}_{a_t}$ denotes the feature of a_t selected for context j. Each arm has a delay distribution $\mathcal{D}_{a_t}$ with unknown mean d_{a_t} supported on positive numbers, and delay $D_{a_t} \sim \mathcal{D}_{a_t}$ is sampled. Because arms have different delays due to exogenous factors, D_{a_t} does not depend on the contexts. Initially, the agent has some budget B. At each round $t \in T$, the remaining budget is b_t. The agent interacts with the environment as follows until the budget is exhausted.

While $b_t > 0$,

- At time t, the agent observes a context from $j \in \mathcal{X}$, where the contexts arrive independent of each other and the set of available arms $\mathcal{A}_t$;
- The agent selects an arm $a_t \in \mathcal{A}_t$ to maximize the weighted reward $r^{(j)}_{t,u,a_t} = \langle \theta_j, \mathbf{f}_{a_t} \rangle$, where u is the latest time in which the reward shall become observable. The agent pays the associated cost c_{a_t} or not if she pulls no arm. The agent is allowed to skip any round, i.e., pull no arm, whenever no arm can be recommended given the remaining budget.

For $u > 0$, $\mathbb{P}(D_{a_t} \leq u) = \tau_{a_t}(u)$. Following [9], we assume $\tau_{a_t}(u)$ satisfying the following inequality. Let $\alpha > 0$,

$$|1 - \tau_{a_t}(u)| \leq u^{-\alpha} \tag{1}$$

By the assumption above, a smaller α implies a lower chance of receiving feedback by u. When we consider the case with heavy-tailed delays, we assume $\mathbb{E}|D_{a_t} - d_{a_t}|^{1+\varepsilon} \leq v_{a_t}$, where $\varepsilon \in (0, 1]$ and v_{a_t} denotes variance. In other words, delays are not assumed to be sub-Gaussian. Also, we assume each arm's mean delay is not larger than a certain portion of the budget, i.e., $d_{a_t} \leq B/4$, because any mean delay larger than the budget is rarely observed and can be discarded. The agent does not know feedback before D_{a_t} exceeds $u - t$. Thus, we define the delayed reward formally as

$$\hat{r}_{t,u,a_t} = r_{t,u,a_t} \mathbb{1}\{D_{a_t} \leq u - t\} \tag{2}$$

where $\mathbb{1}\{D_{a_t} \leq u - t\}$ is an indicator function that returns one if the reward of the decision made at t is observed by u and zero otherwise. The agent's objective is to select the arms sequentially to maximize the total accumulated reward given the budget constraint and delayed feedback. Formally,

$$\begin{aligned} \text{maximize} \quad & U(T, B) = \mathbb{E}\left[\sum_{t=0}^{T} \hat{r}_{t,u,a_t}\right] \\ \text{subject to} \quad & \sum_{t=1}^{T} c_{a_t} \leq B \end{aligned} \tag{3}$$

where the expectation is taken over the distribution of contexts and rewards. Let $U^*(T,B) = \mathbb{E}[\sum_{t=0}^{T} \hat{r}^*_{t,u}]$ denote the total optimal payoff when $\hat{r}^*_{t,u} = \max_{a_t \in \mathcal{A}_t} \hat{r}_{t,u,a_t}$. We measure the performance by the regret, i.e., the difference between the expected gain with hindsight knowledge and the actual gain, and it is defined as

$$R = U^*(T,B) - U(T,B) \tag{4}$$

The agent minimizes the regret by optimal arm selection.

4 Algorithm Design

The proposed algorithm consists of two stages. The first stage is to identify a set of top responsive arms. We describe this stage in Sect. 4.1. The second stage is online allocation with learning; we explain this stage in Sect. 4.2.

4.1 Search for Top Responsive Arms

To mitigate the adverse effects of arm-dependent delays on resource allocation, we need to know the arms' response time in order to determine the cut-off m for contextual learning in the next stage. Hence, we first need to identify the top responsive $A' \leq A$ arms. This task boils down to an identification problem in multi-armed bandits, where the rewards are the average over the number of rounds in which the agent observes delays. Also, ranking arms according to their responsiveness enables us to neglect the ones with rare or no feedback, thus saving scarce resources for the rest.

We build our identification algorithm upon a strategy family known as *Successive Acceptance and Rejection* (SAR) from [5,11]. Different variants of SAR can optimize either the budget for a given confidence level or the quality of exploration for a given threshold. However, in our setting, delays make it challenging to decide on a necessary budget in the first place. To address this issue, inspired by [13], we propose a variant of SAR, namely, *Patient-Racing SAR* (PR-SAR), for a given threshold. Algorithm 1 describes PR-SAR.

Let S_t denote a set of accepting arms, L_t a set of remaining arms at time t, and E_t a set of rejected arms. Till identifying top responsive arms, $|S_t| = A'$, PR-SAR continuously pulls arms from L_t, and the method determines acceptance by comparing confidence bounds. However, due to *delayed response* and possible *non-sub-Gaussian delays*, we need a robust estimation method, namely, *median-of-means* estimator by [4], to measure the responsiveness of arms. The core idea is to prepare several disjoint baskets, calculate the standard empirical mean of received feedback in each basket and take a median value of these empirical means. $X \vee Y \triangleq$ Specifically, for some arm a, $T_a(u) = \sum_t^u \mathbb{1}\{a_t = a\}$ denote the number of times the agent observes arm a by time u, and $h = \lfloor 8 \log(\frac{e^{1/8}}{\delta} \wedge \frac{T_a(u)}{2}))\rfloor$ and $N(u) = \lfloor \frac{T_a(u)}{h} \rfloor$. Let $D_{a,t} = D_{a_t}\mathbb{1}\{a_t = a\}$. Each basket's estimated expected waiting period then yields

$$\hat{d}_{a,1} = \frac{1}{N(u)} \sum_{t=1}^{N(u)} D_{a,t}, \hat{d}_{a,2} = \frac{1}{N(u)} \sum_{t=N(u)+1}^{2N(u)} D_{a,t}, \ldots, \hat{d}_{a,h} = \frac{1}{N(u)} \sum_{t=(h-1)N(u)+1}^{hN} D_{a,t}.$$

Let d_a^M denote the median-of-means estimator of empirical waiting periods. We first need the following lemma to bound d_a^M; the lemma states how the empirical mean behaves when delayed feedback exists. Due to the limited space, the proof is omitted.

Lemma 1. *For some arm a and $\alpha, \delta > 0$, with probability at least $1-\delta-B^{-\alpha}$, $\hat{d}_a \leq d_a + \sqrt{\frac{2B\log\frac{2}{\delta}}{T_a(u)}} + 2d_a T(u)^{-(\alpha\wedge 1/2)}$*

The following theorem states the robust upper confidence bound (UCB) of patient median-of-means estimation.

Theorem 1. *Let $\alpha > 0, \delta > 0$. For some a and any $t > A$ with probability $1-\delta-B^{-\alpha}$,*

$$|d_a^M - d_a| \leq \sqrt{\frac{2\log\left(\frac{16}{1-B^{-\alpha}}\right)}{T_a(u)}} + \frac{B}{2}T_a(u)^{-(\alpha\wedge 1/2)} \tag{5}$$

Proof. Let $Z_l = \mathbb{1}\{\hat{d}_l > d_a + \varepsilon, \forall l \in \{1,\ldots,h\}\}$. According to Lemma 1, Z_l follows a Bernoulli distribution with $p \leq \frac{v_a}{N(u)^\varepsilon \zeta^{1+\varepsilon}} + 2\exp\left(\frac{-2\zeta^2 N(u)}{B^2}\right) + B^{-\alpha}$. If $\zeta = \sqrt{\frac{2\log\left(\frac{16}{1-B^{-\alpha}}\right)}{N(u)}}$, we have $p \leq 1/4 + B^{-\alpha}$. By Hoeffding's inequality, we have

$$\mathbb{P}(d_a^M > d_a + \varepsilon) = \mathbb{P}\left(\sum_{l=1}^{h} Z_l \geq \frac{h}{2}\right) \leq \exp\left(-2h(1/2-p)^2\right) \leq \exp\left(-h/8\right) \leq \delta$$

According to Theorem 1, PR-SAR accepts any arm whose lower confidence bound (LCB) is larger than at least K' of UCB's. Specifically, PR-SAR compares the following delayed robust UCB's and LCB's.

$$UCB_{d_a}(t) = d_a^M(t) + \sqrt{\frac{2\log\left(\frac{16}{1-B^{-\alpha}}\right)}{T_a(u)}} + 2d_a T_a(u)^{-(\alpha\wedge 1/2)} \tag{6}$$

$$LCB_{d_a}(t) = d_a^M(t) - \sqrt{\frac{2\log\left(\frac{16}{1-B^{-\alpha}}\right)}{T_a(u)}} - 2d_a T_a(u)^{-(\alpha\wedge 1/2)} \tag{7}$$

After finding top responsive arms, PR-SAR determines the remaining budget for resource allocation, i.e., $B_{ac} = B - B_{id}$ where B_{id} is the amount of budget spent on identification. PR-SAR can simply decides the cut-off $m = \max_{a\in\mathcal{A}} UCB_{d_a}$. PR-SAR by construction is Hoeffding race method, so it can find top responsive arms.

Algorithm 1. Patient Racing SAR

Input: $\mathcal{A}, B, S_1 = \{\emptyset\}, L_1 = \{1, \ldots, K\}$
Output: A' accepted arms and $m = max\{UCB_{d_a}, \forall a \in \mathcal{A}\}$
  while $|S_t| < A'$ **do**
    for $a \in L_t$ **do**
      pull a and compute UCB_{d_a} using Eq. 6, LCB_{d_a} using Eq. 7
    end for
    for $a \in L_t$ **do** ▷ Update top accepted arms
      if $|\{a' \mid LCB_{d_a} > UCB_{d_{a'}}\}| > A' - |S_t|$ **then**
        $S_t \leftarrow S_t \bigcup \{a\}$
        $L_t \leftarrow L_t \backslash \{a\}$
      end if
    end for
  end while

4.2 Resource Allocation with Delays

We first introduce the decision rule when delayed feedback exists, and then explain online resource allocation with delayed feedback.

Learning Estimators with Delayed Feedback. Waiting indefinitely for feedback results in excessive computation overhead and swift exhaustion of scarce resources. One solution to mitigate the problem is to select a cut-off parameter m. Thus the delayed reward $\hat{r}_{t,u,a}$ can be restated as

$$\tilde{r}_{t,u,a} = r_{t,u,a} \mathbb{1}\{D_{t,a} \leq \min(m, u - t)\} \tag{8}$$

Let $\lambda > 0$ be a regularization parameter. Following [17], we estimate θ using the least square method as

$$\hat{\theta}_t = \left(\sum_{t=1}^{u-1} \mathbf{f}_{a_t} \mathbf{f}_{a_t}^\top + \lambda I \right)^{-1} \left(\sum_{t=1}^{u-1} \tilde{r}_{t,u,a} \mathbf{f}_{a_t} \right) = V_t(\lambda)^{-1} G_t. \tag{9}$$

Let $f_{t,\delta} = \sqrt{\lambda} + \sqrt{2 \log\left(\frac{1}{\lambda}\right) + k \log\left(\frac{k\lambda + t}{k\lambda}\right)}$. Theorem 1 in [17] validates that, after a period of learning, the distance between $\hat{\theta}$ and θ remains bounded with high probability. Hence, we propose the following decision rule to select arms.

– At each round and for each arm a, define the index $\gamma_t(a)$ as

$$\langle \hat{\theta}_t, \mathbf{f}_a \rangle + \left(2f_{t,\delta} + \sum_{t=u-m}^{u-1} \|\mathbf{f}_{a_t}\|_{V_t(\lambda)^{-1}} \right) \|\mathbf{f}_a\|_{V_t(\lambda)^{-1}} \tag{10}$$

The index is the linear upper confidence bound (LinUCB) within the cut-off.
– The agent picks the arm with the highest index, i.e., $a_t^* = \arg\max \gamma_t(a)$, $a \in \mathcal{A}_t$.

Resource Allocation with Delayed Learning. Optimization of resource allocation in (3) with the hard budget constraint is especially challenging with unknown delays. Hence, we approximate the optimal solution with a relaxed budget constraint, i.e., the average budget constraint $\rho = \frac{B}{T}$ at each round. Inspired by the approximation method in [18], we develop an approach for near-optimal delay-oriented allocation. In the following, we drop the parameter m from $\tau_a(m)$ unless it is necessary to avoid ambiguity. For simplicity, we assume that the distribution of contexts is known and static. We have $\tau_a \to 1$ if $m \to \infty$, but such m increases the regret significantly. To simplify the analysis, we assume τ_a's are given. Let $\eta_j^* = \max_{a \in \mathcal{A}} \tau_a \tilde{r}_{j,a}$ the best expected delayed reward the agent can obtain under context j and $\tilde{a}_j^* = \arg\max_{a \in \mathcal{A}} \tilde{r}_{j,a}$ the corresponding arm. Let $\mathbf{p} = (p_1, \ldots, p_j)$ denote a probability vector, and the agent's goal is to solve the following linear programming at each round:

$$\begin{aligned} LP_m \text{ maximize}_{\mathbf{p}} \quad & \sum_{j}^{J} p_j \pi_j \eta_j^* \\ \text{subject to} \quad & \sum_{j}^{J} p_j \pi_j \leq \rho \end{aligned} \tag{11}$$

LP_m maximizes the expected delayed reward with arms with higher probabilities to return, while its constraint considers expected delayed costs to avoid spending resources on arms with no feedback possibilities. The solution of LP_m can be expressed with some threshold $j(\rho)$, which is a function of the average constraint ratio ρ, and the reinterpretation of LP_m can help simplify the regret analysis. Thus,

$$j(\rho) = \max\{j : \sum_{j'}^{J} \pi_{j'} \leq \rho\} \tag{12}$$

$$p_j(\rho) = \begin{cases} 1, & \text{if } 1 \leq i \leq j(\rho) \\ \frac{\rho - \sum_{j'=1}^{j(\rho)} \pi_{j'}}{\pi_{j(\rho)+1}}, & \text{if } i = j(\rho) + 1 \\ 0, & \text{if } i > j(\rho) + 1 \end{cases} \tag{13}$$

and the optimal value of LP_m can be expressed with (12) and (13).

$$v(\rho) = \left(\sum_{j'}^{j(\rho)} \pi_{j'} \hat{\eta}^*_{j'} \right) + p_{j(\rho)+1} \pi_{j(\rho)+1} \eta^*_{j(\rho)+1} \tag{14}$$

With Algorithm 1 and the resource allocation rule (11), we present our proposed decision-making strategy, i.e., delay-oriented resource allocation with learning (DORAL), in Algorithm 2. The algorithm consists of two stages. The first stage spends a portion of the budget B_{id} and time T_{id} to identify top responsive arms, and the second stage explores and exploits with the remaining budget B_{ac} and time T_{ac} using the cut-off obtained in the first stage.

Algorithm 2. Delay-Oriented Resource Allocation with Learning

Input: $\mathcal{A}, S_1 = \{\emptyset\}, B, T, \lambda \in (0, 1)$
  while $|S_t| \leq A'$ **do**
    Algorithm 1
  end while
  while $B_{ac} > 0$ **do**
    Observe context and pick $\tilde{a}^*_j(t)$ with $p_j(\frac{b_t}{t})$ that satisfies LP_m with $\hat{\eta}^*_j$
    Update $b_t, \hat{\theta}_j, V^{(j)}_\lambda(t)$, and $G^{(j)}_t$ if delayed feedback returns.
  end while

5 Experiments

Due to the difficulty of finding a dataset suitable for our scenarios, we evaluate the performance of DORAL with a synthetic dataset with the following settings: The budget is $85,000$ to ensure at least $50,000$ rounds. There are 10 context classes, and the distribution of the contexts is $\boldsymbol{\pi} = [0.09, 0.15, 0.11, 0.05, 0.1, 0.05, 0.08, 0.14, 0.13, 0.1]$. There are 10 arms, and each has a unit cost. Each context and arm has five features, where we represent each feature by some value between $(0, 1)$. We use geometric and Pareto distributions to generate delays for each assigned arm. For the scenario of diverse delays, we have geometric delays for the arms, and their expected delays are $[100, 120, 140, 160, 200, 220, 240, 260, 280, 300]$. For Pareto delays, we set each arm's minimum delay as $[200, 220, 240, 260, 280, 320, 340, 360, 380, 400]$, and the arms share the same shape parameter $\alpha = 2$. For the scenario of similar and short delays, we use [100,110,120,130,140,150,160,170,180,190] for geometric means and Pareto minimum values. We compare our proposed algorithm with the following benchmarks. **Delayed-LinUCB (D-LinUCB)** by [17]: The method selects the arms with the highest reward in each context class. The method selects the arms with the highest reward in each class of context. Because the type of delay distribution is unknown, we choose $m = 500$ for the cut-off m. This method can be interpreted as a greedy method in our scenario. Thus, we

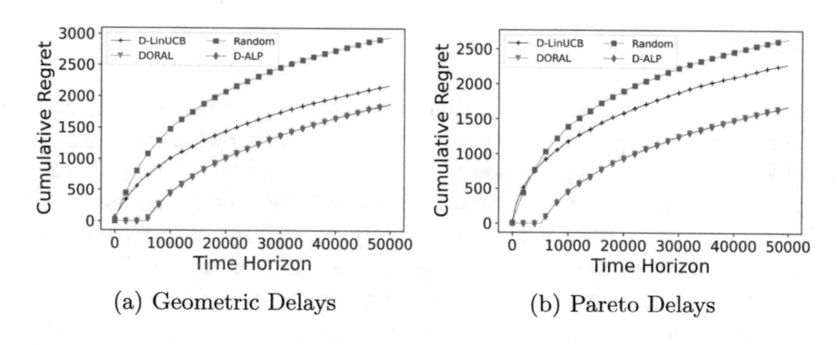

(a) Geometric Delays (b) Pareto Delays

Fig. 1. Similar delays

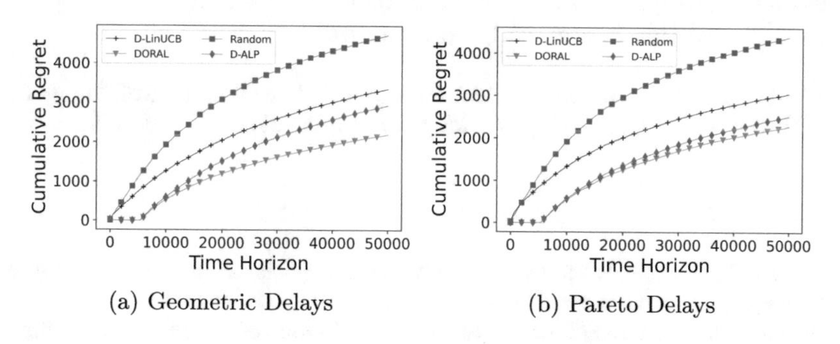

(a) Geometric Delays (b) Pareto Delays

Fig. 2. Diverse delays

have the lowest $\tau_a(m) = 0.81$ in the case of geometric delays, and the lowest $\tau_a(m) = 0.36$ in the case of Pareto delays. **Random Delayed-LinUCB (Random)**: The method is similar to Delayed-LinUCB; Nevertheless, the selection follows with the probability of ρ, i.e., the remaining budget concerning the given budget. It also uses $m = 500$ for the cut-off. **Delayed adaptive linear programming (D-ALP)** of [18]: The method is similar to our proposed method, but we let $\forall a \in \mathcal{A}, \tau_a(m) = 1$. Each figure is the average of 50 runs. As DORAL spends some time identifying top responsive arms, it starts late to accumulate rewards. In Fig. 1, because delays are short, DORAL and D-ALP are overlapping. In Fig. 2 (a), D-ALP identifies a higher $m > 500$, so it's expected to have more regret according to the theorems. Nevertheless, DORAL outperforms D-ALP when facing heavy-tailed delays in Fig. 2 (b) even though both of them identify similar m, while they come close in Fig. 2 (c). These cases indicate that utilizing expected delayed rewards in diverse delays can simultaneously optimize rewards and learning altogether.

6 Conclusion

To tackle the challenges in resource allocation with delayed feedback, we developed a two-stage policy that can efficiently allocate resources while learning with delayed feedback. Also, we proposed a robust method to identify top responsive arms when information delays can be heavy-tailed. Future research involves simultaneously determining cut-off on the fly while ensuring efficient resource allocation. Context-dependent delays are more realistic and challenging when compared to arm-dependent delays in our setting. Also, in our setting, the feedback of different arms is equally important, although often, there exist different levels of urgency to consider in a resource allocation problem, e.g., in sharing limited medical supplies. Hence, another research direction is studying such a hierarchical structure in real-world applications.

Acknowledgement. The work of S.M. was supported by Grant 01IS20051 and Grant 16KISK035 from the German Federal Ministry of Education and Research.

References

1. Amuru, S., Buehrer, R.M.: Optimal jamming using delayed learning. In: 2014 IEEE Military Communications Conference, IEEE (2014), pp. 1528–1533 (2014)
2. Badanidiyuru, A., Langford, J., Slivkins, A.: Resourceful contextual bandits. In: Conference on Learning Theory, PMLR (2014), pp. 1109–1134 (2014)
3. Bastani, H., et al.: Efficient and targeted Covid-19 border testing via reinforcement learning. Nature **599**(7883), 108–113 (2021)
4. Bubeck, S., Cesa-Bianchi, N., Lugosi, G.: Bandits with heavy tail. IEEE Trans. Inf. Theory **59**(11), 7711–7717 (2013)
5. Bubeck, S., Wang, T., Viswanathan, N.: Multiple identifications in multi-armed bandits. In: International Conference on Machine Learning, PMLR (2013), pp. 258–265 (2013)
6. Cesa-Bianchi, N., Gentile, C., Mansour, Y.: Nonstochastic bandits with composite anonymous feedback. In: Conference On Learning Theory, PMLR (2018), pp. 750–773 (2018)
7. Chapelle, O., Manavoglu, E., Rosales, R.: Simple and scalable response prediction for display advertising. ACM Trans. Intell. Syst. Technol. (TIST) **5**(4), 1–34 (2014)
8. Chen, L., Xu, J.: Task replication for vehicular cloud: contextual combinatorial bandit with delayed feedback. In: IEEE INFOCOM 2019-IEEE Conference on Computer Communications, IEEE (2019), pp. 748–756 (2019)
9. Gael, M.A., Vernade, C., Carpentier, A., Valko, M.: Stochastic bandits with arm-dependent delays. In: International Conference on Machine Learning, PMLR (2020), pp. 3348–3356 (2020)
10. Ghoorchian, S., Maghsudi, S.: Multi-armed bandit for energy-efficient and delay-sensitive edge computing in dynamic networks with uncertainty. IEEE Transactions on Cognitive Communications and Networking (2020)
11. Grover, A., et al.: Best arm identification in multi-armed bandits with delayed feedback. In: International Conference on Artificial Intelligence and Statistics, PMLR (2018), pp. 833–842 (2018)

12. Han, B., Gabor, J.: Contextual bandits for advertising budget allocation. In: Proceedings of the ADKDD, vol. 17 (2020)
13. Heidrich-Meisner, V., Igel, C.: Hoeffding and bernstein races for selecting policies in evolutionary direct policy search. In: Proceedings of the 26th Annual International Conference on Machine Learning, pp. 401–408 (2009)
14. Joulani, P., Gyorgy, A., Szepesvári, C.: Online learning under delayed feedback. In: International Conference on Machine Learning, PMLR (2013), pp. 1453–1461 (2013)
15. Thune, T.S., Cesa-Bianchi, N., Seldin, Y.: Nonstochastic multiarmed bandits with unrestricted delays. In: Wallach, H., Larochelle, H., Beygelzimer, A., d' Alché-Buc, F., Fox, E., Garnett, R., eds. In: Advances in Neural Information Processing Systems. Vol. 32., Curran Associates, Inc. (2019)
16. Vernade, C., Cappé, O., Perchet, V.: Stochastic bandit models for delayed conversions. In: Conference on Uncertainty in Artificial Intelligence (2017)
17. Vernade, C., Carpentier, A., Lattimore, T., Zappella, G., Ermis, B., Brueckner, M.: Linear bandits with stochastic delayed feedback. In: International Conference on Machine Learning, PMLR, pp. 9712–9721 (2020)
18. Wu, H., Srikant, R., Liu, X., Jiang, C.: Algorithms with logarithmic or sublinear regret for constrained contextual bandits. In: Cortes, C., Lawrence, N., Lee, D., Sugiyama, M., Garnett, R., eds.: Advances in Neural Information Processing Systems. Vol. 28., Curran Associates, Inc. (2015)
19. Zhou, Z., Xu, R., Blanchet, J.: Learning in generalized linear contextual bandits with stochastic delays. Adv. Neural. Inf. Process. Syst. **32**, 5197–5208 (2019)

AI in Higher Education: Assessing Acceptance, Learning Enhancement, and Ethical Considerations Among University Students

Alexander Griesbeck[1], Jasmin Zrenner[1], Ana Moreira[2,3], and Manuel Au-Yong-Oliveira[1,4,5](✉)

[1] Department of Economics, Management, Industrial Engineering and Tourism, University of Aveiro, 3810-193 Aveiro, Portugal
alexander.griesbeck@st.oth-regensburg.de, jasmin.zrenner@student.hswt.de, mao@ua.pt

[2] APPsyCI - Applied Psychology Research Center Capabilities and Inclusion, ISPA—Instituto Universitário, R. Jardim do Tabaco 34, 1149-041 Lisbon, Portugal
amoreira@ispa.pt

[3] Faculdade de Ciências e Tecnologia, Universidade Europeia, Lisbon, Portugal

[4] INESC TEC, Porto, Portugal

[5] Research Unit on Governance, Competitiveness and Public Policies (GOVCOPP), 3810-193 Aveiro, Portugal

Abstract. This scientific paper aims to investigate the multifaceted impact of Artificial Intelligence (AI) on students at universities, focusing on the three key areas: Acceptance of AI in Higher Education (HE), Learning Enhancement and Ethical Considerations. A survey with 100 valid answers from Graduates and Undergraduates was conducted, to gather insights into their perceptions and experiences with AI-powered tools. The investigation shows that these tools are widely accepted. There is a significant positive correlation between students' comfort level with incorporating AI into their education and their familiarity with them. Regarding the improvement of learning, more than half of the participants think that AI technologies are useful for the understanding of course material, and that their learning efficiency has increased. Also, it could be statistically proven that there is a significant positive correlation between frequency of preparation and performance improvement for exams, essays, and projects. Concerning ethics, most of the students are aware of possible ethical dilemmas and agree that standards are necessary. The study also shows that 75% witnessed other students cheating with the help of AI-tools. Non-parametric Mann-Whitney tests were also used and gender was found to significantly affect the variables under study. This investigation serves as a foundation for informed discussions on harnessing the potential of AI to improve education while addressing the associated concerns.

Keywords: Artificial Intelligence · Higher Education · Learning Enhancement · Ethical Considerations · Student Perceptions

Á. Rocha et al. (Eds.): WorldCIST 2024, LNNS 987, pp. 214–227, 2024.
https://doi.org/10.1007/978-3-031-60221-4_21

1 Introduction

AI systems such as chatbots have recently evolved a great deal and become more accessible on the Internet to more people in a very short period of time. Namely, the interest in AI has increased due to the release and rapid diffusion of the chatbot Chat Generative Pre-Trained Transformer (ChatGPT), a model of language generation published by OpenAI in November 2022 [1]. A chatbot is a computer program designed to simulate conversation with human users [1], typically using natural language processing (NLP). The tool has been labeled as a breakthrough in the academic world [2], due to the ability to produce diverse, original content characterized by its unique, coherent ideas, and profound understanding of existing scientific knowledge. They provide students with a broad array of opportunities for self-directed learning, including tasks such as homework or exam preparation, text translation, and coding. This leads to a concern raised in academia about students using chatbots as a shortcut to success whereby minimal effort may be expended to achieve certain results [3].

The value of AI and its provided support to numerous areas of application is almost undeniable and at the same time it is easy to become dependent on its use for all kinds of tasks regardless of meaningfulness and morality. While AI and its capabilities serve as valuable educational tools, it is essential to note that they cannot replace a teacher's pedagogical ability to guide and support students in a learning process [4].

There is already scientific work that deals with the role in academia and use of AI in education, inter alia, in terms of improvements of the material and the learning surroundings in schools or universities [4–7]. However, there is a need to investigate how students themselves perceive and deal with AI in relation to their coursework and performance. Consequently, its effect on education, students' completion of assignments and their grading needs to be rethought and adapted.

The purpose of this study is to fill in the research gaps in terms of empirical studies assessing the acceptance, learning enhancement, and ethical considerations of AI-powered tools among university students, by examining the current state of acceptance and use of AI in the university context and the exploration of factors that help to enhance the use of AI-powered tools and students' satisfaction. The results of the study provide insights into the perception of AI by university students and may also help with the definition of guidelines for better regulation of the use of AI in academic settings to prevent academic dishonesty. Hence, this study focuses on addressing the following question: **How do University Students Perceive, Utilize, and Ethically Engage with AI-Powered Educational Tools?** To answer this question, three follow-up questions were formulated: How familiar are students with the use of AI-based tools and are they willing to integrate it in education? Do the students that frequently use AI-powered tools perceive performance improvements and in which assignments do they perceive the most improvements (exams, projects, essays)? Are students aware of potential ethical concerns associated with using AI in an educational setting and do they think there should be regulations?

The rest of this article is ordered as follows; the next chapter provides a literature review regarding the role of AI in higher education, with justification for choosing the research hypotheses and the questions for the survey. The Sect. 3 provides a brief explanation of the adopted methodology. Next, Sect. 4 presents and discusses the findings. The Sect. 5 highlights the major conclusions and study implications.

2 Literature Review

2.1 Acceptance of AI in Higher Education

The acceptance of AI in Higher Education is the first topic to be discussed. The challenge is to understand the willingness of students to embrace AI-based technologies as educational aids, examining the factors that influence their acceptance, such as familiarity, ease of use, and perceived benefits.

A Twitter analysis of global perceptions and reactions revealed that, the attitude towards AI in education is twofold [8]. There are the enthusiasts about its potential to support learning and those who are concerned about how students might use it to avoid independent work and thinking or how it contributes to misinformation [8]. As a result of the latter, the use of the tools is not encouraged, and they are not used at all or without the necessary instruction and awareness. Education about the handling of sensitive data is of great relevance so that concerns about privacy and data security can be overcome. It is also helpful to learn how to phrase the instructions to the chatbot precisely and clearly in order to achieve the desired goal or the desired results.

The Twitter study mentioned above [8], also concluded that authority decisions may influence public attitudes. Therefore, teachers should encourage students to the use of AI-based tools such as chatbots. There are already schools that have included learning about AI in their curriculum [9] and surveys at such schools have shown that, on average, students see it as a powerful and useful technology and are enthusiastic and keen to learn about AI. Furthermore, teachers should raise students' intrinsic motivation and redirect them towards tasks that hold personal significance and can only be achieved by humans [10].

In addition to the influence of teachers, the numerous benefits of chatbots contribute significantly to their acceptance. Their ability to learn from previous conversations and adapt to the individual preferences and needs is remarkable. ChatGPT can adjust its responses to the level of knowledge and the learning ability of each student and thus increase the perceived ease of use. Due to the wide range of possible applications, such as supporting with the preparation of homework or presentations, generating exam-style questions, summarizing, or classifying a text, it provides personalized and interactive learning experiences and can therefore improve learning outcomes [5].

Based on the assessment of the literature on the Acceptance of AI in Higher Education, important criteria that influence students' willingness to use AI-based tools were identified and included in the survey. They resulted in the following questions for the survey: Have you ever used AI-powered tools? Do you consider yourself as familiar with the use of AI-based tools for academic purposes, such as AI-powered chatbots (e.g., ChatGPT) or tutoring systems? What factor would influence your willingness to use AI-powered tools for academic purposes more often? Are you comfortable with the idea that AI-based tools will be integrated into education? How frequently do you use AI-powered tools to prepare for exams/projects/essays? Accordingly, the first hypothesis **H1**: *Students who are more familiar with AI-based tools are more comfortable with the idea of integrating AI into their university education*, was derived.

2.2 Learning Enhancement

Students can benefit from AI in many ways. Due to several possible applications, there is a suitable tool for almost every task. In the field of education, these are mostly chatbots, such as ChatGPT. The second part of this literature review therefore focuses on the impact of AI on students Learning Enhancement. There are already numerous studies looking at the ways in which chatbots and AI can be used to improve students' performance and education itself. According to Labadze et al. [6], AI-based tools provide support particularly for homework and as a study aid, for a more personalised learning experience, and as regards the development of various other capabilities.

Moreover, Fütterer et al. [8] point out the benefits of ChatGPT and other large language models in terms of teaching and learning in practice. The potential to enable personalised and adaptive learning and to make assessment and evaluation processes more efficient are mentioned as examples.

However, the risk that chatbots may also generate texts that sound correct but contain incorrect information must not be ignored [1]. Students and educators must therefore learn to recognise misinformation and question the text produced. To achieve this, critical reflection skills must be taught [8]. Another necessary skill is to use suitable tools for a specific task to avoid frustration and thus prevent a negative attitude towards learning with AI [11].

Overall, it can be concluded that AI can support teachers in creating a more effective learning experience for students and, with the right approach, can lead to positive learning outcomes. For instance, in connecting it "with current education reforms such as the digitalization of educational resources, gamification, and personalized learning experiences" [5, p. 2] to create new applications [5].

The aim of this part of the study is to explore the extent of the impact of AI tools, such as ChatGPT, on students' understanding of academic material and their overall learning experience, from the perspective of the students themselves. This includes investigating the impact of AI on learning performance, comprehension, and retention of learning material. The literature review resulted in the following questions: Have you found AI-powered tools to be helpful in enhancing your understanding of course materials? Has the use of AI-powered tools improved your overall learning experience at university? Have AI-powered tools improved your learning efficiency? Have AI-powered tools improved your performance in exams/essays/projects at university? In addition, the hypotheses 2–4 were assumed. **H2**: *Students who use AI-powered tools more frequently for exam preparation also report a greater improvement in their exam performance.* **H3**: *Students who use AI-powered tools more frequently for essay preparation also report a greater improvement in their essay performance.* **H4**: *Students who use AI-powered tools more frequently for project preparation also report a greater improvement in their project performance.*

2.3 Ethical Considerations

Finally, the ethical dimensions of AI usage among students will be explored. In this regard, the present report focuses on the potential for misuse, such as using AI for academic dishonesty and student concerns about the ethical implications of using AI in their academic studies. In addition, students' awareness of these issues and their attitudes towards responsible AI usage will be examined.

There are already scientific analyses of AI in higher education with regard to ethical considerations. These issues are wide-ranging and, in addition to dishonesty and cheating, often deal with questions of data protection and data security. However, this report clearly focuses on the potential of academic dishonesty and cheating during tasks and exams.

Easy access to chatbots like ChatGPT brings new challenges to education. Due to the large number of possible applications, it is tempting to work independently no longer, but to consult AI for every type of task. Students can have a chatbot help them with any assignment or simply let it complete the entire task itself [7, 12]. However, a test of various chatbots concluded that they are not yet as intelligent as is often claimed [13]. According to the study, there are no A- or B-students among the chatbots. GPT-4, for example, performed best, while Bing Chat turned out to be an at-risk student.

Additionally, ChatGPT uses all accessible sources, regardless of copyright [1]. The result is student plagiarism, which is causing increasing concern in education. As a consequence, some teachers have banned the use of ChatGPT from their courses. The solution to this is plagiarism detection applications, which are already being used to detect copied content [7]. However, studies show that ChatGPT has already learnt to circumvent these if it is given the instructions to do so [14]. This problem not only affects academic integrity, but also makes it difficult, if not impossible, to evaluate students fairly [7]. Furthermore, there is the risk that students who complete their assignments without the support of AI-based tools are at a disadvantage compared to those who submit high-quality assignments made by a chatbot [15]. Another problem is that students' comprehension difficulties may not be detected [7].

In order to counteract the problems mentioned, the lecturers need to update their assessments to a time where students have access to chatbots and other AI-based tools. Cotton et al. [15] suggest having students demonstrate their critical thinking, problem-solving, and communication skills in new ways. Group discussions, presentations or other interactive and self-reflective activities are good examples of new types of examination that can promote independent learning and redirect to work practices that not only reproduce.

The results of the literature review were used to formulate the questions for the survey to include students' perceptions in the research: Are you aware of the potential ethical concerns associated with using AI in an educational setting, such as the possibility of using AI for academic dishonesty (e.g., cheating)? Do you think there should be guidelines or policies in place to regulate the use of AI in academic settings to prevent academic dishonesty? If yes, what guidelines would you propose? Have you ever personally witnessed or heard of students using AI or AI-powered tools to cheat on exams or assignments? Are you concerned about the ethical implications of using AI in your academic studies? Afterwards, the last hypothesis **H5**: *There is a positive relationship* *between students' awareness of ethical concerns related to AI in education and their belief in the need for guidelines or policies to regulate AI use in academia*, was put forward.

3 Methodology

First and foremost, the three key areas were searched in scientific databases such as Scopus, Google Scholar, and Springer Link, using keywords or sentences related to the subject, such as "Higher Education", "Artificial Intelligence", "AI", "chatbot*s", "Chat-GPT", "Acceptance", "Learning", "Ethical Considerations", "Cheating". This search was limited to English-language articles. Subsequently, the abstracts of the articles that seemed to be connected to the study's topic were reviewed and assigned to the key areas according to their content. Furthermore, search queries were carried out with different combinations of the mentioned keywords to narrow down the number of search results and at the same time focus more on the main areas of interest.

Besides our analysis of secondary data, we created a survey with the help of Google Forms to meet our research goals. The questionnaire was structured in four parts. The first part contains questions for describing the sample, while the other parts were used to assess Acceptance, Learning Enhancement and Ethical Considerations of AI in HE. In total the survey contained 22 closed questions and one open question. Most of the answers followed the 5-point Likert scale from strongly disagree to strongly agree and were completed by "yes or no" resp. time series. As a sampling method, convenience sampling was chosen due to a lack of time and practical reasons. Although this method is not optimal, because of bias and external influences [16], the endeavor was to extract the maximum value from the available data. To only attract students, the form was shared in a management class at the university of Aveiro and several student WhatsApp groups including an Erasmus group, to also get data from different nationalities and universities. After a period of two weeks (from October 30 to November 13, 2023) the survey was closed, and 104 answers were collected. Due to the limited representation of postgraduate participants, consisting of only two respondents, one incomplete and one inconsistent questionnaire, these instances were excluded from the dataset. Hence, 100 answers could be used as a basis for further analysis.

For Learning Enhancement, which consisted out of six items in the Likert scale format, Cronbach's alpha test was performed. The alpha coefficient of 0.913 indicated acceptable reliability [16]. Because the answers for Acceptance and Ethical Considerations were not completely in this format, Cronbach's alpha coefficient could not be calculated.

To analyze the data descriptive and inferential statistics were used. Descriptive statistics included frequencies, percentages and means. To be able to calculate the means of the Likert scale items, it was assumed that the intervals between the responses are equal. For testing the hypotheses, the Spearman correlation, chi-square test/Fisher's exact test and Cramer's V were used.

Table 1. Sample Description

Demographic Variable	Profile	Percentage
Gender	Female	56%
	Male	44%
Academic stage	Graduate (master's degree)	56%
	Undergraduate (bachelor's degree)	44%
Age	18–20	13%
	21–23	66%
	24–26	19%
	>26	2%
Department	Economics, Management, Industrial Engineering and Tourism	36%
	Electronics, Telecommunications, and Informatics	15%
	Languages and Cultures	14%
	Mechanical Engineering	9%
	Medical Sciences	4%
	Social, Political and Territorial Sciences	3%
	Biology	2%
	Communication and Art	2%
	Education and Psychology	2%
	Physics	2%
	Chemistry	1%
	Civil Engineering	1%
	Environment and Planning	1%
	Other	8%
Nationality	Portuguese	58%
	German	14%
	Italian	6%
	Polish	6%
	French	3%
	Spanish	2%
	Romanian	2%
	Indonesian, Hungarian, Ecuadorian, Czech, US-American, Cape Verdean, Slovakian, Chilean, Swedish	Each 1%

As the Spearman coefficient can be used to calculate the correlation between ordinal data, it was chosen to analyze hypotheses 1–4. Due to the nominal variables in Hypothesis 5, it should be tested using the chi-square test. Because 50% of the cells had expected count less than five, the assumptions for the chi-square test that is - no more than 20% of expected values being less than five - were not met. In the following Cramer's V was performed to measure the strength of the association. Non-parametric Mann-Whitney tests were also used to test whether gender significantly affected the variables under study. The Mann-Whitney non-parametric test was used because the independent variable consists of two groups, and the dependent variable is ordinal. All statistical analyses were performed with Microsoft Excel and IBM SPSS Statistics, version 29. The study had a varied mix of participants, looking at the different nationalities and departments for instance (see Table 1).

4 Results and Discussion

4.1 Acceptance

AI-powered tools have been widely accepted by university students, with a staggering 91% affirming their usage. Also 54% (19% strongly agree, 35% agree) consider them as familiar with the use of AI-based tools for academic purposes and 21% as not that familiar (6% strongly disagree, 15% disagree). This highlights the widespread knowledge of AI-based technologies. When it comes to integration and comfort, a considerable proportion (55%) either strongly agrees (18%) or agrees (37%) with the assimilation of AI tools in education. Only 2% strongly disagree with this idea. Students would use AI-based tools even more often, if learning outcomes would be improved (41%), the usability of these tools would be easier (28%) and concerns about privacy and data security would be less (11%). Analyzing the frequency of AI-powered tool utilization for exam, essay, and project preparation, the data reveals that 13% of participants always or frequently use these tools for exams, 23% for essays, and 31% for projects. Looking at the gender differences, men seem to feel more familiar with AI-tools (mean: 3.86) than women (mean: 3.18). The same trend is observed when integrating AI into education (men: 3.18 vs. women: 3.91). Regarding the same points in connection with the current academic level (graduate vs. undergraduate), the means are very similar (graduate: 3.41 vs. undergraduate: 3.52) and (graduate: 3.52 vs. undergraduate: 3.48). A departmental breakdown of the four departments with the most participants reveals that the department of Electronics, Telecommunications, and Informatics exhibits the highest mean familiarity with AI for academic purposes (4.13), followed by Mechanical Engineering (4), Languages and Cultures (3.5), and Economics, Management, Industrial Engineering, and Tourism (3.13).

The Hypothesis 1 was tested by using the Spearman correlation. With a correlation of 0.473 and a significant p-value (<0.001), there is a moderate positive and significant correlation between students' familiarity with AI-based tools and their comfort with the idea of integrating AI into their university education (see Table 2).

Table 2. Matrix of Spearman correlation for H1

Variables	(1)	(2)
(1) Familiarity with AI-tools for academic use	–	
(2) Comfort of AI integration in academia	.473***	–

Note. *** p < .001

4.2 Learning Enhancement

More than the half of all participants find AI-tools helpful for enhancing understanding of course material (9% strongly agree, 47% agree), whereas 44% reported that these tools improved overall learning experience (5% strongly agree, 39% agree). Furthermore, 52% strongly agree or agree with an improvement of learning efficiency. Opinions are more varied when it comes to improvement of exam, essay, and project performance. Only 22% noticed an improvement in exam preparation. In Contrast, 46% and 55% seemed to have better results in essays and projects. These results underscore varying degrees of effectiveness in different academic areas when integrating AI tools into HE.

The Hypotheses 2, 3, and 4, which suggest a correlation between the frequency of preparation and performance improvement for exams, essays and projects were also tested using the Spearman correlation. The p-value for all Hypotheses of <0.001 shows statistical significance. The observed positive correlation rises ascending for exams (0.521), essays (0.596) and projects (0.741) (see Tables 3, 4 and 5).

Table 3. Matrix of Spearman correlation for H2

Variables	(1)	(2)
(1) AI exam preparation frequency	–	
(2) AI Performance improvement exams	.521***	–

Note. *** p < .001

Table 4. Matrix of Spearman correlation for H3

Variables	(1)	(2)
(1) AI essay preparation frequency	–	
(2) AI Performance improvement essays	.596***	–

Note. *** p < .001

Table 5. Matrix of Spearman correlation for H4

Variables	(1)	(2)
(1) AI project preparation frequency	–	
(2) AI Performance improvement projects	.741***	–

Note. *** $p < .001$

4.3 Ethical Considerations

An overwhelming majority (94%) are aware of the potential ethical issues of AI and recognize its potential role in academic dishonesty (such as cheating). Additionally, 67% of respondents believe that policies or guidelines should be developed to regulate the use of AI in academic settings, underscoring the recognition of the need for an ethical framework. Some of the guidelines, which the study respondents proposed are highlighted in Table 6.

Table 6. Proposals for guidelines, regarding use of AI in HE

Comments
"In the basic or secondary have some classes specific to discuss and teach about this topic"
"Mandatory citing of what AI tools where used and to what extent the tools were used"
"AI system detecting AI use"
"Limits as to in which parts of a project we can use AI as a resource - and define those in which it is prohibited and have a way to check whether it was used or not when the professor corrects the project"
"I guess the AI should not write the full project/essay, but if it's used to do repetitive or supportive working steps (e.g., formatting, spelling) it's fine"

The findings could be clustered to the following points: citing of AI usage, AI control systems to detect cheating, clear rules where and for which tasks to use AI. Additionally, students wanted some training or classes on this topic.

Interestingly, 25% expressed uncertainty about the need for such AI policies, indicating the nuanced views within the student population. When asked if they had ever witnessed or heard a student cheat on a test or assignment using artificial intelligence or AI-powered tools, 75% said "yes." This finding points to some awareness that AI is being misused in academic settings. Additionally, regarding personal concerns about the ethical implications of using artificial intelligence in academic research, 51% agreed (12% strongly agree, 39% agree), 31% were neutral, and 18% disagreed (7% strongly disagree, 11% disagree).

Regarding Hypothesis 5 a chi-square test followed by Fisher's exact test was performed, to test the association between Students who are aware of ethical concerns regarding AI in education and believe that guidelines or policies are needed to regulate AI use. Results show a p-value of 0.014 for the Fisher's exact test, which is under the significance level of 0.05, indicating statistical significance. The Cramer's V of 0.302 implies a moderate positive effect, so the Hypothesis can be accepted (see Tables 7 and 8).

Table 7. Crosstabulation matrix for H5

Variables		Need of AI regulation guidelines in academia							
		No		Unsure		Yes		Total	
		O	E	O	E	O	E	O	E
Awareness of ethical concerns regarding AI in academia	No	2	.5	3	1.5	1	4	6	6
	Yes	6	7.5	22	23.5	66	63	94	94
	Total	8	8	25	25	67	67	100	100

Table 8. Test results for H5

Fisher-Freeman-Halton exact test	Df	p value	Cramer's V
8.420	2	.014*	0.302

Note. * $p < .05$

Non-parametric Mann-Whitney tests were also used to test whether gender significantly affected the variables under study. The results show that gender statistically affects all the variables under study (Table 9). Male participants are more familiar with the academic use of IA, feel more comfortable with its integration into academia, use IA more frequently to prepare for exams, essays, and projects, and that its use helps them improve their performance in exams, essays and projects (Table 9). However, when it comes to awareness of ethical issues related to the use of AI, it is the female participants who have shown greater awareness. AI-powered tools are also seen to be helping male participants more than female participants, regarding: enhancing understanding of course materials, the overall learning experience at university, and improving learning efficiency.

Table 9. Effect of gender on the variables under study

Question	Z	P	Mean Rank	
			Female	Male
Familiarity with AI-tools for academic use	3.17**	.002	42.63	60.52
Comfort of AI integration in academia	3.46***	<.001	41.96	61.36
AI exam preparation frequency	4.33***	<.001	39.79	64.13
AI Performance improvement exams	3.77***	<.001	41.21	62.32
AI essay preparation frequency	2.32*	.020	44.71	57.86
AI Performance improvement essays	2.45*	.014	44.50	58.14
AI project preparation frequency	3.33***	<.001	42.16	61.11
AI Performance improvement projects	2.37*	.018	44.75	57.82
Have you found AI-powered tools to be helpful in enhancing your understanding of course materials?	2.29*	.022	45.01	57.49
Has the use of AI-powered tools improved your overall learning experience at university?	3.54***	<.001	41.91	61.43
Have AI-powered tools improved your learning efficiency?	3.62***	<.001	41.64	61.77
Awareness of ethical concerns regarding AI in academia	-2.29	.022	56.12	43.35

Note. * $p < .05$; ** p .01; *** $p < .001$

5 Conclusion

Based on the findings of the literature analysis, a survey was conducted with 100 valid responses from multifaceted university students, yielding important empirical information about their views and encounters with AI-enabled products. The results offer insightful information to help educators, decision-makers, and interested parties navigate the rapidly changing field of artificial intelligence in higher education.

The survey emphasizes how widely accepted AI-powered technologies are, as 91% of students have used them and more than half are familiar with them in terms of academic use. The accepted hypothesis that comfort with AI integration and familiarity are positively correlated implies that increasing familiarity can improve students' openness to using AI in the university setting.

Regarding learning, AI has a variety of positive effects, including gains in comprehension and learning efficiency. There are notable positive correlations between using

AI tools more frequently and improved performance on exams, essays, and projects. For projects most students use AI-tools and most report improved performance. This underscores the necessity for focused AI applications in various academic contexts.

A strong emphasis is placed on ethical issues, with recommendations for rules and regulations to control the application of AI and deal with issues related to academic dishonesty. As 3/4 of the respondents heard of other students cheating, the need for such standards is stressed. Additionally, 67% would support regulations or procedures to control AI usage. The linkage between the awareness of ethical concerns related to AI in education and the need for guidelines or policies to regulate AI in HE shows that students acknowledge the importance of establishing clear rules.

In summary, this article addresses related issues and provides a basis for educated debates on utilizing AI's potential to enhance education. The advantages, difficulties, and moral conundrums that have been noted offer a road map for developing laws and procedures that will optimize artificial intelligence's beneficial effects in higher education.

Acknowledgments. This work was financially supported by the Research Unit on Governance, Competitiveness and Public Policies (UIDB/04058/2020) + (UIDP/04058/2020), funded by national funds through FCT - Fundação para a Ciência e a Tecnologia.

References

1. OpenAI Introducing ChatGPT. https://openai.com/blog/chatgpt. Accessed 12 Nov 2023
2. Quintans-Júnior, L.J., Gurgel, R.Q., Araújo, A.A.D.S., Correia, D., Martins-Filho, P.R.: ChatGPT: the new panacea of the academic world. Revista da Sociedade Brasileira de Medicina Tropical **56** (2023). https://doi.org/10.1590/0037-8682-0060-2023
3. Boubker, O.: From chatting to self-educating: can AI tools boost student learning outcomes? Expert Syst. Appl. **238**, 121820 (2024). https://doi.org/10.1016/j.eswa.2023.121820
4. Zawacki-Richter, O., Marín, V.I., Bond, M., Gouverneur, F.: Systematic review of research on artificial intelligence applications in higher education–where are the educators? Int. J. Educ. Technol. High. Educ. **16**(1), 1–27 (2019). https://doi.org/10.1186/s41239-019-0171-0
5. Zhai, X., et al.: A review of artificial intelligence (AI) in education from 2010 to 2020. Complexity **2021**, 1–18 (2021). https://doi.org/10.1155/2021/8812542
6. Labadze, L., Grigolia, M., Machaidze, L.: Role of AI chatbots in education: systematic literature review. Int. J. Educ. Technol. High. Educ. **20**(1), 56 (2023). https://doi.org/10.1186/s41239-023-00426-1
7. Lo, C.K.: What is the impact of ChatGPT on education? A rapid review of the literature. Educ. Sci. **13**(4), 410 (2023). https://doi.org/10.3390/educsci13040410
8. Fütterer, T., et al.: ChatGPT in education: global reactions to AI innovations. Sci. Rep. **13**(1), 15310 (2023). https://doi.org/10.1038/s41598-023-42227-6
9. Dai, Y., Chai, C.S., Lin, P.Y., Jong, M.S.Y., Guo, Y., Qin, J.: Promoting students' well-being by developing their readiness for the artificial intelligence age. Sustainability **12**(16), 6597 (2020). https://doi.org/10.3390/su12166597
10. Overono, A.L., Ditta, A.S.: The rise of artificial intelligence: a Clarion Call for higher education to redefine learning and reimagine assessment. Coll. Teach. **71**(4), 1–4 (2023). https://doi.org/10.1080/87567555.2023.2233653

11. Ijaz, K., Bogdanovych, A., Trescak, T.: Virtual worlds vs books and videos in history education. Interact. Learn. Environ. **25**(7), 904–929 (2017). https://doi.org/10.1080/10494820.2016.1225099
12. Qasem, F.: ChatGPT in scientific and academic research: future fears and reassurances. Library Hi Tech News **40**(3), 30–32 (2023). https://doi.org/10.1108/LHTN-03-2023-0043
13. Rudolph, J., Tan, S., Tan, S.: War of the chatbots: Bard, Bing Chat, ChatGPT, Ernie and beyond. The new AI gold rush and its impact on higher education. J. Appl. Learn. Teach. **6**(1), 1–26 (2023). https://doi.org/10.37074/jalt.2023.6.1.23
14. Khalil, M., Er, E.: Will ChatGPT get you caught? Rethinking of plagiarism detection. arXiv: 2302.04335 (2023). https://doi.org/10.48550/arXiv.2302.04335
15. Cotton, D.R., Cotton, P.A., Shipway, J.R.: Chatting and cheating: ensuring academic integrity in the era of ChatGPT. Innov. Educ. Teach. Int. **61**, 1–12 (2023). https://doi.org/10.1080/14703297.2023.2190148
16. Saunders, M.N.K., Lewis, P., Thornhill, A.: Research Methods for Business Students, 8th edn. Pearson Education, New York (2019)

Intellectual Capital and Information Technologies in Education: Some Insights

Óscar Teixeira Ramada(✉)

Instituto Superior de Ciências Educativas do Douro, ISCE Douro, Penafiel, Portugal
oscarramada@gmail.com

Abstract. This paper has the goal verify what insights can be drawn from the relationship between intellectual capital and information technologies, in education. In fact, it is clear that there is an omission of the first in the second. These are only possible because, previously, there exists that one. And this importance can be seen in the design, development and use. It is the latter that is the most visible face of information technologies, in education. The most cited examples in the literature review are the Google Classroom platform (not used massively) and videos. The ultimate aim is to develop tools and platforms that facilitate teaching (by teachers) and learning (by students). It should be noted that they can be desktop and/or mobile. In the state-of-the-art, in the 21st century, it appears that research on the 2 topics, especially the second, has an embryonic development, with the use of complex models (UTAUT2 in Kumar and Bervell (2019)) that make it difficult its use, even if not research at all. What can be expected, in future developments of information technologies, in education, is to facilitate teaching, in terms of content available and in terms of reducing the distance at which it is taught.

Keywords: Intellectual Capital · Information Technologies · Education

1 Introduction

In the 21st century, the topic of technology, information technology, in particular, has played an increasingly important role in several areas, facilitating communications between people. However, in the area of education, it assumes special prominence. This is the case, whatever the level of education, of learning, such as foreign languages (Saritepeci et al. (2019)) and online courses, in particular, which may have requirements to make face-to-face teaching less necessary, removing the distances between who learns and who teaches, namely.

Examples of technologies include videos (Engeness et al. (2019)), tools and platforms, specifically for teaching, such as Google Classroom (Kumar and Bervell (2019)), Microsoft Teams (Martin and Tapp (2019)), Google Meet (Singh and Awasthi (2020)). Saritepeci et al. (2019), in particular, state that a tool with potential, in the context of mobile learning, is WhatsApp. Thus, any and all mobile tools can serve as a learning aid.

Á. Rocha et al. (Eds.): WorldCIST 2024, LNNS 987, pp. 228–233, 2024.
https://doi.org/10.1007/978-3-031-60221-4_22

Knowing what possible relationships exist between, on the one hand, intellectual capital (Andriessen (2004)) and, on the other, information technologies, in the context of education, is of added interest, as a means of transmitting knowledge, in addition to be the result of knowledge. Without intellectual capital, there are no information technologies applied to education (and other areas) (Secundo et al. (2018), Shehzad et al. (2014)), because the latter are the product of the former. According to the aforementioned (Andriessen (2004)), intellectual capital constitutes a set of skills that allow you to do something, hence the justification.

However, it is extremely important to know what the literature refers to and relates about these 2 topics. Thus, the research question is the following: *The literature related to intellectual capital and information technologies (restricted to education) implicitly and/or explicitly, what insights does it convey?*

Therefore, is there a gap between the topics, which are approached in a divergent way when in reality they are intrinsically convergent?

The brief literature review aims to make a contribution, with some answers and identify some insights, which constitute the ultimate goal of this paper.

The choice of papers was based on Google Scholar, which met the criteria of scientific interest and being recent: both from 2019. Therefore, their content incorporates the most current information available, and the length of the paper restricts extensions with a broader literature review. In any case, it is noted that scientific production is still scarce, because the technologies used are also recent and complex, in particular, those of an empirical (rather than theoretical) nature.

2 Literature Review

Kumar and Bervell (2019), are 2 authors who studied the platform called Google Classroom and how it serves for mobile learning, in higher education and how the initial perceptions of students can be modeled. To do this, they use a model called modified *Theory of Acceptance and Use of Technology2* (UTAUT2).

In fact, technologies, such as Learning Management Systems (Moodle, Blackboard, Edmodo, Schoology, Skai and Google Classroom), according to several authors, are not radically adopted in classrooms, contrary to what one might think. The goal of Google Classroom is to reduce paper work, share resources and improve interaction between teachers and students, in addition to managing a greater number of students in classes. It is easier to use, can be used via mobile app and has similarities with the Facebook platform. However, the most recent models, to study the usage behavior of Google Classroom (and others) rarely consider the effects of non-linear relationships, in addition to using linear techniques. Using the UTAUT2 model, allows us to uncover their effects and predict user behavior (from Google Classroom). Therefore, this paper aims to make a contribution to understanding these (non-linear and linear).

The general framework of this model focuses on identifying the relationships, linear and non-linear, between the 2 constructs: prediction of Behavioral Intentions (BI – degree to which students intentionally formulate plans to carry out activities in Google Classroom) which, in turn, make it possible to predict behaviors that predict voluntary Use Behavior (UB). These relationships are based on 7 other constructs:

- Performance Expectancy (PE – represents the perceived belief that using Google Classroom is beneficial in learning activities);
- Effort Expectancy (EE – represents the perception that using Google Classroom is easy);
- Social Influence (SI – represents the idea that, according to the point of view of others, the use of Google Classroom is important for the purposes of teaching activities);
- Facilitating Conditions (FC – represents the idea that using Google Classroom, there is associated technical support and resources);
- Hedonic Motivation (HM – represents the pleasure gained when using Google Classroom) and;
- Habit ((H – represents the level at which Google Classroom is used automatically).

It should be noted that BI, FC and H are used to predict Google Classroom usage behavior. The price value (PV – which represents the benefit obtained from the use of technologies compared to their financial cost) was excluded, because Google Classroom is free for students to use.

Regarding the methods used, the authors used a quantitative design to materialize students' perceptions when using Google Classroom. It covered first-year students, undergraduates, of the teaching initiation course, covering areas such as technological instructional practices, philosophy of education and skills in 21st century classes. The sample size was 206 students (of which only 163 answered) associated with studies related to Geography, Humanities, History and Sciences. By gender, 30 were male and 133 were female. Regarding education, 150 came from the Education area and 13 from Humanities. Regarding the year, 150 students attended the first year, 11 the second and 2 the third. Generally, the use of smartphones was the main device used to access Google Classroom to access learning activities. Data collection was obtained using Google Forms and questionnaires whose items, 26, were associated with the aforementioned PE, EE, SL, FC, HM and BI. The questionnaires were presented to students at the end of the semester, after using Google Classroom. Answers to the questionnaires, followed a 5-point Likert Scale classification, where 5 meant "strongly agree", 4 "agree", 3 "neither agree nor disagree", 2 "disagree" and 1 "strongly disagree". The method used for analysis was the *Partial Least Squares-Structural Equation Modeling* (PLS-SEM).

Regarding the results found, the authors concluded that 3 constructs proved to be the most important in explaining students' BI in relation to Google Classroom: PE, HM and H. They begin to be enthusiastic about using it, mainly, in the most pleasant aspects, derived from the characteristics, very similar to Facebook, which leads to the development of positive intentions in pedagogical use, as well as for other purposes, in the classroom. The greater the use, the greater the scope of learning and habit is an important factor if students accept Google Classroom using it as a mobile technology for learning purposes. Habit works as a factor with positive effects and gains for students.

Regarding implications for theory, the authors highlight that, all non-linear relationships, well explained the complex relationships between several BI predictors in relation to Google Classroom. This enables further studies of other antecedents of the main BI predictors and, on the other hand, it is imperative to include non-linear relationships in models that study Google Classroom based on UTAUT2.

In the implications, specifically, for the purposes of policies and practices, the authors emphasize that, for Google Classroom to be used, it is necessary to design activities that instill interest and encourage the use of it, via mobile devices, to create a habit in everyone the students. This is based on hedonic motivations, which lead to learning content and interactions online, with effects on the ease and usefulness of learning via mobile devices. Thus, the most important factor, with regard to BI, is that content and activities that have a playful nature are embedded in Google Classroom.

Regarding future avenues of research, the authors suggest a more in-depth study of habit, as a facilitator of conditions and BI, in determining it (or not). The relationship between facilitation conditions and hedonic motivation, as well as the habit of using Google Classroom, can also be an object of study. And even a comparative study, between the perceptions of learning via Google Classroom and its acceptance. The moderating effects of factors such as age, gender, learning content and location may be the subject of future research.

Finally, regarding limitations, the authors mention difficulties in generalizing the results obtained, because they only covered undergraduate students and only a single university was considered. Other courses and universities would perhaps allow us to obtain other results. The introduced moderators did not allow us to know the origins of the changes when using Google Classroom. Finally, a final limitation can be mentioned and concerns the fact that only the students' side and not the teachers' side were taken into account.

Engeness et al. (2019), are other authors who, within the scope of the same 2 topics, focus on the role/support of videos in information and communication technologies (ICT) in the context of teaching online courses. In this sense, they seek to research some insights into learning with the aim of developing digital transformation and improving learning capacity, in Norway, with teachers, both in internship and already permanent.

These authors begin by highlighting that, following Zimmerman (2002), students, with a view to improving learning through self-learning, can be achieved by defining objectives, strategies and even carrying out their own self-assessment and that ICT can play an important role in this area together with teachers. However, in the case of videos, in the authors' opinion, the question arises as to how videos can facilitate this entire self-learning process. There is evidence, in several studies consulted by the authors, of a positive correlation between self-learning and achievement of objectives. We need to know how students interact with videos and how they make a valid contribution (to self-learning).

With regard to the sample used in the paper, the authors used a questionnaire to the 2 types of teachers mentioned above. It was administered online, with 22, in total, covering some questions about length of service, learning experiences and facilitator of learning processes and others of a qualitative nature. The sample period was from 2014 to 2018. The teachers in question worked at various levels of education, such as kindergarten, schools, primary, secondary, universities and other establishments. Their countries of origin were mainly Norway and also Sweden.

Regarding the results obtained, the authors divided the analysis into 2 perspectives: teachers' preferences regarding the duration and their interaction with the videos. In preferences, the majority revealed that they were extremely satisfied. In the interaction, the videos served 3 purposes in their use: orienting support, executive support, controlling support and other support. Executive support provided the most support to students: 52.11%. Followed by orienting support: 18.12%.

As the main conclusions of the paper, the authors highlight 4 aspects provided by the videos. First, they created a global perspective of the teaching material, provided commitment to the recommended activities and clarified the criteria for learning outcomes and their assessment. Second, if videos provide orienting, executive and controlling support, they enable students to commit to the learning process, carry out activities and validate the results achieved. Third, if videos provide the 3 aforementioned supports, they contribute to improving digital transformation. Fourth, the majority of participants expressed a preference for videos lasting between 5/6 min and 7/10 min.

As main implications for the design of the videos, in supporting the learning of the courses, the interactions revealed orienting, executive, controlling and other types of support. On the other hand, interactions with videos may have repercussions on their design. Future research could bring benefits and expansion of knowledge with longitudinal studies. All these implications could provide insights for pedagogical improvement by the professionals involved, of any nature: teachers, designers, videographers, in particular.

3 Conclusions

This paper deals with 2 topics of relevance today: intellectual capital and information technologies, in education. Combining them, the goal to discover some insights that a Literature Review, confined to 2 papers, displays within the scope of scientific production.

Therefore, it is clear from the outset that, in terms of quantity, it is not abundant. On the other hand, given that information technologies, in education, are only possible due to the prior existence of intellectual capital, it was found that this appeared completely omitted in the papers displayed.

The research question, remember, is: *The literature related to intellectual capital and information technologies (restricted to education) implicitly and/or explicitly, what insights does it convey?*

In response, one of the insights that can be highlighted is that the initial factor, from which the elaboration of information technologies, in education is possible, does not appear associated, so its dual relationship is non-existent.

Another insight is that technologies such as Google Classroom are not used massively in an educational environment and therefore it cannot be said that there is a complete paradigm shift in teaching-learning (higher education). This is due, in the case of mobile technologies, to the fact that habit, that is, daily use, has not yet been assimilated by both students and teachers, being an observable reality.

In the particular case of the use of videos, their use was the basis of digital transformation and the improvement of learning (on the part of students) and teaching (on the part of teachers), in the particular case in Norway (period from 2014 to 2018). One approach is worth highlighting in this context: videos and other information technologies in the educational context, encourage motivation and self-learning behaviors, with the achievement of pre-defined objectives. The videos must be short: between 5 and 10 min, to prove to be an effective tool. Consequently, even though one of the topics is omitted, this does not take away its usefulness but only limits the scope of the joint approach. There is a gap that denotes divergence when in reality there is convergence.

As contributions of this research, it is worth highlighting that the development and use of information technologies, in education, constitutes a learning facilitator and a stimulus factor, in addition to reducing teaching time (the same, but in less time).

The most obvious limitations of the paper include the lack of similar studies that could serve as a comparison, especially, in terms of results (in addition to insights). Countries such as Sweden, Denmark and Finland are Scandinavian countries in which the teaching structure would be interesting to compare.

As more pronounced implications, it is worth mentioning that information technologies, in education, have development potential, to the point that, in the not very near future, physical presence can be dispensed with and replaced by online, which would have profound impacts on frequency of teaching, because it would attenuate or limit the factor, distance, between the home and teaching environment.

In future research avenues, it is worth mentioning the need for studies extended to used platforms, such as Zoom, Microsoft Teams, among others.

References

Kumar, J., Bervell, B.: Google classroom for mobile learning in higher education: modelling the initial perceptions of students. Educ. Inf. Technol. **24**(2), 1793–1817 (2019)

Castro, R.: Blended learning in higher education: trends and capabilities. Educ. Inf. Technol. **24**(4), 2523–2546 (2019)

Saritepeci, M., Duran, A., Ermis, U.: A new trend in preparing for Foreign Language Exam (YDS) in Turkey: case of WhatsApp in mobile learning. Educ. Inf. Technol. **24**(5), 2677–2699 (2019)

Engeness, I., Nohr, M., Singh, A.: Use of videos in the information and communication technology massive open online course: insights for learning and development of transformative digital agency with pre- and in-service teachers in Norway, Special Issue: Dataphilosophy and Transcurricular Praxis in the Digital Society and Education, Policy Futures in Education, pp. 1–20 (2019)

Zimmerman, B.: Becoming a self-regulated learner: an overview. Theory Pract. **41**(2), 64–70 (2002)

Martin, L., Tapp, D.: Teaching with teams: an introduction to teaching an undergraduate law module using Microsoft Teams. Innov. Pract. High. Educ. **3**(3), 1–9 (2019)

Singh, R., Awasthi, S.: Updated Comparative Analysis on Video Conferencing Platforms – Zoom, Google Meet, Microsoft Teams, WebEx Teams and GoToMeetings, EasyChair Preprint, vol. 4036, pp. 1–10 (2020)

Secundo, G., Lombardi, R., Dumay, J.: Intellectual capital in education. J. Intellect. Cap. **19**(1), 2–9 (2018)

Shehzad, U., Fareed, Z., Zulfiqar, B., Shahzad, F., Latif, H.: The impact of intellectual capital on the performance of universities. Eur. J. Contemp. Educ. **10**(4), 273–280 (2014)

Andriessen, D.: Making Sense of Intellectual Capital - Designing a Method for the Valuation of Intangibles, Elsevier, Butterworth-Heinemann (2004)

Integrating an Advanced Evolutionary Intelligence System into the Internal Contracting Operations at Centro Hospitalar Universitário Lisboa Central

João Melo e Castro[1(✉)], José Neves[2,3], Teresa Magalhães[1], and Magda Reis[4]

[1] Escola Nacional de Saúde Pública - Universidade Nova de Lisboa, Lisbon, Portugal
joaomelo.ecastroemelo@gmail.com
[2] Algoritmi Center/LASI@, Universidade do Minho, Braga, Portugal
[3] IA&SAÚDE@IPSN, CESPU, Famalicão, Portugal
[4] Centro Hospitlar Universitário Lisboa Central, Lisbon, Portugal

Abstract. The research outlined aims to advance and integrate an Evolutionary Intelligence Framework within the Internal Contracting Unit of Centro Hospitalar Universitário Lisboa Central (CHULC). This initiative is expected to streamline and amplify the effectiveness of the hospital's internal processes, especially concerning Outpatient Consultations and Surgical Operations. By adopting this technology, enhanced analytical insights will be accessible to middle management, including hospital administrators and department heads, and subsequently to the CHULC Board of Directors. Such insights will be instrumental for CHULC in optimizing its operational capacities in accordance with the anticipated volume of healthcare activities, and in identifying any underutilized resources or services.

Keywords: Evolutionary Intelligence · Artificial Intelligence · Healthcare management

1 Introduction

1.1 Internal Contracting Process

Healthcare Institutions must continually evolve and adapt to the changing landscape of healthcare needs and delivery methods, a complex challenge due to their inherent nature [1]. Contractualization serves as a mechanism for these institutions to meet the demand for agility and adjustment. This approach involves a negotiation process aimed at optimizing the allocation of public health resources by establishing, monitoring, and evaluating goals [2]. In Portugal, the coordination between supervisory bodies (both regional and central) and hospital facilities is facilitated through a contracting process. This process determines the expected levels and mix of services, sets objectives in agreement with the supervisory authorities and employs indicators to assess institutional performance in

Á. Rocha et al. (Eds.): WorldCIST 2024, LNNS 987, pp. 234–237, 2024.
https://doi.org/10.1007/978-3-031-60221-4_23

terms of accessibility, quality, and financial efficiency [3, 4]. Therefore, adhering to the contracting process mandates strong coordination and professional motivation that aligns with overarching strategic objectives. Internal contracting becomes pivotal, as it translates external commitments into actionable objectives within the healthcare institutions, promoting alignment and coherence across different management levels toward a unified purpose [5]. To ensure the efficacy of internal contracting, prerequisites include a well-defined policy, robust management control, internal auditing mechanisms, collaborative management, and an information system based on actual healthcare activities.

1.2 Evolutionary Intelligence

Evolutionary Intelligence (EI) is a subset of Artificial Intelligence (AI) that employs evolutionary algorithms to enhance the optimization of complex and non-linear issues, especially in unpredictable environments. Essentially, EI adopts biological evolution, particularly Darwinian principles, to refine problem-solving strategies [6]. EI evolves as it incorporates better-quality information into the predicates that define its domain. It enables the forecasting of behavior across a spectrum of failure possibilities, especially when faced with partial information. EI extends logic programming into a symbolic system that swiftly adjusts to environmental shifts [7]. In EI, computational elements are depicted as network nodes where logical deductions, influenced by the information quality within predicate extensions, take place [8]. These logical deductions, guided by information quality and modulated by confidence levels, help in correlating erroneous data with potential performance in various considered conditions. The proposed computational model, termed virtual intellect, is conceptualized as a network of 'symbolic neurons'. Each neuron processes a series of sub-problems relevant to different scenarios.

2 General and Specific Objectives

The overarching goal of this research is to devise and implement an Evolutionary Intelligence Framework in the Internal Contracting Unit. The objective is to gauge the available capacity by medical specialty, focusing solely on the medical staff (doctors) involved in Outpatient Consultations and Operating Room procedures, both conventional and outpatient. This will allow for an evaluation of the existing capacity versus the projected care activity potential. The detailed objectives of the study are, viz.

- To benchmark the existing capacity by specialty, anchored on the medical staff (doctors), against the actual healthcare activities, including the number of consultations and surgeries conducted;
- To compare the healthcare activities carried out by the medical staff (doctors) in outpatient consultations with the average duration of initial and follow-up consultations by specialty;
- To examine the healthcare activities conducted by the medical staff (doctors) in surgical settings against the average duration of various surgery types (ambulatory, conventional, basic, and additional) by specialty;
- To contrast the available structural capacity (operating rooms) with the potential healthcare activities (based on medical staff availability);

- To analyze the outpatient care demand using the Consultation Waiting List relative to the available and utilized capacities; and
- To scrutinize the demand for surgical care via the Surgery Waiting List in light of the available capacity and the extent to which it has been utilized.

3 Methods

To achieve the main goal of this research, a hybrid exploratory and practical approach was selected. Initially, the study adopts an exploratory stance, employing a dual method: a bibliographic survey for theoretical grounding and field research for practical insights. Subsequently, the study shifts to applied research. To develop a potent Evolutionary Intelligence model that adds value in the given context, a six-step methodology will be implemented, viz.

- Data Collection and Preparation: This initial phase involves gathering data to feed into the model and processing it to establish a strong basis for subsequent steps;
- Attribute Selection: Post data preparation, redundant attributes are identified and removed to retain only those pertinent to the model's construction [9].
- Algorithm Selection: Optimal algorithms are chosen to address the research question effectively;
- Model and Parameter Selection: Parameters are fine-tuned in alignment with the chosen algorithms;
- Training: The prepared data is employed to train the model; and
- Performance Evaluation: Finally, the trained model's efficacy is assessed using specific metrics [9].

The Centro Hospitalar Universitário Lisboa Central encopasses six hospital facilities. The study zeroes in on the medical staff assigned to twenty-two medical areas and forty specialties within the Outpatient Consultation (OC) and Operating Room (OR) sectors across CHULC. The research examines several variables, viz.

- The distribution of medical hours between OC and OR activities;
- The healthcare services provided (consultations and surgeries);
- Average durations for initial and follow-up consultations; average times for surgeries, categorized as outpatient, conventional, basic, and additional;
- The allocation of operating block times to each medical area/specialty; and
- The demand for services, as indicated by the Waiting Times for Consultations (WTC) and the Surgery Enrollment List (SEL).

4 Conclusion

By integrating the Evolutionary Intelligence (EI) system, CHULC will gain valuable insights for its middle management, including local Hospital Administrators and their respective services and departments. This will also inform the CHULC Board of Directors. Such insights will enable CHULC to align its available resources with the expected care activities, and pinpoint areas that are underutilized due to unmet needs. Identifying these areas will be crucial for restructuring and enhancing overall effectiveness in achieving the objectives outlined in a Performance Contract.

References

1. Douglas, J., Brooks, R.: The future hospital: the progressive case for change. Institute for Public Policy Research, London (2007)
2. Escoval, A.: O processo de contratualização na saúde em Portugal (1996–2005). Revista Portuguesa de Saúde Pública **9**, 7–24 (2010)
3. Buchelt, B., Frączkiewicz-Wronka, A., Dobrowolska, M.: The organizational aspect of human resource management as a determinant of the potential of Polish hospitals to manage medical professionals in Healthcare 4.0. Sustainability **12**(12), 5118 (2020). https://doi.org/10.3390/su12125118
4. Matos, T., Ferreira, C., Lourenço, A., Escoval, A.: Contratualização interna vs. contratualização externa. Revista Portuguesa de Saúde Pública **9**, 161–180 (2010)
5. Gruson, B.: L'expérience d' une structure de gestion décentralisée aux hôpitaux universitaires de Genève. Gestion Hospitalières **416**, 353–356 (2002)
6. Fathi, M., Bevrani, H.: Artificial intelligence and evolutionary algorithms-based optimization. In: Optimization in Electrical Engineering. Springer, Cham (2019). https://doi.org/10.1007/978-3-030-05309-3_7
7. Neves, J., Ribeiro, J., Pereira, P., et al.: Evolutionary intelligence in asphalt pavement modeling and quality-of-information. Prog. Artif. Intell. **1**, 119–135 (2012). https://doi.org/10.1007/s13748-011-0003-5
8. Ribeiro, J., Abelha, A., Machado, J., Marques, A., Neves, J.: The inference process with quality evaluation in healthcare environments. In: 9th International Conference on Computer and Information Science 2010 IEEE/ACIS, Yamagata, Japan, pp. 183–188 (2010). https://doi.org/10.1109/ICIS.2010.160
9. Alzubi, J., Nayyar, A., Kumar, A.: Machine learning from theory to algorithms: an overview. J. Phys: Conf. Ser. **1142**, 012012 (2018)

Language Policy in the Social Media - Efforts to Revitalize Indigenous Languages in Peru

Frederick Schruba[1], Marcelo Leon[2(✉)], Rene Faruk Garzozi-Pincay[3], and Darlys Sares[2]

[1] Ruhr Bochum University, Bochum, Germany
[2] Universidad ECOTEC, Samborondon, Ecuador
marceloleon11@hotmail.com
[3] Universidad Estatal Peninsula de Santa Elena, La Libertad, Ecuador

Abstract. The context of this work is the revitalization of indigenous languages in Peru. The main research question is what concrete measures can be observed in practice to revitalize the indigenous languages Quechua and Mochica, in which contexts they occur, and what are the main motives of the actors involved. To this end, the current social and legal situation of indigenous languages in Peru is described, and their historical development is traced. Two languages have been deliberately chosen to understand the variety of problems and challenges faced by indigenous languages in Peru, given their extreme different language situations. The current threat situation in which many indigenous languages find themselves and the consequent need for more protection and revitalization make this topic a relevant research field within linguistics.

The work is articulated in three parts. In the first part, the theoretical basics of language policy, multilingualism, and language revitalization are covered. In the second part, the language situation and historical development of language policy in Peru are explained. In the third part, exemplary revitalization aspirations for the Quechua and Mochica languages are shown.

Keywords: Indigenous · Peru · Language policy · Social media

1 Introduction

UNESCO estimates (2017) suggest that up to 50% of the approximately 6,000 existing languages could disappear by the end of the 21st century. Additionally, the Society for Threatened Languages warns that one-third of all languages will die out in the next few decades. Klein (2013, p.9) states that one language disappears every two weeks, and more than 98% of the existing languages are poorly documented, with limited knowledge about them. There has long been a consensus among linguists that the linguistic diversity of the planet is severely endangered.

Latin America, which is known for its linguistic diversity, particularly regarding language families, is no exception. In Peru, dozens of languages have died out in the last few centuries, and according to the Peruvian Ministry of Education, 21 of the 47

Á. Rocha et al. (Eds.): WorldCIST 2024, LNNS 987, pp. 238–250, 2024.
https://doi.org/10.1007/978-3-031-60221-4_24

indigenous languages still spoken are currently threatened with extinction (Ministry of Education of Peru 2013, p.57). In recent decades, there has been a growing awareness of the so-called "original languages," and more and more people and institutions are working towards protecting and utilizing these languages.

In the theoretical foundations, an overview of the concept of language policy is given, and the dimensions and actors of language policy are explained. Additionally, the mission of language policy in social media is discussed in more detail, as it forms the basis for the third part of the analysis. In connection with this, the goals and motivations of language policy are explained, and two models described by Nahir (1984) and Vikør (1994) are discussed. The role of ethnicity and nationalism in language planning is also examined, as it provides important motives for revitalizing indigenous languages in Peru.

Various aspects of multilingualism and language contact are then presented. An overview of the concept of transculturality/lingualism is provided, as it is fundamental to understanding the reality of Latin American societies. The meaning of language contacts for their conflict potential, as well as the terms of the minority language, the threat to the very existence of languages, and language revitalization, are also explained.

2 Theoretical Foundations: Language Policy, Multilingual Ability and Language Revitalization

Language planning - in its broadest sense - has a millennia-old tradition and, according to Baldau and Kaplan (1997, p. IX), probably exists since the beginning of human historiography. Already after the spread of the Roman Empire in the Mediterranean area, the authorities used various language planning measures to establish the positions of Latin and Greek as common languages and to strengthen them (ibid.). In the scientific literature, there exists a variety of different definitions, concepts, and models that deal with the concepts of language planning and language policy. First, the most important of these aspects are explained from a theoretical perspective, and the development process of this debate is illuminated. For a further understanding of this area, this overview and the demarcation of terms are fundamental.

At the beginning, the central term of language is defined more precisely, since the acknowledgment of a linguistic (en)political approach deviates from the traditional linguistic understanding (Ricento 2006, p. 5). Unlike in linguistics, the concept of language is frequently translated into linguistic (en)political discussions as a clearly definable expression. According to Blommaert (2010), this is indeed a "rough simplification," since the assumption that languages exist separately is a distortion of linguistic reality. Rather, in the most common multilingual contexts, a code-switching and mixing can be observed (cf. Marten 2016, p.16).

According to Marten (2016, p.15), the terms language planning, linguistics, and language policy are often used in parallel and synonymously, making a clear demarcation difficult to define. Historically, the terms have developed through a winding run that is briefly sketched in this text. During the war, the idea of directly influencing language in the German-speaking area was treated very carefully, as it was heavily influenced for propaganda purposes (cf. Marten 2016, p.18). Initially, the concept of language planning related mainly to the activities of linguists dealing with the standardization of orthography, grammar, and dictionaries (cf. Dovalil and Šichová 2017, p.11).

The expression language planning, as we use it today, was largely shaped in the 1950s by the Norwegian language scientist Einar Haugen. In 1959, he published his work on the planning of a modern standard Norwegian language. The four-tier language planning model is one of the most significant and still influences the scientific debate. Not only linguistic aspects but also social aspects, such as the dissemination of a language, are significant.

This distinction between linguistic and social planning objects was significantly influenced by Heinz Kloss in 1969, who established corpus and status planning of languages. In the historical process, the planning objects were expanded to include further areas. This includes, for example, prestige, language acquisition, use, and discourse planning, which are explained in more detail in Chapter 2.1.2 (cf. Marten 2016, p. 24).

Terms such as language cultivation or language care contain a rather traditional, general understanding of language policy, and relate above all to measures that affect only one language (see Marten 2016, p.19 et seq.). Frequently, the concept of language criticism and the language as a "good worth protecting" is also invoked, which actively protects against a supposed loss of language (ibid.).

Since the 1980s, the terms language policy and linguistic planning have become prevalent over the above terms, and have developed to be understood as they are today (ibid.). Non-governmental actors are also able to act on language policy (see Marten 2016, p. 23). Companies, hospitals, schools, clubs, music groups, families, or individuals can all influence language behavior in their environments (cf. Ibid.). This form of language policy can also be open - with clearly formulated intentions and goals - or covert - implicitly through language action (cf. Ibid.). While open language policy always represents a conscious action, covert language politics can also occur unconsciously. This depends on whether the actors knowingly and intentionally influence language action or if they are not even aware of it (see Shipman 1996, p. 13ff) (Fig. 1).

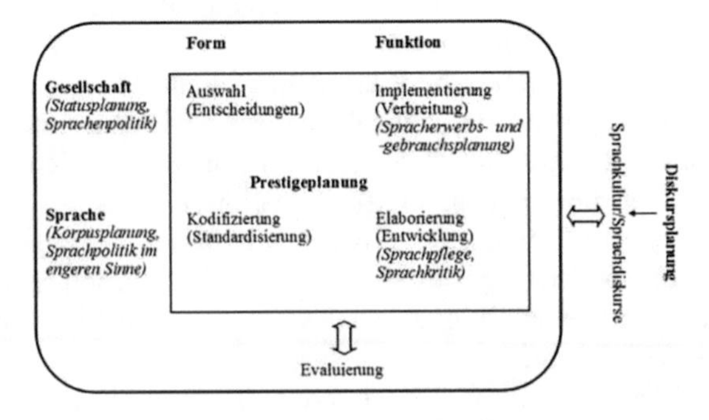

Fig. 1. Marten 2016, p.27

3 Language Policy in the (Social) Media

Media, especially social media, play an important role in language policy debates and can be a driving force in the upgrading and revitalization of minority languages. For the purposes of this work, these aspects are of great interest, and therefore, the importance of media for linguistic policy activities is being represented briefly at this point.

In democracies, the media are independent of the state, and according to Baldauf and Kaplan, they are either quasi-governmental organizations or are assigned to their respective organizations. In dictatorships, the media are controlled by the state and consequently, the government can exercise their language policy direction over the media (see Wilke 2012).

Media serve the dissemination of languages and their varieties, influence material language-political discourses, and at the same time, reflect them. Especially for the strengthening of minority languages, activists often call for "active promotion of media that use the smaller languages" (Marten 2016, p.98). Of particular importance is the question of whether the language in the media is also perceived as modern and attractive by younger generations, there by increasing the probability that the number of speakers will develop positively (ibid.). As mentioned above, this is significant in order for languages to not become "museum pieces" and to develop into modern everyday languages (ibid., p.100).

Also, new media such as the internet and mobile phones, and online networks like Facebook, Instagram, or TikTok have an enormous influence on the development of languages and the implementation of language policy activities. Since social networks are now accessible to almost everyone, people can easily find and network with others, and by uploading videos or blog posts, reach large numbers of people very quickly. In this way, campaigns for the rights of speaker groups and minority languages can spread easily. Social networks thus simplify language policy activities "from below". Gugenberger (2019, p.611) also takes the view that new media have become an "important instrument in the service of language revitalization" because they create new communicative spaces. This way, minority languages can be made visible in the digital space, and language policy activities can be coordinated and made accessible to a wide audience (cf. Ibid.).

In the next chapter, the goals and motivations of language policy must be explained to understand why people want to influence the development of languages. This is important for the analysis of language-political aspects of interpretation.

4 Language Revitalization

There are definitely voices that criticize multilingualism and see it as a handicap for international communication (cf. Marten 2016, p.65). Some critics declare multilingualism as a danger for political and social stability (see ibid.). Such voices usually come from the majority population rather than from the minority language groups.

These critical voices are countered by many arguments in favor of protecting endangered languages. In scientific and language-political discussions, measures to revitalize and protect minority languages are hardly questioned anymore (see Marten 2016, p.66). Hairman (2001, p.142) emphasizes that the protection of minorities and regional cultures is already enshrined in the 1948 Universal Declaration of Human Rights. Thus, the recognition of general human rights is also the acknowledgment of the right to the mother tongue and home culture (cf. Ibid.). He also stresses that the protection of minority languages includes components of a "general ecological awareness". Respecting biological and cultural diversity goes hand in hand (ibid. p.146). Against this background, the "preservation of minority languages, their language communities, and cultures is an integrating element in the network of global ecological awareness" (ibid.).

Marten (2016, p.66) compares efforts to protect endangered languages with efforts to "conserve biodiversity". He argues that the loss of a language "takes away a small part of mankind's cultural inheritance" and that this way, knowledge "about the facets and possibilities of human language ability" is lost (ibid.)

However, it should also not be concealed that there are also cases in which parents actively choose not to pass on their traditional languages to their children. This is often done with the prospect of better economic and social position and the desire for social advancement of their children.

The concept of revitalization has already been mentioned several times in this work and requires a closer look at this point. Language revitalization is closely related to the concept of cultural revitalization. Hirschmann (2010, p.25) understands cultural revitalization as "the process of bringing back cultural symbols and practices that were hidden for a while or were not executed." Gugenberger and Kirchmair (2020, p.11) emphasize that revitalization "does not mean the restoration of partially or entirely lost culture in its primal original 'pure form,' but rather involves processes of transformation, adaptation, and recontextualization." Thus, a revitalized culture is reinterpreted, adapted, and integrated into new contexts according to the contemporary cultural circumstances of the actors (ibid.). As already described in the concept of transculturality, traditional and modern elements can flow together, creating new hybrid forms.

5 The Revitalization of Quechua in the Music and the Social Media

Music, especially modern genres such as pop or rap, are a particularly effective way to contribute to the valorization and revitalization of indigenous languages (see Gugenberger and Kirchmair 2020, p.10). Since these genres are popular among young people, they can be used to spark their interest in using traditional indigenous languages. Artists can help transform these languages, which are often associated with tradition and backwardness, by integrating them into an urban context and combining them with modern music genres, giving them a new role and value" (Gugenberger and Kirchmair 2020, p.10).

This transformation can be observed in the work of Peruvian singer Renata Flores and rapper Liberato Kani. Flores describes this process on her Instagram profile as follows:

> How do I transform the traditional into something modern? Well, I listen to and still enjoy my traditional musical roots, and then I blend them with more modern genres such as pop, rap, and trap. By doing so, I am able to reconnect the ancestral and traditional with the new. (Flores 2021)

In addition to using the Quechua language, in her videos and on social media platforms, it also incorporates further traditional and historical symbols from Inca culture.

For example, the video for his Michael Jackson cover was filmed in front of the historical ruins of an Inca building, and is commented on as follows: 'Recorded in Vilcashuamán, an architectural jewel of the Incas in Ayacucho - Peru; with the aim of revaluing our ancestral language, 'Quechua" (Flores 2015b) (see Fig. 2).

Fig. 2. Flores 2015b, 00:34 at least

Moreover, traditional clothing is also used as a sign of cultural identity. In the video for the song "Tijeras", the singer is proudly dressed in traditional attire, which is a symbol of identity for the descendants of the Inca culture (see Fig. 3). She is accompanied by two musicians who also wear traditional clothes and play traditional instruments. Additionally, the video features a beautiful Andean mountain landscape in the background.

Fig. 3. Flores 2018b, 00:14 at least

The picture describes Renata Flores' efforts to promote and revitalize Quechua language and culture through her music and social media presence. In her works, she combines traditional musical roots with modern genres like pop, rap, and trap, creating a connection between the old and the new. Her covers of classic pop songs in Quechua, as well as her own compositions in the language, have garnered millions of views on YouTube.

In addition to showcasing traditional architecture and clothing in her music videos, Flores also uses her platform to address issues of discrimination faced by Quechua speakers. Many struggle with the Spanish language due to differences in phonetics, such as the absence of the "e" sound. Flores works to combat this discrimination and promote language teaching, demonstrating an open language policy that aims to integrate Quechua as a modern and hip language in society.

A further artist who also promotes Quechua and Quechua-speaking people is the rapper Liberato Kani, whose real name is Ricardo Flores Carrasco. He is known for his contribution to the genre of ethnorap, which involves the incorporation of modern rap music into the culture of an indigenous people (Gugenberg & Kirchmair, 2020, p.10). Unlike Renata Flores, Liberato Kani only performs his own compositions and uses both Quechua and Spanish languages. In an interview, he explains that using Spanish is a way to promote Quechua and to reach people who do not yet speak the language (cf. Flores Carrasco 2017).

6 The Revitalization of Quechua in Academic Contexts

The education sector plays a central role in the transmission and generational transfer of languages. According to the model by Soon and Chaplain (1997, p.5ff), which examines the various groups of language policy actors, educational institutions form their own group. [22] Two examples will be described here that illustrate how and in what form Quechua is taught and used at universities. The following post by the YouTube channel TVPerú Noticias from 2016 sheds light on this topic.

The video "Cada vez más jóvenes revaloran el idioma quechua" ("More and more young people value the Quechua language") discusses the increasing interest of students from the Pontificia Universidad Católica del Perú in Quechua language courses (TVPerú Noticias 2016). In the video, the linguistic teacher describes that students from various courses are participating in Quechua language courses and that there is a growing demand (see ibid. 1:27–1:53). Furthermore, the video presents teaching materials developed specifically by the university. These are three textbooks for teachers, arranged according to different skill levels and based on different language levels. According to the coordinator of the Programa de Lenguas Peruanas, the university developed these materials according to international standards (see ibid. 2:10–2:54).

As a result of this university initiative, a news program exclusively in Quechua was launched on Radio Nacional in 2016, which further increases the language's reach. People can now listen to news in Quechua on the radio from Monday to Friday (cf. Radio Nacional 2021).

7 Aspirations to the Revitalization of Mochica

The Mochica language was already the most important language in the coastal region of northern Peru between the years 100–800 AD, before the expansion of the Inca Empire (cf. Calderón et al. 2019, p.78 & Jarque 2015). Although the last native speaker of

Mochica died in the 1970s, there are numerous studies and research publications dealing with the language, including records of its grammar, phonetics, and vocabulary that have helped to preserve the language from complete extinction (cf. Serrepe Ascencio 2018). One of the oldest and most important records is the work "El Arte de la lengua Yunga" by the Spanish priest Fernando de la Carrera, which includes both grammatical and phonetic elements that are fundamental for the recovery of the Mochica language (ibid.). Another significant researcher who has contributed to the preservation of the language is the German Ernst Middendorf, who conducted various studies on the Mochica at the end of the 19th century and published a dictionary titled "Diccionario en la lengua Muchik" (ibid.). Similarly, the German researcher Heinrich Bruning also conducted investigations on the Mochica and made sound recordings of Mochica speakers, although unfortunately these recordings were lost before they could be analyzed (see Calderón et al. 2019, p.78).

In 1974, the history professor Américo Herrera Calderón took recordings of probably the last Mochica speaker - Simón Quesquén (cf. Calderón 2019, p.86). There is over an hour of sound material preserved, which was published in 2019 on the website of the Ibero-American Institute of Prussian Cultural Heritage and is available for free (Ibero-American Institute of Prussian Cultural Heritage 2019). These audio recordings of a Mochica speaker are important original documents, as they provide unique insights into the phonetics of the language and are essential for a possible revival of the language.

Although the language is officially considered extinct, it is in fact in a process of revitalization by a group of pobladores in the Lambayeque region "who are reviving the language through teaching and use" (ServirTV PERU 2017, 00:34 min). The language is being taught especially through the mediation in schools and, according to Serrepe Ascension (2018), in 2018 approximately 80 Mochica teachers taught at 38 schools in the Lambayeque region (Fig. 4).

Fig. 4. ServirTV PERU 2017, 00:42 at least

The school has been teaching the Mochica language and Isidora de Jesus since 2004 Ruiz Pena, coordinator of committee en identity Cultural on the School, told, that. First of all, the teachers who - also for them - had to learn a new language (Ser- virTV PERU 2017, 04:48–06:30 min). The initiative was initiated by the reit above mentioned anthropologist Jorge Sachún Cedeño, which directly on the school received is and at the establishment of Mochica on the school contributed Has (see. Ibid.). Next to the language courses will also the traditional crafts on passed on to the student body. For example, the school offers courses for the work with pottery or the crafts of *Mate burilado* [23] offered.

Also, the regional government in Lambayeque publicly confesses to the revalorization of the history, language, and customs of the Mochica culture (Ser-virTV PERU 2017, 06:36 - 08:38). As outlined in Chapter 2.1.5.3, the (social construction of) identity plays an important role in language-political activities aimed at revitalizing the Mochica culture. The close relationship between language and identity is evident in these activities. Additionally, the common story, exhibited as an identity feature, and the belief in a shared history also contribute to the creation of collective identities in a significant way [24].

According to Guevara Paico, the regional education manager of Lambayeque, the regional government's goal is "to recover the Mochica language not only by speaking it, but also by writing it" (ServirTV PERU 2017, 06:36 - 08:38). Therefore, the focus is primarily on improving and expanding the written form of the language, since Mochica was only transmitted orally for centuries and writing was only introduced by Spanish conquerors. Hence, the corpus work of the Mochica language is still underdeveloped, and standardization and normalization are still pending, similar to other indigenous languages. The release of the first Mochica textbook in 2011 was an important step towards this direction (see Chero Zurita et al. 2011). Adequate teaching materials are necessary for the dissemination of the language, and only through writing can the language gain official recognition. Researchers, museum directors, Luis Enrique Chero Zurita, and the Ministry of Education of Peru are involved in supporting these efforts (cf. Ibid. p.8; Andina - Peruvian News Agency 2012). Local, regional, and national institutions cooperate to convey the Mochica culture consciously and to strengthen collective identity.

The case of the Mochica language represents a unique process of language revitalization on the South American continent. Although the language has been officially classified as extinct since the 1970s, revitalization efforts have been observed since the beginning of this century. These efforts aim to recover the preserved remnants of the Mochica language and develop it as an active language through educational measures among the population in the Lambayeque region. It is clear how successful language policies can be achieved through the cooperation of different cultural and educational institutions, as well as the support of regional and national governments and the United Nations. Efforts to write down the language, such as the creation of teaching materials, have driven the standardization and normalization of the language, and have opened up the possibility for people to use Mochica in more official contexts in the future. The revitalization of the language goes hand in hand with a revival of the traditions, customs, and folk practices of the Mochica culture. This highlights the tight relationship between language and identity politics in these activities. However, the language-political activities described in this context show a clear reference to traditional themes and folklore, giving the language a museum-like character for now, as stated by Marten (2016, p.100). This indicates that, unlike Quechua, Mochica has not yet developed into a modern everyday language and has not yet penetrated all areas of everyday life, due to the still very low number of speakers and the fact that there are no native speakers of Mochica. Whether the language will acquire a more symbolic character in the future or develop into a modern everyday language depends on a variety of factors, including the appreciation and pride towards one's own cultural identity and origin. In any case, the results of these revitalization measures are exceptional. They show how coordinated efforts in language policy from above and below have led to a language being spoken again by parts of the population without active speakers, while promoting an appreciative and respectful treatment of one's own cultural heritage.

8 Discussion

Language revitalization is a frequently mentioned goal of language policy activities, particularly in multilingual states and regions where language conflicts prevail due to unequal linguistic prestige and where some languages are threatened with extinction. Language conflicts are often just a symptom of underlying social conflicts, and language policy measures, including revitalization measures, may not necessarily aim at preserving the language itself, but may be instrumentalized to achieve political, economic, or social goals. Language policy activities need not necessarily emanate from state institutions, but can be carried out by any person or institution. A central criterion is whether the measures are top-down or bottom-up. For a successful language policy, it is fundamental that top-down and bottom-up activities work together in a cooperative way.

Regarding indigenous languages, one of the most important motives for language revitalization and language policy action is identity. Identity can be understood as a social construct through which individuals distinguish themselves, feel part of a group, and through which collective identities can emerge. Language acts as a central feature in the formation of identity, and the belief in a common lineage and history also plays an important role in the construction of collective identities. Due to centuries-long economic, social, political, and cultural oppression and discrimination, language-political activities often play a significant role in meeting the legal requirements of indigenous peoples.

Under the rule of the Bourbons at the end of the 18th century, a relatively tolerant policy was replaced by a new language policy aimed at complete assimilation of the indigenous population and the eradication of all indigenous languages to maintain political control. Many indigenous languages fell victim to this repressive policy and died out.

After Peru gained independence from Spain and became a republic, little changed initially for the situation of indigenous languages, which remained associated with poverty and marginalization. To this day, indigenous languages and their speakers suffer from centuries-long oppression and discrimination. This often leads to discrimination against speakers and the shift to the dominant and more prestigious Spanish language.

In addition to language, other symbols such as traditional clothing or architecture from the Inca period are used in the analyzed videos and images, embedding them in modern contexts. This makes the transcultural activities of the artists visible and accessible to a broad audience. These examples demonstrate that tradition and modernity do not have to exclude each other, but can be shaped into something new. This can help the Quechua language develop into a modern everyday language.

The examples from the Pontifical Catholic University of Peru and the National University of San Marcos show the contributions that academic institutions can make to the revitalization of Quechua. In addition to teaching Quechua, the Pontifical Catholic University of Peru has also involved itself in the development of appropriate teaching materials, which can be used to aid language acquisition processes and contribute to the normalization and standardization of the language. Furthermore, the university helped promote the language in society through the co-development of a radio program in Quechua. In 2019, the first doctoral thesis in Quechua was completed, which promoted the prestige of the language and indicated the slow advancement of the language in academic circles, which had been locked out for centuries.

9 Conclusions

Mochica is in a completely different situation than Quechua. Although the language was officially declared extinct in the 1970s, it was documented in written and auditory forms before the death of the last speaker. In the beginning of this century, a group of researchers and residents from the Lambayeque region of northern Peru revitalized the language through school education and cultural programs. Teachers were trained to teach Mochica, and in 2011 the first Mochica textbook was published. The language is now taught in 38 schools, and initiatives by museums in the region also offer holiday workshops. Along with the language, traditional crafts and customs of the Mochica culture are also taught. The revitalization of Mochica shows that a growing number of people in Lambayeque are returning to their cultural identity of the Mochica culture. The successful language policy and revitalization efforts are based on cooperation and collaboration between different educational and cultural institutions, as well as support from regional and national governments. Moreover, this example shows the importance of documenting endangered languages, so that knowledge is not lost and future generations have the opportunity to learn these languages.

The revitalization efforts for Quechua demonstrate how the language can develop into a modern everyday language used in urban contexts and areas of life, while Mochica remains clearly tied to traditional themes and folklore. With over 10 million speakers, Quechua has become well-represented on social media. Revitalization efforts for Mochica, on the other hand, are limited to cultural and educational areas and have limited relevance to daily life. Both languages share a growing respect and pride for one's own origins, language, and culture of the ancestors, which is the most important motive for revitalization efforts. Whether Mochica will ever have an active speaker remains to be seen, as universal social use is not yet recognizable. Despite this, revitalization has high value as it demonstrates respect and appreciation for linguistic and cultural diversity, especially for indigenous populations and their languages and cultures. Language revitalization efforts should not aim to make extinct languages suitable for everyday use in all areas of life but should aim to preserve linguistic and cultural diversity in Peru. Further research could investigate the use of indigenous languages on social media and explore language revitalization efforts in other areas and for other languages. The linguistic diversity of Peru and Latin America provides almost limitless research material.

References

Aguirre Nuñez, A.: Hablando de Discriminación (2010). Accessed 27 Apr 2021. https://scc.pj.gob.pe/wps/wcm/connect/3ad7328043eb964c942ff40365e6754e/Hablando_de_discriminaci%C3%B3n_Aroldo_Aguirre_Nu%C3%B1ez.pdf?MOD=AJPERES&CACHEID=3ad7328043eb964c942ff40365e6754e

Andina – agency Peruana de noticias. En Lambayeque revivirán lengua de civilización mochica para perpetuarla (2012). Accessed 13 May 2021. https://andina.pe/agencia/noticia-en-lambayeque-reviviran-lengua-civiliza-cion-mochica-para-perpetuarla-434583.aspx

Assmann, J.: The cultural memory. Writing, memory and political identity in early civilizations Munich. Beck (2005)

Baldauf, R.B., Jr., Kaplan, R.B.: Language Planning from practice to theory. Multilingual Matters, Bristol (1997)

Bochmann, K.: Language and identity in multilingual regions eastern europe - theoretical and methodological starting position. In: Bochman, K., Dumbrava, V. (eds.) Linguistic Individualization in Multilingual Regions of Eastern Europe, pp. 13–43. Leipzig University Publishing House, Leipzig (2007)

Bush, B.: Multilingualism. Facultas, Vienna (2013)

Calderón, A.H., Ziemendorff, M., Ziemendorff, S.: Grabaciones del extinto idiom mochica. Indiana **36**(1), 77–108 (2019)

Cooper, R.L.: Language Planning and Social Change. CambridgeUniversity Press, Cambridge (1989)

Devetak, S.: Ethnicity. In: Goebl, H., et al. (eds.) Contact Linguistics/ Contact/Linguistics, pp. 203–209. Mouton en Gruyter, Berlin/New York (1996)

Dovalil, V., Šichová, K.: Language Policy, Language Planning and Language Management. Winter, Heidelberg (2017)

Fishman, J.A.: Concluding comments. In: Fishman, J.A. (eds.) Handbook of Language and Ethnic Identity, pp. 444–454. OUP, Oxford/New York (1999)

Fishman, J.A.: DO NOT Leave Your Language Alone: The Hidden status agendas Within body Planning in Language Policy. Law rence Erlbaum Associates, New York (2006)

Flores, R.: The House of the rising sun. The Animals version Quechua. [Video] Youtube (2015a). Accessed 14 May 2021. https://www.youtube.com/watch?v=eX-9Pb-QlZs

Flores, R.: The way you make me feel. Michael Jackson - Version in Quechua. [Video] Youtube (2015b). Accessed 14 May 2021. https://www.youtube.com/watch?v=BvT9y0HqItE

Flores Carrasco, R.: Liberato Kani: El Quechua Es resistencia. [Video] Youtube (2017). Accessed 16 May 2021. https://www.youtube.com/watch?v=_jXnZNK0O94

Flores Carrasco, R.: [liberatokani] "Kaynata hallarispa umayta kuyuchispa..."[Instagram post], Instagram (2021). Accessed 14 May 2021. https://www.inst-a-gram.com/p/COA7KT9Bm5W/

Fritz, S.: Afterword. Indigenous women in Peru: between marginalization and departure. In: Supa Huamán, H. (ed.) Awayu. The Quechua Woman Hilaria Supa Huamán Talks About Her Life, pp. 125–137. Fuerth: Heiko Beyer & Markus Frederick GdbR (2005)

García, C.: La lengua de los incas conquista el espacio académico (2019). Accessed 10 May 2021. https://news.un.org/es/story/2019/12/1467351

Society for Endangered Languages. Accessed 17 May 2021. http://gbs.uni-koeln.de/wordpress/

Gugenberger, E.: Identity and language conflict in a pluriethnic Company. One sociolinguistic study above Quechua speakers and -speak- cherinnen in Peru. WUV University Press, Vienna (1995)

Gugenberger, E., Sartingen, K.: Foreword. In: Gugenberger, E., Sartingen, C. (eds.) Hybridity – Transculturality – Creolization: In- innovation and Change in Culture, Language and Literature Latin America, pp. 7–9. LIT, Vienna (2011)

Henk, E.: Indigenous Languages in Peru – Status of Languages and Rights the Speaker. Johannes Herrmann Verlag, Giessen (2016)
Hirschman, B.: Del native al Maya: identity politics the maya movement in Guatemala. Investigations: researches to Latin America. LIT, Vienna (2010)
Ibero-American Institute of Prussian Cultural Heritage. Grabaciones Mochica (2019). Accessed 11 May 2021. https://publications.iai.spk-berlin.de/receive/riai_mods_00003043
López-Hurtado, L.E.: Planificación para la Revitalization Lingüística en América Latina (Dr. Luis Enrique López-Hurtado) [Video] Youtube (2016). Accessed 10 May 2021. https://www.youtube.com/watch?v=NzoHqp7LGiw
Ministerio de Cultura del Peru (2018) Lengua quechua: ISO. Lima: Ministerio de Culture. Accessed on 05/01/2021 at https://bdpi.cultura.gob.pe/sites/default/files/archivos/lenguas/Ficha%20de%20lengua%20-%2034%20Quechua_0.pdf
Ministerio de Educación del Peru. Documento Nacional de Lenguas Originals (DNLO). Lima: Ministerio en Education (2013). Accessed 01 May 2021. https://centroderecursos.cultura.pe/es/registrobibliografico/documento-nacional-de-lenguas-originarias-del-perú
Ministerio de Educación del Perú. Lenguas Originarias del Perú. Lima: Ministerio deEducación (2018). Accessed 1 May 2021. https://centroderecursos.cultura.pe/sites/default/files/rb/pdf/Lenguas%20Originarias%20del%20Peru%20%282018%29_7_MB.pdf
Mader, E.: Cultural entanglements. Hybridization and identity in Latin America. In: Borsdorf, A., Kromer, G., Parnrider, C. (eds.) Latin America in the Upheaval. Spiritual Currents in the Globalization stress, pp. 77–85. StudiaStudentenförderungs-GmbH, Innsbruck (2001)
Marten, H.F.: Linguistic(en)politics - An introduction. Fool Franke Attempt, Tubingen (2016)
Chisel, J.M.: The bilingual child. In: Bhatia, T.K., Richie, W.C. (eds.) The Manual of Bilingualism, pp. 91–113. Blackwell, Malden (2004)
Mumm, P.-A.: Language community, ethnicity, identity. In: Mumm, P.A. (ed.) Languages, People and Phantoms – Linguistic and Culture- Scientific Ethnicity Studies, pp. 1–96. De Gruyter, Berlin (2018)
Nahir, M.: Language planning goals: a classification. Lang. Prob. Lang. Plan. **8**(3), 294–327 (1984)
Rubin, J.: Evaluation and language planning. In: Rubin, J., Jernudd, B.H. (eds.) Can Language be Planned? – Sociolinguistics Theory Other Practice for Development Nation, pp. 217–252. University Press of Hawaii, Honolulu (1971)
Schiffman, H.F.: Linguistic Culture and Language Policy. Routledge, London (1996)
Spolsky, B.: Conditions for language revitalization: a compared of the cases of Hebrew other Maori. Curr. Issu. Lang. Other Soc. **2**(3), 177–201 (1995)
Thomason, S.G.: Language Contact—An Introduction. George town University Press, Washington (2001)
UNESCO. Endangered languages (2017). Accessed 15 May 2021. http://www.unesco.org/new/en/culture/themes/endangered-languages/atlas-_of-languages-in-danger/
Universal Declaration of Linguistic Rights (1998). Accessed 16 Mar 2021. https://culturalrights.net/descargas/drets_culturals389.pdf
Vikor, L.S.: Språkplan leggings. principle above practice. Novus, Oslo (1994)
Wegener, C.: Media, Appropriation and Identity - "Stars" in Everyday Life youthful fans. Publisher for social sciences, Wiesbaden (2008)
Welsch, W.: Transculturality. On the changed constitution of today cultures. In: Cutter, I., Tomson, C.W. (eds.) Hybrid Culture Door: Media, Networks, Arts, pp. 67–90. Wienand, Cologne (1997)
Wodak, R.: Linguistics analyses into language policies. In: Ricento, T. (ed.) An Introduction to Language Policy: Theory and Method, pp. 170–183. Blackwell, Malden (2006)
Zimmermann, K.: Política del lenguaje y planificación para los pueblos Amerindios: Ensayos en ecología linguistica. Ibe roamericana/Spoiled, Madrid/Frankfurt at the Main (1999)

A Graph-Based Approach for Searching and Visualizing of Resources and Concepts in Data Science

David Morales-Quezada and Janneth Chicaiza(✉)

Departamento de Ciencias de la Computación, Universidad Técnica Particular de Loja, Loja, Ecuador
{drmorales4,jachicaiza}@utpl.edu.ec

Abstract. The paper presents a resource search and visualization approach for learning data science fundamentals. We implement a web prototype that uses a knowledge graph built from the topics of the domain and a metadata repository of resources related to these topics. The application allows the graphical and interactive visualization of the graph and its semantic relationships, as well as the exploration, search, and visualization of concepts and resources. The results of the preliminary validation show its potential to improve the understanding of data science topics and promote free access to educational resources such as datasets, notebooks, and multimedia material.

Keywords: Learning data science · knowledge graphs · search engine · visualization of educational resources

1 Introduction

One of the most popular services on the Web is providing access to learning material. Through initiatives inspired by the Open Access movement, different institutions and authors share, every day, thousands of educational materials, such as books, academic articles, presentations, multimedia material, etc. For learning topics related to Data Science (DS), there are also other types of resources available online, such as notebooks, code scripts, and datasets.

In recent years, data science has emerged as an academic and professional discipline that has become one of the most exciting fields today [5] and a differentiator factor for many organizations. That is reflected in the growing demand for experts capable of processing and analyzing large volumes of data [13]. Additionally, salaries for data science professionals typically exceed those of web and JavaScript developers, and other types of professionals in the computer science field [6]. But being a new profession, organizations cannot meet the demand since companies from all sectors or fields have begun to hire data scientists as a crucial component to carry out their business strategy, as they allow companies to make better decisions, optimize their processes, and increase their potential.

Thus, considering the high demand for specialists in the field of DS and the deficit to cover this demand, in this paper we present a prototype for searching

Á. Rocha et al. (Eds.): WorldCIST 2024, LNNS 987, pp. 251–260, 2024.
https://doi.org/10.1007/978-3-031-60221-4_25

and displaying resources, such as concepts, notebooks, datasets, and videos, for learning the fundamentals of data science. Through the web application, we hope to provide free access to information and learning material to those who are taking their first steps in data science.

Therefore, the contribution of this paper can be summarized as: 1) supporting students and interested people to find material to learn topics in the domain of DS; and 2) facilitating the understanding of the topics of DS foundations, taking advantage of the structure and semantics provided by graph technologies.

In the next Sect. 2, we present the background of this research. Next, in Sect. 3, we present the methodology used in our proposal, which includes the creation of a graph of concepts about the DS domain, the creation of the resource metadata repository, and the implementation and preliminary evaluation of the search prototype. Finally, Sect. 4 contains the conclusion.

2 Background

2.1 An Introduction to Knowledge Graphs

Nowadays, due to the explosion of digital information, information management has become an increasingly challenging task [8]. The use of information retrieval methods incorporated into web search engines has allowed users to access heterogeneous and distributed information that would otherwise be impossible to discover. However, a simple search, based on keywords and textual content, suffers from some limitations, such as the inability to understand the context or meaning of the information. To overcome this obstacle, Tim Berners-Lee proposed the idea of the Semantic Web [3,4].

To define the structure of the content and describe real-world events, a graph-based model is used in the semantic web. Knowledge graphs (KG) provide a common structure and interface for connecting data through named relationships. According to [7], a knowledge graph can describe any fact-information or data about people, places, and all kinds of things-, and the way all these entities are connected. By establishing connections between data and understanding their interactions, graphs provide rich information context, helping us make decisions more accurately and reduce errors.

The semantic web provides a stack of technologies to describe, connect, and query data in a graph. The core technology used to describe and exchange data is RDF. Then, ontologies and vocabularies allow the creation of open and consensual models. These technologies are the reference for describing real-world entities and resources. Finally, SPARQL is the query language you use to retrieve RDF data and build valuable applications.

RDF is an extension of the XML markup language, and it is used to create directed and labeled graphs, where edges represent named relations between two nodes, which in turn represent entities. This view of the graph is the intuitive mental model often used in easy-to-understand visual explanations. By using a markup language with the syntax and semantics necessary to record entity properties, it significantly improves information retrieval on the web. For this reason, we adopted RDF as the basis for the creation of the graph used in this proposal.

2.2 Related Work

To identify related works, we performed a search in the scientific database Scopus. The document search was carried out using keywords such as *("knowledge graph" OR "linked data" OR "semantic web") AND (visualization OR "information retrieval" OR search) AND (education OR student)*. Then, using the Scopus search interface, we applied three inclusion criteria: 1) the range from 2017 to 2022 was established as publication year; 2) as document types, we chose *Conference Paper, Article, Book Chapter, Review and Book*; and 3) as study area, we limited it to Computer Science. Applying these criteria, 88 documents were recovered. Then, we discarded one document because it was written in the Chinese language; therefore, the literature corpus was made up of 87 documents.

From this corpus, the title and abstract of each document were scanned, with the aim of finding more specific proposals in the area. For the selection of documents, three criteria were considered: 1) as an application domain, we verified that each study is oriented towards education or students; 2) that the implemented application is for visualization, exploring or searching for learning content; and 3) that the technologies used are RDF, ontologies, or knowledge graphs. Applying these criteria in the document review, we found that 8 documents met all these requirements. Table 1 presents the main tasks performed by each selected proposal.

In our proposal, similar to [1,2,9], we use a graph structure to make it easier for students to understand the relationships between concepts. By browsing a domain-specific graph, both students and teachers can start from any specific concept and reach any other related concept, discovering associated resources and information. Furthermore, for knowledge graph creation, our proposal aims to enrich the domain graph by connecting topics with similar entities of an external graph; thus, the enrichment feature is similar to [10]. In conclusion, although there are already some works similar to ours, we believe that our proposal tries to gather the best features to create a useful application for learning a topic that is constantly growing.

3 Methodology

This section describes the methodology used to build our proposal to search for and visualize resources and concepts related to data science. Figure 3 presents the three main phases: 1) creation and enrichment of the KG for the DS domain; 2) metadata extraction from resources' repositories; and 3) implementation of the prototype.

In order to provide an overview of what was done in each of the stages, below we describe the tasks of each phase.

Table 1. Description of related work

Ref.	Performed tasks
[11]	- Creation and validation of an ontology that considers the Body of Knowledge (BOK) description model. - Publication and visualization of data related to the academic offer of a higher education center by using the description model provided by the BOK
[9]	- Construction of a knowledge graph based on the textual content of resources related to computing. - Visualization of the knowledge graph to make it easier for students to understand the relationship between keywords
[12]	- Application of supervised machine learning and rules to automate classification decisions for unstructured information. - Development of a software agent for context-sensitive searching of new knowledge as a toolkit for ontologies enrichment
[10]	- Retrieval and semantic enrichment of theses data sets stored on open educational repositories. - Implementation of a semantic web service that retrieves theses from a graph and extends the user's search query based on keywords
[14]	- Application of knowledge graph technologies in teaching circuit courses by helping teachers carry out targeted instruction and helping students to implement personalized learning. - Construction of an interdisciplinary and multimodal knowledge graph that supports the retrieval and visualization of knowledge units, the recommendation of personalized learning paths, and the analysis of intelligent teaching
[2]	- Application of named entity recognition (NER), named entity linking (NEL), and knowledge graph techniques to enable semantic exploration of a domain. - Development and evaluation of ARCA, a system based on visual search that allows the semantic exploration of concepts
[1]	- Implementation of an incremental adaptive learning model based on knowledge graphs to reduce some issues in China's online learning system. The knowledge graph is used to describe entities (concepts) and their relationships
[15]	- Modeling a study plan through linked data by using an ontology which maps different elements of the curriculum to learning objectives

3.1 Creation and Enrichment of the KG

This phase starts with the identification of terms in fundamentals in data science and the creation of the graph of topics related to this domain.

Definition of the Topics of Interest. This task aims to identify the relevant terms or topics corresponding to DS fundamentals. As a reference for the identification of terms, the contents of the syllabus of the homonymous subject taught in an engineering major of a higher education center were analyzed. Furthermore, to obtain a more complete vision of the relevant topics in the field of introduction to data science, we carried out the analysis of other syllabuses and meshes of DS courses taught by other institutions and organizations (Fig. 1).

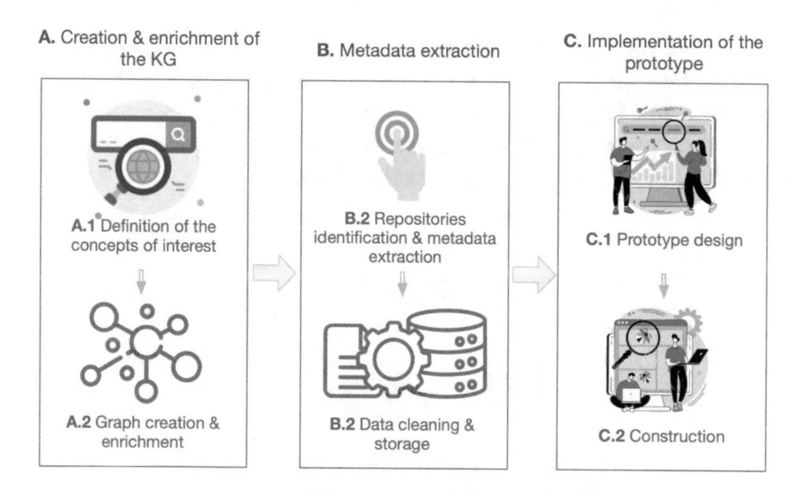

Fig. 1. Phases and activities proposed for the prototype creation.

Based on the identified terms, we filter and refine the list to leave the most appropriate terms for the vocabulary. Thus, we discard some topics or subtopics for having a very general or broad scope, such as *Data Science Concepts*; we also rename certain terms; for example, instead of *Python*, we rename to *Python language*, thus we avoid meaning issues. This step of topics definition is important because the extraction of metadata (see 3.2) of academic resources uses the terms of the vocabulary as keywords to search for related resources.

Another important aspect here is to create associations between themes and subtopics. For example:

- *Exploratory data analysis* $\Longrightarrow$ *Univariate analysis*, *Multivariate analysis*,...

This example means that two specific topics (*Univariate analysis* and *Multivariate analysis*) should be connected to the more general topic (*Exploratory data analysis*) on the graph.

Graph Creation and Enrichment. Based on the vocabulary of terms, we create a knowledge graph of related concepts to describe relevant topics in DS domain.

Using the OpenRefine tool[1], we created a project to generate the RDF triples from the list of terms. Additionally, we apply the reconciliation functionality provided by OpenRefine to find related DBpedia resources. As a result, from this last step, we create mappings between the concepts of our graph and the equivalent or related DBPedia resources. In the educational context, DBPedia[2] can be used as a source for development of innovative applications that take advantage of the richness of structured data from Wikipedia.

[1] https://openrefine.org/.
[2] https://www.dbpedia.org/.

Figure 2 presents a partial view of the graph created and stored in GraphDB Free. In this view, we can notice that the relationship *skos:broader* was used to connect topics and their sub-topics. This form of concept organization creates a structure that the application will use to make it easier for users to navigate search results and understand a set of related topics.

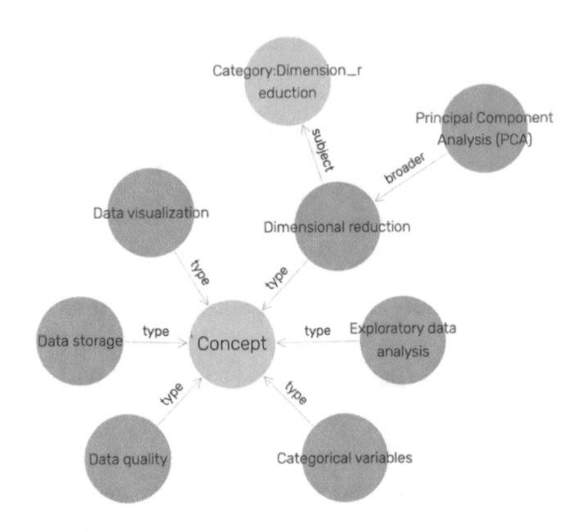

Fig. 2. A partial view of the graph for describing concepts of the DS domain

To enrich the domain graph, we use the URI of the DBPedia resources to retrieve, from the DBPedia SPARQL Endpoint, additional information on each topic, such as description and image URL.

3.2 Metadata Extraction from Resources Repositories

The second phase includes the identification of repositories for the metadata extraction of resources related to the domain topics.

Repositories Identification and Metadata Extraction. To learn data science topics, we consider three types of resources of interest: datasets, notebooks, and multimedia material. For this reason, we identified two repositories that offer this type of material and that facilitate data extraction through APIs:

- Kaggle: with access to a multitude of features and tools to help a DS project accomplish its goals, Kaggle is regarded as the greatest community in the field of data science. This platform hosts competitions and offers a ton of publicly accessible datasets for educational reasons and freely accessible notebooks for experimenting with and testing machine learning and data science models.
- YouTube: this platform allows us to find and share multimedia content, such as audio and video. Many educators, educational institutions, and organizations use this tool to share their lessons, tutorials, and other educational content for free.

Once the data sources are identified, we implement different functions for data extraction. Thanks to Python libraries such as *Requests*, we send HTTP requests and use the APIs offered by the two aforementioned platforms to search for resources and obtain their metadata.

The data extraction process is based on the use of each of the domain topics, which we send as a parameter to each extraction function. In summary, after running the process on each source, we obtained more than 32 thousand resources between datasets, kernels, notebooks, and videos.

Data Cleaning and Storage. From the data extracted in the previous step, we cleaned the data by eliminating duplicate resources and eliminating irrelevant columns (such as constants or with a large number of missing values). In addition, we applied functions to integrate and transform the data. After cleaning the data, we automated the process of storing the data in the NoSQL database, MongoDB.

3.3 Implementation of the Prototype

In this last stage, the prototype is built and a preliminary evaluation is carried out.

Prototype Design. As a reference for the search and visualization prototype, we established two basic design criteria:

- Intuitive design. The search should be based on easy-to-use inputs and filters, thus the user will be able to interact with the application and get personalized results.
- Clear presentation of relevant information. The application must take advantage of the structure of the domain graph to facilitate the exploration and learning of the topics of interest in the field of data science.

Prototype Construction. The construction of the prototype was carried out through the use of the Streamlit library, a Python library focused on creating Data Science applications and it is characterized by its ease of use and its ability to create graphical user interfaces quickly and efficiently.

Regarding the operation of the application, the process begins when the user enters a query based on keywords. From this base information, three fundamental actions occur:

1. Retrieval of related topics. The user's query is directed to the knowledge graph in GraphDB to obtain all the related topics that could be of interest to the user. For example, if the user enters the word "Regression", the application retrieves related terms, such as Linear Regression, Simple Linear Regression, Multiple Linear Regression, and Nonlinear Regression. Thanks to the use of the SPARQLWrapper library, we execute the query 1.1 accessing the SPARQL Endpoint of GraphDB. Additionally, we define another SPARQL

query to extract metadata of the topic queried by the user, such as image and description. This basic information can help users better understand the fundamentals of the topic of interest.
2. Retrieval of resources related to the user's topic from the MongoDB database, in which the metadata extracted from the repositories were stored. Metadata of resources are extracted through queries carried out with the help of the PyMongo library.
3. Through the front end, created in Streamlit, all relevant resources (both concepts and resources) related to the topic asked by the user are visually presented. The presentation of the related topics, as a graph, is done through the use of the Python Pyvis library. From the graph, the user can more deeply explore the topic of interest and have a better understanding of its relationship with other topics.

Listing 1.1. SPARQL query to get related topics

```
SELECT ?broader ?subject ?sameAs WHERE {
  ?concept rdfs:label ?label.
  OPTIONAL {?concept skos:broader ?broader}
  OPTIONAL {?concept dcterms:subject ?subject}
  OPTIONAL {?concept owl:sameAs ?sameAs}
  FILTER (LCASE(str(?label)) = LCASE(''{query_term}''))}
```

Figure 3 shows the results that the application generates after a query; In this case, the user searched for the term *Correlation*. As can be seen in the figure, the output is organized into three parts: 1) *Topic description* shows the title, image, and description of the topic, information that was obtained from the graph; 2) *Related topics* presents the graph of related topics graphically and interactively, which significantly enriches the user experience by allowing them to explore and learn about the topic of interest; and 3) *Results* shows a table containing the main metadata of the resources related to the user's topic of interest, including the link to the resource source (Kaggle or Youtube).

After construction of the prototype, code and datasets can be downloaded from a GitHub repository[3].

Through some test cases made for students of the Data Science Fundamentals matter, we were able to verify that the application meets the established design requirements and that its operation is acceptable because:

- Allows users to search for resources on specific topics related to DS fundamentals.
- Users can apply filters and advanced search options to personalize their search experience for more specific results.
- Facilitates the understanding of a topic because it provides a description of the topic searched by the user and presents related topics through an interactive visualization.

[3] https://github.com/drmorales4/data-scienceTIC.

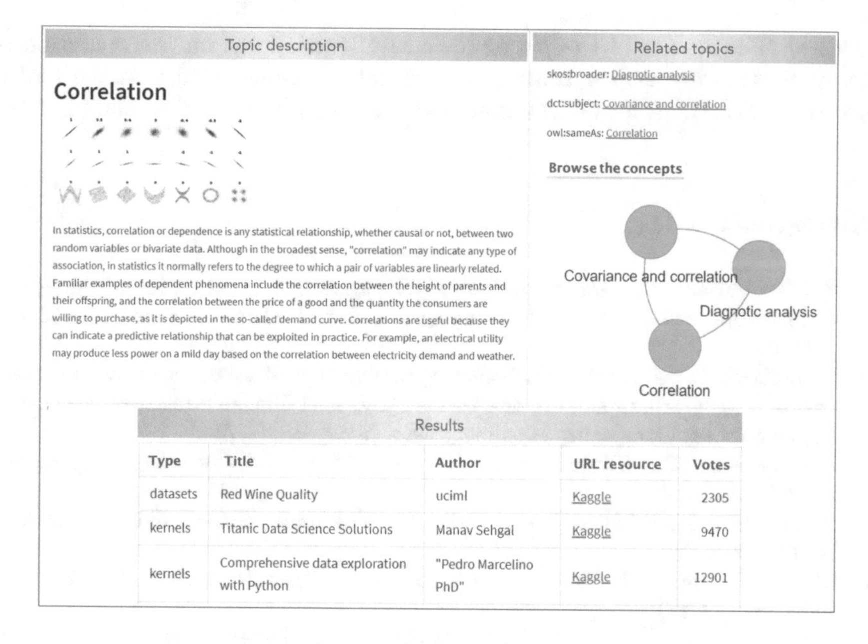

Fig. 3. Output generated by the prototype.

In summary, the results of the application's operation indicate that it is a useful tool so that students can easily find relevant resources and can improve their understanding of the topic related to the DS domain. In future work, metrics and a questionnaire will be used to evaluate the relevance of the results provided to students and to validate users' experience when interacting with the application.

4 Conclusion

The current demand for data science professionals is constantly increasing due to the increase in data production around the world. This has led organizations to seek experts in this field to help them make data-driven decisions and find solutions to business problems. Through this proposal, we hope to contribute to students and practitioners being encouraged to explore and search for resources related to this industry that is constantly growing and changing.

The methodology used in this project was useful to create a graph based on concepts of data science fundamentals. The graph structure was used for the visual representation of the relationships between the topics of the subject, which facilitates the understanding of the Data Science topics. This graph can be expanded and used as a controlled vocabulary in the DS domain.

Also, using the API of three repositories, we gathered and prepared metadata of resources for learning DS topics. This repository feeds a prototype that allows users to search for resources and concepts through an intuitive and friendly interface. The results of the preliminary tests carried out on the operation of the application are very encouraging since they demonstrated that the tool is effective in searching for educational resources relevant to topics related to the DS.

References

1. Bai, Z.: Variable incremental adaptive learning model based on knowledge graph and its application in online learning system. Int. J. Comput. Appl. **44**, 650–658 (2021)
2. Bernasconi, E., et al.: ARCA. Semantic exploration of a bookstore. In: Proceedings of the International Conference on Advanced Visual Interfaces. AVI 2020. Association for Computing Machinery, New York, USA (2020)
3. Berners-Lee, T.: Linked data (2006). https://www.w3.org/DesignIssues/LinkedData.html
4. Berners-Lee, T., Hendler, J.A., Lassila, O.: The semantic web. Sci. Am. **284**(5), 34–43 (2001)
5. Blei, D.M., Smyth, P.: Science and data science. Proc. Natl. Acad. Sci. **114**(33), 8689–8692 (2017)
6. Chen, J., Ayala, B.R., Alsmadi, D., Wang, G.: Fundamentals of data science for future data scientists (2018)
7. Hogan, A., et al.: Knowledge graphs. ACM Comput. Surv. **54**(4), 71 (2021)
8. Konstantinou, Nikolaos, Spanos, Dimitrios-Emmanuel.: Materializing the Web of Linked Data. Springer, Cham (2015). https://doi.org/10.1007/978-3-319-16074-0
9. Li, Y., Zhao, J., Yang, L., Zhang, Y.: Construction, visualization and application of knowledge graph of computer science major. In: Proceedings of the ICBDE 2019, p. 43–47. Association for Computing Machinery, New York (2019)
10. Nieto, M., et al.: Web service to retrieve and semantically enrich datasets for theses from open educational repositories. IEEE Access **8**, 171933–171944 (2020)
11. Quezada-Sarmiento, P.A., et al.: Body of knowledge model and linked data applied in development of higher education curriculum. In: Arai, Kohei, Kapoor, Supriya (eds.) CVC 2019. AISC, vol. 943, pp. 758–773. Springer, Cham (2020). https://doi.org/10.1007/978-3-030-17795-9_57
12. Telnov, V., Korovin, Y.A.: Machine learning and text analysis in the tasks of knowledge graphs refinement and enrichment. In: International Conference on Data Analytics and Management in Data Intensive Domains (2020)
13. UCM: ¿Por qué estudiar Data Science? 5 razones que harán que te decidas (2020). https://www.masterdatascienceucm.com/por-que-estudiar-data-science/
14. Xing, X., Dou, J., Xiangjun, W., Xiaolin, Y.: Knowledge graph based teaching analysis on circuit course. In: 2020 International Conference on Modern Education and Information Management (ICMEIM), pp. 767–771 (2020)
15. Zablith, F., Azad, B.: Reconciling instructors' and students' course overlap perspectives via linked data visualization. IEEE Trans. Learn. Technol. **14**(5), 680–694 (2021). https://doi.org/10.1109/TLT.2021.3118902

Choosing a Data Model for a Data Warehouse from a Non-experienced End-User Perspective

L. Botha and E. Taylor(✉)

North-West University, Potchefstroom, South Africa
estelle.taylor@nwu.ac.za

Abstract. A critical decision that will determine if the implementation of a data warehouse is a success or a failure is selecting the data modelling approach. This research was done in the Design Science Research (DSR) Paradigm. As part of the first phase (awareness of the problem), questionnaires were completed by 112 respondents at companies in South Africa. The results of these showed that data models and the use thereof still present problems in many of these companies in South Africa. The aim of this paper is to assist an end user or company to choose a suitable data model. The Data Modelling Selection Framework (DMSF) is introduced in this paper. This framework considers: business information needs, data parameters, enterprise size, business processes, current data architecture and environment, as well as data strategy. The DMSF includes 15 guidelines that can be followed by an end-user or a company when having to decide which data model will be suitable for a data warehouse.

Keywords: Data warehousing · data model · data modelling · business intelligence

1 Introduction

Data is the most fundamental piece of information that a system uses. Data models can help database designers, application programmers, and end-users communicate more effectively. A well-designed data model can even help you gain a better understanding of the data requirements and why the database is being developed. Determining which data modelling approach will suit a company's data, information and business intelligence needs best can be a complex process. Many companies have not yet adopted the correct data modelling approach, which has a direct impact on the decision making in the company. The need for information contributing to decision making within a company is the only constant in an ever-changing world of business. The aim of this paper is to assist in choosing a suitable data model for a data warehouse, especially if a user does not have an immense amount of knowledge regarding data models. The research question for this study is: What aspects or factors should be considered when selecting an appropriate data modelling approach in a data warehouse? To answer the research question, the main research objective of this study is the creation of a framework that guides the selection of a suitable and feasible data model for a data warehouse that has a direct impact on

Á. Rocha et al. (Eds.): WorldCIST 2024, LNNS 987, pp. 261–270, 2024.
https://doi.org/10.1007/978-3-031-60221-4_26

the business intelligence system. This framework should take characteristics, such as business information needs, data parameters, enterprise size, business processes, current data architecture and environment, as well as data strategy into account.

In the next section of this paper substantiation and background which led to identifying the problem are discussed, followed by a description of the methodology followed for this research, the literature review, results and discussion. A conclusion to the paper follows in the last section of this paper.

2 Problem Substantiation and Background

The importance and use of data warehouses and business intelligence in businesses today cannot be overemphasised [1, 2]. Business intelligence enables organisations to better understand internal business processes, as well as business competitors, through systematic acquisition, collation, analysis, interpretation and exploitation of information gained from data [3–5]. Business intelligence can deliver enormous benefits, but also has enormous risks [6]. Business users use the information gained from business reports as facts, and incorrect information can lead to poor business decisions [6]. Large and complex businesses are urged to make use of data warehouses to ensure that decision making in critical times are influenced by relevant data [1, 2, 7, 8]. Data quality and integrity are improved when a data warehouse is modelled correctly. This also lowers the chances of incorrect information in the business intelligence system.

However, some businesses do not fully utilise the power that a data warehouse holds if implemented correctly [7]. Data models should be used to build quality data warehouses in business intelligence systems [9]. As global data continues to proliferate, budgets and time that businesses have to spend on data model design and development are not keeping pace and businesses need support and assistance in this area [6, 10]. Companies are thus faced with difficult choices when attempting to implement a data warehouse and choosing data modelling techniques that will benefit their business intelligence the most.

A data model serves as a map to understanding and managing data in a structured environment [11]. When referring to the structured environment, it means that the environment contains a considerable volume of complex data that can hold a couple of possibilities to organise and arrange the data and the data analyst(s) can shape the data according to their needs. In this paper, the term data model is defined as a model that organises and describes data and the relationships between the data based on informational needs of a company or end-user, whilst data modelling is defined as the act of designing and developing a data model.

The problem or gap will be further substantiated and placed in the South African context, by the results of questionnaires completed by companies in South Africa (reported in the Results section of this paper).

3 Methodology

For this study, the design science research process model by [12] is used, suggesting a design science research framework as illustrated in Fig. 1. This framework includes five steps; the first step is the awareness of a problem. The main goal of this is to identify the problem and to define it. The second step is to identify a preliminary suggestion(s) or solution for the problem identified in the first step of the framework. The suggestion(s) or solution(s) is based on existing theory or knowledge or developed by using a suitable research methodology. The third step involves the actual development. When the development of the artefact is completed, the fourth step is to evaluate the artefact according to the functional specification. The fifth and last step is to terminate the project and conclude.

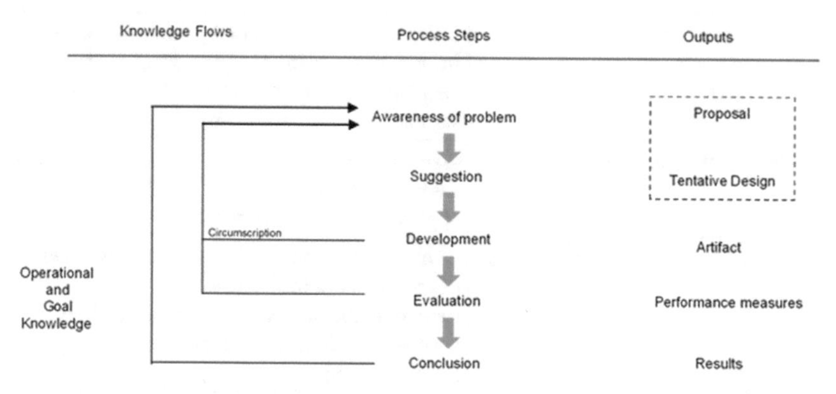

Fig. 1. Design science research framework [12]

This study includes the development of a framework. A framework can be described as an artefact and because the application domain of these artefacts is organisational systems, the researcher makes use of design science research cycles and guidelines as developed by [13]. These guidelines assist the researcher in conducting and evaluating design science research. In Table 1, the seven guidelines of design science research are given, as well as a description of each guideline according to [13]. A brief description of how the guideline is applied in this study is also given.

Table 1. Design science research guidelines according to [13] and application to this study

Guideline	Application to this study
Guideline 1: Design as an artefact	A framework is produced during this study by using design science research. This framework can be implemented in other research projects, as well as the information technology industry
Guideline 2: Problem relevance	The artefact is a proposed solution to assist a business or organisation when selecting a data modelling approach
Guideline 3: Design evaluation	The researcher uses descriptive design evaluation methods, by using information from the knowledge base to build an argument for the artefact's utility
Guideline 4: Research contributions	This study contributes to the fields of not only information system design science research, but also knowledge of business intelligence, data warehouses, data vaults, etc
Guideline 5: Research rigor	To characterize the stated and constructed artefact rigor is assessed in terms of the artefact's applicability and generalisability
Guideline 6: Design as a search process	Using feedback the design of the artefact and development of the framework is revisited to determine the most suitable data modelling approach while data parameters, rules and processes are considered
Guideline 7: Communication of research	The results of the study are communicated to technology- and management-oriented audiences through related articles

4 Results and Discussion

In this section the applicable sections of a questionnaire completed by different companies in South Africa are described, followed by the discussion of the case study and development of the framework.

4.1 Results of Questionnaires (Phase 1 of DSR, Awareness of the Problem)

Questionnaires were sent out in the form of an online questionnaire. The questionnaire had an introduction that explained the purpose of the questionnaire. Respondents were informed that their responses would be kept confidential and that taking part in the questionnaire was entirely voluntary. Businesses approached were in South Africa and included: information technology companies specialising in software development, telecommunication companies and information technology departments at banks. Approximately 450 prospective respondents were identified and contacted. A total of 112 respondents completed the questionnaire.

Almost half of the respondents had more than 4 years' experience. The results obtained regarding the months or years of experience in the information technology industry are presented in Table 2.

Table 2. Results obtained regarding experience in the information technology industry

Experience	Frequency	%
3 months – 1+ year(s)	15	13.40%
2 years to 4+ years	42	37.50%
5 years to 5+ years	33	29.46%
10+ years	22	19.64%

Respondents were asked to indicate how challenging they find choosing a data model. A 4-point Likert scale, ranging from not challenging at all to extremely challenging, was used to determine how challenging each of the items was. Table 3 presents the item addressed in the current data warehouse environment, the sample size (n), frequency, percentage, and mean associated with each of the listed items. From these results 50% of the respondents found choosing a data model challenging while 25.89% indicated that, it is extremely challenging.

Table 3. Descriptive statistics regarding challenges faced in the current data warehouse environment

Item in current data warehouse	n	Scale	Frequency	%	Mean
Choosing a data model	112	Not challenging at all	0	0.00%	3.02
		Hardly challenging	27	24.11%	
		Challenging	56	50.00%	
		Extremely challenging	29	25.89%	

Respondents were asked to state their agreement level with two statements that concern an element that could be improved within the data warehouse environment. A Likert scale was used to determine the respondents' agreement level with the statement. This scale ranged from strongly disagree to strongly agree. The results obtained are presented in Table 4 and includes a statement regarding an element that needs improvement, the sample size (n), frequency, percentage and mean associated with each of the listed items.

From these results in can be seen that users from companies in South Africa indicated that data models and the data modelling process require improvement.

Table 4. Descriptive statistics regarding elements that could be improved within the data warehouse environment

Statement	n	Scale	Frequency	Percentage	Mean
Data models used require improvement	112	Strongly disagree	0	0.00%	3.01
		Disagree	24	21.43%	
		Agree	63	56.25%	
		Strongly agree	25	22.32%	
Data modelling process requires improvement	112	Strongly disagree	0	0.00%	3.09
		Disagree	12	10.71%	
		Agree	78	69.64%	
		Strongly agree	22	19.64%	

4.2 Development of the Data Modelling Selection Framework (DMSF)

To address this need, a framework was developed based on literature studied as well as case studies done for companies. For this article one case study is discussed to explain how certain guidelines were identified during the development of the framework. For context, the case study was done based on data of a South African travel agency that provides individual components of travel packages or complete packages to future travellers. This business currently has 10–49 employees and can be described as a small enterprise. This business is continuously trying to improve experiences for its customers. The company's data set contains data from 2013 to 2021. Only one data set was provided containing data that included different package options that had been sold to travellers, accommodation options that had been selected, touring information, as well as travel insurance data. The data is currently being captured by the company's system being used daily. There is currently no database and data are stored from the company's system into flat files. The data set provided by the company is structured, text data and no-predefined data models exist for the data set. The data set is a.zip folder that is downloaded via a shared path on Google Drive. After extracting the data set from the.zip folder, the flat file is imported into Microsoft SQL Server 2019 Developer edition as a staging table. This is easy to do, as the flat file contains the headers in the first row of the flat file and Microsoft SQL Server can identify the column headings during the import. The quality of the data is measured based on definition conformance, completeness, validity or business rule conformance, accuracy to source, precision, non-duplication, derivation integrity, accessibility and timeliness.

Understanding the context and structure of the company's data allows the researcher to also assist in identifying the requirements that the company and end users have with regards to the information needed from the data. The company used in this case study only had one specific requirement, to improve the travel experience of a customer. Three of the employees of the company were involved during the development of the requirements. Two of the employees were travel agents and the third employee was a financial manager. The researcher identified the following requirements that can help the company to improve travel experiences for its customers:

- A report indicating events available for travellers that can be pulled for a specific date range and location. These events can include any restaurants, festivals, and other activities.
- A daily report that serves as a reminder to contact customers that are travelling within the coming week to ensure that everything has been organised and to answer any other questions that the customer might have.
- A report that can be pulled that contains information about customers that travelled in the last month which can then be used to follow-up with the customers on their experience and how it can be improved.
- A report that can be pulled that contains information about customers that have travelled in the last two years that can be used to do marketing of new packages available.

The researcher chose different data models based on their complexity level to implement, the first data model chosen was a relational data model and the second a hybrid data model that included the combination of a relational data model and a dimensional data model. Before the hybrid data model could be designed, a dimensional data model was chosen to be incorporated with the relational data model developed to then form the hybrid data model developed. All data models were developed by using Microsoft SQL Server 2019 Developer edition. Data was easy to import from staging tables into the relational data model with T-SQL (Transact-SQL) INSERT INTO statements. After developing the hybrid data model, the basic structure of the relational data model was used and most of the tables had already been populated with data. For this case study the implementation aspects of the hybrid design were looked at, but one of the main values of the dimensional design is that it is easy for those who are not database experts to understand. The hybrid design not only loses this ease of understanding but also produces the most complex design. While addressing the requirements identified from the data set, the researcher found that with the hybrid data model data it is easier to analyse and retrieve historical data due to the tables added to the hybrid design from the dimensional data model. T-SQL (Transact-SQL) queries were used to address the requirements set by the company.

After the design and development of the data models, employees at the company who are dependent on the data and who were responsible for setting requirements were asked to complete a questionnaire that would assist in gaining feedback regarding the data models. The data models were shared and explained to these end-users and the data that was extracted from the T-SQL queries were also shared. The questionnaire was used as an indication of whether the set requirements were met and if the end-users agreed that the proposed data model was of good quality. The level of the Capability Maturity Model integrated framework addressed during the development of the data models for the company included maturity level 3, by setting the requirements and providing a technical solution.

Based on the findings of the development of this company's data models, the following initial guidelines were identified which assisted in identifying data models to use in other case studies that the researcher did:

- A hybrid data model approach takes up more time to design and develop, especially when first designing the two or more data models before deciding on the design of the hybrid data model.
- A hybrid data model can be complex to read and understand and can make the process of retrieving data complex as well.
- A relational data model is easy to read and to understand the relationships between tables.
- A dimensional model can be easy to understand, especially when a dimensional model is presented as a star schema.

After doing more case studies the researcher also identified the following initial guidelines:

- A dimensional data model is easy to present to an audience.
- To retrieve information from a dimensional data model can be complex, especially when multidimensional information is required. A company team implementing a data model should ensure that the chosen data model can be designed, developed and implemented, based on the team's skills.
- Machine learning techniques should be investigated and implemented if any requirements regarding forecasting form part of the user expectations.
- Communication is essential during the process of data modelling and different iterations of the data model should be done. After each iteration, feedback from the end-user should be received. This will ensure that requirements and scope creep are addressed.
- When deciding to use a data vault model, ensure that enough storage is available for future changes in the data model or additional data that can be.
- The time taken to design, develop and implement a data vault model is not necessarily the same as when designing and developing a dimensional data model, as it is a more complex process due to record sources and load dates being recorded.
- A data vault model requires a greater number of joins when querying data than a dimensional data model, but by using link entities this can be resolved.
- A company might be hesitant to move to a new data model if the company already has a data model and data warehouse in place.

The researcher identified three different themes within the guidelines identified in literature and the initial guidelines identified in the case studies: project management and communication, data modelling approaches and data model qualities, and data warehouse and business intelligence elements. This was done by following six steps of thematic analysis: familiarization of the data, coding, generating initial themes, reviewing the themes, defining, and naming the themes and the last step, writing up (Caulfield, 2019). Based on the three themes, the researcher divided the guidelines and started selecting and developing a revised list of guidelines.

The researcher could also determine which of the initial guidelines were of importance to all three themes, which initial guidelines occurred in only two of the themes and which of the initial guidelines could only be found in one of the identified themes. Initially 33 different guidelines were identified. Five of the initial guidelines were associated with all three themes, while twelve of the initial guidelines were associated with two themes and sixteen of the initial guidelines were only associated with one of the themes. A couple of the initial guidelines that concern data modelling approaches and data model qualities are use case specific.

Based on the initial guidelines identified, the following revised guidelines were identified:

1. Identify the business information needs, data parameters, business processes, current data architecture and data strategy.
2. Identify and evaluate the data sources.
3. Discuss the identified business information needs, data parameters, business processes, current data architecture and data strategy with management to identify any missing information.
4. Create test data that represents the data in the data.
5. Create mock reports based on the test data.
6. Create and visualise data warehouse tables, identify possible keys and relationships.
7. Have a discussion with management and end-users and present the outputs of guidelines 4, 5 and 6. This will assist in determining any other data needs.
8. Based on the data requirements and the complexity of the data within the data source, choose two possible data models.
9. Use data modelling software and create the conceptual and logical data models for the two chosen data models.
10. Present the conceptual and logical data models to the appropriate audiences and retain feedback.
11. Based on the feedback received in guideline 10, make the necessary adjustments.
12. Choose one of the two data models.
13. Design, develop and implement the data model.
14. Identify any possible issues and maintain the data model to ensure optimisation is achieved.
15. Re-evaluate the data model in the event of a change to a business process.

The framework consists of 15 guidelines and serves as a guideline or tool for end user's or companies that are struggling to decide on a suitable data model for a data warehouse. This framework is referred to as the Data Modelling Selection Framework (DMSF). A company should keep in mind that even though some of the guidelines in the framework are repeated, quality time spent on the design and development of the data model architecture will save time at a later stage and ensure that most of the data needs have been addressed.

5 Conclusion

The aim of this paper is to assist in choosing a suitable data model for a data warehouse, especially if a user does not have an immense amount of knowledge regarding data models. In the substantiation and background section of this paper, substantiation and background literature which led to the problem identification was discussed. A comparison between different data models were done following the problem substantiation and background section. The data models compared include hierarchical, network, relational, entity-relationship, object-oriented, hybrid, dimensional, NoSQL, as well as data vaults. A framework was developed based on literature studied as well as case studies done for companies and was also discussed. During the development of the framework three different themes were identified within literature and initial guidelines identified in case studies: project management and communication, data modelling approaches and data model qualities, and data warehouse and business intelligence elements. The developed framework consists of 15 guidelines and serves as a guideline or tool for end user's or companies that are struggling to decide on a suitable data model for a data warehouse and is referred to as the Data Modelling Selection Framework (DMSF).

References

1. Makele, P., Doss, S.: A survey on data warehouse approaches for higher education institution. Int. J. Innov. Res. Appl. Sci. Eng. **1**, 223–227 (2018). https://doi.org/10.29027/IJIRASE.v1.i11.2018.223-227
2. Shahid, M.B., et al.: Application of data warehouse in real life: state-of-the-art survey from user preferences' perspective. Int. J. Adv. Comput. Sci. Appl. **7**, 415–426 (2016). https://doi.org/10.14569/IJACSA.2016.070455
3. Imhoff, C., Galemmo, N., Geiger, J.G.: Mastering Data Warehouse Design - Relational and Dimensional Techniques. Indiana Wiley Publishing Inc., Indianapolis (2003)
4. Joseph, M.V.: Significance of data warehousing and data mining in business applications. Int. J. Soft Comput. Eng. **3**, 329–333 (2013). https://www.ijsce.org/wp-content/uploads/papers/v3i1/A1391033113.pdf
5. Olszak, C.M.: Dynamic Business Intelligence and Analytical Capabilities in Organizations (2014)
6. Azeroual, O., Theel, H.: The effects of using business intelligence systems on an excellence management and decision-making process by start-up companies: a case study. Int. J. Manag. Sci. Bus. Administ. **4**, 30–40 (2018). https://arxiv.org/ftp/arxiv/papers/1901/1901.10555.pdf
7. Guerra, J., Andrews, D.: Why You Need a Data Warehouse. http://magnitude.com/wp-content/uploads/2014/01/2013-03-Why-You-Need-a-Data-Warehouse.pdf
8. Mullins, C.S.: Big Data Guidance for Relational DBAs (2016). www.dbta.com/Columns/DBA-Corner/Big-Data-Guidance-for-Relational-DBAs-113356.aspx
9. Blaha, M.: Data Models Have Many Benefits. Here Are 10 of Them (2014). https://www.dataversity.net/data-models-many-benefits-10/
10. Korada, M.: Why Organizations Can't Afford Not to Have a Data Model (2012). https://www.ibmbigdatahub.com/blog/why-organizations-can-t-afford-not-have-data-model
11. Inmon, W.H., Linstedt, D.: Data Architecture: A Primer for the Data Scientist. Elsevier Inc., Waltham (2015)
12. Vaishnavi, V.K., Kuechler, W.L.: Design Science Research Methods and Patterns. Auerbach Publications, Boca Raton (2007)
13. Bichler, M.: Design science in information systems research. Wirtschaftsinformatik **48**(2), 133–135 (2006). https://doi.org/10.1007/s11576-006-0028-8

Middle Management, Higher Education and Open Innovation: A Systematic Review of Qualitative Literature

Milton Labanda-Jaramillo[1,2(✉)], Henry Paul Saraguro Calle[1], and Luis Miguel Gutierres Camacho[1]

[1] Facultad de Educación, Universidad Nacional de Loja, Loja, Ecuador
{miltonlab,henry.saraguro,luis.gutierres}@unl.edu.ec
[2] Universidad Nacional Yacambú, Barquisimeto, Venezuela

Abstract. The growing relationships between Higher Education and Open Innovation highlight the importance of middle management, the challenges and opportunities in university education, as well as the role of open innovation in the advancement of these educational institutions. This research aims to determine the relationships between middle management, universities and open innovation through a systematic literature review of qualitative studies. A quali-quantitative research approach and the PRISMA 2020 guidelines in the four phases of Systematic Literature Reviews (identification, screening, eligibility and inclusion) are used to select relevant studies. The results, from an initial set of 177 studies, identified 48 eligible studies, of which only 8 showed a direct relationship between the three categories. Due to the scarcity of binding literature, we conclude the importance of understanding the interactions and implications of middle management in promoting open innovation in higher education. It also highlights the need for further research to explore the role of middle managers in higher education and to understand how open innovation can influence these academic institutions.

Keywords: Middle management · higher education · open innovation

1 Introduction

Middle management in higher education institutions plays a key role in promoting open innovation, an approach that seeks to foster collaboration and knowledge transfer between academia, industry and other relevant stakeholders. As the business and educational environment becomes increasingly dynamic and competitive, it is crucial to understand how middle management in higher education can drive innovation and promote a culture of open collaboration. Therefore, nowadays universities can be seen as innovation hubs that allow to articulate initiatives and entrepreneurship in society as well as to promote the development of technology-based solutions that can materialize in new products/services and business models driving the creation of new startups and spin-offs. [1].

Á. Rocha et al. (Eds.): WorldCIST 2024, LNNS 987, pp. 271–283, 2024.
https://doi.org/10.1007/978-3-031-60221-4_27

The comprehensive analysis of the existing qualitative literature in these fields will provide a solid knowledge foundation and perspectives to inform future research and guide mid-level management stakeholders, as defined by [2] as those operating in specific departmental leadership roles such as registrars, finance managers, government officials, deans, directors, and librarians; and who are responsible for making strategic decisions related to promoting open innovation in the realm of higher education.

This investigative work related to the outcomes of the doctoral thesis titled 'A Theoretical Approach to Middle Management in Higher Education Institutions from an Open Innovation Perspective' will enable a better understanding of the key factors influencing the success of implementing open innovation strategies in higher education institutions, as well as the obstacles and barriers encountered in this process [3]. In this context, the research focuses on conducting a systematic review of the existing qualitative literature, aiming to answer the question: What is the current state of literature concerning the relationship between middle management, higher education, and open innovation? Therefore, the goal is to identify and analyze the main findings, trends, and challenges related to this topic, based on previous research conducted in various contexts and disciplines.

1.1 Higher Education and Its Academic Leadership

In the current context of higher education, characterized by tumultuous changes driven by globalization, public universities are compelled to face new challenges and meet numerous emergent obligations while remaining within traditional structural and organizational frameworks. This transition from a conventional to a modern university has become increasingly difficult and disruptive. Therefore, it is imperative to adapt university management through models aligned with current and future requirements [4].

On the other hand, the digital era has introduced new dynamics in higher education, where technology and connectivity play a central role. In this regard, universities must be prepared to fully leverage the opportunities offered by this era, both in terms of teaching and learning, as well as administrative management. To achieve this, it is essential to adopt a systematic and continuous management approach, enabling agile and effective adaptation to constant environmental changes.

In conclusion, higher education is facing unprecedented challenges today. The transition towards a modern university requires university management that aligns with present and future demands. Open communication among administrators and managers, along with the implementation of a systematic and continuous management system, are key elements to ensure success in this transformative process. The digital era demands constant adaptation and efficient utilization of technological tools available in academic and administrative realms.

Relevant studies emphasize that decision-making in a university is one of the most remarkable facets of management, and to achieve success in their objectives, managers must pay special attention to the performance of key managerial functions such as planning, organizing, directing, and controlling [5]. Hence, it is crucial that mid-level academic leaders acquire an in-depth understanding of the challenges facing higher education today and in the future, as well as the best management practices. This will enable them to develop effective and adaptive strategies to foster open innovation and promote institutional growth. Moreover, possessing a strong foundation of managerial knowledge and skills will provide them with the confidence necessary to address challenges and make well-informed decisions in an ever-evolving educational environment.

1.2 Open Innovation as an Enabler of University Managerial Transformation and Innovation

Open innovation has transcended its initial technological scope and expanded its reach into new fields. It has focused on topics such as the effects of open innovation on corporate performance, including competitiveness and conquering new markets. Furthermore, it has explored the role of managerial capabilities, diversity, and intellectual capital in the context of open innovation, as well as the influence of external sources of knowledge on non-technological innovations [6].

In the realm of higher education, open innovation also drives learning approaches that embrace innovative methodologies to enhance creative education and equip students for the real world. Additionally, it empowers research management through communities of practice and innovation laboratories, creating spaces for motivation and interaction between researchers and entrepreneurs, where the protection of intellectual property plays a crucial role in ensuring development and collaboration [7].

A prominent study in this context [8] emphasizes the need to steer the institutional transformation of universities towards a global, holistic, forward-looking, and humanistic management approach. This entails adopting more open, flexible, decentralized, and horizontal frameworks, characteristic of organizations based on high-performance teamwork. Likewise, a pluridisciplinary and transdisciplinary approach is required to promote the alignment between disciplines and the fundamental functions of the university, enabling the effective addressing of the existential challenges in university education.

2 Methodology

The methodology of this research was based on a quali-quantitative approach, the main objective being to conduct a systematic review of qualitative literature related to three conceptual categories: middle management, higher education and open innovation. Scientific articles published between 2016 and 2023 were selected from journals, congresses and book chapters, extracted from seven different databases. The work was divided into two phases, one carried out in 2022 and the second in early 2023. In the systematic review, specific search strings were used for each of the seven databases used, as detailed in Table 1.

The primary selection criterion was that the articles should have a qualitative or mixed research methodology, excluding quantitative approaches, as shown in Table 2.

Table 1. Databases and search strings used in systematic review.

N°	Databases	Search strings
1	Emerald	Middle Management + Higher Education + Open Innovation
2	Jstor	1) “Middle Management” and “Higher education” and “Open Innovation” 2) “Middle Management” and Open Innovation and University
3	Scielo	(Innovación)AND(Educación)AND(Superior)
4	Proquest	(Middle Management + Higher Education + Open Innovation) AND (stype.exact(“Scholarly Journals”) AND la.exact(“SPA”) AND pd(20180101–20230127) AND PEER(yes))
5	Econstor	“Middle Management” “Higher education” “Open Innovation”
6	Iresie	Educación superior + innovacion abierta
7	La Referencia	1° “middle management” and open innovation Filtro: Año de publicación de 2018 hasta 2022 2° “Middle Management” and Open Innovation and University” Filtro: Año de publicación de 2018 hasta 2022

In the screening phase, two independent reviewers assessed compliance with the primary selection criteria during the first review phase in 2022. The inclusion criteria analyzed in the eligibility phase focused on the study population, which had to be linked to higher education institutions and address topics related to educational management, academic management, and innovation. Exclusion criteria were established so as not to include studies conducted in primary or secondary educational institutions, such as schools or high schools.

Table 2. Inclusion and Exclusion Criteria used in systematic review.

N°	Inclusion Criteria	Exclusion Criteria
1	Years 2016–2023	Studies with a quantitative methodological approach
2	Sources: Scientific journals, conference proceedings, book chapters	Studies whose population is schools
3	Languages: Spanish, English, Portuguese	Studies whose population is secondary education
4	Access: Open access and downloadable	

Five eligibility criteria (shown in Table 3) were established to determine in a weighted manner whether an article met the necessary requirements to be considered in the corresponding phase of the systematic literature review.

Table 3. Eligibility Criteria

N°	Código	Criteria description
1	CC1	Has the study proposal been clearly and adequately described?
2	CC2	Were the methods or techniques used in the primary studies clearly reported?
3	CC3	Was there an adequate description of the context in which the research was conducted?
4	CC4	Was the study proposal evaluated/validated?
5	CC5	Were the results clearly reported?

Applying the phases of a systematic literature review in conjunction with the PRISMA 2020 guidelines, through a thorough analysis we were able to develop the systematic review process according to the flow and results plotted (see Fig. 1):

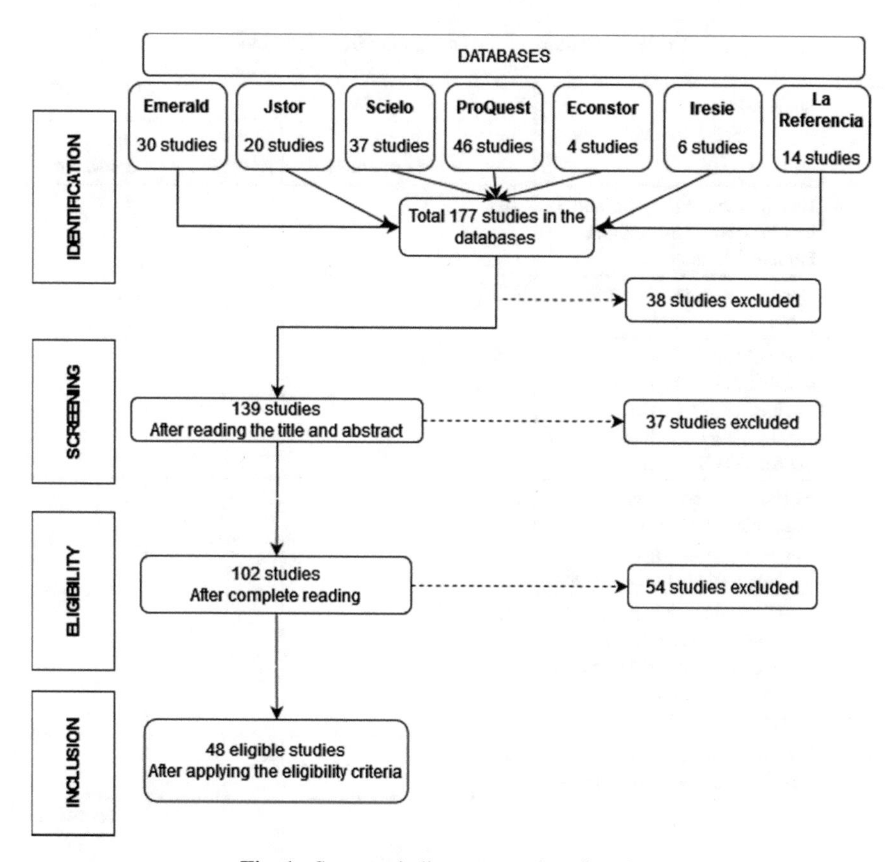

Fig. 1. Systematic literature review flowchart

3 Results

In the identification phase, a total of 177 articles were collected from the 7 databases using the search strings described in Table 1, in addition to applying the inclusion/exclusion criteria established in Table 2. After an initial review of the title and abstract in the screening phase, 38 articles considered not relevant were eliminated, resulting in a total of 139 studies. In the eligibility phase, an exhaustive reading of each of the studies was carried out, which made it possible to exclude 37 of them, leaving a total of 102 selected studies. Subsequently, the 5 quality criteria in Table 3 were applied, which resulted in the exclusion of 54 studies, leaving a total of 48 studies considered suitable for the investigation, as detailed in Table 4.

In application of the good practices of open science, finally, and with the aim of making this research reproducible, the tools used and the data generated throughout the process of the systematic literature review are made available at the following site https://cutt.ly/V8e45E3.

Table 4. Results of the systematic literature review

N°	Study title	Year	Categories		
			Middle Management	Higher Education	Open Innovation
1	User innovation rings the bell for new horizons in e-health: A bibliometric analysis	2022	X		X
2	Open Universities: The next phase	2018	X	X	
3	Innovational duality and sustainable development: Finding optima amidst socio-ecological policy trade-off in post-COVID-19 era	2022	X		X
4	Inclusive leadership and extra-role behaviors in higher education: Does organizational learning mediate the relationship	2021	X	X	X
5	Local competence building and international venture capital in low-income countries: Exploring foreign high-tech investments in Kenya's Silicon Savanna	2018	X		X
6	Recent trends of research in open and distance education in India	2020		X	X

(continued)

Table 4. (*continued*)

N°	Study title	Year	Categories		
			Middle Management	Higher Education	Open Innovation
7	Subsidiarity as secret of success: "Hidden Champion" SMEs and subsidiarity as winning HRM configuration in interdisciplinary case studies	2020	X		X
8	Has the innovative city pilot policy improved the level of urban innovation?	2021	X		X
9	Balancing between accountability and autonomy: The impact and relevance of public steering mechanisms within higher education	2021		X	
10	Can blockchain be a strategic resource for ODL?: A study	2020		X	X
11	Socialist political economy with Chinese characteristics in a new era	2020	X		X
12	Contribution of developed countries towards MOOCs: An exploration and assessment from a representative platform Coursera	2020	X	X	X
13	A framework of online-merge-offline (OMO) classroom for open education: A preliminary study	2019	X	X	X
14	Incorporating the Gender Perspective in Engineering Curricula: The Case of École Centrale Marseille	2022		X	X
15	The role of leadership in business model innovation: A case of an entrepreneurial firm from India	2019	X	X	
16	Individual innovativeness, self-efficacy and e-learning readiness of students of Yenagoa study centre, National Open University of Nigeria	2020	X	X	X

(*continued*)

Table 4. (*continued*)

N°	Study title	Year	Categories		
			Middle Management	Higher Education	Open Innovation
17	Boundaries that Matter: Workforce diversity in the STEM field in Germany	2019		X	X
18	Disruptive innovation, the episteme and technology-enhanced learning in higher education	2021		X	X
19	Arquitectura de horizontes en emprendimiento social: Innovación con tecnologías emergentes	2022		X	X
20	Tres experiencias sobre clases invertidas para promover el compromiso por el aprendizaje. Percepciones de estudiantes universitarios	2019		X	X
21	Study of the implementation of Help Desk software in an institution of higher education	2018		X	X
22	Las voces de los conocedores y conocedoras de los pueblos originarios en la formación docente	2017		X	
23	Un canal aberto no ensino superior? MOOC e REA no mundo digital	2017	X	X	X
24	Implementación de Recursos Educativos Abiertos (REA) a través del portal TEMOA (Knowledge Hub) del Tecnológico de Monterrey, México	2010	X		X
25	Proyecto oportunidad: Integrar principios y estrategias de educación abierta para la innovación docente en las universidades latinoamericanas	2016		X	X
26	Evolution of Blended Learning and its Prospects in Management Education	2022		X	X

(*continued*)

Table 4. (*continued*)

N°	Study title	Year	Categories		
			Middle Management	Higher Education	Open Innovation
27	Bridging the qualitative-quantitative divide in knowledge transfer studies: The use of QCA in the exploration of university-industry relationships	2022		X	X
28	Generation of University Spin Off Companies: Challenges from Mexico	2021		X	X
29	Artificial-Intelligence-Based Fuzzy Comprehensive Evaluation of Innovative Knowledge Management in Universities	2022		X	X
30	Artificial-Intelligence-Based Fuzzy Comprehensive Evaluation of Innovative Knowledge Management in Universities	2022		X	X
31	Construction of Incentive Mechanism for College Students' Innovation and Entrepreneurship Based on Analytic Hierarchy Process	2022		X	X
32	On the Development of a Model to Increase Innovative Capabilities by Using Knowledge Management in the Educational Field within Majmaah University: An Applied Study	2019		X	X
33	Developing knowledge creation capability: The role of big-five personality traits and transformational leadership	2019	X		X
34	Innovación abierta entre universidades y empresas locales: Condiciones, complejidades y retos	2021		X	X
35	Universidad Nacional Abierta ¿Una estampa para educar sin fronteras	2020		X	X

(*continued*)

Table 4. (*continued*)

N°	Study title	Year	Categories		
			Middle Management	Higher Education	Open Innovation
36	Análisis de datos abiertos de instituciones de educación superior colombianas como apoyo a la relación Universidad-Entorno	2020		X	X
37	Teaching profiles and their association or dissociation with education 4.0 construct elements	2020		X	X
38	La educación superior en el contexto de la innovación: Higher education in the context of innovation	2020		X	X
39	La Ciencia y Educación Abierta como movimientos articuladores de la investigación, la tecnología y la innovación	2020	X	X	X
40	Del maestro al educador profesional. Bases para la profesionalización	2022		X	
41	De la innovación docente universitaria a su transferencia a la escuela: Una experiencia educativa desde la perspectiva de género	2022		X	X
42	Factores de éxito para los programas tecnológicos en el área de negocios en la modalidad b-learning para el contexto de la educación pública	2022		X	X
43	Las investigadoras en el Sistema Nacional de Investigadores: Tan iguales y tan diferentes	2022		X	
44	La colaboración Universidad-Empresa como elemento motivador. Un estudio de caso	2019	X	X	X
45	Competencias y educación para los trabajos y desafíos del mañana: La perspectiva de una empresa	2019	X	X	X

(*continued*)

Table 4. (*continued*)

N°	Study title	Year	Categories		
			Middle Management	Higher Education	Open Innovation
46	Evaluación del Programa Doctoral No Escolarizado en el CEPIES como una Innovación Educativa	2019		X	X
47	Beneficios y oportunidades de mejoramiento para la internacionalización de empresas comercializadoras de lulo en Colombia	2020	X		X
48	Los Estilos de Apego, las Redes Personales y la Satisfacción con la Carrera Profesional del Personal Directivo de Quebec, Canadá	2019	X		X

The resulting studies address aspects such as leadership, innovation, and decision-making in educational institutions. For instance, the article [10] underscores the importance of subsidiarity as a successful configuration of human resources in SMEs, while the study [11] analyzes public management mechanisms and their impact on autonomy and accountability in higher education.

The findings also reveal articles focusing on higher education that explore topics such as organizational learning, inclusion, student engagement, and classroom technology implementation. Likewise, Aboramadan et al. investigate the relationship between inclusive leadership and extra-role behaviors in higher education in [12], and [13] examines the innovative duality and sustainable development in the post-COVID-19 era.

Regarding open innovation, the results of the SLR highlight the interest in this approach within the field of higher education. Several studies investigate the implementation of open educational resources (OER), the use of emerging technologies, and the management of innovative knowledge in universities, specifically. For example, [14] explores the contribution of developed countries through massive open online learning platforms (MOOCs), while [15] analyzes the role of MOOCs and OER in the digital domain.

In summary, the research findings demonstrate the intersections of these categories, enabling the identification of a consolidated state of the art in Open Innovation in Higher Education, quantified in 24 interrelated studies, as depicted in Fig. 2. Secondly, and no less importantly, the 10 studies on Open Innovation in Middle Management denote a growing field of study. Thirdly, there is a virtually unexplored space with only 2 studies on Middle Management in Higher Education.

Finally, the interrelationship of the three categories, supported by a total of 8 studies, reveals a remarkable opportunity that could be an outstanding indication of a future significant body of scientific literature in an evolving area of research.

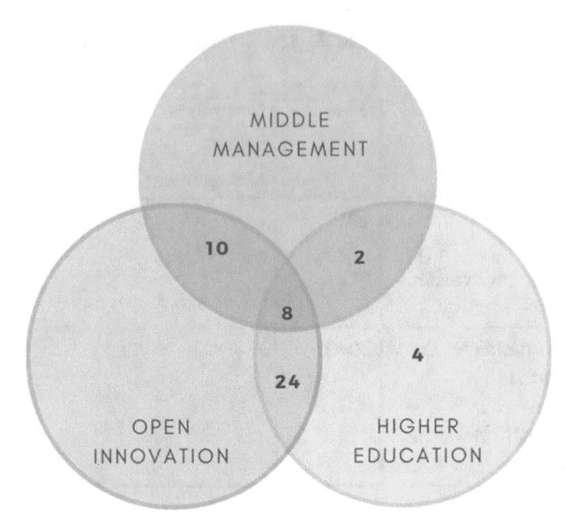

Fig. 2. Quantitative relationship of the categories and the results of the SLR.

4 Conclusions

The current state of scientific literature regarding the existing relationship among middle management, higher education, and open innovation is comprised of 48 valid studies. Notably, the intersection of these three categories constitutes an emerging and under-explored area, with only 8 studies directly addressing them. This underscores the significance of comprehending the interactions and implications of middle management in promoting open innovation within higher education.

The connection between higher education and open innovation stands out as a widely addressed and established theme, evident in the 24 analyzed studies encompassing both categories. This unequivocally demonstrates that open innovation is a topic of great interest and relevance within the framework of higher education. Universities actively seek external collaborations and adopt knowledge from external sources to drive internal innovation and enhance competitiveness in an ever-changing environment.

These findings reveal a substantial research gap within the existing literature regarding the diverse interrelationships and combinations among the three selected categories for this systematic literature review. This highlights the need for further research to explore the role of middle managers in the realm of higher education and to understand how open innovation can influence these academic institutions.

References

1. Miranda, J., Rosas-Fernández, J.B., Molina, A.: Collaborative networking to enable innovation and entrepreneurship through open innovation hubs: the entrepreneurship learning centre of Mexico city. In: Camarinha-Matos, L.M., Afsarmanesh, H., Ortiz, A. (eds.) PRO-VE 2020. IAICT, vol. 598, pp. 311–323. Springer, Cham (2020). https://doi.org/10.1007/978-3-030-62412-5_26
2. Mande, W.M., Nambatya, A.K., Nsereko, D.N.: Contribution of middle management to enhancement of quality education in Ugandan universities. In: SSRN, p. 36 (2016). https://ssrn.com/abstract=2721387

3. Rosalía, O.C., Serpa, R., Morell, R.: Modelo teórico–metodológico para perfeccionar la gestión educativa universitaria semipresencial en UNIANDES- Babahoyo. Uniandes Episteme **6**(1), 14 (2019). http://45.238.216.13/ojs/index.php/EPISTEME/article/view/1252/554
4. Aguilar, F.V.: La necesidad de un nuevo modelo de gestión para las Universidades Públicas2. Anales (2015). https://dspace.ucuenca.edu.ec/bitstream/123456789/22929/1/4.pdf
5. Sirilak, S., Wannasri, J.: The management model for Higher Education Institutions in the digital era. J. Educ. Innov. **25**(1), 46–52 (2023)
6. Valdez Reyes, G.: Management and leadership performance of academic middle managers and the attainment of their trilogic functions: an input to an enhancement program. Asia Pac. J. Multidisc. Res. **7**(3), 2019 (2022). https://www.academia.edu/download/62409060/APJMR-2019.7.03.0120200319-81496-qmaus3.pdf
7. Mignon, S., Ayerbe, C., Dubouloz, S., Robert, M., West, J.: Managerial innovation and management of open innovation. J. Innov. Econ. Manag. **32**(2), 3–12 (2020)
8. Chamba-Eras, L., Labanda-Jaramillo, M.: Contextualización preliminar de la Gerencia Intermedia Universitaria desde un enfoque de Innovación Abierta: Caso Universidad Nacional de Loja. In: 2022 17th Iberian Conference on Information Systems and Technologies (CISTI) (2022)
9. Niño, L., Piñero, M.L.: Significados sociales de la gerencia universitaria en un contexto de transformación institucional. Revista Chakiñán **1**, 101–117 (2016). https://chakinan.unach.edu.ec/index.php/chakinan/article/view/15/12
10. Mear, F., Werner, R.A.: Subsidiarity as secret of success: 'Hidden Champion' SMEs and subsidiarity as winning HRM configuration in interdisciplinary case studies. Empl. Relat. **43**(2), 524–554 (2020)
11. Kallio, T.J., Kallio, K.-M., Huusko, M., Pyykkö, R., Kivistö, J.: Balancing between accountability and autonomy: the impact and relevance of public steering mechanisms within higher education. J. Public Budg. Account. Financ. Manag. **34**(6), 46–68 (2022)
12. Aboramadan, M., Dahleez, K.A., Farao, C.: Inclusive leadership and extra-role behaviors in higher education: does organizational learning mediate the relationship? Int. J. Educ. Manag. **36**(4), 397–418 (2022)
13. Sinha, A., Adhikari, A., Jha, A.K.: Innovational duality and sustainable development: finding optima amidst socio-ecological policy trade-off in post-COVID-19 era. J. Enterp. Inf. Manag. **35**(1), 295–320 (2022)
14. Ayoub, A., Amin, R., Wani, Z.A.: Contribution of developed countries towards MOOCs: an exploration and assessment from a representative platform Coursera. Asian Assoc. Open Univ. J. ()2020
15. Mallmann, E.M., Ferreira Nobre, A.M.: Um canal aberto no ensino superior? MOOC e REA no mundo digital. Apertura **9**(2), 24–41 (2017)

Sustainable Crowdfunding: Value-Adding or Greenwashing?

Nuno Rosário[1], Nuno Melão[2(✉)], and João Reis[3,4]

[1] Instituto Politécnico de Viseu, Campus Politécnico, 3504-510 Viseu, Portugal
[2] CISeD - Research Centre in Digital Services, Instituto Politécnico de Viseu, Campus Politécnico, 3504-510 Viseu, Portugal
nmelao@estgv.ipv.ptn
[3] Center for Research, Development and Innovation (CINAMIL), Portuguese Military Academy, Rua Gomes de Freire 203, 1169-203 Lisbon, Portugal
[4] Industrial Engineering and Management, Faculty of Engineering, Lusofona University and EIGeS, Campo Grande, 1749-024 Lisbon, Portugal

Abstract. Sustainable crowdfunding is becoming increasingly popular among entrepreneurs, but there is a lack of research about the economic, environmental, and social impacts of the projects financed via this source. This paper aims to examine the post-funding phase of sustainability-oriented projects and determine whether they provide the actual sustainability benefits or engage in greenwashing practices. After developing a framework to evaluate the sustainability level of crowdfunding projects, a content analysis is conducted on a sample of 50 projects financed by six European sustainability-oriented crowdfunding platforms. The results indicate that the sample projects have positive impacts on sustainability, but they rarely create value across the entire triple bottom line of economic, environmental, and social impacts. Overall, this study contributes to the literature that seeks to uncover the post-financing outcomes of sustainable projects funded by crowdfunding platforms.

Keywords: Digital platforms · crowdfunding · sustainability

1 Introduction

Recent advances in digital technologies have transformed how financial products and services are produced, delivered, and consumed. In this context, crowdfunding has emerged as a growing financing approach. Crowdfunding is a source of alternative finance for individuals or organizations to fund their projects or ventures, usually through online platforms, from many people, to which each individual typically contributes a small amount [3]. It is believed that crowdfunding enables the democratization of access to capital and promotes innovation and entrepreneurship [14, 26].

Given the urgency of addressing global challenges, such as climate change, hunger, and poverty, sustainable crowdfunding is a particular form of crowdfunding that has received significant interest. It refers to using the "crowd" to fund projects that create environmental and social value, along with economic impacts [4]. Prior research

Á. Rocha et al. (Eds.): WorldCIST 2024, LNNS 987, pp. 284–296, 2024.
https://doi.org/10.1007/978-3-031-60221-4_28

highlights that sustainable crowdfunding has the potential to facilitate sustainable development since it can finance projects that contribute to achieving the sustainable development goals (SDGs) [23]. It also emphasizes that sustainable crowdfunding presents challenges, such as the need for project initiators to show the environmental and social impacts and avoid greenwashing issues [27]. Greenwashing refers to misleading claims about the environmental or social performance of an organization or project [17]. However, most existing literature on this topic has focused on the phase leading up to the funding campaign (pre-financing). A few exceptions are [12] and [5], who analyzed the post-financing phase, i.e., when project initiators are expected to implement their projects and deliver results. Still, there is little empirical evidence on the actual impacts of sustainable crowdfunding projects.

This paper examines the post-financing of sustainable crowdfunding projects and determines whether they deliver sustainability impacts or lead to greenwashing. To achieve this, we develop a multi-stage, multi-dimensional framework to evaluate the sustainability level of crowdfunding projects in the post-financing stage and apply it to a sample of 50 projects funded by six European crowdfunding platforms. This paper contributes to the theory and practice of sustainable crowdfunding by providing a framework to evaluate the sustainability level of crowdfunding projects in post-financing, as well as empirical evidence on the post-financing outcomes of sustainable projects funded by crowdfunding platforms.

The remainder of this paper is structured as follows. Section 2 presents background concepts and reviews relevant literature. The following section describes the research methodology and explains the sustainability assessment framework. Section 4 presents and discusses the results of applying the framework to 50 projects. The final section concludes the paper and provides implications and suggestions for future research.

2 Background and Related Work

This section defines crowdfunding, explains why crowdfunding is important for sustainable development, and identifies previous literature conducted in this field.

2.1 Crowdfunding

Crowdfunding refers to a group of people who come together to finance a particular project, regardless of its nature (entrepreneurial, social, philanthropic, or other), or conversely, when someone with an idea or project seeks funding from various individuals [3, 20]. The Internet plays a crucial role in crowdfunding in two ways [2]: by reducing transaction costs, the Internet serves as a means to capture small amounts of capital from large groups of investors (the "crowd" effect as a catalyst); the Internet enables a direct connection between backers and fundraisers, eliminating the need for intermediaries. Internet-based platforms have played a crucial role in the globalization of funding sources, thus contributing to the democratization of entrepreneurship and access to business financing; besides, the crowd can be invited to evaluate the project or provide feedback on the product to be produced [20, 24]. The focus of this paper is on loan-based crowdfunding or crowdlending.

From an operational standpoint, the crowdfunding process typically involves the participation of three stakeholders: the project promoter seeking to raise funds, the platform acting as an intermediary, and the backers who invest in specific projects. Any individual or organization needing financial resources to develop a new product or service can initiate a fundraising campaign through one of the many available crowdfunding platforms [3, 20].

A crowdfunding project unfolds through two stages [12]: pre-financing and post-financing. The pre-financing stage encompasses the period leading up to the completion of financing on the crowdfunding platform (including campaign preparation, communication, marketing, and the actual funding period). The post-financing stage begins after the crowdfunding campaign ends and corresponds to the period in which project initiators must communicate successes, failures, and results achieved to the backers who expect to receive the promised returns. Implementing the previously announced actions is equally important in this phase.

2.2 Sustainability and Crowdfunding

Corporate sustainability refers to a company's ability to manage its activities to create long-term value while providing social and environmental benefits to its stakeholders [18]. It incorporates three dimensions: economic, social, and environmental sustainability. Economic sustainability focuses on finding ways of development based on sustainable economic growth with reduced pressure on the environment. The social dimension emphasizes the importance of creating a society where all individuals have equal opportunities and access to resources, ensuring that development benefits everyone and does not marginalize certain groups. The environmental dimension emphasizes the need to protect and preserve the natural environment, acknowledging its finite capacity and the importance of sustainable practices to maintain its integrity for future generations.

The United Nations (UN) included in its 2030 Agenda for Sustainable Development [23] the use of digital technologies to fund projects aligned with the SDGs. One of its groups of experts identifies crowdfunding as a relevant form of financing for SDGs [16]. Therefore, comparing the actual contributions of successfully funded projects through crowdfunding with global sustainability frameworks, such as the SDGs, becomes particularly interesting.

Some authors, such as [6], suggest that crowdfunding offers greater benefits to sustainability-focused initiatives compared to traditional or commercial ones. The authors investigated the effect of social and environmental sustainability orientation in new crowdfunding initiatives and found that a sustainability orientation positively impacts the financing success of crowdfunding projects. Moreover, it enables entrepreneurs to raise capital from other sources.

Within the domain of crowdfunding and its impact on sustainability-focused initiatives, it is essential to recognize the key role of entrepreneurship. Entrepreneurs with a sustainability-oriented vision often turn to crowdfunding platforms not only for financial support but also to engage a community of like-minded backers [5]. This synergy between entrepreneurship and crowdfunding reflects the growing importance of sustainable entrepreneurship ecosystems. These ecosystems offer a favorable environment for sustainable ventures, offering access to resources, mentorship, and collaborative opportunities [10, 25]. In addition, they serve as a bridge between the financial support gained through crowdfunding campaigns and the broader context of sustainable development [22]. In this context, successful crowdfunding initiatives not only raise capital but also contribute to corporate sustainability.

2.3 Sustainable Crowdfunding: Value Creation or Greenwashing?

The contemporary nature of crowdfunding and its progress through collaborative financing platforms has led to a significant number of studies in this area. Research in crowdfunding has delved into various aspects, including general concepts, motivations and typologies (e.g., [13]), models (e.g., [8]), platforms (e.g., [4]), success factors (e.g., [15]), and outcomes (e.g., [7]).

In recent years, more focused research has explored niche areas of crowdfunding, particularly investments in sustainable development following the triple-bottom-line approach. However, there remains a shortage of studies in this area, especially those addressing the post-financing stage [27]. There are a few exceptions, though. For example, [12] examines the post-financing phase for 52 sustainable projects and finds that the announced measures are generally implemented, but only a few projects disclose the actual environmental impacts. [5] find that more than 70% of sustainable ventures survived at least one year of operations after the crowdfunding campaign, concluding that crowdfunding is an adequate funding source for such ventures.

Evaluating the success of sustainable projects financed via crowdfunding goes beyond merely raising the necessary funds. Particular attention should be given to the information conveyed to crowdfunding platforms and backers in post-financing [21]. Several authors [8, 12] emphasize the importance of disclosing whether the effects announced during the campaign are actually realized in post-financing, confirming the allocation of funds to the previously announced cause. Transparent and accurate information about project development is crucial to avoid potential greenwashing practices. Consequently, disclosing whether the effects reported during the campaign are actually realized in post-financing is an essential step in assessing the contribution of crowdfunding to sustainable development, avoiding the emergence of greenwashing phenomena.

The main objective of this research is to analyze the post-financing of projects announced as having a sustainable impact, relating them to the triple bottom line, and determining whether crowdfunding genuinely supports sustainable projects or leads to greenwashing practices.

3 Methodology

The unit of analysis of this research was the crowdfunding project (CP). To better understand the phenomenon being studied, several methods were used embedded in a qualitative research design. The investigation started with two exploratory semi-structured interviews with representatives from two European crowdfunding platforms. They served the dual purpose of understanding the motivations of these platforms for targeting this market niche and providing insights into the path leading up to the selection of projects offered on the market, the communication provided to investors, the funding process, and the subsequent monitoring.

The next step consisted of contacting and registering with several European crowdfunding platforms for several purposes: 1) to understand the communication processes between the platform and investors in post-financing; 2) to know how the platforms controlled and monitored project execution in post-financing; 3) to grasp the nature and extent of information provided in post-financing about the projects. The platforms made an apparent effort to monitor the projects and provide investors with as much information as possible. For example, investors could contact the project initiators, read project news, and access periodical reports comparing the predetermined standards with what is executed.

An evaluation framework was needed to ascertain whether sustainable projects added value or were a phenomenon of greenwashing. Therefore, the third step involved developing a framework to assess the sustainability level of a project after the crowdfunding campaign. The framework is described in more detail in Sect. 2.1 and builds on existing literature and the SGDs.

The next step involved conducting a pilot study to test the framework and plan data collection. A sample of 10 projects meeting the following criteria was chosen: 1) projects that were fully funded through leading European, loan-based crowdfunding platforms dedicated exclusively to sustainable projects; 2) availability of adequate information on both the crowdfunding platforms and the project initiators websites to facilitate the application of the framework and subsequent classification of each project. Two projects were funded by the Portuguese platform GoParity, two were funded by the French platform Lendosphere, two projects were funded by the German platform Econeers, one project was funded by the Italian platform Ener2crowd, and one project was funded by the Spanish platform La Bolsa Social. Some minor adjustments and improvements were incorporated into the framework to enhance its objectivity and applicability to a broader sample.

The framework was subsequently applied to a final convenience sample of 50 projects from six crowdfunding platforms meeting the abovementioned criteria between October 1 and November 30, 2022. The data corpus contained a wide range of text types, including newsletters, management reports, business plans, corporate and project videos, investor information sheets, proposals for impact measures, among others. The primary data source came from the crowdfunding platforms, while the project initiators' website was used as a complementary source. The analysis of data was carried out using content analysis. Through this technique, the researchers examined the data to identify the presence, meanings and relationships of the ideas and concepts within the previously developed framework, and drew inferences based on their findings [19].

3.1 Evaluation Framework

The sustainability evaluation framework employs a multi-stage, multi-dimensional approach to assess sustainability comprehensively. The initial step involved conducting a literature search on the Web of Science database for articles using the term "sustainability assessment framework" or "sustainability assessment model" or "sustainability evaluation framework" or "sustainability evaluation model" in the topic field. The analysis of the results revealed that no existing framework could be directly applied to address the research question; therefore, a new one was developed. Three references were selected as the basis for developing this framework, drawing from the SDGs, sustainability dimensions, and sustainable business ecosystems.

The evaluation framework has five stages. The first stage is based on the work of [11], and it seeks to exclude projects that undermine the SDGs, as shown in Table 1.

Table 1. Stage 1 of the framework (adapted from [11]).

SDG	The Crowdfunding Project harms SDG?	Yes/No/Neutral?
1. No poverty	The CP supports poverty	
2. Zero hunger	The CP The FI supports hunger, deteriorates food security, nutrition, and sustainable agriculture	
3. Good health and well-being	The CP is detrimental and deteriorates well-being	
4. Quality education	The CP has a negative impact on equitable quality education and interferes lifelong learning opportunities	
5. Gender equality	The CP promotes gender inequality and interferes with female empowerment	
6. Clean water and sanitation	The CP deteriorates the availability and sustainable management of water and sanitation	
7. Affordable and clean energy	The CP interferes access to affordable, reliable, sustainable and modern energy	
8. Decent work and economic growth	The CP interferes sustained, inclusive and sustainable economic growth, full and productive employment and decent work for all	
9. Industry, innovation and infrastructure	The CP impedes resilient infrastructure, inclusive and sustainable industrialization and innovation	
10. Reduced inequalities	The CP promotes inequality within and among countries	
11. Sustainable cities and communities	The CP interferes inclusiveness, safety, resilience and sustainability of cities and human settlements	
12. Responsible consumption and production	The CP deteriorates sustainable consumption and production patterns	
13. Climate action	The CP promotes climate change and its impacts	
14. Life below water	The CP impedes a gently and sustainable use of the oceans, seas and marine resources or sustainable development	
15. Life on land	The CP damages the terrestrial ecosystem and fosters deforestation, desertification, land degradation and biodiversity loss	
16. Peace, justice and strong institutions	The CP impedes the development of peaceful and inclusive societies, deteriorates access to justice and interferes effective, accountable and inclusive institutions	
17. Partnerships for the goals	The CP weakens the means of implementation and revitalizing the Global Partnership for sustainable development	
	Crowdfunding as a means of financing harms sustainability?	

If the SDGs are not infringed, the project advances to the second stage, which allows to assess it in terms of the 17 SDGs. Subsequently, the SDGs were grouped into three dimensions [9]: economic (affordable and clean energy; decent work and economic growth; industry, innovation, and infrastructure; sustainable cities and communities; responsible consumption and production); social (no poverty; zero hunger; good health and well-being; quality education; gender equality; reduced inequalities; peace, justice and strong institutions; partnerships for the goals); environmental (clean water and sanitation; climate action; life below water; life on land).

The third stage evaluates the project according to 13 aspects of sustainability identified by [11] (Table 2). These aspects were derived from the SDGs, but certain redundant elements of the SDGs were eliminated to maintain the manageability of the assessment [11]. Generally, the more aspects identified, the greater the project's sustainability orientation.

After confirming that the project does not violate any of the SDGs and that it covers various dimensions and aspects of sustainability, the fourth stage aims to assess whether a project has the potential to be classified as a sustainable business ecosystem. This stage is based on the work of [10], who emphasize that new businesses increasingly need to align a dual orientation, one directed towards sustainability and entrepreneurship, i.e., seeking business opportunities that contribute to the ecosystem in a sustainable manner. Such a dual orientation is pursued when four conditions are met: 1) the sustainability orientation of actors, 2) the recognition of sustainable opportunities and resource mobilization, 3) collaborative innovation of sustainability opportunities, and 4) markets for sustainable products. The favorable presence of these conditions creates a favorable environment for sustainable entrepreneurs to interact and engage in entrepreneurial experimentation, focusing on sustainable solutions, products, and innovations.

Table 2. Stage 3 of the framework (adapted from [11]).

1	Wealth	
2	Health	
3	Education	
4	Inclusion	
5	Equality (incl. justice)	
6	Working opportunities	
7	Sustained economic growth	
8	Resilient infrastructure	
9	Foster innovation	
10	Combat climate change	
11	Sustainable use of resources (incl. conserving natural resources and sustainable energy supply)	
12	Sustainable industry patterns (incl. sustainable agriculture, sustainable economic growth and sustainable industry)	
13	Sustainable living (incl. sustainable cities and sustainable consumption)	
		0

The final stage assesses the project's level of sustainability using an adapted version of [1]'s framework. According to this framework, a fully sustainable organization integrates sustainability into its corporate strategy and communicates it both internally and externally; its organizational focus extends beyond the organization itself and includes the entire supply chain; it recognizes the inspiration sustainability can bring to innovation and entrepreneurial efforts; a sustainable organization takes an active role in shaping sustainability regulations and adheres to them, emphasizing the crucial balance among the economic, environmental, and social elements. Based on these aspects, the framework (Fig. 1) has five dimensions, each with an equal weight, and four levels of sustainability, ranging from 'reduced' (1 point) to 'very good' (4 points). The final project classification is obtained by weighting the five indicators by their respective points.

Table 3. Summary of the projects by platform (n = 50).

Crowdfunding platform	No. of projects	Capital raised			Number of backers		
		Average	Min	Max	Average	Min	Max
GoParity	11	134000	9000	300000	821,82	246	1441
La Bolsa Social	4	222500	100000	355000	97	72	145
Lendosphere	8	1500537,5	50000	3900000	794,25	41	1647
Bettervest	12	416487,5	83850	913050	391,25	122	739
Econeers	11	795560	305750	1598660	481,55	213	789
Ener2crowd	4	185943	130071	278261	*	*	*
Total	50	577221,64	9000	3900000			

*not reported by the platform

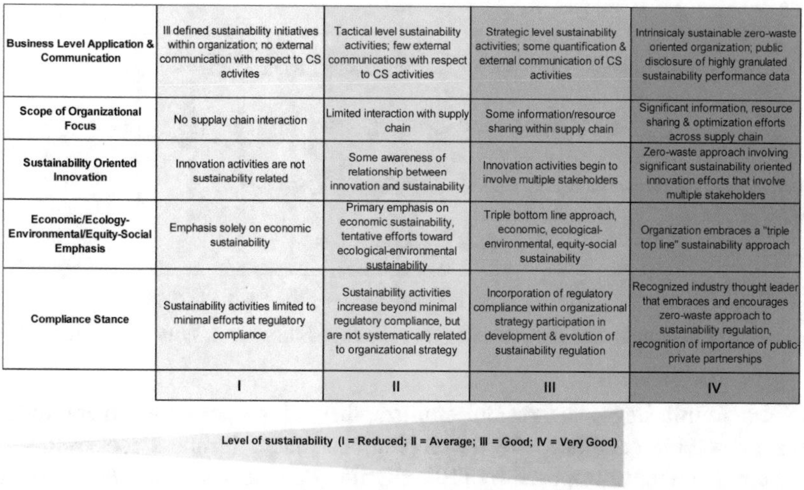

	I	II	III	IV
Business Level Application & Communication	Ill defined sustainability initiatives within organization; no external communication with respect to CS activites	Tactical level sustainability activities; few external communications with respect to CS activities	Strategic level sustainability activities; some quantification & external communication of CS activities	Intrinsicaly sustainable zero-waste oriented organization; public disclosure of highly granulated sustainability performance data
Scope of Organizational Focus	No supply chain interaction	Limited interaction with supply chain	Some information/resource sharing within supply chain	Significant information, resource sharing & optimization efforts across supply chain
Sustainability Oriented Innovation	Innovation activities are not sustainability related	Some awareness of relationship between innovation and sustainability	Innovation activities begin to involve multiple stakeholders	Zero-waste approach involving significant sustainability oriented innovation efforts that involve multiple stakeholders
Economic/Ecology-Environmental/Equity-Social Emphasis	Emphasis solely on economic sustainability	Primary emphasis on economic sustainability, tentative efforts toward ecological-environmental sustainability	Triple bottom line approach, economic, ecological-environmental, equity-social sustainability	Organization embraces a "triple top line" sustainability approach
Compliance Stance	Sustainability activities limited to minimal efforts at regulatory compliance	Sustainability activities increase beyond minimal regulatory compliance, but are not systematically related to organizational strategy	Incorporation of regulatory compliance within organizational strategy participation in development & evolution of sustainability regulation	Recognized industry thought leader that embraces and encourages zero-waste approach to sustainability regulation, recognition of importance of public-private partnerships

Level of sustainability (I = Reduced; II = Average; III = Good; IV = Very Good)

Fig. 1. Stage 5 of the framework (adapted from [1]).

4 Results

The sample comprises 50 projects selected from six European crowdfunding platforms (Table 3). From a geographical standpoint, the projects' activities primarily focus on Europe (43%), but there is also a significant incidence in regions such as Africa (36%), Asia (14%), and South America (7%). The platforms raised varying amounts of capital per project, ranging from €9,000 to €3,900,000. The number of backers per project also shows significant variability, from 41 to 1,647.

In the first stage of the framework, which involves the "SDGs violation" verification process, the analysis of the projects indicated that none of them were detrimental to sustainability within the scope of any of the SDGs. This suggests that sustainability-oriented crowdfunding platforms effectively fulfill their role, ensuring that they do not select projects for funding that infringe on the SDGs.

Regarding the second stage, the results reveal that all projects have an impact on the economic dimension; 48% have an effect on the environmental dimension in addition to the economic one; 32% achieve the triple bottom line, impacting the economic, social, and environmental dimensions; 16% impact the economic and social dimensions only; and 4% impact only the economic dimension. In other words, 68% of the sample still does not generate impact across all three sustainability dimensions (Fig. 2).

The examination of the sustainability aspects (stage 3 of the framework) shows that the projects present a lower predominance of the social dimension (Fig. 3). Among the 13 categorized sustainability aspects, 5 are present in 50% or more of the projects, relating to the economic and environmental dimensions. Aspects related to the social dimension, such as Education, Equality, Inclusion, Employment Opportunities, and Health, show a lower prevalence, highlighting that there is still much work to be done regarding the social dimension of sustainability.

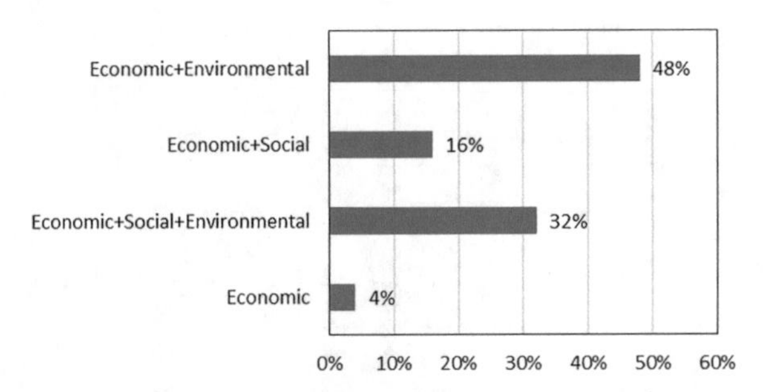

Fig. 2. Impacts of the projects on the sustainability dimensions (n = 50).

In the fourth stage, the results confirm that all projects implement measures to create favorable conditions for developing sustainable business ecosystems. They meet the requirements proposed by [10], aligning various actors, available resources, identification of opportunities, and solution development towards sustainability.

Moving to the conclusion of the sustainability classification process in stage five, it is evident that the sustainability initiatives are relatively well-defined within the organizations (90%), and their communication is carried out with external stakeholders (98%). However, the external communication process is not consistent across all organizations. Specifically, 14% engage in comprehensive external communication, 46% at a reasonable level, and 40% to a limited extent. Concerning the interaction within the supply chain, there is still room for improvement. In 48% of the sample, the level of interactivity within the supply chain is either non-existent or limited. Regarding the project's sustainability orientation, 38% exhibit Sustainable Awareness, 44% involve various parties in sustainable development, and 18% have already reached a level of excellence with a complete focus on sustainability among all stakeholders. Regarding the targeted sustainability dimensions, there is still progress to be made to ensure that all three dimensions are covered, which is observed in only 36% of the projects. In addition, 46% focus on the Economic dimension and direct efforts towards the Environmental dimension, 16% focus on the Economic dimension and direct efforts towards the Social dimension, and 2% focus solely on the Economic dimension. The regulatory compliance verification process reveals maturity. Half of the analyzed projects incorporate it into their organizational strategy, 36% adhere to the minimum mandatory compliance, 8% are thought leaders in the sector, and 6% exceed the minimum compliance, although it's not always directly related to the organizational strategy.

The five evaluated indicators (Business Level Application & Communication; Scope of Organizational Focus; Sustainability-oriented Innovation; Economic, Environmental, and Social Emphasis; Compliance Stance) score on average above "2" (Medium level), three of which even exceed 2.5, moving towards the "Good" level (Graph 6). Overall, 50% of the projects receive a Sustainability Level rating in the "Medium" category (ranging between 2 and 3), 30% receive a "Good" rating (ranging between 3 and 4), and 20% receive a "Reduced" rating (below 2).

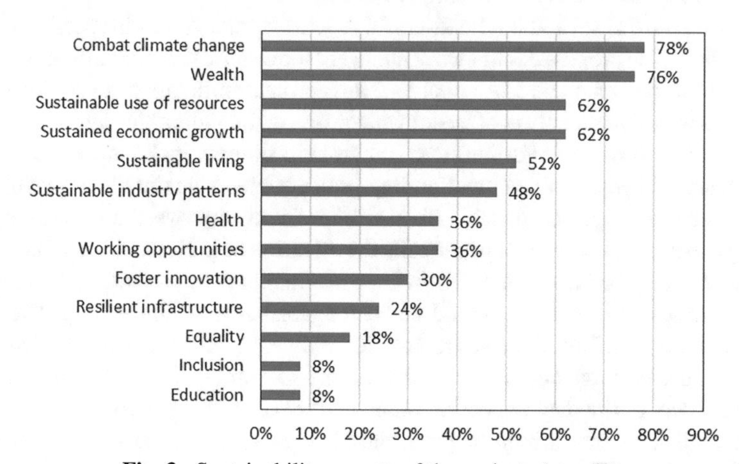

Fig. 3. Sustainability aspects of the projects (n = 50).

Overall, the research confirms the existence of projects with added value and evident sustainability aspects. However, some projects lacked sufficient information, both in terms of quality and quantity, preventing their classification with an acceptable level of sustainability. This finding is consistent with previous studies [5, 12] in that crowdfunding is an effective financing source for sustainability ventures and that project initiators are implementing the measures advertised during the crowdfunding campaigns.

In addition, some projects tended to overstate the sustainability dimensions they aimed to achieve, as the available information did not validate these claims. On a scale ranging from 1 to 4, the representative sustainability level of these projects is 2.57 (average), with minimum values of 1.4 and maximum values of 3.8 (a range of 2.4); 70% of the projects have a reduced or medium sustainability level. This work went further than previous research by making a first attempt at analyzing the sustainability effects of crowdfunding projects in post-financing.

5 Conclusion

This paper analyzed the post-financing phase of sustainable projects funded through crowdfunding platforms and determined whether crowdfunding genuinely supports sustainable development or leads to greenwashing practices. To do this, a five-stage evaluation framework was developed based on extant literature and then applied to a sample of 50 projects funded by six European crowdfunding platforms. The findings revealed that most projects have a positive impact on sustainability, but there is still room for improvement in terms of covering all the different aspects of sustainability.

The paper contributes to the theory of crowdfunding and sustainability by developing and testing a comprehensive framework that integrates the SGDs, sustainability aspects, and sustainable business ecosystems. The framework enables to evaluate the sustainability level of crowdfunding projects in the post-financing stage, addressing a gap in the literature. The paper also provides empirical evidence on the post-financing outcomes of sustainable crowdfunding projects.

This research has several implications for practitioners involved in sustainability and crowdfunding. The proposed framework identifies the criteria that crowdfunding projects should meet to be considered sustainable. Thus, project initiators can use this to guide the design and implementation of sustainable projects, including the communication of their progress and impact. Furthermore, it can help funders make informed decisions about investing in sustainable crowdfunding projects and evaluating their performance. Platforms can leverage the insights obtained herein to improve their monitoring and reporting processes. Lastly, this paper calls the attention of policymakers to the need for more transparency and accountability in the post-financing of crowdfunding projects.

The findings of this paper should be seen against several limitations. The sample size is relatively small and may not represent the diversity of sustainable crowdfunding projects. Future research could expand the sample size and include more crowdfunding platforms. Second, the data collection relies mainly on self-reported data from project initiators and platforms, which may be biased or incomplete. Researchers could use more reliable data sources such as third-party audits, impact assessments, or stakeholder feedback. Third, the framework used is qualitative in nature and relies on the researchers' interpretations, which may not capture all the aspects of the impacts of sustainability. Future studies could refine the framework via the development of a scale and analyze the results using quantitative methods such as structural equation modeling.

Acknowledgments. This work is funded by National Funds through the FCT—Foundation for Science and Technology, I.P., within the scope of the project Ref. UIDB/05583/2020. We would like to thank the Research Center in Digital Services (CISeD) and the Polytechnic of Viseu for their support.

References

1. Amini, M., Bienstock, C.: Corporate sustainability: an integrative definition and framework to evaluate corporate practice and guide academic research. J. Clean. Prod. **76**, 12–19 (2014)
2. Baumgardner, T., Neufeld, C., Huang, P., Sondhi, T., Carlos, F., Talha, M.: Crowdfunding as a fast-expanding market for the creation of capital and shared value. Thunderbird Int. Business Rev. **59**(1), 115–126 (2017)
3. Belleflamme, P., Lambert, T., Schwienbacher, A.: Crowdfunding: tapping the right crowd. J. Bus. Ventur. **29**(5), 585–609 (2014)
4. Belleflamme, P., Omrani, N., Peitz, M.: The economics of crowdfunding platforms. Inf. Econ. Policy **33**, 11–28 (2015)
5. Bento, N., Gianfrate, G., Thoni, M.: Crowdfunding for sustainability ventures. J. Clean. Prod. **237**, 117751 (2019)
6. Calic, G., Mosakowski, E.: Kicking off social entrepreneurship: how a sustainability orientation influences crowdfunding success. J. Manage. Stud. **53**(5), 738–767 (2016)
7. Chan, C.R., Parhankangas, A.: Crowdfunding innovative ideas: how incremental and radical innovativeness influence funding outcomes. Entrep. Theory Pract. **41**(2), 237–263 (2017)
8. Cumming, D., Leboeuf, G., Schwienbacher, A.: Crowdfunding models: Keep-it-all vs. all-or-nothing. Finan. Manage. **49**(2), 331–3 60 (2020)
9. D'Adamo, I., Gastaldi, M., Imbriani, C., Morone, P.: Assessing regional performance for the sustainable development goals in Italy. Sci. Rep. **11**(1), 1–10 (2021)
10. DiVito, L., Ingen-Housz, Z.: From individual sustainability orientations to collective sustainability innovation and sustainable entrepreneurial ecosystems. Small Bus. Econ. **56**, 1057–1072 (2021)
11. Dressler, A., Bucher, J.: Introducing a sustainability evaluation framework based on the sustainable development goals applied to four cases of South African frugal innovation. Business Strategy Develop. **1**(4), 276–285 (2018)
12. Hörisch, J.: Take the money and run? Implementation and disclosure of environmentally-oriented crowdfunding projects. J. Clean. Prod. **223**, 127–135 (2019)
13. Hossain, M., Oparaocha, G.: Crowdfunding: Motives, definitions, typology and ethical challenges. Entrep. Res. J. **7**(2), 1–14 (2017)
14. Kim, K., Hann, I.: Crowdfunding and the democratization of access to capital—An illusion? evidence from housing prices. Inf. Syst. Res. **30**(1), 276–290 (2019)
15. Koch, J., Siering, M.: The recipe of successful crowdfunding campaigns: an analysis of crowdfunding success factors and their interrelations. Electron. Mark. **29**(4), 661–679 (2019)
16. Kukurba, M., Waszkiewicz, A., Salwin, M., Kraslawski, A.: Co-created values in crowdfunding for sustainable development of enterprises. Sustainability **13**(16), 8767 (2021)
17. Lyon, T., Maxwell, J.: Greenwash: corporate environmental disclosure under threat of audit. J. Econom. Manage. Strat. **20**(1), 3–41 (2011)
18. Maehle, N., Otte, P., Drozdova, N.: Crowdfunding Sustainability. In: Shneor, R., Zhao, L., Flåten, BT. (eds) Advances in Crowdfunding. Palgrave Macmillan, Cham (2020)
19. Miles, M., Huberman, A.: Qualitative data analysis: A methods sourcebook. Sage, Thousand oaks (2014)
20. Mollick, E.: The dynamics of crowdfunding: an exploratory study. J. Bus. Ventur. **29**(1), 1–16 (2014)
21. Petruzzelli, A., Natalicchio, A., Panniello, U., Roma, P.: Understanding the crowdfunding phenomenon and its implications for sustainability. Technol. Forecast. Soc. Chang. **141**, 138–148 (2019)
22. Tenner, I., Hörisch, J.: Crowdfunding sustainable entrepreneurship: what are the characteristics of crowdfunding investors? J. Clean. Prod. **290**, 125667 (2021)

23. United Nations: Transforming our world: the 2030 Agenda for Sustainable Development. Resolution adopted by the General Assembly on September 25 2015, New York (2015)
24. Vismara, S.: Information cascades among investors in equity crowdfunding. Entrep. Theory Pract. **42**(3), 467–497 (2018)
25. Volkmann, C., Fichter, K., Klofsten, M., Audretsch, D.: Sustainable entrepreneurial ecosystems: an emerging field of research. Small Bus. Econ. **56**(3), 1047–1055 (2021)
26. Wachs, J., Vedres, B.: Does crowdfunding really foster innovation? evidence from the board game industry. Technol. Forecast. Soc. Chang. **168**, 120747 (2021)
27. Wehnert, P., Beckmann, M.: Crowdfunding for a sustainable future: a systematic literature review. IEEE Trans. Eng. Manage. **70**(9), 3100–3115 (2023)

Bridging the Digital Divide: A Study on the Feasibility of Smart University Integration in Timor-Leste

Rita Pires Soares[1], Ramiro Gonçalves[1,2](✉), Ana Briga-Sá[1,3](✉), José Martins[4](✉), and Frederico Branco[1,2](✉)

[1] Universidade de Trás-os-Montes e Alto Douro, Vila Real, Portugal
al79819@alunos.utad.pt, {ramiro,anas,frederico.branco}@utad.pt
[2] INESC TEC, Porto, Portugal
[3] CQ-VR: Center of Chemistry of Vila Real, Vila Real, Portugal
[4] Instituto Politécnico de Bragança, Bragança, Portugal
jose.l.martins@inesctec.pt

Abstract. Education is vital in fostering economic growth and societal development, particularly in developing countries like Timor-Leste. As technology has revolutionised education in the digital transformation era, the concept of a smart university, driven by advanced technologies and data analytics, has gained prominence globally. Timor-Leste, amid its progress in institutional structures and public infrastructure, is also exploring integrating smart technologies in higher education. This underscores a commitment of The East Timor National Education Strategic Plan (NESP) 2011–2030 to meet national and international standards, positioning the country at the forefront of educational innovation. This study aims to assess the feasibility of implementing a Smart University in Timor-Leste to evaluate the readiness of the country to embrace digital technologies and integrate them into higher education practices. The research employs a Design Science Research methodology where qualitative and quantitative data are gathered through interviews, surveys, and document analysis. Design artefacts, including system architecture and an evaluation framework, are developed to comprehensively understand the technological and informatics aspects of implementing a Smart University in Timor-Leste. The findings will contribute to decision-making and inform the implementation plan, offering valuable insights into stakeholders' perspectives and perceptions, and will support the advancement of the educational landscape in Timor Leste by integrating smart technologies and innovative practices in higher education.

Keywords: Smart University · Higher Education in Timor Leste · Design System Architecture · Evaluation Framework

1 Introduction

Education is crucial for economic growth in developing countries, providing social, economic, and political benefits [1]. Suggests that investment in education is essential for long-term economic development, with a 9% return on investment. Timor Leste, a

Á. Rocha et al. (Eds.): WorldCIST 2024, LNNS 987, pp. 297–305, 2024.
https://doi.org/10.1007/978-3-031-60221-4_29

democratic country, has made significant progress in institutional structures and public infrastructure since gaining independence. The East Timor National Education Strategic Plan (NESP) 2011–2030 aims to achieve universal conditions of primary education, eliminate illiteracy, and ensure gender parity through priority programs. The Minister of Higher Education, Science and Culture (MESCC) oversees the education sector, focusing on infrastructure development, curriculum development, teacher training, access and equity, research, innovation, and ensuring institutions meet national and international standards [2].

The smart-university concept, based on smart technologies, is gaining global attention, particularly in developing countries like Timor Leste, under the supervision of The MESCC is integrating smart technologies into its education system, aligning with national and international standards also fostering knowledge exchange and technology transfer through collaborations with international institutions. It could enhance education quality, provide innovative learning opportunities, and drive global progress. By integrating technologies and digital platforms, Timor-Leste can foster creativity, interdisciplinary research, and collaboration, ensuring more people have access, leading to the development of new ideas and solutions, quality higher education, and strengthening its international reputation. However, the higher education landscape in Timor-Leste is still in its early stages of development. It faces numerous challenges in its higher education sector, including limited resources, inadequate infrastructure, relatively low access to higher education compared to other countries, and a shortage of qualified faculty. Timor-Leste's universities also need to prioritise developing skills and competencies relevant to the digital age, preparing graduates for the demands of the modern workforce.

In the digital age [3], a smart university is a technology-driven institution that utilises advanced technologies and data analytics to provide quality education and personalised learning experiences for each student, enhancing engagement and resource efficiency. Smart University is a complex concept that integrates technology, sustainability, and efficiency. Key features include intelligent resource utilisation, environmentally conscious practices, and the integration of cutting-edge technologies in educational and administrative processes. These universities are driven by systems such as smart devices, online platforms, and data-driven decision-making processes. These systems optimise administrative operations and student support services, fostering a holistic understanding of the university's role in promoting sustainability and efficiency.

A robust and well-designed system architecture is essential to create a smart university. This architecture is the foundation for all the technological components of a smart university, including the infrastructure, applications, and devices. The design system architecture of a smart university should be scalable, flexible, and secure, ensuring that it can accommodate the evolving needs and demands of both students and faculty. With a scalable system architecture, a smart university can quickly adapt and expand as the student and faculty population grows. An evaluation framework is crucial for assessing the effectiveness of smart universities. It helps measure the impact of their initiatives, identify strengths and weaknesses, and make informed decisions [4, 5].

To assess the feasibility of implementing a smart university model in Timor-Leste, it is crucial to consider infrastructure, technology readiness, funding, and human resources. Engaging with government, academia, and industry stakeholders is essential to ensure that Smart University aligns with the country's development goals and meets the needs of its citizens. The findings of this research will contribute to educational innovation, assist Timor-Leste in building a technologically advanced higher education system that supports students and aligns with national development goals, and provide valuable insights and recommendations for decision-making processes regarding establishing a smart university in Timor-Leste.

2 Purpose and Research Question

This study aims to develop and implement an effective design system architecture and evaluation framework tailored for a smart university in Timor Leste. By focusing on the unique context of Timor Leste University, the research aims to enhance the quality and impact of higher education through technological advancements and systematic evaluation.

The research focuses on the current state of the higher education system at Timor Leste University, addressing institutional needs and challenges for implementing a smart university framework. It also explores the components of an efficient design system architecture for a smart university and how it can be customised to suit its unique requirements. The evaluation framework is essential for assessing the effectiveness and impact of smart university initiatives, including metrics and indicators to measure success.

The research questions will enable a focused investigation into the challenges and opportunities related to implementing smart universities in Timor-Leste and the effective integration of technology to enhance teaching and learning experiences in the country's universities. The research questions encompass the main ideas of designing a system architecture and evaluation framework for the implementation of a smart university in Timor Leste, which are:

RQ 1: What are the challenges and opportunities related to implementing smart universities in Timor Leste, and how can integrating smart universities effectively enhance teaching and learning experiences in the country?

RQ 2: How can a system architecture be designed for smart university implementation in Timor Leste, considering the unique needs and constraints of the context?

RQ 3: What key components, functionalities, and evaluation metrics are required in the system architecture and evaluation framework to ensure a comprehensive and integrated smart-university environment and assess its effectiveness and impact in Timor Leste?

3 Literature Review

The literature review encompasses a thorough examination of topics such as "smart university," "technology integration," "digital solutions in education", "design system architecture", and "evaluation framework". Drawing from diverse and reputable sources, including research papers, articles from esteemed publications like IEEE and ACM Digital Library, and various international journals, the review utilises academic databases such as Google Scholar, Scopus, Semantic Scholar, Web of Science, and Crossref from 2017–2022 that classifies in the Fig. 1. That can be described as follows:

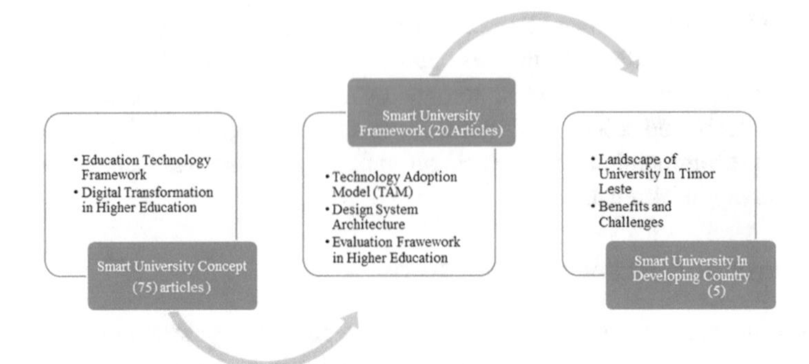

Fig. 1. Literature review.

The literature review provides a comprehensive overview of the theoretical framework and literature review for this study that is divided into three main points: "Smart Universities Concept", "Smart University Framework", and "Smart University in Developing Country", selected from 100 articles related to smart universities.

The literature review found that the concept of a smart university is an institution that utilises advanced technologies and digital solutions to enhance the quality of education, research, and administrative processes. [6] also, insights into the definition and essential characteristics of smart universities from [7] highlight smart universities' key characteristics, including technologically enabled infrastructure, personalised learning experiences, collaborative learning environments, digital tools for teaching and research, and efficient administrative processes.

The review sources that support the discussion on digital transformation in higher education and technological advancements in smart universities provided by [8] discuss leveraging emerging technologies to enhance teaching and learning experiences, improve access to education, and foster innovation through digital transformation in higher education. Villegas-Ch [9] explores the reimagining of traditional educational practices by integrating digital technologies such as blended learning, online assessments, and learning analytics to support personalised and data-driven learning experiences. Caţă [10] emphasises how incorporating various technologies in education, including LMS, mobile learning, big data analytics, IoT, cloud computing, and AI, can provide personalised and context-aware learning experiences, improving engagement and performance. These authors explore the concept of digital transformation in higher education and its impact on smart universities. It emphasises the integration of digital technologies and strategies to reshape traditional educational practices, enhance learning outcomes, and improve institutional efficiency.

The literature review for the smart university framework defines the Technology Acceptance Model (TAM) as a widely recognised and applied framework for assessing the adoption and acceptance of technology. Its relevance in integrating smart university initiatives in Timor-Leste lies in its ability to provide a systematic approach to understanding the factors that influence the acceptance and use of technology in an educational setting. Davis (1989) provided a theory about the TAM that is crucial for understanding user adoption of new technologies in higher education [11, 12]. Mishra and Koehle

(2006) [13] propose a conceptual framework called the Technological Pedagogical Content Knowledge (TPCK) framework that captures essential teacher knowledge for technology integration. By identifying and addressing TAM in the context of smart university implementation, TAM helps analyse factors influencing the acceptance and adoption of technological solutions for Timor Leste universities.

Design System Architecture for Smart Universities: The authors [14–16] provide insight into the design system architecture model for developing smart universities. Senge proposed Leveraging systems thinking by emphasising the interconnectedness of elements within an organisation. Applying this perspective aids in designing a holistic system architecture that considers the interdependencies of various components in a smart university, ensuring a coherent and effective implementation [17].

Research by [9, 18–20] explores the implementation of smart initiatives in developing countries, providing contextual relevance to the study in the specific case of Timor Leste University. The works [21–24] offer insights into the higher education landscape in Timor Leste, providing a foundational understanding of the challenges and opportunities faced by institutions in the region.

By synthesising these theoretical frameworks and literature, the study aims to construct a comprehensive foundation for designing and evaluating the implementation of smart university initiatives at Timor Leste University. Although the literature recognises the potential for Smart Universities to be revolutionary, it does not adequately examine how they will specifically affect Timor Leste's distinct socioeconomic and cultural setting. Future studies and real-world applications will be directed by filling in these gaps.

4 Research Methodology

The methodology employed in this research is Design Science Research (DSR) proposed by Peffers et al. [25], which is highly appropriate for investigating technology-related topics. DSR can effectively be applied to research focused on digital transformation, incorporating insights from behavioural science and design. This methodology will guide the model development process, which will subsequently undergo validation in selected universities. The DSR methodology proposed comprises six key steps: Problem identification and motivation, defining objectives for the solution, design and development, Demonstrating, Evaluating, and Communicating with the activities described in Table 1.

Designing comprehensive system architecture artefacts is crucial to implement smart university initiatives in Timor-Leste effectively. These artefacts will serve as the blueprint for developing and integrating various technologies and systems within educational institutions. These artefacts provide valuable documentation and insights to ensure the success and effectiveness of the smart university initiative. The system architecture artefacts include a System Architecture Diagram, Component Diagram, Deployment Diagram, Data Flow Diagram, Use Case Diagram, Sequence Diagram, Communication Diagram, Entity-Relationship Diagram, Security Architecture, and Scalability and Flexibility assessment. These diagrams provide a comprehensive understanding of the system's structure, components, and interactions, ensuring its effective implementation.

An evaluation framework is a valuable tool for organisations and researchers as it provides a structured and systematic approach to assess the impact and efficiency of initiatives. It aids in making informed decisions, enhancing accountability, and driving improvements based on evidence-based insights, as shown in Fig. 2. The evaluation framework consists of an evaluation plan, survey questionnaires, interview protocols, data collection instruments, data analysis framework, evaluation report, lessons learned document, presentation materials, and feedback incorporation plan. The plan outlines the overall approach to evaluating smart university implementation, including goals, objectives, KPIs, data collection methods, and timeline. The questionnaires gather feedback from students, faculty, and staff, while the interview protocols involve in-depth discussions with stakeholders. The evaluation report presents findings, strengths, weaknesses,

Table 1. Design Science Research Methodology.

Activity	Description
Problem Identification and Motivation	Identify the research focus and its relevance in addressing the problem through a literature review on smart universities. Clearly articulate the problem of the educational divide in Timor-Leste, emphasising the need for smart university integration to address existing challenges. Identify specific educational disparities, technological limitations, and cultural considerations that contribute to the educational divide in Timor-Leste
Define Objectives for the Solution	Specify the requirements for a clear solution and assess existing solutions for comparison. Define the objectives of evaluating Timor-Leste's readiness for smart university integration and establish the scope of the research
Design and Development	Break down the main research problems and use theoretical knowledge to construct the final artefact, focusing on building an explanatory model. Develop a design for evaluating readiness, considering existing literature, best practices, and innovative approaches to smart university integration. Propose a comprehensive evaluation framework that considers the unique context of Timor-Leste, incorporating insights from the literature review and addressing identified gaps
Demonstrate	Demonstrate that the developed artefact solves all or part of the problem through a case study. Required resources include knowledge of how to use the artefact to solve the problem. In this phase, interviews/surveys will be conducted with university administrators, faculty, students, and relevant stakeholders to verify and validate the developed artefact

(*continued*)

Table 1. (*continued*)

Activity	Description
Evaluate	Validate the efficiency of the results using appropriate methods, refine the artefact if necessary, and enhance understanding during this phase. Implement the designed evaluation framework in a real-world context and evaluate its effectiveness in assessing Timor-Leste's readiness. Conduct a pilot assessment using the developed framework, gather data on infrastructure, policies, and cultural factors, and evaluate how well it captures the readiness for smart university integration
Communicate	Communicate the findings and outcomes of the readiness assessment to relevant stakeholders. Present the final solution by writing the doctoral thesis and disseminating the results to researchers and professionals in the field through publications in relevant journals or conferences. Share the results with educational institutions, policymakers, and the community in Timor-Leste, emphasising the areas of strength and improvement in preparation for smart university integration

and recommendations for improvement. Ethical considerations, such as informed consent and confidentiality, will be prioritised. Acknowledging limitations in generalizability due to Timor-Leste's specific context and resource constraints, the research aims to provide a comprehensive understanding of technological and informatics aspects related to Smart University implementation.

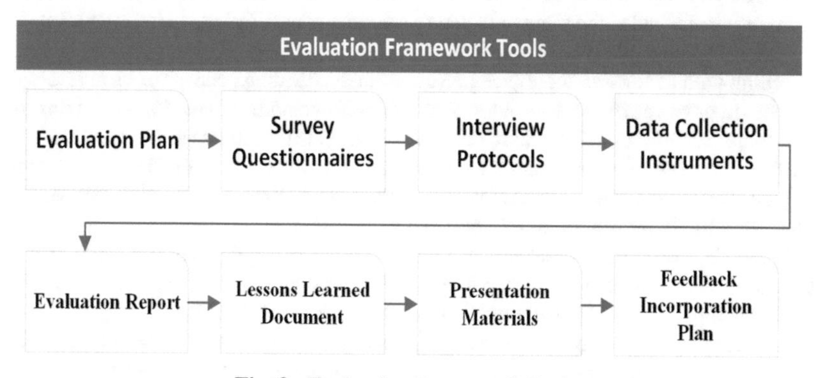

Fig. 2. Evaluation Framework Tools.

5 Conclusion

The findings of this study will provide valuable insights for Timor Leste University's smart university initiative, focusing on the design system architecture and evaluation framework artefacts. The evaluation report, with its findings and recommendations, serves as a guide for improvement and helps identify areas for further attention. The design system architecture will provide a blueprint for developing and integrating various technological components, ensuring a cohesive user experience. The evaluation framework artefacts establish a systematic approach to assess the performance and impact of the smart university initiative, identifying strengths, weaknesses, and areas for enhancement. The combination of a well-designed system architecture and a robust evaluation framework ensures that the smart university evolves and adapts to meet the needs of its stakeholders, contributing to the overall success and effectiveness of the smart university environment.

References

1. Mutton, D., Ciriello, R.: The Impact of Education on Economic Growth in Developing Countries (2021). https://doi.org/10.13140/RG.2.2.32332.54400
2. Ministério do Ensino Superior, Ciência e Cultural|Jornal da República. https://www.mj.gov.tl/jornal/?q=node/7045. Accessed 13 Nov 2023
3. Uskov, V.L., Bakken, J.P., Howlett, R.J., Jain, L.C. (eds.): Smart Universities: Concepts, Systems, and Technologies. Springer, Cham (2018). https://doi.org/10.1007/978-3-319-59454-5
4. Apanaviciene, R., Vanagas, A., Fokaides, P.A.: Smart building integration into a smart city (SBISC): development of a new evaluation framework. Energies **13**, 2190 (2020). https://doi.org/10.3390/en13092190
5. Teaching Performance Evaluation in Smart Campus|IEEE Journals & Magazine|IEEE Xplore, https://ieeexplore.ieee.org/document/8552406. Accessed 13 Nov 2023
6. Fernández-Caramés, T.M., Fraga-Lamas, P.: Towards next generation teaching, learning, and context-aware applications for higher education: a review on Blockchain, IoT, Fog and Edge Computing enabled smart campuses and universities. Appl. Sci. **9**, 4479 (2019). https://doi.org/10.3390/app9214479
7. Teaching with technology in higher education: understanding conceptual change and development in practice: Higher Education Research & Development, vol. 36, no. 1. https://www.tandfonline.com/doi/full/10.1080/07294360.2016.1171300. Accessed 16 Nov 2023
8. Bisri, A., Putri, A., Rosmansyah, Y.: A systematic literature review on digital transformation in higher education: revealing key success factors. Int. J. Emerg. Technol. Learn. (iJET). **18**, 164–187 (2023). https://doi.org/10.3991/ijet.v18i14.40201
9. Villegas-Ch, W., Palacios-Pacheco, X., Luján-Mora, S.: Application of a smart city model to a traditional university campus with a big data architecture: a sustainable smart campus. Sustainability **11**, 2857 (2019). https://doi.org/10.3390/su11102857
10. Cață, M.: Smart university, a new concept in the Internet of Things. In: 2015 14th RoEduNet International Conference - Networking in Education and Research (RoEduNet NER), pp. 195–197 (2015). https://doi.org/10.1109/RoEduNet.2015.7311993
11. Alismaiel, O.A., Cifuentes-Faura, J., Al-Rahmi, W.M.: Social media technologies used for education: an empirical study on TAM model during the COVID-19 pandemic. Front. Educ. **7**, 882831 (2022). https://doi.org/10.3389/feduc.2022.882831

12. Scherer, R., Siddiq, F., Tondeur, J.: The technology acceptance model (TAM): a meta-analytic structural equation modeling approach to explaining teachers' adoption of digital technology in education. Comput. Educ. **128**, 13–35 (2019). https://doi.org/10.1016/j.compedu.2018.09.009
13. Mishra, P., Koehler, M.J.: Technological Pedagogical Content Knowledge: A Framework for Teacher Knowledge (2006)
14. Zhi, J.: Systems approach model: the application research of UML for instructional design. Presented at the Computational Intelligence, 28 December 2009. https://doi.org/10.1109/CISE.2009.5366826
15. He, Y., Wang, J.: Study on system architecture design of university students quality evaluation. Int. J. Wirel. Microwave Technol. **2**, 23–28 (2012). https://doi.org/10.5815/IJWMT.2012.06.04
16. Zhang, L.-M., Xia, H.-X.: School administration management information system in university based on UML. J. Wuhan Univ. Technol. **31**, 139–142 (2009)
17. Mohammed, A.R., Kassem, S.S.: E-learning system model for university education using UML (2020). https://doi.org/10.1109/ECONF51404.2020.9385482
18. John, T.M., Ucheaga, E.G., Badejo, J.A., Atayero, A.A.: A framework for a smart campus: a case of Covenant University. In: 2017 International Conference on Computational Science and Computational Intelligence (CSCI), pp. 1371–1376 (2017). https://doi.org/10.1109/CSCI.2017.239
19. Fernando Raguro, Ma.C., Lagman, A.C., Juanatas, R.: Technology management framework for smart university system in the Philippines. In: Proceedings of the 2021 9th International Conference on Information Technology: IoT and Smart City, pp. 372–380. Association for Computing Machinery, New York, NY, USA (2022). https://doi.org/10.1145/3512576.3512642
20. Bond, M., Marín, V.I., Dolch, C., Bedenlier, S., Zawacki-Richter, O.: Digital transformation in German higher education: student and teacher perceptions and usage of digital media. Int. J. Educ. Technol. High. Educ. **15**, 1–20 (2018). https://doi.org/10.1186/s41239-018-0130-1
21. Ximenes, P.B.: Higher education in Timor-Leste. In: Symaco, L.P., Hayden, M. (eds.) International Handbook on Education in South East Asia, pp. 1–29. Springer, Singapore (2021). https://doi.org/10.1007/978-981-16-8136-3_19-1
22. Ximenes, I.F.A.: Exploring the clarity of the head of department role in a Timor-Leste higher education institution (2019)
23. King, M., Forsey, M., Pegrum, M.: Southern agency and digital education: an ethnography of open online learning in Dili, Timor-Leste. Learn. Media Technol. **44**, 283–298 (2019). https://doi.org/10.1080/17439884.2019.1639191
24. Burns, R.: Education in Timor-Leste: envisioning the future. J. Int. Comp. Educ. (JICE) **6**, 33–45 (2017). https://doi.org/10.14425/JICE.2017.6.1.3345
25. Peffers, K., Tuunanen, T., Rothenberger, M., Chatterjee, S.: A design science research methodology for information systems research. J. Manag. Inf. Syst. **24**, 45–77 (2007)

Towards Safer and Efficient Dowry Transactions: A Blockchain-Based Approach

Marianus M. S. Bria[1], Ramiro Gonçalves[1,2(✉)], José Martins[2,3(✉)], Carlos Serôdio[1,4,5(✉)], and Frederico Branco[1,2(✉)]

[1] Universidade de Trás-os-Montes e Alto Douro, Vila Real, Portugal
al79828@alunos.utad.pt, {ramiro,cserodio,fbranco}@utad.pt
[2] INESC TEC, Porto, Portugal
jose.l.martins@inesctec.pt
[3] Instituto Politécnico de Bragança, Bragança, Portugal
[4] Center ALGORITMI, Universidade do Minho, Guimarães, Portugal
[5] CITAB, Universidade de Trás-os-Montes e Alto Douro, Vila Real, Portugal

Abstract. The dowry payment system is used in the cultural context and tradition of certain financial transactions related to marriages and engagement. However, disputes, fraud, and financial gaps in exploitation occur in these systems, which affect user confidence. This study uses an exploratory approach to identify the main weaknesses of current traditional dowry payment systems and analyses the benefits that blockchain technology and smart contracts can provide. The proposed data security framework combines blockchain security features such as decentralisation, cryptography, and automatic verification through smart contracts to ensure the integrity and reliability of dowry payment transactions.

In this study, we adopt the Design Science Research (DSR) methodology to propose producing and developing artefacts that support solving problems in the existing dowry payment system more efficiently. We will disseminate new ideas or concepts developed to indigenous communities in Timor-Leste using the Diffusion of Innovation (DOI) and Technology Acceptance Model (TAM) frameworks to ensure that the technological framework developed can be used safely and efficiently.

Keywords: Data Security Framework · Blockchain · Smart Contract · Dowry Payment System · User Trust

1 Introduction

Culture is the heritage of every nation's ancestor and should be nurtured and preserved. A country's beautiful and exciting cultural diversity provides it with a unique identity [1]. As a developing country, Timor-Leste consists of 13 districts with different cultural diversities under the supervision of the *Secretaria Estado Arte e Cultura* (SEAC) [1]. It was established in the formulation of Article 6 paragraph (g) of the Constitution of Timor-Leste [2]. One of the traditional cultures that has attracted the attention of researchers is the problem of payments based on the dowry system in an engagement ceremony with the

Á. Rocha et al. (Eds.): WorldCIST 2024, LNNS 987, pp. 306–313, 2024.
https://doi.org/10.1007/978-3-031-60221-4_30

patrilineal system [3]. Dowry (barlake) is the amount of money or jewellery of a specific value given by the groom's family to the bride's family as a form of respect for the bride's extended family [3–5]. Before marriage, this tradition is the first step, or betrothals [3, 6]. The payment of a dowry is an essential factor in traditional weddings before marriage [3]. The tradition of engagement begins with the meeting of the two parties to decide on the time and the dowry, which is given by the groom's family to the bride's family at the time of the engagement ceremony [7]. According to the researcher's observations, in Timor-Leste, dowry is still traditionally paid at the place of engagement, according to a mutually agreed date. However, conflicts often arise during this process because the unilateral agreement of the bride's family changes when an engagement ceremony occurs [8]. In dowry practices, disputes over the amount of dowry, fraud in determining the amount of dowry, and financial disparities for exploitation [9]. Therefore, this study explores the potential of blockchain smart contracts based on a data security framework to support the payment process in a dowry system.

Blockchain technology and smart contracts can support the payment process in the dowry system [10] where blockchain technology provides a distributed ledger that can record financial payments in a transparent, immutable, and traceable manner [10]. This will facilitate the distribution of dowries to eligible parties such as mothers, uncles, chiefs, and church thanksgiving ceremonies, while smart contracts are automated protocols that regulate and execute the terms of the contract where both parties enter into a digital contract agreement for dowry payments [11]. The development of this framework model will provide the best solution to solve the problem of convenience and trust in the ongoing traditional dowry payment in Timor-Leste.

Some researchers, such as [12], define data security from an economic perspective to protect large amounts of data so that it is more secure, private, and effective.

Research conducted by [13] defines data security by analyzing eight elements of data security contained in cloud storage systems, such as "data confidentiality, data integrity, data availability, detailed access control, secure data sharing in dynamic groups, anti-leakage, complete data deletion, and privacy protection."

This study aims to evaluate the potential of blockchain technology and smart contracts in the context of dowry payments by developing a data security framework based on the activities of indigenous people in making marriage dowry payments in Timor-Leste. Data security in marriage dowry payments is vital for maintaining the trust of indigenous people against fraud, theft, and forgery of marriage dowry payments.

This article is organized as follows: Sect. 1 discusses the themes of marriage dowry, blockchain smart contracts, and data security; the objectives and research questions in Sect. 2, where the main objectives are focused on solving and where each review question is specified; Sect. 3 presents the theoretical framework and literature review reporting the exploratory work in the initial literature review; Sect. 4 presents the research methodology to be used; and finally, in Sect. 5, we conclude that a better data security framework will increase user confidence and efficiency towards indigenous people.

2 Purpose and Research Questions

This research aims to evaluate the potential of blockchain technology and smart contracts in the context of dowry payments in Timor-Leste. This is an alternative or solution that could be applied to overcome problems or conflicts related to dowry payments and to explore innovations or new practices that can reduce conflicts, injustices, or economic difficulties that arise.

2.1 Objective

This research focuses on the data security framework developed based on the activities of indigenous people in making marriage dowry payments in Timor-Leste. The problem to be solved in this research is to measure the level of trust of indigenous people in Timor-Leste towards implementing the developed data security framework. Data security in marriage dowry payments is fundamental to maintaining the trust of indigenous people against fraud, theft, and transparency of marriage dowry payments.

Therefore, we will develop a model that can overcome the problems encountered in the community and increase the confidence of indigenous people in more efficient dowry payments.

2.2 Research Questions

Based on the formulation of the article 6th paragraph (g) of the Timor-Leste Constitution, "Affirm and enhance the personality and cultural heritage of the people of Timor-Leste" [2].

According to its Strategic Development Plan, Timor-Leste's millennium development aims to protect and preserve the nation's diverse and unique cultural, historical, and heritage [14].

Based on the above background, the issues discussed in this research include security, fraud, and efficiency. The research question for this study is: How much does the trust of indigenous people in Timor-Leste in implementing the marriage dowry payment system method increase when the blockchain smart contract is applied in a data security framework?

3 Theoretical Framework and Literature Review

It is important to define the theoretical framework and review the literature to demonstrate the importance of this research and the problems it seeks to solve. An initial literature review was conducted using the Google Scholar, ScienceDirect, and Scopus databases, published between 2011 and 2023. A total of 211 articles related to dowry payments, data security, blockchain, smart contracts, and Timor-Leste were analyzed. Some of the articles obtained were not used as written references. Figure 1 provides information on the number of articles obtained.

The articles were selected based on author categorisation, title, year of publication, theoretical approach, context, and relevant information to contribute to knowledge and

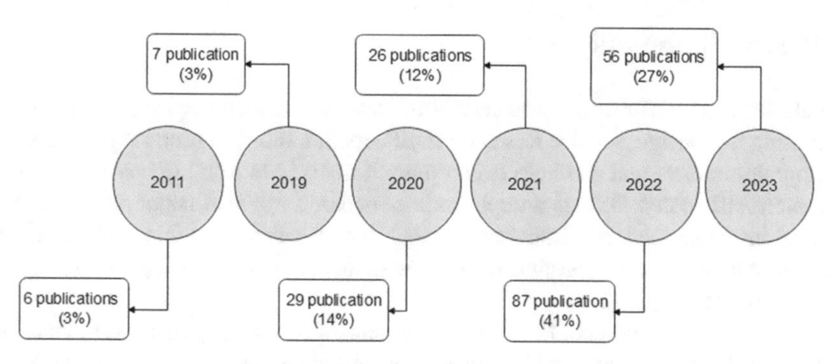

Fig. 1. Articles included in the study.

research on blockchain smart contracts and financial payments in dowry systems in the future.

The research results conducted by [15] on the causes of the continued practice of marriage dowry are due to solid community customs and culture. This study used data from Understanding the Lives of Teens and Young Adulthoods (UDAYA) in two Indian states, Uttar Pradesh, and Bihar. From the data obtained, they collected information from 5206 married adolescent girls. Their research focused on dowry as an observed variable. They conducted univariate, bivariate, and multivariate logistic regression analyses to examine the factors associated with dowry payments during marriage.

Research by [16] explained that the high cost of marriage or dowry creates a substantial financial burden on families, especially in rural households. This study aimed to analyse the relationship and role of marriage distance. This study adopts Logit, Tobit, and SUR models with a 70-year database in rural China from a nearly national representative sample. The results show that couples matched for love are 10.7% and 10.3% less likely to pay dowry and dowry, respectively, compared to couples matched by parents, and they pay fewer dowries and dowries. There is an inverted U-shaped relationship between marriage distance and dowry, and arranged marriage has the most significant correlation with marriage payment behaviour among couples with long marriage distances.

Descriptive research was conducted by [17] in the Lotha community with purposive sampling of the 50 couples interviewed. The Lotha community calls the marriage dowry "Hanlam". *Hanlam* in Lotha has specific differences because it benefits newly married couples. This study aims to determine the relevance of *Hanlam* practice among the Lotha community and its impact on women's rights.

Research conducted by [18] in rural areas of India is related to dowry payments by brides to grooms, where women who give dowry to men become poor after marriage.

Based on the analysis of published data on dowry payments made by indigenous peoples in several countries, such as India, Pakistan, China, Indonesia, and Timor-Leste:

a. The amount of dowry that must be paid in a marriage will cause one of the brides and groom to suffer losses [18]
b. Strong community customs and culture cause dowry payments to continue throughout the marriage [19].

4 Research Methodology

To make the research more focused and follow academic principles, we propose a framework using the Design Science Research (DSR) methodology to propose producing and developing artefacts that can help overcome problems in existing dowry payment systems more efficiently. DSR is an appropriate methodology that supports intuitive and practical investigations. It is important to note that research using this method does not mean that a product will be produced and presented, but rather a concept, model, or what is called an artefact [20–22].

The research will be conducted in an indigenous community in a traditional house (Uma Lisan Lakmeta, Kab Lia Nai, Resleo) in the Ainaro District, Sub-District Ainaro. How to disseminate new ideas or concepts developed to indigenous communities in Timor-Leste, we will adopt the Diffusion of Innovation (DOI) model framework [23] and the Technology Acceptance Model (TAM) to Ensure that the developed technology can be used safely and efficiently to increase public confidence in the model being developed [24] (Table 1).

Table 1. Design Science Research applied to this work (adapted from [22])

Activity	Description
1. Identify Problem & Motivate	Identify the focus of the study and its value in response to the problem. The resources needed in this step are the problem's state of the art and the solution's relevance. To this end, it will be necessary to conduct a survey of the state of the art (using an appropriate protocol) regarding dowry payments to indigenous communities in Timor-Leste and to develop a model for a data security framework to increase user confidence
2. Define Objectives of a Solution	Specify the requirements (quantitative and/or qualitative) to present a possible solution and describe how it can be implemented. The resources needed at this stage will be the state-of-the-art of the problem and the knowledge of possible solutions previously presented. At this stage, it is also essential to know what solutions have already been implemented, if any, to understand their effectiveness and to serve as a comparison
3. Design & Development	Focus on the search for knowledge for the construction of the final artefact, which can be achieved by separating the main problems encountered during research into something simple. The necessary resource is the knowledge of the theory that can give rise to a solution. The artefact is built in this phase, from its definition to its design and development. The artefact to be developed comprises the construction of an explanatory model

(*continued*)

Table 1. (*continued*)

Activity	Description
4. Demonstration	Demonstrate that the developed artefact solves all or part of the problem through a case study. The required resources include knowledge of how to use artefacts to solve the problem. In this phase, interviews and surveys will be conducted with people in Timor-Leste, specifically employees from different municipalities, to verify and validate the developed artefact
5. Evaluation	Validate whether the results are efficient by focusing on methods that can determine the feasibility of the final solution. If it is necessary to improve the artefact, return to Step 3 (design and development) or Step 4 (demonstration). At this stage, it is essential to increase the understanding of the research conducted, as this will allow those with access to the artefact to understand and learn from what has been presented. Construction activities should be evaluated regarding the applicability and generalizability of artefacts
6. Communication	Present the final solution by writing the thesis. It is essential to disseminate the results obtained to researchers and professionals in the field, and this work must be published in journals or conferences related to the research area. In this manner, publications will be submitted to conferences and scientific journals for dissemination throughout the work

According to Peffers, DSR is a structured guide that is practical and oriented such that it can be used to investigate a problem and provide the best solution.

5 Conclusion

This research contributes to knowledge by creating a data security framework to increase the confidence of indigenous communities in Timor-Leste in conducting marriage dowry payments.

The data security framework built will guide the development of applications to protect user data when conducting dowry payment transactions.

The conclusion drawn from this research is that application development based on the built data security framework will provide efficient and secure services, and there will be no fraud in marriage dowry payments.

The proposed data security framework also incorporates blockchain security features, such as decentralisation, cryptography, and automatic verification through smart contracts, to ensure the integrity and reliability of dowry payment transactions.

References

1. Ramos-Horta, J.: timor-leste_national_strategic_development_plan_2011-2030 (2011)
2. Governu, D., et al.: Viii governo constitucional, pp. 1–11 (2020)
3. Chen, J., Pan, W.: Bride price and gender role in rural China. Heliyon **9**, e12789 (2023). https://doi.org/10.1016/j.heliyon.2022.e12789
4. Chiplunkar, G., Weaver, J.: Marriage markets and the rise of dowry in India. J. Dev. Econ. **164**, 103115 (2023). https://doi.org/10.1016/j.jdeveco.2023.103115
5. Mehdi, R.: Danish law and the practice of mahr among Muslim Pakistanis in Denmark. Int. J. Sociol. Law **31**, 115–129 (2003). https://doi.org/10.1016/j.ijsl.2003.02.002
6. Niner, S., Fellow, R.: Changes to Barlake in Timor-Leste (2011)
7. Niner, S.: Barlake: an exploration of marriage practices and issues of women's status in Timor-Leste. Local-Global: Identity Secur. Community **11**, 138–153 (2012). Traversing customary community and modern nation-formation in Timor-Leste
8. Ullah, A., et al.: The ubiquitous phenomena of dowry practice and its relation with women prestige in District Swabi-Pakistan. PalArch's J. Archaelogy Egypt/Egyptology **17**, 14176–14189 (2020)
9. Menon, S.: The effect of marital endowments on domestic violence in India. J. Dev. Econ. **143**, 102389 (2020). https://doi.org/10.1016/j.jdeveco.2019.102389
10. Chatterjee, K., Singh, A., Neha: A blockchain-enabled security framework for smart agriculture. Comput. Electr. Eng. **106**, 108594 (2023). https://doi.org/10.1016/j.compeleceng.2023.108594
11. Kaafarani, R., Ismail, L., Zahwe, O.: An adaptive decision-making approach for better selection of blockchain platform for health insurance frauds detection with smart contracts: development and performance evaluation. Procedia Comput Sci. **220**, 470–477 (2023). https://doi.org/10.1016/j.procs.2023.03.060
12. Tao, H., et al.: Economic perspective analysis of protecting big data security and privacy. Futur. Gener. Comput. Syst. **98**, 660–671 (2019). https://doi.org/10.1016/j.future.2019.03.042
13. Yang, P., Xiong, N., Ren, J.: Data security and privacy protection for cloud storage: a survey. IEEE Access **8**, 131723–131740 (2020). https://doi.org/10.1109/ACCESS.2020.3009876
14. Plano-Estrategico-de-Desenvolvimento_PT1 (2011)
15. Srivastava, S., et al.: Banned by the law, practiced by the society: the study of factors associated with dowry payments among adolescent girls in Uttar Pradesh and Bihar, India. PLoS One **16**, e0258656 (2021). https://doi.org/10.1371/journal.pone.0258656
16. Lyu, Q., Zhang, L.: Love match, marriage distance, and marriage payment: evidence from rural China. Sustainability (Switzerland) **13**, 13058 (2021). https://doi.org/10.3390/su132313058
17. Mhabeni, W., Pongen, M.: Hanlam: marriage payment in the 21st century: a case study from Wokha Town. J. Humanit. Educ. Dev. **5**, 53–56 (2023). https://doi.org/10.22161/jhed.5.1.8
18. Calvi, R., Keskar, A.: Dowries, resource allocation, and poverty. J. Econ. Behav. Organ. **192**, 268–303 (2021). https://doi.org/10.1016/j.jebo.2021.10.008
19. Srivastava, A., Mukherjee, S., Jebarajakirthy, C.: Aspirational consumption at the bottom of pyramid: a review of literature and future research directions. J. Bus. Res. **110**, 246–259 (2020). https://doi.org/10.1016/j.jbusres.2019.12.045
20. Peffers, K., Tuunanen, T., Rothenberger, M.A., Chatterjee, S.: A design science research methodology for information systems research. J. Manag. Inf. Syst. **24**, 45–77 (2007). https://doi.org/10.2753/MIS0742-1222240302

21. Vaishnavi, V., Kuechler, B.: Design Science Research in Information Systems (2013)
22. Hevner, A.R., March, S.T., Park, J., Ram, S.: Design science in information systems research. MIS Q. **28**, 75 (2004)
23. Cipolla, C.M.: The Diffusion of Innovations in Early Modern Europe (1962)
24. Davis Jr., F.D.: A technology acceptance model for empirically testing new end-user information systems: theory and results, pp. 1–291 (1985)

A Taxonomy on Human Factors that Affect DevOps Adoption

Juanjo Pérez-Sánchez[1(✉)], Saima Rafi[2], Juan Manuel Carrillo de Gea[1], Joaquín Nicolás Ros[1], and José Luis Fernández Alemán[1]

[1] University of Murcia, C. Campus Universitario, Edificio 32, 30100 Murcia, Spain
juanjose.perez9@um.es

[2] Edinburgh Napier University, Merchiston Campus, 10 Coliton Road, Edinburgh EH10 5DB, Scotland, UK

Abstract. DevOps is a software development methodology created to reduce or even remove the division between the Development (Dev) and Operations (Ops) teams. However, DevOps adoption requires overcoming several impediments, and between them, culture change and human factors have the biggest impact. Therefore, this paper addresses the challenge of DevOps adoption from the perspective of DevOps culture and human factors. A systematic mapping study was carried out to create a taxonomy of human factors affecting DevOps adoption. A total of 21 studies were selected and 59 human factors were included in the taxonomy after the extraction and synthesis processes.

1 Introduction

DevOps is a Software Engineering (SE) paradigm that focuses on the reduction of the division between the Development (Dev) and Operations (Ops) teams, that are traditionally separated [29]. Software development companies are currently facing a major challenge, on the one hand, they must quickly adapt to ever-changing market conditions while delivering quality software, on the other hand, the systems that make these deliveries possible are more complex than ever and they have to maintain a high degree of maintainability and availability [7].

DevOps aligns the objectives of both development and operations teams through cultural, methodological and technical change, in order to improve quality and reduce delivery time. Organizations are switching to DevOps motivated by different reasons. While some organizations focus on improving collaboration or enabling team and process efficiency, most of them adopt DevOps to reduce time-to-market while delivering high-quality products [6].

Adopting DevOps is, consequently, a task that involves a variety of aspects of software development, which is considered non-trivial and requires overcoming several impediments. DevOps itself, as a concept, is a multifaceted phenomenon which definition still requires more attention from the scientific community [26].

This paper focuses on the cultural aspect of DevOps, in particular it focuses on DevOps human factors. While the soft side of software development is always present in literature, it is not always confessed among DevOps practitioners and

Á. Rocha et al. (Eds.): WorldCIST 2024, LNNS 987, pp. 314–324, 2024.
https://doi.org/10.1007/978-3-031-60221-4_31

researchers. Culture is very related to human factors [27], and human factors are, in fact, the greatest source of opportunity to improve productivity [5]. Human factors are often used as a broad term, in this paper, we consider the term "human factor" as the set of psychological, behavioral, sociological, and contextual aspects of humans that are related to the software process. The importance of human factors research has been growing steadily recently [3], as the amount of papers dealing with this topic increased in SE research [1]. Since human factors and team culture are more complex in nature than technical and process issues, they have been identified as the main difficulty for a successful DevOps adoption [28]. However, Restrepo-Tamayo and Gasca-Hurtado [22] did not identify any research focusing on DevOps human aspects in a recent mapping study regarding human aspects of SE. While there is a lack of DevOps studies regarding human factors, DevOps includes a number of sub-methodologies [11] that have been subject to research for a longer time (e.g., Agile, Lean, or Automation). This study builds on of these more mature fields adapting their knowledge to shape a DevOps human factor taxonomy. The approach of adapting knowledge from more mature fields or disciplines has been encouraged and applied successfully in Information Systems research [15]. In contrast, Software Engineering methodologies have suffered from a lack of cumulative research, as they have been reported to keep "reinventing the wheel" [10]. The overall objective of this paper is to characterize DevOps culture from the perspective of human factors, and to identify which human factors affect DevOps adoption, and how do they affect it.

The rest of this document is structured as follows: Sect. 2 presents the methodology followed in this study, Sect. 3 reports the results given by the methodology applied, Sect. 4 elaborates over the possible threats to validity of this study's methodology, Sect. 5 reports the conclusions of this paper and provides the lines of future work.

2 Methodology

In order to address the problems detailed in the previous section, a Systematic Mapping Study was performed [20]. A Systematic Mapping Study is a methodology for secondary studies with which to extract an overview from a research field, and classify the information regarding the topic selected. While it shares some commonalities with a Systematic Literature Review, they do not share the main purpose, which in the case of a Systematic Mapping Study is the establishment of the structure from a research field, without the synthesis of evidence that systematic literature reviews require. To the extent of our knowledge, no other secondary studies have been performed to identify human factors that affect DevOps adoption. We chose the Systematic Mapping Study methodology as opposed to a Systematic Literature Review in order to provide background knowledge and structure for a research topic (DevOps human factors) that has not been explored yet. This paper follows the Systematic Mapping Study guidelines provided by Petersen et al. [21], complemented by the guidelines for secondary studies proposed by Kitchenham et al. [14].

Table 1. Research questions and motivation.

ID	Research question	Motivation
RQ1	In which years, venues and sources have the selected papers been published?	To identify time tendencies and the different venues and sources (journals, conferences, books)
RQ2	Which research types and methods were used in the selected papers?	To determine the frequency of research strategies and types of empirical studies adopted for evaluation
RQ3	In which contexts have human factors been identified?	To identify tendencies in research fields (agile, lean, software process improvement, others)
RQ4	Which human factors affect DevOps adoption?	To extract the human factors that have an impact on DevOps adoption
RQ5	How do human factors affect DevOps adoption?	To understand the effect of human factors on DevOps adoption

2.1 Research Questions

Table 1 presents the research questions along with their motivation.

2.2 Search Strategy

The main search strategy selected for this review was an automated search, which was complemented with the snowballing technique. The search was carried out on September, 28, 2021.

2.3 Database Selection

ACM Digital Library (https://dl.acm.org), IEEE Xplore (https://ieeexplore.ieee.org), Scopus (https://www.scopus.com) and Web of Science (https://www.webofscience.com) were the four online databases selected for the search process. Following the guidelines of Petersen et al. [21], two publishing databases and two indexing databases were selected, with the inclusion of ACM Digital Library and IEEE Xplore as the two publishing databases.

2.4 Search String

The search string was proposed using keywords extracted from PICO(C) (Population, Intervention, Comparison, Outcomes, Context), relevant studies previously known by the authors, and a thesaurus. The different versions of the search string were iteratively improved and discussed until the preliminary results satisfied every author of this paper. The following search string was used for every database with only syntactic modifications:

("human factor" OR "human behavior" OR "human aspect") AND ("implementation" OR "adoption" OR "acceptance") AND ("software engineering" OR "software development" OR "devops" OR "agile" OR "continuous practices" OR "software process" OR "continuous development" OR "continuous deployment")

The search scope in every database was title, abstract and keywords. Since DevOps emerged in 2009 [7], a search filter excluding studies published before 2010 was applied when possible to avoid papers that were published before DevOps was introduced into scientific literature. The search string was built based on three main semantic blocks, "human factors" AND "adoption" AND "DevOps or DevOps-related methodologies", where every block includes as much relevant synonyms as possible.

2.5 Selection Criteria

The selection process was carried out according to the following criteria:

- Inclusion criteria
  - IC1: Studies in the field of software engineering concerning methodologies, techniques, processes, tools or practices used in DevOps contexts.
  - IC2: Studies published in journals or conferences or as book chapters.
  - IC3: Studies written in English.
  - IC4: Studies published from 2010 to the date of the search.
- Exclusion criteria
  - EC1: Papers that did not include any human factors or did not focus on software engineering and the adoption of technology.
  - EC2: Papers available only in the form of abstracts, guidelines, reports, or PowerPoint presentations.
  - EC3: Duplicate papers.
  - EC4: Studies for which the full text was not accessible.

2.6 Quality Assessment Criteria

The quality of the papers selected was assessed by employing the following Quality Criteria (QCs) based on the studies by Khan et al. [13] and Shameem et al. [25]:

- QC1: Are the aims/objectives of the research clearly stated?
- QC2: Are the threats to the validity of the methodology specified?
- QC3: Did the study provide sufficient details for it to be repeatable?
- QC4: Are the results and findings clearly reported?
- QC5: Are the conclusions supported by the results of the study?
- QC6: Is the data analysis clearly described and systematic?

The first two authors of this paper assessed each one of the QCs assigning a score of 0 in case the criteria was not fulfilled, 0.5 if it was partially fulfilled, or 1 if the criteria was completely fulfilled. After calculating the sum of every QC for every paper, only papers with a total score of 3 or more were included in the final papers. A full-text review was done for every paper in order to assess its quality. The quality criteria assessment form can be accessed online (https://bit.ly/3LpOrh1).

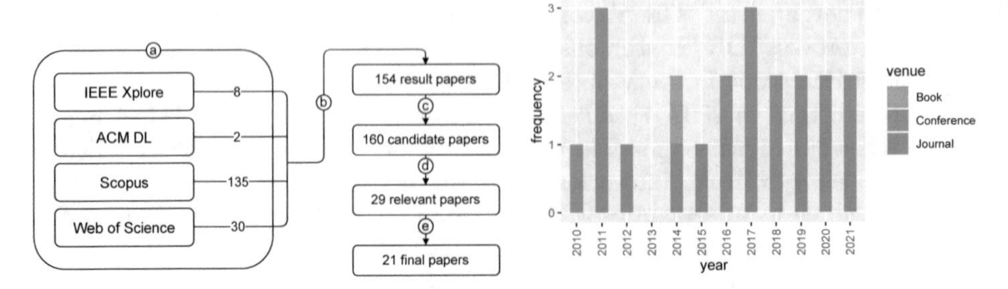

Fig. 1. Systematic mapping study process diagram. Legend: a) Perform the automatic search using the search string. b) Collect the results returned by the automatic search. c) Snowballing. d) Apply the selection criteria to the candidate papers. e) Apply the quality criteria to the relevant papers.

Fig. 2. Publication venues over 11 years tendency.

2.7 Data Extraction

For every paper included in the final selection for the review, the following information was collected: 1) the authors, 2) title, 3) publication venue, 4) venue name, 5) year of publication, 6) research type of the paper, 7) research method of the paper, 8) list of human factors identified, 9) the description of each human factor, and 10) the category where every human factor was included. The data extraction form can be accessed online (https://bit.ly/3EUKA9a).

3 Results

A set of 154 candidate papers were returned from the databases listed previously. Figure 1 details the selection process. Table 2 shows the final list of selected papers after the process detailed in Sect. 2.

3.1 RQ1. In Which Years, Venues, and Sources Have the Selected Papers Been Published?

Figure 2 shows the selected papers divided by venue type from January 2010 to September 2021, when the search was conducted. A stable publication tendency over the last 11 years can be observed, with a maximum of 3 papers published in 2011 and 2017, and a minimum of 0 papers published in 2013. A 57% of the selected papers (12) were published in journals, followed by 38% (8) in conferences, and 5% (1) in books. A slight increase in journal publication can be seen in the last 3 years.

Table 2. List of final papers. The complete reference details can be consulted in https://bit.ly/3EUKA9a.

ID	Title
1	Understanding Conflicts in Agile Adoption through Technological Frames
2	Software Process Improvement Programs: What happens after official appraisal?
3	A conceptual lean implementation framework based on change management theory
4	Empirical Investigation on Agile Methods Usage: Issues Identified from Early Adopters in Malaysia
5	How to Make Lean Cellular Manufacturing Work? Integrating Human Factors in the Design and Improvement Process
6	Lean continuous improvement to information technology service management implementation: Projection of ITIL framwork
7	Lean Implementation Strategies: How are the Toyota Way Principles addressed?
8	Scared, frustrated and quietly proud: Testers' lived experience of tools and automation
9	Software process improvement: A organizational change that need to be managed and motivated
10	How human aspects impress Agile software development transition and adoption
11	From Empowerment Dynamics to Team Adaptability: Exploring and Conceptualizing the Continuous Agile Team Innovation Process
12	Lean manufacturing and operational performance
13	Persuading Software Development Teams to Document Inspections: Success Factors and Challenges in Practice
14	Controlled Experiments as Means to Teach Soft Skills in Software Engineering
15	Human Aspects of Agile Transition in Traditional Organizations
16	Gamification solutions for software acceptance: A comparative study of Requirements Engineering and Organizational Behavior techniques
17	Getting the Best out of People in Small Software Companies
18	Uncertainty, personality, and attitudes toward devops
19	Agile methods and organizational culture: Reflections about cultural levels
20	Sustainable Risk Management in IT Enterprises
21	The unified theory of acceptance and use of technology (UTAUT): A literature review

3.2 RQ2. Which Research Types and Methods Were Used in the Selected Papers?

Figure 3 presents the research method employed in the selected papers according the study by Petersen et al. [21]. Figure 4 details the research method used in each selected paper according to Wieringa et al. [30] and Petersen et al. [21]. Petersen et al. did not include the categories "Practitioner targeted interview" nor "Literature review", but we have employed them in order to better categorize the data.

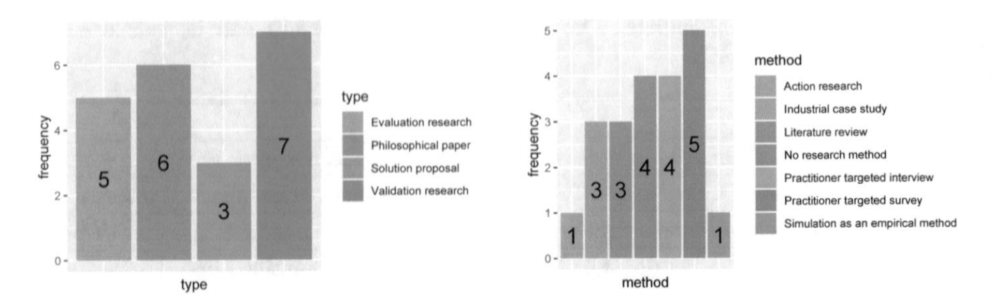

Fig. 3. Research type tendency.

Fig. 4. Research method frequency.

Figure 3 presents a 57% of empirical studies (Evaluation research and Validation research) while 43% did not include an empirical component (Philosophical papers and Solution proposals).

As can be seen in Fig. 4, the majority (57%) of the papers employed research methodologies that are closely related to a validation in industry (Industrial case study, Practitioner targeted interview, Practitioner targeted survey).

3.3 RQ3. In Which Contexts Have Human Factors Been Identified?

Table 3 shows every context in which the selected papers focused that is related to DevOps, along the description of that particular context's relationship with DevOps. Regarding the context where the human factors were identified, 33% of the papers (7) identified human factors in the context of Agile software development, followed by 24% of papers (5), which were related to Lean Management. Only one paper directly addressed DevOps.

3.4 RQ4. Which Human Factors Affect DevOps Adoption?

A total of 59 human factors were identified in the selected papers. The factors were synthesized based on the names and descriptions available on the selected papers. Merging factors that referred to the same concept was the only modification that was applied to the original data. The list of identified factors, the synthesized definitions, their original definitions, and their paper of origin, can be accessed online (https://bit.ly/3OGZVi9). A simpler version with the list of synthesized human factors and definitions is available at https://bit.ly/46d6Kjc.

3.5 RQ5. How do Human Factors Affect DevOps Adoption?

Based on the synthesized definitions provided in the previous research question results, the human factors could be separated into those that are positive if present (and negative if absent, e.g. "Motivation for change"), those that are negative if present (and positive if absent, e.g. "Negative experiences"), and those that can be positive or negative depending on how are they applied in

Table 3. DevOps human factor contexts.

Context	Relation with DevOps	Pap.
Agile methodology	DevOps extends the agile methodology in terms of the principles involved, as DevOps can provide a pragmatic extension of current agile activities [11]. Agile methods can be considered as enablers for the adoption of DevOps thinking [9]	1, 4, 10, 11, 15, 19, 20
Lean Management (LM)	DevOps and lean methodology share similar goals and can effectively complement each other. DevOps and lean are key elements that achieve rapid software development and validated learning [24]	3, 5, 6, 7, 12
Software Process Improvement (SPI)	Software process improvement (SPI) will help DevOps to enhance quality of software product [19]. This program will assist DevOps organizations to successfully manage and improve their processes so as to achieve the actual benefits of DevOps [2]	2, 9
Software development teams	DevOps practices have improved the performance of software development teams by establishing cross-functional teams and providing closer collaboration with customers [17]. DevOps is promoting collaboration between development and operation teams [18]	13, 14
Technology adoption	The adoption of DevOps is an initiative toward the adoption of new technology [8]. DevOps has led to an innovation from a technology perspective [16]	16, 21
Process management	Process management helps in DevOps adoption. Challenges as regards managing current changes and DevOps adoption can be resolved by managing the DevOps process [12]	17
Automation	DevOps encourages automation, which helps improve the quality of releases [23]. DevOps has attempted to use automated systems to bridge the information gap between project team entities and to enforce rigorous processes to ensure real-time communications [4]	8
DevOps	The focus of this paper is DevOps human factors	18

an organization (e.g. "Social influence"). We labeled them as Positive human factors, Negative human factors, and Ambivalent human factors, respectively. A total of 42% of the human factors identified (25) were classified as positive human factors, 39% (23) were classified as ambivalent, and 19% (11) were classified as negative. The effect of each human factor extracted from its definition its available at https://bit.ly/3MHGsyw.

4 Threats to Validity

The threats to validity have been assessed according the validity dimensions identified by Wohlin et al. [31]. In order to address possible threats to validity, the search string was built and tested iteratively, the search was conducted in multiple databases (Construct validity), the selection process was documented and the results given were reviewed by every author of this paper and the data extraction was properly documented and thoroughly reviewed and agreed (Internal validity), the analysis was performed directly from the data extracted without any selection nor modification of the data involved, and the data was made publicly available (Conclusion validity), and the subject and results of the study are timely, relevant, and could be applied in a practical industrial environment (External validity).

5 Conclusions and Future Work

This paper contributes to DevOps state of art extending the knowledge regarding DevOps human factors by adapting the knowledge available from other DevOps-related contexts. For researchers, this knowledge extends what is known about DevOps culture, which has been partially defined from the perspective of human factors. For practitioners, the knowledge regarding DevOps human factors can improve the process of tailoring a DevOps adoption process, as we provide a detailed list of human factors that should be taken into account.

We find the additive and extensive empirical research the main line of future work. In addition, the DevOps human factors research could benefit from the knowledge adaptation from other disciplines, such as Psychology, which may provide a much deeper understanding of the information that is gathered regarding human behaviour.

Acknowledgements. This research is part of the OASSIS-UMU project (PID2021-122554OB-C32), supported by the Spanish Ministry of Science and Innovation and the European Regional Development Fund.

References

1. Amrit, C., Daneva, M., Damian, D.: Human factors in software development: on its underlying theories and the value of learning from related disciplines. Inf. Softw. Technol. **56**, 1537–1542 (2014)
2. Badshah, S., Khan, A.A., Khan, B.: Towards process improvement in DevOps: a systematic literature review. In: ACM International Conference on Proceedings Series, pp. 427–433 (2020)
3. Capretz, L.F.: Bringing the human factor to software engineering. IEEE Softw. **31**, 104–104 (2014)
4. Cois, C.A., Yankel, J., Connell, A.: Modern DevOps: optimizing software development through effective system interactions. In: IPCC, pp. 1–7 (2014)
5. Curtis, B., Krasner, H., Iscoe, N.: A field study of the software design process for large systems. Commun. ACM **31**, 1268–1287 (1988)
6. Díaz, J., López-Fernández, D., Pérez, J., González-Prieto, Á.: Why are many businesses instilling a DevOps culture into their organization? EMSE **26**, 25 (2021)
7. Debois, P.: DevOps: a software revolution in the making? Cutter IT J. **24**, 1–41 (2011)
8. Gill, A.Q., Loumish, A., Riyat, I., Han, S.: DevOps for information management systems. VINE J. Inf. Knowl. Manag. Syst. **48**, 122–139 (2018)
9. Hosono, S.: A DevOps framework to shorten delivery time for cloud applications. Int. J. Comput. Sci. Eng. **7**, 329–344 (2012)
10. Hron, M., Obwegeser, N.: Why and how is scrum being adapted in practice: a systematic review. J. Syst. Software **183**, 111110 (2022)
11. Jabbari, R., bin Ali, N., Petersen, K., Tanveer, B.: What is DevOps? In: XP, vol. 24-May-201, pp. 1–11 (2016)
12. Jones, S., Noppen, J., Lettice, F.: Management challenges for DevOps adoption within UK SMES. In: International Workshop on Quality-Aware DevOps, pp. 7–11 (2016)

13. Khan, A.A., Keung, J.: Systematic review of success factors and barriers for software process improvement in global software development. IET Software **10**, 125–135 (2016)
14. Kitchenham, B., Brereton, P.: A systematic review of systematic review process research in software engineering. Inf. Softw. Technol. **55**, 2049–2075 (2013)
15. Kjærgaard, A., Vendelø, M.T.: The role of theory adaptation in the making of a reference discipline. Inform. Organ. **25**, 137–149 (2015)
16. Koilada, D.K.: Business model innovation using modern DevOps. In: IEEE Technology & Engineering Management Conference, pp. 1–6. IEEE (2019)
17. Lwakatare, L.E., et al.: Towards DevOps in the embedded systems domain: why is it so hard? In: HICSS, vol. 2016-March, pp. 5437–5446 (2016)
18. Lwakatare, L.E., Kuvaja, P., Oivo, M.: Dimensions of DevOps. In: Lassenius, C., Dingsøyr, T., Paasivaara, M. (eds.) XP 2015. LNBIP, vol. 212, pp. 212–217. Springer, Cham (2015). https://doi.org/10.1007/978-3-319-18612-2_19
19. Mishra, A., Otaiwi, Z.: DevOps and software quality: a systematic mapping. Comput. Sci. Rev. **38**, 100308 (2020)
20. Petersen, K., Feldt, R., Mujtaba, S., Mattsson, M.: Systematic mapping studies in software engineering. In: EASE, pp. 1–10 (2008)
21. Petersen, K., Vakkalanka, S., Kuzniarz, L.: Guidelines for conducting systematic mapping studies in software engineering. Inf. Softw. Technol. **64**, 1–18 (2015)
22. Restrepo-Tamayo, L.M., Gasca-Hurtado, G.P.: Human aspects in software development. In: Wong, L.H., Hayashi, Y., Collazos, C.A., Alvarez, C., Zurita, G., Baloian, N. (eds.) CollabTech. LNCS, vol. 13632, pp. 1–22. Springer, Cham (2022)
23. Riungu-Kalliosaari, L., Mäkinen, S., Lwakatare, L.E., Tiihonen, J., Männistö, T.: DevOps adoption benefits and challenges in practice: a case study. In: Abrahamsson, P., Jedlitschka, A., Nguyen Duc, A., Felderer, M., Amasaki, S., Mikkonen, T. (eds.) PROFES 2016. LNCS, vol. 10027, pp. 590–597. Springer, Cham (2016). https://doi.org/10.1007/978-3-319-49094-6_44
24. Rodríguez, P., Mäntylä, M., Oivo, M., Lwakatare, L.E., Seppänen, P., Kuvaja, P.: Advances in using agile and lean processes for software development. Adv. Comput. **113**, 135–224 (2019)
25. Shameem, M., Kumar, C., Chandra, B., Khan, A.A.: Systematic review of success factors for scaling agile methods in global software development environment. In: APSEC, vol. 2018-January, pp. 17–24 (2017)
26. Smeds, J., Nybom, K., Porres, I.: DevOps: a definition and perceived adoption impediments. In: Lassenius, C., Dingsøyr, T., Paasivaara, M. (eds.) XP 2015. LNBIP, vol. 212, pp. 166–177. Springer, Cham (2015). https://doi.org/10.1007/978-3-319-18612-2_14
27. Sánchez-Gordón, M., Colomo-Palacios, R.: Characterizing DevOps culture: a systematic literature review. In: Stamelos, I., O'Connor, R.V., Rout, T., Dorling, A. (eds.) SPICE 2018. CCIS, vol. 918, pp. 3–15. Springer, Cham (2018). https://doi.org/10.1007/978-3-030-00623-5_1
28. Toh, M.Z., Sahibuddin, S., Mahrin, M.N.: Adoption issues in DevOps from the perspective of continuous delivery pipeline. In: ICSCA, vol. F1479, pp. 173–177 (2019)
29. Wettinger, J., Breitenbücher, U., Leymann, F.: DevOpSlang – bridging the gap between development and operations. In: Villari, M., Zimmermann, W., Lau, K.-K. (eds.) ESOCC 2014. LNCS, vol. 8745, pp. 108–122. Springer, Heidelberg (2014). https://doi.org/10.1007/978-3-662-44879-3_8

30. Wieringa, R., Maiden, N., Mead, N., Rolland, C.: Requirements engineering paper classification and evaluation criteria. Req. Eng. **11**, 102–107 (2006)
31. Wohlin, C., Runeson, P., Höst, M., Ohlsson, M.C., Regnell, B., Wesslén, A.: Experimentation in Software Engineering. Springer, Heidelberg (2012). https://doi.org/10.1007/s00766-005-0021-6

Creating Textual Corpora Based on Wikipedia and Knowledge Graphs

Janneth Chicaiza[1(✉)], Mateo Martínez-Velásquez[1], Fabian Soto-Coronel[1], and Nadget Bouayad-Agha[2]

[1] Universidad Técnica Particular de Loja, Loja, Ecuador
{jachicaiza,msmartinez12,fasoto}@utpl.edu.ec
[2] Universitat Oberta de Catalunya, Barcelona, Spain
nbouayad_agha@uoc.edu

Abstract. Information overload reduces the capability of machines to find relevant information. Furthermore, when dynamic topics or emerging events occur that arouse the interest of the community, unofficial or unreliable sources of information quickly emerge that, instead of satisfying the information needs of users, increase misinformation. To address this issue, this paper proposes a method to create domain-specific corpora of text that can offer immediate answers on a particular topic. The approach involves creating a vocabulary of the domain and then creating a textual corpus from Wikipedia pages related to the different terms of the domain. The authors tested this method by creating a specialized corpus for the pollution domain and implementing a process to answer queries about the domain. Preliminary results show that the Q&A system could provide accurate and up-to-date information on the topic, based on Wikipedia, a free-content platform that users continuously feed.

Keywords: Controlled vocabulary · DBPedia · knowledge graphs · NLP applications · SKOS · Wikipedia

1 Introduction

On the Web, one of the most popular applications for finding information is search engines. This type of tool is effective for indexing large amounts of information. However, using a general search engine when trying to find answers to domain-specific questions can be a bad decision because the engines return large amounts of results, many of them irrelevant. One of the reasons this problem occurs is due to the inability of machines to understand users' information needs and to understand natural language. Today, thanks to advances in artificial intelligence, and specifically deep learning and natural language processing (NLP), the understanding of human language by machines has improved several tasks.

To provide accurate responses to users' requests for information, another aspect to consider is that, when it comes to dynamic topics or emerging events, that arouse the immediate interest of the community, unofficial or unreliable sources of information quickly emerge that, instead of satisfying the information needs of users, increase misinformation. Therefore, in this proposal, we focus

Á. Rocha et al. (Eds.): WorldCIST 2024, LNNS 987, pp. 325–337, 2024.
https://doi.org/10.1007/978-3-031-60221-4_32

on the construction of domain-specific corpora of text that are suitable to offer specific and immediate answers on a particular topic, which can change over time and therefore, it is crucial that the knowledge base be kept updated. The central sources of the proposal are knowledge graphs and the most used encyclopedia in the world, Wikipedia [9].

Wikipedia is a widely used data source for corpus creation in various areas of computational linguistics and natural language processing, such as information extraction [18], text classification [8], information extraction [22], named entity recognition [11], text summarization [19], question answering (Q/A) systems [4], coreference resolution [2], evaluation and benchmarking [15] among others tasks and applications. It is also widely used for training large language models (LLM) such as OpenAI's GPT models, Google's PaLM, Meta's LLaMa, BLOOM, Ernie 3.0 Titan, and Anthropic's Claude, among others. This is because, despite the biases, errors, or criticisms, Wikipedia contains reliable content because it is created, debated, and curated by people [6].

Due to recent advances in generative artificial intelligence and LLM, machines can gain a better understanding of natural language, thus making better predictions and creating responses that mimic humans. However, using LLM for NLP tasks carries some challenges. First, to address the hallucination phenomenon. According to [12], hallucination occurs because LLM can only use the information they have been trained on, and when there is uncertainty, machines invent facts. Second, LLM can get worse performance over time if they don't receive a steady supply of original content written by humans. These two issues have made Wikipedia and other sources of human-generated content even more valuable [6].

Therefore, to face these limitations, there are at least two strategies: i) improving LLM with external knowledge and automated feedback [17], and ii) updating embedding representations of text, created with a particular language model, from new knowledge acquired from new textual content. In this proposal, we focus on the second strategy.

This paper describes a proposal for creating a close-domain corpus from Wikipedia pages. By using such an open and collaboratively created source, we attempt to meet the demand for information from a particular domain, expressed as questions that the community may have and concerns that may arise for a topic or during an event of global interest.

The proposal guides through the creation of a document store and includes two main stages: i) creation of a controlled vocabulary of a domain from a knowledge graph (DBPedia), and ii) extraction and contextualized chunking of content from a subset of Wikipedia pages that are related to terms of the controlled vocabulary. In Sect. 3, after describing the motivation and background of this research in Sect. 2, we present our proposal. Later, in Sect. 4 we present a concept test where the specialized corpus created for the pollution domain is used for a Q/A task, getting promising results. Finally, in the last section, we present the conclusion.

2 Motivations and Background

Nowadays, Wikipedia is one of the largest, and most dynamic and accessible sources of information online; for these reasons, it has become an invaluable source of data for various NLP applications. However, in certain contexts or usage scenarios, it is necessary to create specialized corpora from Wikipedia pages related to a specific topic. In this section, we will explore the reasons and benefits of creating specialized corpora based on Wikipedia, and how these resources can power various NLP applications, specifically Q/A systems.

2.1 Why Create a Specialized Wikipedia Corpus?

Among the reasons why we select Wikipedia, as a valuable source of information, is because it is constantly evolving as it contains updated information on events and topics in which people are interested [3]. Wikipedia is also interesting as a source of information because it can be leveraged [13] to provide answers to any factual question [3] even in open and large-scale domains. Finally, another reason to select Wikipedia as a source of information for different downstream applications is the ease of automatically monitoring and detecting changes in the content of the pages. In [20], the authors propose a natural disaster monitoring system based on Wikipedia. This approach can be adapted to monitor any event of community interest.

Given the general characteristics of Wikipedia, the following arguments support the idea of creating a specialized corpus based on it:

1. Context and Relevance. According to [14], context and relevance are essential to understanding the meaning of words and structures in a text. By focusing a corpus on a given topic, a richer and more specific context is provided compared to the entire Wikipedia corpus.
2. Noise Reduction. Wikipedia is based on an open-editing model that allows users from around the world to contribute and edit its content, and even though the project strives to maintain a high standard of quality and reliability, it is not immune to issues related to low-quality information [21]. Since data quality plays a critical role in the performance of NLP models [7], by creating a specialized corpus, noise or incorrect information can be identified and reduced because editors, domain experts, and machines can collaborate to ensure the accuracy of the corpus information.
3. Training language models. In some usage scenarios, NLP models are trained on specialized corpora for performing specific tasks, such as machine translation, medical information extraction, or legal document classification. Training models on domain-specific data can improve their performance because this action enables models to be adapted to specific tasks. Also, specialized corpora are essential for the development and evaluation of highly effective NLP models in real-world applications.

In conclusion, domain-specific language models, tuned or trained on specialized corpora contain language and terminology specific to this domain, thus these models can better understand and generate text relevant to that domain.

2.2 Use of Domain-Specific Corpora for Question-Answering Systems

As was mentioned before, specialized corpora, in addition to improving the performance of language models for a specific domain, can be useful to adapt NLP applications, like QA systems, to specific domains.

According to [10] in closed-domain Q/A systems, the goal is to find answers to questions within a specific domain, and the challenge is to train models with limited data. In this sense, creating a domain-specific corpus based on open sources like Wikipedia can alleviate the lack of information, and in contrast to open-domain systems, the quality of answers is expected to be higher [1].

Other advantages of these systems are the ease of removing noise and ambiguities and the ability to provide detailed responses or explanations because they have a deeper knowledge of the domain. These advantages make closed-domain systems ideal for applications and use cases where accuracy and consistency are crucial. Their domain-specific focus allows them to provide high-quality answers and improve user experience in a variety of contexts.

Using a specialized corpus based on Wikipedia pages, a Q/A system can exhibit other advantages, such as ease of personalization for a particular domain, catching information about new events, and keeping previous information updated. These features are possible to reach by using the Wikipedia API and querying the knowledge graphs built around Wikipedia.

3 Creation Method for a Specialized Corpus

To create a text corpus based on Wikipedia pages and specialized in a particular topic or domain, we propose a method composed of two main phases: 1) creation of a terminological vocabulary of the domain, defined from published knowledge organization systems as RDF graphs; and 2) creation of a textual corpus from the Wikipedia pages that are related to the different terms of the domain. That is, to identify the pages that deal with a domain, we take advantage of the terms or concepts that are connected in graphs such as DBPedia.

The following sections describe each of the phases that we propose.

3.1 Creation of the Controlled Vocabulary

The goal of this phase is to create a semantic base of interrelated topics or concepts. In this paper, we propose using knowledge graphs as a source of terminological information related to the domain of interest.

Knowledge graphs described by vocabularies such as SKOS[1] facilitate the identification of terms or concepts in a domain. This set of concepts can then be used as a reference to find Wikipedia pages related to terms or topics in a particular domain.

When we use RDF-based knowledge graphs, such as DBPedia or Wikidata, we can query the underlying data using the SPARQL language[2], which is a

[1] https://www.w3.org/TR/swbp-skos-core-spec/.
[2] https://www.w3.org/TR/rdf-sparql-query/.

standard query language recommended by W3C to recover RDF data. Taking the SPARQL EndPoint service provided by each graph provider as a gateway, it is possible to automatically obtain information related to the terms of interest.

Figure 1 shows our proposal to create a vocabulary of terms that represents a particular domain.

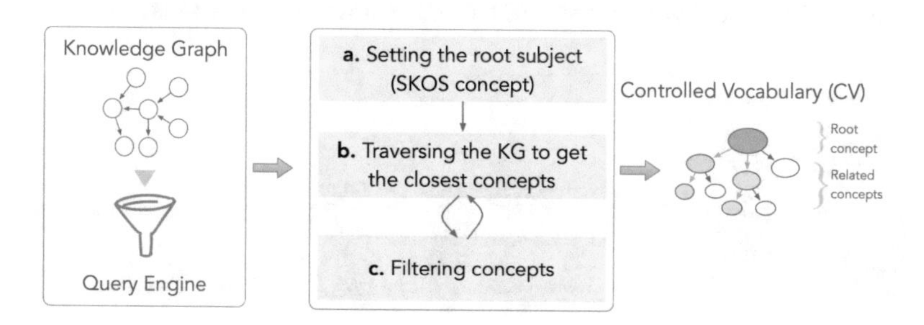

Fig. 1. Components of the creation method of a controlled vocabulary.

As shown in Fig. 1, the creation method consists of three steps:

1. The first step is to define a root node, that is, the broader SKOS concept that defines the domain.
2. Starting from the root node, the query engine uses SPARQL to traverse the knowledge graph and obtain the concepts related to the root concept. The traversing of the graph must be performed using relations that lead to the terms of the domain. For SKOS-based datasets like DBPedia Categories, the *skos:broader* property is used to traverse the graph. This property and others allow us to apply an iterative process of traversing the graph, starting from the root node. The traversing path process is based on the Breadth First Search (BFS) algorithm and allows us to collect the terms close to the root node, which are at a maximum distance. This second step can be carried out for several moments until an acceptable distance is reached, which according to authors like [5] is recommended to do up to three iterations.
3. The third step is interactive because it allows the user to choose which concepts interest them or are relevant to the domain. This step is carried out in conjunction with the previous step, that is, with each iteration carried out to obtain more sub-concepts, it is recommended that the user select those that are important, before expanding the terminological network.

In this paper, DBPedia[3] was leveraged as the source of knowledge to create the controlled vocabulary. DBPedia was one of the first projects focused on publishing linked open data in a structured and readable format for both humans and machines. Furthermore, for this phase, we take advantage of the *DBPedia Categories* subdomain defined by SKOS to identify the relevant terms of a domain.

[3] https://www.dbpedia.org/about/.

3.2 Creation of the Document Store

To create the specialized corpus, we extracted textual information from Wikipedia.

Figure 2 describes the three fundamental stages of the process followed to create the repository of indexed documents. For each stage, the tasks performed are indicated and the outputs generated are shown for each task.

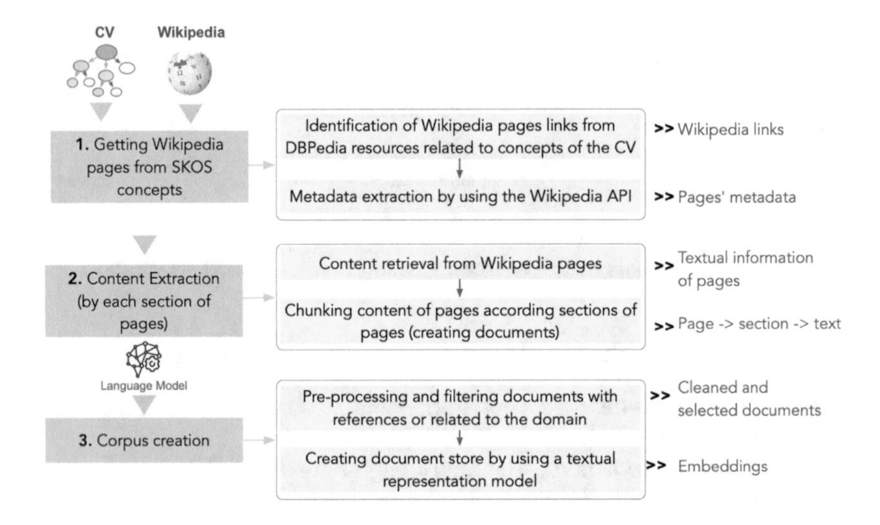

Fig. 2. Creation process for the document store.

The most important aspects of the process are described below:

1. Getting Wikipedia pages from SKOS concepts. This stage aims to identify the Wikipedia pages specialized in a domain and get their metadata. To obtain this result, we propose two steps:
   (a) Starting from the set of terms or SKOS concepts that are part of the controlled vocabulary (CV), created during the previous phase (see 3.1), we design a query to traverse the graph and reach all the Wikipedia pages connected to the DBPedia resources which are related to the SKOS concepts. At this point, we use the relation *dcterms:subject* to find all the DBpedia resources connected to the concepts, and the relation *foaf:isPrimaryTopicOf* to reach the links of the Wikipedia pages from the DBPedia resources. The query 1.1 summarizes this step.
   (b) Once we identify the links of the Wikipedia pages related to one of the domain's topics, it is possible to use the Wikipedia API to obtain their metadata, such as page ID, creation date, modification date, title, etc. The first three metadata mentioned will be used later to detect possible changes in the content of the pages.

2. Extraction of content by sections. After selecting the Wikipedia pages that are related to a domain, during this stage, we propose to carry out the following tasks:
   (a) Extraction of the content of each page. Using libraries based on the Wikipedia API we obtain the textual content of each page.
   (b) Identification of the outline or table of contents of each page. Based on this information, the content of each page is divided according to its sections. The objective of disaggregating the content of each page by sections is to create text documents within a specific context. Thus, tasks such as content filtering and parsing using Information Retrieval and NLP methods will be more precise.

Listing 1.1. Query to retrieve the Wikipedia page links from a given concept

```
SELECT DISTINCT ?resource ?wikipage
WHERE {
    VALUES ?concept {dbc:Pollution}
    ?resource dcterms:subject ?concept.
    ?resource foaf:isPrimaryTopicOf ?wikipage}
```

3. Corpus creation. To create a semantic representation of the selected documents we perform two final tasks:
   (a) Pre-processing and filtering documents consist of cleaning the text, selecting the documents that are related to the topics of the domain, and discarding those texts that do not contain related information. To identify relevant documents from those that are not, we can apply two strategies: i) apply some clustering method to determine documents that belong to isolated clusters, or ii) verify that any of the domain terms exist in the content of the document, that is, use the list of vocabulary terms that was built in the previous stage 3.1.
   (b) Creation of the document store. This offline task aims to create the vector representations associated with each text document. To carry out this task we could use any language model, such as Bag-of-words, word embeddings or based on transformers [16].

4 Concept Testing: Specialized Corpus to Provide Answers for Pollution Domain

In this section, we test the defined process by implementing a web prototype, which enables the creation of a specialized corpus in a particular domain. The online interface guides the user during the creation of the specialized corpus, according to the process defined in Fig. 2.

For testing, we created a corpus specialized within the domain of pollution domain, then, we used the corpus to feed a Q/A system and answer queries related to that domain.

For the prototype implementation, we employed some Python frameworks and libraries. Specifically, we utilized Django to build the web application, SPARQLWrapper to make efficient queries in semantic databases like DBpedia, Pandas to conduct efficient analysis and manipulation of the acquired data, and the Wikipedia library to simplify access to Wikipedia's knowledge base.

Below, we describe important aspects related to corpus creation.

4.1 Creation of the Controlled Vocabulary

As we explained in 3, the terminological vocabulary defines the topics of interest within a domain, therefore, the first step was to create the CV. Figure 3 illustrates the steps that the user can perform in the web interface of the created prototype.

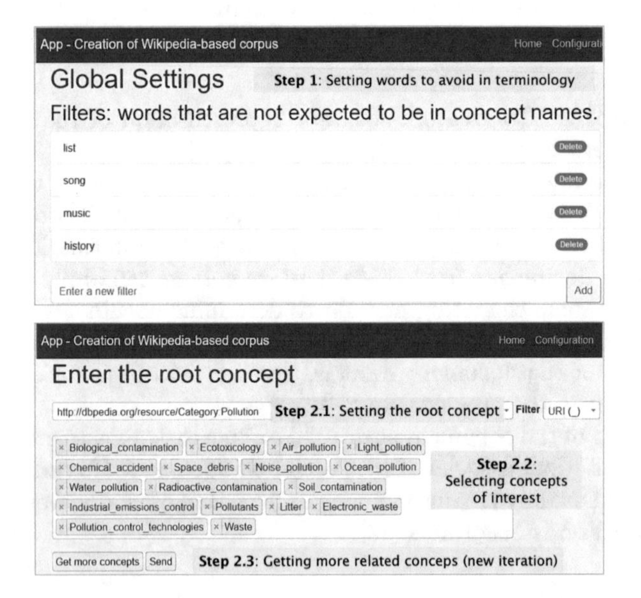

Fig. 3. Web interface that allows the user to construct a controlled vocabulary.

The top of the Fig. 3 shows that the user must perform a previous step, before creating the CV: set certain terms or words (the equivalent of stop words in NLP) that should not be present in the names of concepts or topics. In DBPedia, the source used as a basis to create the CV, there are SKOS concepts that are not very informative, that is, they do not group pages of content relevant to an academic domain such as the word *list* that is included in the concept *dbc:Pollution-related_lists*.

After establishing the stop words, the user must set a root node (Step 2.1), that is, the SKOS concept that represents the most general topic within the domain of interest. To define the root concept, the interface allows the user to specify it in two ways: i) enter the name of a topic, or ii) indicate the URI of the DBPedia concept. In Fig. 3 it is observed that the user uses the second option, which facilitates the work of locating the related concept.

Once the user sets the root node, below, the application will connect to DBPedia SPARQL EndPoint[4], to obtain its subconcepts. The prototype uses the relationship *skos:broader* and through SPARQL queries obtain the related concepts of the domain. When the found concepts are displayed, the user has the possibility of removing concepts that are not of interest (Step 2.2).

In the next step (Step 2.3), the user, iteratively, can interact with the application to choose the depth to which they want to traverse the DBPedia graph and obtain the concepts connected to the concepts found during the previous iteration. In this way, the pilot application automatically obtains information related to the topic of interest and thus builds the controlled vocabulary of the domain.

In this case, after setting the concept *dbc:Pollution* as the root node and traversing the graph up to three levels deep, we obtained the following resources:

- 17 concepts during the iteration 1. For example: *dbc:Biological_contamination & dbc:Air_pollution*
- 73 concepts during the iteration 2. For example: *dbc:Hazardous_air_pollutants*
- 233 during the iteration 3. For example: *dbc:Chlorine*

As a result of this phase, the controlled vocabulary was made up of around 280 unique concepts related to pollution.

4.2 Creation of the Document Store

Based on the topics or concepts of the CV, the application uses the *dct:subject* predicate to find the DBPedia resources associated with those concepts. During the testing, we found more than 6 thousand resources associated with the 280 SKOS concepts. Finally, through the URIs of the DBPedia resources, the application identified 6,557 links to Wikipedia pages available online.

Once the application identifies the Wikipedia pages that describe some topic in the domain, it begins a batch process to obtain and pre-process the textual content of each page. Then, as we explained in 3.2, the sections of each page are taken as a reference to segment the content and thus create the corpus documents.

Finally, and before creating the vector representations of the text, we use the Haystack framework[5] to segment the text of each section up to 100 tokens. This

[4] http://dbpedia.org/sparql.

[5] https://haystack.deepset.ai.

length considers the framework recommendation: *we recommend that documents contain significantly fewer words. We have found decent performance with Documents around 100 words long*[6]. Table 1 presents an extract of the metadata and content of each document created after chunking the text of each section.

Table 1. A sample of the documents created for the Pollution corpus. Each line describes the metadata and the content of each document.

Wikipage ID	Title	Last Modification	Section	Sub-section	Split ID	Text
6389531	Environmental effects of aviation	2023-09-07 T23:41:36Z	Climate change	Factors	1	Airplanes emit gases (carbon dioxide, water vapor, nitrogen oxides or carbon monoxide
6389531	Environmental effects of aviation	2023-09-07 T23:41:36Z	Climate change	Factors	2	Airport ground vehicles, those used by passengers and staff to access airports,
6389531	Environmental effects of aviation	2023-09-07 T23:41:36Z	Climate change	Volume	1	By 2018, airline traffic reached 4.3 billion passengers with 37.8 million departures,

From the textual corpus, the prototype was configured to use the LLM *bigscience/bloom* and thus create the semantic representations of each document. Finally, Pinecone was used as a repository to store the vectors and thus facilitate the search and retrieval of information.

4.3 Potential Application: Q/A System

To demonstrate the potential use of the Pollution corpus created from Wikipedia pages, we built a function that receives a user query and returns the most relevant answer(s).

Starting from the document repository stored in Pinecone, the implemented function provides at least one relevant answer to the user. Using *google/flan-ul2* as a language model, the function is capable of answering questions focused directly on the collection of embeddings in Pinecone.

To have a preliminary idea about the value provided by the new corpus created, we compare the response given by the LLM and the other one that is given by the implemented function connected to the created store. Table 2 shows five questions and the output that is obtained, i) using the specific-corpus created, and ii) only using the LLM, i.e. only using the knowledge of the LLM without the update.

[6] https://docs.haystack.deepset.ai/docs/optimization.

Table 2. Answers that were provided for potential user's questions about Pollution.

Question	Answer without the new corpus	Answer with the new corpus
What is the Aerosol size-selective sampling?	Aeronef	To measure the respirable fraction of particles in air, a pre-collector is used with a sampling filter
How is the control of Noise pollution?	It is a good idea to use earplugs when you are sleeping	The Hierarchy of Controls concept is often used to reduce noise in the environment or the workplace
What are the effects of ozone depletion?	It causes an increase in skin cancers	Sunburn, skin cancer, and cataracts
What is Marine debris?	Marine debris is any man-made material that is discarded into the ocean	Particulates suspended in a liquid, usually water on the Earth's surface

As can be seen in Table 2, the best answers (second column) are those that are generated when the Q/A function is connected to the document store created from updated information from Wikipedia. Currently, we are working on a mechanism to detect changes in the content of Wikipedia pages related to the domain and their subsequent update in the document store.

5 Conclusion

In this paper, we propose a method to create a specialized document store based on Wikipedia. According to our proposal, to identify pages related to a specific domain, the first step is to create a controlled vocabulary for the domain from DBPedia categories. DBPedia-based KG offer a way to retrieve updated content for a domain, unlike LLMs, which may not work with online content.

As a source of textual information, we propose to use Wikipedia because, in this repository, content on various topics is continually updated, even more so on events that arouse the interest of the community. We are currently designing a method to update the content of a domain based on changes in Wikipedia pages over time.

The creation of domain-specific corpora based on Wikipedia is essential to driving the development and continuous improvement of a variety of natural language processing applications. These corpora offer context, precision, and relevance, resulting in more effective and applicable models in diverse fields. As the field of NLP advances, the importance of specialized Wikipedia corpora becomes even more evident in creating NLP systems that can accurately and consistently understand and generate content in specific domains.

During the concept testing, we created a specialized corpus for the pollution domain by using a web prototype; its interface eased the process and guided the users to interactively build the corpus. From this repository, we implement a function to answer queries about the domain. Preliminary results, with a small set of queries, demonstrate that the answers provided by a Q/A system, which we are working on, could provide accurate and up-to-date information on a topic. In future work, we will evaluate the system by using metrics and reviews of human users to improve the usability and reliability of the proposal.

Acknowledgements. The authors would like to thank the *Universidad Técnica Particular de Loja* for sponsoring this research. The work is supported by the project PROY_PROY_ARTIC_CE_2022_3693.

References

1. Almotairi, M., Fkih, F.: A review on question answering systems: Domains, modules, techniques and challenges. In: 38th International Business Information Management Association (IBIMA) (11 2021)
2. Azerkovich, I.: Employing Wikipedia data for coreference resolution in Russian. In: Filchenkov, A., Pivovarova, L., Žižka, J. (eds.) Artificial Intelligence and Natural Language, pp. 107–112. Springer International Publishing, Cham (2018)
3. Chen, D., Fisch, A., Weston, J., Bordes, A.: Reading wikipedia to answer open-domain questions. In: Proceedings of the 55th Annual Meeting of the Association for Computational Linguistics (ACL), vol. 1, pp. 1870–1879. ACM (Mar 2017)
4. Chicaiza, J., Bouayad-Agha, N.: Enabling a question-answering system for COVID using a hybrid approach based on wikipedia and Q/A Pairs. In: Nagar, A.K., Jat, D.S., Marín-Raventós, G., Mishra, D.K. (eds.) Intelligent Sustainable Systems, pp. 251–261. Springer Nature Singapore, Singapore (2022)
5. Chicaiza, J., Piedra, N., Lopez-Vargas, J., Tovar-Caro, E.: Domain categorization of open educational resources based on linked data. In: Klinov, P., Mouromtsev, D. (eds.) Knowledge Engineering and the Semantic Web, pp. 15–28. Springer International Publishing, Cham (2014)
6. Deckelmann, S.: Wikipedia's value in the age of generative ai. Tech. rep., Wikimedia (Jul 2023). https://wikimediafoundation.org/news/2023/07/12/wikipedias-value-in-the-age-of-generative-ai/
7. Frąckiewicz, M.: The importance of data quality in nlp. Tech. rep., TS2-Space (May 2023). https://ts2.space/en/the-importance-of-data-quality-in-nlp/
8. Han-Joon, K., Jiyun, K., Jinseog, K., Pureum, L.: Towards perfect text classification with wikipedia-based semantic naïve bayes learning. Neurocomputing **315**, 128–134 (2018). https://doi.org/10.1016/j.neucom.2018.07.002
9. Jemielniak, D.: Wikipedia: Why is the common knowledge resource still neglected by academics? Gigascience **8**(12) (2019)
10. Kia, M.A., Garifullina, A., Kern, M., Chamberlain, J., Jameel, S.: Adaptable closed-domain question answering using contextualized CNN-attention models and question expansion. IEEE Access **10**, 45080–45092 (2022). https://doi.org/10.1109/ACCESS.2022.3170466
11. Krishnan, A., Ziehe, S., Pannach, F., Sporleder, C.: Employing Wikipedia as a resource for named entity recognition in morphologically complex under-resourced languages. In: Proceedings of the 14th Workshop on Building and Using Comparable Corpora (BUCC 2021), pp. 28–39. INCOMA Ltd. (Sep 2021)
12. Lee, M.: A mathematical investigation of hallucination and creativity in gpt models. Mathematics **11**(10) (2023). https://doi.org/10.3390/math11102320
13. Lymperopoulos, P., Qiu, H., Min, B.: Concept Wikification for COVID-19. In: Proceedings of The 2020 Conference on Empirical Methods in Natural Language Processing (Aug 2020)

14. Manning, C.D., Raghavan, P., Schütze, H.: Introduction to information retrieval. Cambridge University Press (2008). https://doi.org/10.1017/CBO9780511809071
15. Mernyei, P., Cangea, C.: Wiki-CS: A Wikipedia-Based Benchmark for Graph Neural Networks. arXiv e-prints (Jul 2020). https://doi.org/10.48550/arXiv.2007.02901
16. Neelima, A., Mehrotra, S.: A comprehensive review on word embedding techniques, pp. 538–543 (2023). https://doi.org/10.1109/ICISCoIS56541.2023.10100347
17. Peng, B., et al.: Check your facts and try again: Improving large language models with external knowledge and automated feedback (Feb 2023). https://doi.org/10.48550/arXiv.2302.12813
18. Rodriguez-Ferreira, T., Rabadán, A., Hervás, R., Díaz, A.: Improving information extraction from wikipedia texts using basic english. In: International Conference on Language Resources and Evaluation (2016)
19. Sankarasubramaniam, Y., Ramanathan, K., Ghosh, S.: Text summarization using wikipedia. Inform. Process. Manage. **50**(3), 443–461 (2014). https://doi.org/10.1016/j.ipm.2014.02.001
20. Steiner, T., Verborgh, R.: Disaster Monitoring with Wikipedia and Online Social Networking Sites: Structured Data and Linked Data Fragments to the Rescue? (Jan 2015)
21. Sugandhika, C., Ahangama, S.: Assessing information quality of wikipedia articles through google's e-a-t model. IEEE Access **10**, 52196–52209 (2022). https://doi.org/10.1109/ACCESS.2022.3172962
22. Wu, F., Weld, D.S.: Open information extraction using wikipedia. In: Proceedings of the 48th Annual Meeting of the Association for Computational Linguistics, pp. 118–127. ACL '10, Association for Computational Linguistics, USA (2010)

Innovation of the Logistics Process of Grupo Logistico Especializado S.A.S. Through the Development and Implementation of the aGILE Platform

Carlos Hernán Fajardo-Toro[1] and Sandra Rodríguez Ardila[2]

[1] Escuela Superior de Administración Pública – ESAP, Bogotá, Colombia
Carlosh.fajardo@esap.edu.co
[2] Grupo Logístico Especializado S.A.S, Bogotá, Colombia
Sandra.rodriguez@glecolombia.com

Abstract. The implementation of the aGILE tool has had an impact on all areas of the company's logistics operations, leading to better organisational practices based on the identified opportunities and process optimization. aGiLE effectively integrates operational processes with the software of allied carriers and customers.

In addition to a brief conceptual background on digitalization, Industry 4.0 and information systems, this article will provide a brief description of the impact that the implementation of 4.0 technologies and the aGILE platform has had on the company.

Keywords: Industry 4.0 · Information Systems · process improvement · Cost reduction

1 Introduction

Organisations define their business processes on the basis of the differentiation strategies with which they seek to compete. For these processes to add value, the flow of information to synchronise and control them, and the management and knowledge based on the collection, processing and analysis of data to generate information for decision making, are fundamental.

Digitization is therefore becoming a necessity for organizations as part of their strategy to engage with customers, manage their internal processes and align and manage their supply chains [1–3], and is becoming an imperative for organizations across industries for a number of compelling reasons:

- **Efficiency and productivity**: Digital tools streamline processes, automate tasks and improve operational efficiency. They enable faster data processing, reduce the time it takes to complete tasks and increase overall productivity.
- **Cost reduction**: Adopting digital technologies often results in cost savings. It eliminates paperwork, reduces manual labour, minimizes errors and optimizes resource utilization, thereby reducing operational costs.

Á. Rocha et al. (Eds.): WorldCIST 2024, LNNS 987, pp. 338–346, 2024.
https://doi.org/10.1007/978-3-031-60221-4_33

- **Improved customer experience**: Digitalization enables companies to offer personalized services, faster responses and smoother interactions, leading to improved customer satisfaction and loyalty. This includes everything from online customer support to tailored recommendations based on user behavior.
- **Data-driven decision making**: Digitization is generating vast amounts of data. By leveraging analytics and data-driven insights, organizations can make informed decisions, predict trends and strategise effectively to drive better outcomes and stay competitive.
- **Global reach and market expansion**: Digital platforms facilitate global reach, enabling companies to transcend geographical boundaries and enter new markets. Online presence, e-commerce and digital marketing expand customer bases and diversify revenue streams.
- **Adapting to changing trends**: The digital landscape is evolving rapidly. Organizations need to adapt to new technologies to stay relevant. Embracing digital fosters agility, allowing companies to pivot quickly in response to market changes.
- **Improved collaboration and communication**: Digital tools enable seamless communication and collaboration between teams, regardless of their physical location. This fosters teamwork, knowledge sharing and innovation within organizations.
- **Sustainability and eco-friendliness**: Going digital often reduces the need for physical resources such as paper, leading to a more environmentally friendly approach. It promotes sustainability by reducing waste and energy consumption.
- **Regulatory compliance**: Many industries face evolving regulatory requirements. Digitalization helps ensure compliance by facilitating accurate record keeping, traceability and adherence to industry standards.

In summary, digitalization is essential for organizations to thrive in the modern landscape, offering improved efficiency, better customer experience, adaptability to change, and a competitive edge in the marketplace.

The main objective of this thesis is to show the benefits of digitization for an organization whose main business is to provide logistics services. This digitization will be the basis to transform the business processes, improve the service and reduce the operational costs.

Firstly, an introduction to Industry 4.0 is presented, and then an introduction to enterprise information systems and the importance of enterprise architecture as the basis for their design and implementation.

All this has been possible thanks to the incentives granted by the Colombian government.

2 Industry 4.0

Industry is currently considered to be in the midst of the so-called Fourth Industrial Revolution. This revolution began in Germany in 2011 to meet the challenges of changing environmental conditions and the development of new technologies [12–14]. These environmental changes include globalization, market volatility, competitive intensity and complexity. On the other hand, there is the technology developed for industry, which enables production automation, digitalization and networking between machines, products and users.

In a first historical review, the industrial revolutions can be distinguished as follows: The first industrial revolution (1780) was characterized by mechanization, the second (1870) by the use of electricity, the third (1960) by automation and the fourth -Industry 4.0- (2011) by the fusion between digital and physical systems, generating intelligent production systems [10, 11] to face the challenges generated by the changes in environmental conditions and the development of new technologies [14]. These changes in the environment include globalization, market volatility, competitive intensity and complexity. On the other hand, there is the technology developed for industry, which enables production automation, digitalization and networking between machines, products and users [15–18].

A key concept that has enabled the development of Industry 4.0 is the Internet of Things (IoT), which is associated with the terms M2M (machine-to-machine communications) and CPS (cyber-physical systems). From the above, a definition of IoT is the physical interconnection based on telecommunications for the exchange of data between two M2M compatible entities such as: devices, gateways and network infrastructure (ETSI, 2013).It is considered one of the six disruptive civilian technologies according to the US National Intelligence Council [19, 20].

Industry 4.0 is still a trending topic in all discussions about the near future, especially in developing countries. As already mentioned [21], the concept first appeared in Germany in 2011 and was presented to the general public by the German Manufacturers of Machines and Production Equipment (Verband Deutscher Maschinen- und Anlagenbau - VDMA)) at the Hannover Messe 2013, where it combined three technological innovations (automation, IoT and artificial intelligence) by creating innovative business models and going beyond product and process innovation [22].

The digital technologies that form the backbone of Industry 4.0, the Fourth Industrial Revolution. These technologies revolutionize industrial processes, manufacturing, and operations, leading to increased efficiency, connectivity, and innovation across various sectors. Here are some key technologies underpinning Industry 4.0 [14, 21–26]:

- **Internet of Things (IoT)**: IoT refers to the network of interconnected devices embedded with sensors and software, allowing them to collect and exchange data. It enables real-time monitoring, predictive maintenance, and automation in industries.
- **Big Data and Analytics**: Big Data involves the processing and analysis of large and complex datasets to extract valuable insights. Advanced analytics help in decision-making, predictive modeling, and optimization of processes.
- **Artificial Intelligence (AI)**: AI encompasses machine learning, neural networks, and cognitive computing. It enables machines to learn from data, recognize patterns, make decisions, and perform tasks that traditionally required human intelligence.
- **Cloud Computing**: Cloud technology provides on-demand access to computing resources like servers, storage, and applications over the internet. It facilitates scalability, flexibility, and cost-effectiveness for businesses.
- **Cyber-Physical Systems (CPS)**: CPS integrates computational and physical processes, enabling the interaction between the digital and physical worlds. These systems control and monitor physical processes through computer algorithms.

- **Additive Manufacturing (3D Printing):** Additive manufacturing builds objects layer by layer based on digital models. It offers customization, rapid prototyping, and reduced material waste in manufacturing.
- **Robotics and Automation**: Robotics involves the use of machines programmed to perform tasks autonomously. Automation, through robotics, streamlines production, improves precision, and enhances efficiency in various industries.
- **Augmented Reality (AR) and Virtual Reality (VR):** AR superimposes digital information on the physical world, while VR creates immersive simulated environments. Both technologies have applications in training, design, maintenance and customer experience.
- **Blockchain:** Blockchain is a distributed ledger technology that ensures a secure, transparent and tamper-proof record of transactions. It has applications in supply chain management, financial transactions and data security.
- **Edge computing**: Edge computing involves processing data closer to its source, rather than relying on a centralized cloud. It enables faster processing, reduced latency and more efficient use of network resources.

Together, these technologies are driving the digital transformation seen in Industry 4.0, enabling organizations to achieve greater efficiency, agility, innovation, and competitiveness in the modern era. This new industrial paradigm is based on the use of technologies related to Artificial Intelligence (AI), as well as on the interconnection or relationship of physical elements through sensors that allow the connection between them (M2M and CPS), which is the basis of the Internet of Things [23]. In industrial terms, the use of these technologies allows flexibility and optimization of processes and services, both in terms of planning with accurate information, control of the supply chain and its logistics, real-time information on operations and changes in the skills and competences of people [24, 25].

However, all of the above technologies serve to support the 4 pillars of Industry 4.0, as they do not make an organization 4.0 on their own. These pillars are Interconnectivity, Information transparency, Technical support: Integration, Decentralized decision-making and all 4 must be fulfilled in order to achieve a 4.0 level.

In order to achieve the correct implementation and use of these technologies, it is necessary to have an appropriate IT infrastructure, which means that there has been implemented all the software and hardware necessary for business management, such as Enterprise Resource Planning - ERP, Warehouse Management Systems - WMS, Manufacturing Enterprise Systems - MES or any other software adapted to the business processes.

3 Enterprise Architecture and Information Systems

Enterprise architecture (EA) provides a high-level view of an organization's business and IT systems and their interrelationships [26]. Enterprise architecture is defined as "as a discipline capable of providing the overview and insight needed to translate strategy into execution, enabling senior management to make key decisions about the design of the future enterprise" [27] and defines the current and desired future states of an organization's processes, capabilities, application systems, data and IT infrastructure, and provides a roadmap to get from the current state to the desired state [28].

As a result of the processes developed, information systems emerge, where by definition an information system is the set of elements and agents related to each other and the flow of information between them to achieve a goal [29].

On the other hand, these information systems use information technology as the basis for processing, transmitting and communicating information. These technologies, as mentioned before, are the software and hardware required to enable information management. Among these technologies are the ERP, MES, WMS, EDI, etc., and this brings functions and benefits such as [29]:

- **Integration**: EIS integrates different functions, processes and departments, enabling a consistent flow of information throughout the organization.
- **Efficiency and Productivity**: Streamline operations, automate routine tasks and reduce redundancy, improving overall efficiency and productivity.
- **Data Centralization**: Provide a single source of truth for data, ensuring consistency and accuracy in decision making.
- **Improved decision making**: Provide rapid access to real-time data and analysis, supporting informed and strategic decision-making at all levels.
- **Improved collaboration**: Foster better communication and collaboration between departments and stakeholders by providing a common platform for information.
- **Scalability and adaptability**: Support organizational growth and adaptability to changing business needs and market conditions.

4 Logistics Process Improvement at Grupo Logistico Especializado S.A.S.: aGILE

GRUPO LOGISTICO ESPECIALIZADO S.A.S, is a company located in Bogota - Colombia, dedicated to the commercialization of logistics services, which focuses its efforts to fulfill commercial agreements, satisfying the requirements and expectations of customers and stakeholders. It provides parcel services, document return, bulk cargo and design of customized operations.

To remain in the market, transport logistics companies must be increasingly competitive and need to meet the expectations of their clients, who require high levels of quality in service at the most competitive prices.

The company had a need to improve the logistical process, optimize time and improve control of the company's processes. As a result, the needs of the decision makers were basically, in terms of management, the creation and consultation of guides, the creation and consultation of manifests, the processing of concerns, petitions, complaints and grievances, management indicators, the generation of reports, reverse logistics, the automation of product reception, picking and shipping.

This involved the development of a software application, aGILE, which, as well as being the basis for more efficient management of the various processes, would allow the implementation of digital technologies that would take the company into a 4.0 environment.

For this reason, the platform developed should allow the control and reading of customer guides and products through electronic devices, the intelligent control of logistics operations, which can be done in real time, the processing of concerns, petitions, complaints and grievances, the generation of reports and management indicators. All this is based on the automation and digitalization of operational processes.

To identify and design the software requirements, the current processes were analyzed and modelled to understand how activities involving multiple functions and the intervention of several people were carried out.

During the implementation phase, procedures and roles were modified, and a requirements workshop was held with the staff of each department responsible for the activities to be managed, resulting in a list of functional requirements from which the technical requirements and architecture of the solution were designed.

The development of some of the functionalities of the aGiLE platform was based on the use of a unified and scalable Business Intelligence (BI) platform, in order to support decision making by displaying current and historical data of the company's operation in its business context. Basically, this application is based on two main processes: (i) the logistics process and (ii) the customer service process.

The logistics process was made up of 66 activities and the implementation of the functionalities in aGiLE reduced by more than 85% the number of activities or tasks carried out by the logistics operation (creation of guides, consultation of guides, creation of manifests, consultation of manifests, creation of reports), reduction of the current process activities to eight (8), given the degree of automation achieved and the reduction of the number of activities or tasks performed by the logistics operation (creation of guides, consultation of guides, creation of manifests, consultation of manifests, generation of reports), reduction of the current process activities to eight (8), given the degree of automation achieved and the reduction of the number of activities or tasks performed by the logistics operation.

The customer service process (handling and managing complaints and enquiries) saw a reduction of over 83% in the number of activities initially performed - from thirty-five (35) tasks to just six (6) after the implementation of Agile.

Based on the above, it can be said that the benefits obtained by implementing the application were:

- Reduced cost overruns in logistics operations by at least 30%.
- 100% processing of compensation within the timeframe required by the regulations.
- Adoption of two 4.0 technologies: RPA (process automation) for the operational processes of receiving, preparing and dispatching shipments, data analysis for management indicators and control panels.
- Easy access to information for the customer and centralized information.
- Customer self-management via the web, from any electronic device (indicators, reports, logistics management).
- Centralization of information and logistics operations on a single platform.
- Access to reports that allow effective analysis of the most relevant developments in the operation at any time.
- Direct, measurable and quantifiable results obtained with the development of each of the specific objectives, indicating the characteristics of the new knowledge generated, the new product, process or service, prototypes, designs, technological packages, etc.
- Similarly, for the customer service process (Management Indicators functionality), which was initially carried out manually in Excel, with the implementation of this

functionality in aGiLE, they are now generated automatically and can be consulted by any customer via the web or any electronic device.

- aGiLE has been adopted by 6 transport partners who have developed integration models for their aGiLE platforms, thereby improving their own IT infrastructure and processes.
- A web API has been developed, shown below, to allow customers to integrate and obtain information such as the creation, consultation and printing of waybills, traceability consultation (electronic tracking of consignments). This functionality is necessary in GLE, developed in REST architecture using the http protocol for data exchange with XML.
- Reverse logistics - Documentary control: A functionality has been implemented for the tracking, control and return of documents that our customers send with their orders.

5 Conclusions

This work basically tries to show the impact that the development and implementation of a platform, aGILE, has had as part of the process that allows us to move to a 4.0 environment, but it is only the basis to be able to comply with the principles that allow us to say that we are in Industry 4.0.

On the other hand, it can be seen how the development does not only study the requirements based on functionalities, but also on a reduction of activities, which implied an in-depth study of the processes in question and defining which ones could be eliminated without affecting the end of it, resorting to RPAs as well as the use of intelligent tools for better planning.

The above shows the favorable result in terms of costs, customer relations, time, service and paper reduction, but perhaps the most important thing is that the design is based on what was sought in the process and not as a tool that solves the problem in itself, so it is decided to do a personalized development, i.e. the IT responds to the needs of the process and not the other way round.

References

1. Direction, S., Limited, E.P.: Understanding digital platform strategies. Strateg. Dir. **35**, 34–36 (2019). https://doi.org/10.1108/SD-01-2019-0001
2. Sehlin, D., Truedsson, M., Cronemyr, P.: A conceptual cooperative model designed for processes, digitalization and innovation **11**, 504–522 (2019). https://doi.org/10.1108/IJQSS-02-2019-0028
3. Bouwman, H., Nikou, S., Molina-Castillo, F.J., Reuver, M.D.E.: The impact of digitalization on business models **20**, 105–124 (2018). https://doi.org/10.1108/DPRG-07-2017-0039
4. Bertola, P., Teunissen, J.: Fashion 4.0. innovating fashion industry through digital transformation. Res. J. Text. Appar. **22**, 352–369 (2018). https://doi.org/10.1108/RJTA-03-2018-0023
5. Hardwig, T., Klötzer, S., Boos, M.: Software-supported collaboration in small- and medium-sized enterprises. Meas. Bus. Excell. **24**, 1–23 (2020). https://doi.org/10.1108/MBE-11-2018-0098

6. Khin, S., Ho, T.C.: Digital technology, digital capability and organizational performance. Int. J. Innov. Sci. **11**, 177–195 (2019). https://doi.org/10.1108/IJIS-08-2018-0083
7. Liboni, L.B., Cezarino, L.O., Jabbour, C.J.C., Oliveira, B.G., Stefanelli, N.O.: Smart industry and the pathways to HRM 4.0: implications for SCM **1**, 124–146 (2019). https://doi.org/10.1108/SCM-03-2018-0150
8. Shaughnessy, H.: Creating digital transformation: strategies and steps. Strategy Leadersh. **46**, 19–25 (2018). https://doi.org/10.1108/SL-12-2017-0126
9. Tripathy, S., Aich, S., Chakraborty, A., Lee, G.M.: Information technology is an enabling factor affecting supply chain performance in Indian SMEs. J. Model. Manag. **11**, 269–287 (2016). https://doi.org/10.1108/JM2-01-2014-0004
10. Anand, P., Nagendra, A.: Industry 4.0: India's defence industry needs smart manufacturing. Int. J. Innovative Technol. Exploring Eng. **8**, 476–485 (2019). https://doi.org/10.35940/ijitee.K1081.09811S19
11. Hocaoğlu, M.F., Genç, İ.: Smart combat simulations in terms of industry 4.0. In: Gunal, M.M. (ed.) Simulation for Industry 4.0 Past, Present, and Future, pp 247–273. Springer, Cham (2019). https://doi.org/10.1007/978-3-030-04137-3_15
12. Ghobakhloo, M.: Corporate survival in industry 4.0 era: the enabling role of lean-digitized manufacturing **31**, 1–30 (2020). https://doi.org/10.1108/JMTM-11-2018-0417
13. Ghobakhloo, M.: The future of manufacturing industry: a strategic roadmap (2018). https://doi.org/10.1108/JMTM-02-2018-0057
14. Ghobakhloo, M.: Industry 4.0, digitization, and opportunities for sustainability. J. Clean Prod. (2019). https://doi.org/10.1016/j.jclepro.2019.119869
15. Alcácer, V., Cruz-Machado, V.: Scanning the industry 4.0: a literature review on technologies for manufacturing systems. Eng. Sci. Technol. Int. J. **22**, 899–919 (2019). https://doi.org/10.1016/j.jestch.2019.01.006
16. Bendul, J.C., Blunck, H.: The design space of production planning and control for industry 4.0. Comput. Ind. **105**, 260–272 (2019). https://doi.org/10.1016/j.compind.2018.10.010
17. Reischauer, G.: Industry 4.0 as policy-driven discourse to institutionalize innovation systems in manufacturing. Technol. Forecast Soc. Change **132**, 26–33 (2018). https://doi.org/10.1016/j.techfore.2018.02.012
18. Sharpe, R., van Lopik, K., Neal, A., et al.: An industrial evaluation of an industry 4.0 reference architecture demonstrating the need for the inclusion of security and human components. Comput. Ind. **108**, 37–44 (2019). https://doi.org/10.1016/j.compind.2019.02.007
19. Sahay, R., Meng, W., Estay, D.A.S., et al.: CyberShip-IoT: a dynamic and adaptive SDN-based security policy enforcement framework for ships. Futur. Gener. Comput. Syst. **100**, 736–750 (2019). https://doi.org/10.1016/j.future.2019.05.049
20. Chatterjee, S., Kar, A.K., Gupta, M.P.: Success of IoT in smart cities of India: an empirical analysis. Gov. Inf. Q. **35**, 349–361 (2018). https://doi.org/10.1016/j.giq.2018.05.002
21. Lele, A.: Industry 4.0 (2019)
22. Fajardo-Toro, C.H., Mula, J., Poler, R.: Adaptive and hybrid forecasting models—a review. In: Ortiz, Á., Andrés Romano, C., Poler, R., García-Sabater, J.-P. (eds.) Engineering Digital Transformation. LNMIE, pp. 315–322. Springer, Cham (2019). https://doi.org/10.1007/978-3-319-96005-0_38
23. Zona-Ortiz, A.T., Fajardo-Toro, C.H., Pirachicán, C.M.A.: Proposal for a general framework for the deployment of smart cities supported in the development of IOT in Colombia | Propuesta de un marco general para el despliegue de ciudades inteligentes apoyado en el desarrollo de IOT en Colombia. RISTI - Revista Iberica de Sistemas e Tecnologias de Informacao **2020**, 894–907 (2020)
24. Büchi, G., Cugno, M., Castagnoli, R.: Smart factory performance and industry 4.0. Technol. Forecast Soc. Change **150**, 119790 (2020). https://doi.org/10.1016/j.techfore.2019.119790

25. Lins, T., Oliveira, R.A.R.: Cyber-physical production systems retrofitting in context of industry 4.0. Comput. Ind. Eng. **139**, 106193 (2020). https://doi.org/10.1016/j.cie.2019.106193
26. Gong, Y., Janssen, M.: The value of and myths about enterprise architecture. Int. J. Inf. Manage. **46**, 1–9 (2019). https://doi.org/10.1016/j.ijinfomgt.2018.11.006
27. van den Berg, M., Slot, R., van Steenbergen, M., et al.: How enterprise architecture improves the quality of IT investment decisions. J. Syst. Softw. **152**, 134–150 (2019). https://doi.org/10.1016/j.jss.2019.02.053
28. Shanks, G., Gloet, M., Someh, I.A., Frampton, K.: Achieving benefits with enterprise architecture. J. Strateg. Inf. Syst. **27**, 139–156 (2018). https://doi.org/10.1016/j.jsis.2018.03.001
29. Laudon, K.C., Laudon, J.P.: Management Information Systems: Managing the Digital Firm (2004)

Exploring Transparency in Decisions of Artificial Neural Networks for Regression

José Ribeiro[1(✉)], Ricardo Santos[1], Cesar Analide[2], and Fábio Silva[1]

[1] CIICESI, ESTG, Politécnico do Porto, Felgueiras, Portugal
{jfr,rjs,fas}@estg.ipp.pt
[2] ALGORITMI Research Centre / LASI, University of Minho, Braga, Portugal
analide@di.uminho.pt

Abstract. The interpretation and explanation of machine learning models are important tasks to ensure transparency and credibility. Various methods have been developed to achieve this goal, including Shap, Lime, and Explanatory Dashboards. The application of these explainability methods aims not only at interpreting the machine learning model but also at incorporating transparency into the implementation of a system that uses artificial intelligence algorithms. By leveraging part of a an initial manufactoring project, we implemented techniques to enhance the understanding of the model's time prediction through explainability methods. This improves trust among the involved entities and contributes to the overall effectiveness of the system in various application contexts. Our objective is to apply those strategies to other projects with in industry 4.0 and manufactoring to increase the adoption of new and innovative understandable intelligent systems.

Keywords: Machine Learning · Shap · Lime · Explainer Dashboard

1 Introduction

At the intersection of data analysis, machine learning, and industry [6], a deep understanding of decisions made by complex models is imperative to boost confidence and effectiveness in business applications. This study builds upon previous research where data was utilized to explore the decisions from "black box" algorithms for a regression problem.

The adopted approach focused on applying advanced interpretability methods to these machine learning models. The project employed techniques such as SHAP (SHapley Additive exPlanations) [2,4], LIME (Local Interpretable Model-Agnostic Explanations) [8], and Explainer Dashboard [1]. These were strategically chosen to provide profound insights into the decisions of the regression neural network. In the context of the intial project, focused on competitiveness in the industry, the emphasis was on the use of Artificial Intelligence (AI) techniques to ensure tangible predictions and strengthen trust between the company

Á. Rocha et al. (Eds.): WorldCIST 2024, LNNS 987, pp. 347–356, 2024.
https://doi.org/10.1007/978-3-031-60221-4_34

and its clients. This involved creating a prediction model using ML methods to anticipate time duration, as well as developing an automatic project management system. The significant progress achieved in the intial project facilitated the implementation of explainability techniques in the XGBoost, Random Forest Regressor, and Gradient Boosting Regressor algorithms. These tools not only highlight the most influential variables in predictions [5] but also provide a detailed insight into the decision-making process of AI [7].

Interpretation and explanation have become essential elements to maximize the benefits of these implementations [8]. The techniques applied enabled an in-depth understanding of the weight of decisions, contributing to a more informed and transparent decision-making [1], which are fundamental in a scenario where trust, effectiveness and competitiveness play central roles [3].

2 Literature Review

Explainable AI (xAI) algorithms play a fundamental role in providing transparency and interpretability in the decisions of artificial intelligence models. Shapley Additive Explanations (SHAP) [4] is a common approach to interpret machine learning models, providing insights into the importance of each feature in both global and local predictions [7]. Based on cooperative game theory, this technique calculates the Shapley value for each feature, representing its expected contribution to model predictions [5]. Applicable to various model types, SHAP [2] offers explanations at different levels of detail, from summary graphs to understand overall feature importance to local charts explaining individual predictions. SHAP [3] is a powerful method used in various machine learning tasks, contributing to transparency and confidence by detailing how each feature influences model decisions.

Local Interpretable Model-Agnostic Explanations (LIME) [8] is a popular approach in interpreting machine learning models. Similar to SHAP, LIME is model-agnostic and applicable to different model types. It generates local explanations by adjusting the model's behavior near predictions, making models simpler and more interpretable. The methodology perturbs input data around the instance of interest, observing resulting changes in the model's output. Despite its flexibility and applicability to any model, including deep learning, LIME has computational limitations, especially in high-dimensional data. Its approximate descriptions provide valuable insights into understanding the model's behavior in specific cases, making it a powerful tool for interpreting and explaining machine learning models. The Explainer Dashboard [1] is a visualization tool designed to interpret and explain machine learning models. Developed by IBM as part of the open-source AI Explainability toolkit, the dashboard allows interaction with model inputs, demonstrating how changes in these inputs affect predictions. With components such as Data Explorer, summary view, and detailed view, the dashboard handles complex models, including those with multiple inputs and outputs, as well as non-numeric inputs such as text and images. Utilizing techniques such as feature importance, partial dependence plots, and local explanations, this tool is valuable for understanding model predictions, identifying areas for improvement, and transparently communicating model behavior to stakeholders, contributing to building trust in model predictions.

Recent machine learning research explores enhancing business profitability through strategic use of corporate disclosures [2]. The study emphasizes the profitability potential of trading based on corporate disclosures, accounting for transaction costs and liquidity. It underscores the importance of factoring in these considerations when developing investment strategies. The authors utilize shap to assess feature importance in corporate disclosures, aiding the accuracy of machine learning models. Shap is leveraged to clarify each feature's contribution to model predictions, with graphs illustrating their significance in evaluating machine learning algorithms. The article [3] introduces a modern approach to tree boosting that strikes a balance between predictive accuracy and interpretability, making it ideal for applications relying on fragmented data. The approach is aptly named Locally Interpretable Tree Boosting (LitBoost) and is applied to predict real estate prices. The interpretability provided by LitBoost values the use of shape functions to create specific local explanations for predictions, illustrating how each attribute contributes to the final prediction.

The research in [7] highlights a new approach, the SAFE ML framework. Its goal is to enhance the interpretability of machine learning models without compromising their performance. By utilizing complex models to generate innovative features, the framework enables the creation of more transparent and straightforward models. The researchers tested the effectiveness of this framework on various datasets and observed that, in many cases, it offered comparable performance to complex models but with greater transparency. This suggests that implementing the framework could be the solution to address the trade-off challenge between transparency and performance in the field of machine learning.

In [8], a study on the development of a machine learning model aimed at identifying patients with inflammatory arthritis and other non-inflammatory conditions during the screening process. This pioneering study used machine learning to optimize the patient screening process from primary to secondary care. Instead, the LIME method (Local Interpretable Model-Agnostic Explanations) was adopted, providing accurate explanations for each patient's prediction to enhance the medical decision-making.

In [1] an example of a study on optimizing the integrated home care problem through the use of heuristic methods to enhance routing and scheduling decisions. The proposed methods, named ConstructByCaregiver and ConstructbyUser, are carefully designed to find high-quality solutions, considering a wide range of constraints and objectives. Real data from home care providers in Barcelona, Spain, were used to test these methods, and the results showed that both methods were able to achieve efficient solutions in a reasonable computational time. Additionally, a decision support system, called Explained Dashboard, was developed to assist in decision-making processes.

The pursuit of explainability, transparency, and interpretability in machine learning models is essential for building trust and understanding. Previously referenced methods play roles in this scenario, providing insights that go beyond the black box of algorithms [1,4]. This approach not only enhances the reliability of predictions but also strengthens the ability to make informed decisions in various sectors [8].

3 Case Study

In the Portuguese business scenario, small and medium-sized enterprises still rely on employees' experience to calculate delivery times without resorting to reliable tools that guarantee accurate estimates. The common practice of processing orders in a First In First Out (FIFO) format, although widely used, may not be the most effective, as it does not consistently deal with possible interruptions, such as a temporary lack of raw materials or labor.

We decided to develop explainability strategies based on a manufactoring process which predicted the time duration for different stages. Recognizing that the transparency of the predictions made may be insufficient to assure trust in the decision process, a new layer of explainability was added to make product owner understand the cause of time predictions.

Thus, we grounded our investigation in the use of a production database, which included 5 production stages in the manufactoring process, namely: cutting, bending, welding, treatment, and finishing. This database stems from the project despicted in [6] conducted in collaboration with a company in the metalworking sector. This databse simulates a manufactoring plant based with generated data.

In the current phase of the research, we have directed our focus to a specific subset of this database, narrowing it down to two stages, namely the cutting and finishing stages. This approach aims to provide a detailed insight into the variables that directly impact the production time in these two stages.

After analyzing this sample, we direct our attention to selected variables, considering their operational relevance and impact on performance. Our goal is to deepen the understanding of the behavior of XGBoost, Random Forest Regressor, and Gradient Boosting Regressor models in this specific context, exploring how each algorithm responds and contributes to predictions related to production time in the focused sector. To enhance the explicability of the machine learning decisions we will leverage the capabilities of Shap, Lime, and the Explainer Dashboard algorithms.

The implementation of advanced explainability methods on this specific data sample should allow the identification od patterns, relationships, and specific contributions of each variable to the algorithms' decisions. These advanced techniques were applied to the problem, aiming to gain clear insights into the inner workings of the "black box" and understand how time each stage is influenced by the input variables.

4 Study of Explainabilty Algorithms

Tools such as Shap, Lime, and Explainer Dashboard were incorporated to provide clear insights into model decisions, highlighting influential variables and offering an explanation for the machine learning decision-making process. To assess the interpretability of machine learning models, specifically the algorithms for regression problems (XGBoost, Random Forest Regressor, and Gradient Boosting Regressor), we applied explainability techniques to explore the impact of input variables on model decisions.

As mentioned in Sect. 3, we focused on examining two phases of the production process: the cutting phase and the finishing phase. In the cutting phase, we analyzed three variables: the number of simple cuts, the number of complex cuts, and the number of employees involved in this phase. Meanwhile, in the finishing phase, we expanded our analysis to five variables: the number of simple and complex finishing faces, the type of finishing (brushed or polished), the type of cleaning (easy, medium, or difficult), and the number of employees in the finishing phase.

The generated graphs become powerful tools to identify areas for optimization, promoting efficiency, and enhancing quality in both phases of the production process. This study exemplifies the importance of combining technology and data analysis to drive significant improvements in complex production environments.

4.1 Cutting Phase

Analyzing the XGBoost, Random Forest Regressor, and Gradient Boosting Regressor models in the cutting phase using the Shap technique provides valuable insights. In XGBoost, the graph Fig. 1 shows that the first variable has an impact between −4 to 3, the second between −1.9 to 1.9, and the third between −1.9 to 2.2. The mean Shap values highlight specific values for each variable: 1.4, 0.57, and 0.47, respectively. For the Random Forest Regressor, the impact on the model varies from −3 to 1.8, −1 to 1.5, and −0.9 to 1 for the three variables, with mean shap values of 0.85, 0.38, and 0.36. Lastly, in the Gradient Boosting Regressor, the impact on the model ranges from −4 to 2.9, −1.4 to 1.9, and −1.1 to 1.7, with mean SHAP values of 1.26, 0.39, and 0.34 for the three variables.

Red points, representing higher values, indicate a greater contribution to the algorithms' decisions, while the blue points, associated with lower values, have a comparatively lower influence. Furthermore, it is noted that variables extending further to the right on the graphs, have a more significant impact on predictions, whereas those leaning to the left, have a lesser influence on decision impact. It is noteworthy that the "Simple number" variable had the most impact on the decision among the three algorithms used.

In the cutting phase, for XGBoost, using the lime technique, in the graph Fig. 2 the first variable contributes 1.70, the second variable 0.36, and the third variable 0.14, with contributions specified in the graph. The prediction interval ranges from -0.66 to 2.95, with key values such as Intercept (5.235), Prediction local (3.03), and Right (2.95). For the Random Forest Regressor, we identified specific contributions to the variables, with the prediction interval ranging from -0.34 to 2.95 and key values including Intercept (5.232), Prediction local (3.04), and Right (2.94). Analyzing the Gradient Boosting Regressor in the cutting phase, contributions are highlighted, with a prediction interval between -0.66 and 2.95. Important indicators include Intercept (5.235), Prediction local (3.04), and Right (2.94). These results provide valuable insights into the influences of variables in each model, contributing to a deeper understanding of decisions made in the cutting phase of the production process.

The most significant impact on the predicted value during the cutting phase is attributed to the "Simple number" variable in the XGBoost model, with a negative contribution of 1.70. This negative contribution suggests that negative variations in "Simple number" are associated with a substantial reduction in the prediction, highlighting the crucial importance of this variable in the model's decision-making process.

In the cutting phase, considering the XGBoost model and the explainer dashboard, the mean impact on prediction (mean absolute SHAP value) reveals significant contributions from the variables. The first variable stands out with 1.43, followed by the second variable with 0.4, and the third variable with 0.38, graph Fig. 3. For the Random Forest Regressor in the same phase, the influences are also noteworthy. The first variable contributes with 1, the second variable with 0.3, and the third variable with 0.25. Regarding the Gradient Boosting Regressor in the cutting phase, the mean impact on prediction emphasizes the importance of the variables. The first variable has an impact of 1.39, the second variable contributes with 0.38, and the third variable with 0.24. We can conclude that the "Simple number" variable with values of 1.43, 1, 1.39 had a preponderant influence on the model's prediction and, consequently, its relevance to the decisions during this phase.

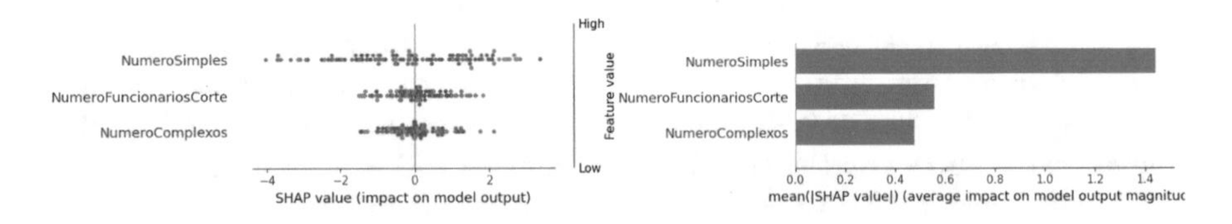

Fig. 1. Cutting Step - XGBoost Algorithm - Slap Technique

4.2 Finishing Phase

In the finishing phase, using the SHAP method on the models, the XGBoost algorithm reveals intervals for the five analyzed variables. The mean SHAP values specify distinct values for each variable: 0.25, 0.13, 0.06, 0.05, and 0.04. For the Random Forest Regressor, the graph Fig. 4 presents varied intervals, and the mean SHAP values highlight specific values: 0.12, 0.10, 0.07, 0.0043, and 0.039.

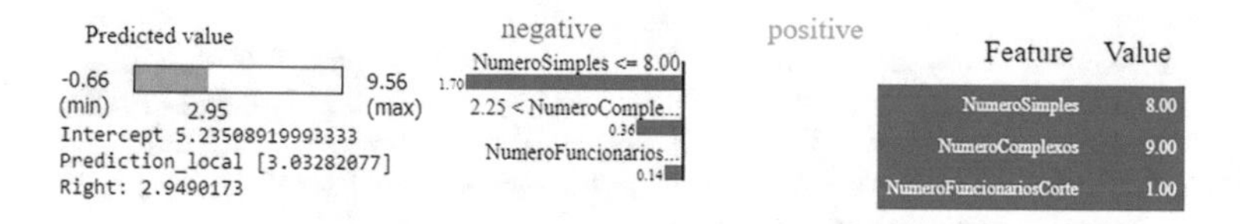

Fig. 2. Cutting Step - XGBoost Algorithm - Lime Technique

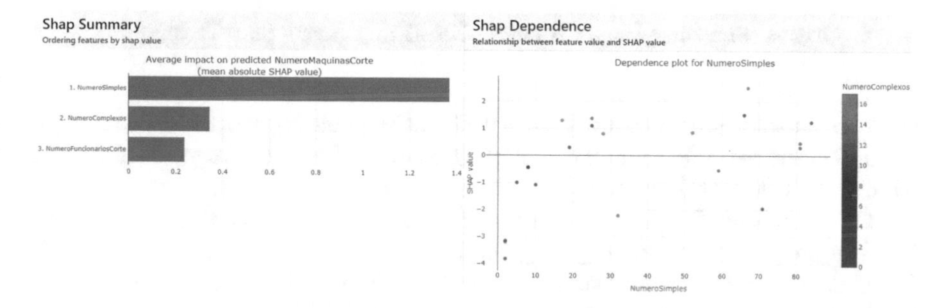

Fig. 3. Cutting Step - Gradient Boosting Regressor Algorithm - Technical Explainer Dashboard

In the Gradient Boosting Regressor, the mean SHAP values present specific values: 0.25, 0.9, 0.07, 0.0047, and 0.01 for the analyzed variables. Concluding that the "Simple number" variable also has a significant influence during this phase, being the variable with the most weight in the decision of the algorithms.

Analyzing contributions using the Lime method reveals valuable insights into various variables. Notably, the type of finish and the number of employees exert significant influence. The prediction interval ranges from 0.58 to 1.90, with Intercept (1.50), Prediction local (1.41), and Right (1.90) as key values. Applying the RandomForestRegressor Fig. 5, with a prediction interval between 1.02 and 1.94, highlights essential values like Intercept (1.38), Prediction local (1.63), and Right (1.94). In the XGBoost model for finishing, contributions are detailed, with a prediction interval between 0.58 and 1.73 and significant values such as Intercept (1.48), Prediction local (1.49), and Right (1.72). One can verify in the graph Fig. 5 that the variable with the most significant weight in the algorithm's outcome was the "type of finish" variable, with a positive contribution.

Applying the Explainer Dashboard method to the models, specifically with XGBoost, the mean impact on prediction stands out, with the variable (number of simple finishing stages) contributing 0.24, the variable (number of complex finishing stages) contributing 0.2, the variable (type of finish) contributing 0.06, the variable (number of employees in the finishing stage) contributing 0.049, and the last variable (type of cleaning) contributing 0.04. For the Random Forest Regressor 6, the mean impact on prediction reveals significant influences, with the variable (number of simple finishing stages) contributing 0.24, the variable (number of

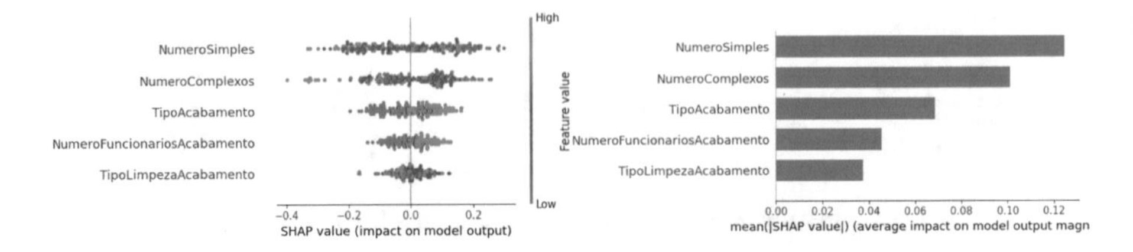

Fig. 4. Finishing Stage - Random Forest Regressor - Slap Technique

complex finishing stages) contributing 0.11, the variable (type of finish) contributing 0.077, the variable (number of employees in the finishing stage) contributing 0.05, and the last variable (type of cleaning) contributing 0.039.

The Gradient Boosting Regressor in the finishing phase presents mean impacts on prediction, highlighting the relevance of the variables: the variable (number of simple finishing stages) with 0.24, the variable (number of complex finishing stages) with 0.08, the variable (type of finish) with 0.07, the variable (number of employees in the finishing stage) with 0.046, and the last variable (type of cleaning) with 0.01. We can infer that the variable "Simple number" exerted a decisive influence on the model's prediction, thus highlighting its significant relevance to the decision-making process (Fig. 6).

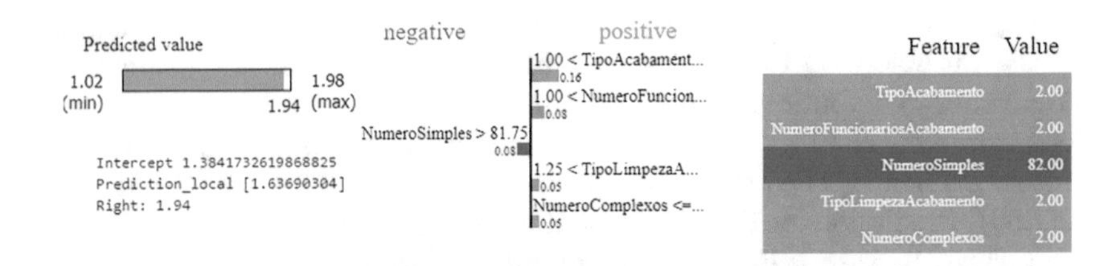

Fig. 5. Finishing Stage - Random Forest Regressor Algorithm - Lime Technique

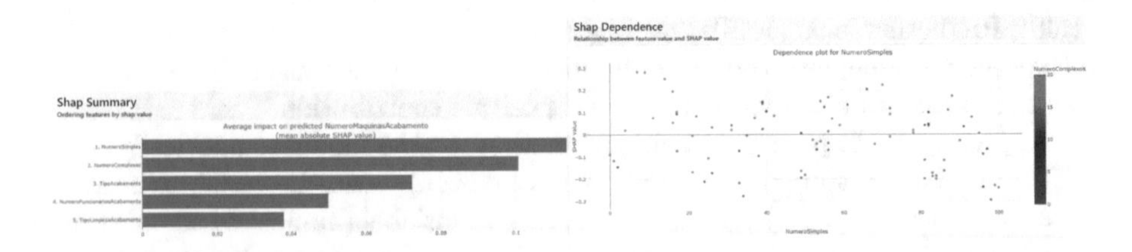

Fig. 6. Finishing Stage - Random Forest Regressor - Technical Explainer Dashboard

4.3 Analysis of Results

The results emphasize the variation in variable contributions across different algorithms. In the cutting phase, there are distinct impact ranges and mean

SHAP values for variables in each algorithm, indicating differences in variable contributions to model decisions. In the finishing phase using XGBoost, specific intervals and mean SHAP values for each variable highlight the diversity of influences. The Random Forest Regressor exhibits varied intervals, and the Gradient Boosting Regressor reveals specific patterns in model impact graphs and mean SHAP values. These findings provide a comprehensive understanding of each variable's contribution to the models, offering insights to interpret and optimize operational decisions in the finishing phase of the production process. The analyzed graphs using the Lime Technique reveal distinct contributions to the variables, categorized as "negative" and "positive". Overall, these results provide an understanding of variable influences in the applied models during the analyzed stages of the production process, contributing to the optimization and interpretation of operational decisions.

Utilizing the Explained Dashboard technique and examining the graphs in different stages of the production process, we identified valuable insights. In the cutting phase, the models exhibited distinct contributions, highlighting specific variables with significant impacts. In the finishing phase, the detailed analyses of these models revealed the relative influence of different variables, providing an in-depth understanding of decisions in each context. These results are to improve understanding of the production process and optimizing decision-making based on the applied models. The present analysis of the applied techniques allows inferring that the variable "Simple number" exerted a significant influence on the models' predictions, highlighting its preponderant relevance for the modeled decisions.

5 Conclusions

Model interpretation tools stand out as powerful instruments for understanding and explaining machine learning models. Each method offers a unique perspective on the model's behavior, assisting in identifying the features driving its predictions. Model interpretation is important to understand the underlying complexities of algorithms. Methods such as Shap, Lime, and Explainer Dashboard are usefull tools in this process.

Shap ranks the importance of features, providing an analysis of each input variable's contribution to the model's predictions. On the other hand, Lime takes a more localized approach, generating specific interpretations for particular data points. By assigning importance to features in specific instances, it highlights influences on specific predictions and helps diagnose situations where the model may fail. The Explainer Dashboard, in turn, provides an interactive approach to explore the model's behavior. By visualizing feature importance values, partial dependence plots, and explanations for individual predictions, the dashboards allow for in-depth analysis of the model in different contexts. This functionality is particularly useful for identifying unexpected behaviors or patterns in the data that may not be immediately apparent. Together, these methods offer insights into machine learning models, highlighting the driving features of predictions and assisting in diagnosing cases of incorrect forecasts. As for future work, we consider applying these methods to different datasets from various contexts, expanding the scope and validating the results. In the future, we aim to compare existing results with other explainability techniques, such as Grad-CAM (Gradient-weighted Class Activation Mapping), and additional feature importance techniques, further enriching the understanding of machine learning models, providing a benchmark for explainablity algorithms.

Acknowledgements. This work has been supported by FCT - Fundação para a Ciência e Tecnologia within the RD Units Project Scope: UIDB/00319/2020.

References

1. Burgas, L., Melendez, J., Colomer, J., Massana, J., Pous, C.: N-dimensional extension of unfold-PCA for granular systems monitoring. Eng. Appl. Artif. Intell. **71**, 113–124 (2018). https://doi.org/10.1016/J.ENGAPPAI.2018.02.013
2. Gosiewska, A., Kozak, A., Biecek, P.: Simpler is better: lifting interpretability-performance trade-off via automated feature engineering. Decis. Support Syst. **150**, 113556 (2021). https://doi.org/10.1016/J.DSS.2021.113556
3. Hjort, A., Scheel, I., Sommervoll, D.E., Pensar, J.: Locally interpretable tree boosting: an application to house price prediction. Decis. Support Syst., 114106 (2023). https://doi.org/10.1016/J.DSS.2023.114106
4. Kraus, M., Feuerriegel, S.: Forecasting remaining useful life: interpretable deep learning approach via variational bayesian inferences. Decis. Support Syst. **125**, 113100 (2019). https://doi.org/10.1016/J.DSS.2019.113100
5. Li, Y., Chan, J., Peko, G., Sundaram, D.: An explanation framework and method for AI-based text emotion analysis and visualisation. Decis. Support Syst., 114121 (2023). https://doi.org/10.1016/J.DSS.2023.114121, https://linkinghub.elsevier.com/retrieve/pii/S0167923623001963
6. Ribeiro, J., Ramos, J.: Forecasting system and project monitoring in industry. Lect. Notes Netw. Syst. **583**, 249–259 (2023). https://doi.org/10.1007/978-3-031-20859-1_25
7. Schmitz, H.C., Lutz, B., Wolff, D., Neumann, D.: When machines trade on corporate disclosures: using text analytics for investment strategies. Decis. Support Syst. **165**, 113892 (2023). https://doi.org/10.1016/J.DSS.2022.113892
8. Wang, B., Li, W., Bradlow, A., Bazuaye, E., Chan, A.T.: Improving triaging from primary care into secondary care using heterogeneous data-driven hybrid machine learning. Decis. Support Syst. **166**, 113899 (2023). https://doi.org/10.1016/J.DSS.2022.113899

Predictive Process Mining a Systematic Literature Review

Eduardo Silva[1(✉)] and Goreti Marreiros[1,2]

[1] Institute of Engineering, Polytechnic of Porto, Porto, Portugal
{1180881,mgt}@isep.ipp.pt
[2] Research Group on Intelligent Engineering and Computing for Advanced Innovation and Development, Polytechnic of Porto, Porto, Portugal

Abstract. As business process complexity reaches an all-time high, the competitive and rapidly changing environments where organisations operate created the need for business processes to be continuously analysed, improved, and supported by adequate tools and techniques, which led to the conception of Process Mining (PM). Recently, Predictive Process Mining (PPM) emerged as a novel field that integrates PM with predictive techniques, such as Machine Learning, allowing the future behaviour of ongoing processes to be predicted and actions to be taken to steer them in favour of business interests. This allows PM to evolve from a reactive to a proactive tool for process enhancement, which helps reduce costs, time, and necessary resources. Nevertheless, the impact and value of PPM remain mostly unproven and the number of related studies is relatively low. A systematic review has therefore been elaborated to support future studies, based on a search carried out in three databases and according to the PRISMA statement. This review aims to identify the methods and techniques used in the implementation of PPM use cases and to understand the perceived business impact and value. In total, 411 articles were initially identified, with 24 meeting the defined criteria and being selected for discussion. From these articles, next-event, outcome, and suffix prediction have been identified as the most common use cases of PPM, in addition to the predictive techniques (e.g. XGBoost, LSTM, BERT, GAN) and optimisation methods (e.g. trace bucketing) being applied in their development. Finally, although limited, the perceived impact and value of PPM in organisations have been recorded, according to stakeholders actively using PPM in their daily work.

Keywords: Process Mining · Business Processes · Process Enhancement · Predictive Techniques · Machine Learning · Deep Learning

1 Introduction

Process complexity is at an all-time high. A survey conducted by Forrester Consulting in 2022, with more than 800 business leaders, has shown that 71% of companies are using 10 or more applications (e.g., Enterprise Resource Planning

Á. Rocha et al. (Eds.): WorldCIST 2024, LNNS 987, pp. 357–378, 2024.
https://doi.org/10.1007/978-3-031-60221-4_35

(ERP) and Customer Relationship Management (CRM) software) to execute a single business process [1].

To address this issue, organisations must improve and support their processes with the right tools and techniques [2]. Process Mining (PM) emerged as a solution, proposed in 2004 by Dr. Wil van der Aalst in the article "Process mining: a research agenda" [3].

PM combines Data Mining and Business Process Management to continuously analyse, learn, and enhance processes by leveraging the growing amount of event data in information systems. This approach often reveals complexities and deviations that were not anticipated during the process design.

In recent years, Predictive Process Mining (PPM) emerged as a novel field that combines predictive techniques, such as Machine Learning (ML), with PM to enhance ongoing processes in favor of business interests.

PPM has been successfully applied in various contexts, such as healthcare [4,5], logistics [6], and finance [7], and is likely to become increasingly relevant as organizations continue to collect large amounts of data.

Nevertheless, its full potential has not yet been proven to organisations as there is some resistance to its adoption and integration with traditional PM. According to a systematic review of PM, published in 2019 [8], predictive techniques represent less than 5% of the PM-related scientific articles published as of 2019.

Organisations are primarily using PM as a reactive tool to analyse the past and make changes to improve the process in the future.

However, with PPM, organisations can use PM proactively by anticipating the outcome of ongoing processes and following the best course of action to change them in favour of business interests. This optimises processes and reduces the costs, time, and resources necessary to execute them, potentially changing the way business processes are designed and decision-making is made.

For PPM's value to be proven for organisations, real-life use cases must be developed and studied, measuring the impact on the operational performance and costs of the organisation. In addition, more interviews should also be held with stakeholders to understand their perception of PPM, if they are considering using it, and what their needs are.

The following systematic review has been elaborated to identify the current state of PPM and the mechanisms being applied in the field, as well as the perceived value of PPM from stakeholders and organisations already working with it in their daily work, aiming to produce reference documentation for researchers and developers to develop future case studies.

The remainder of this article is structured in 4 sections, the first detailing the explicit and reproducible methodology defined for the systematic review, based on the PRISMA checklist, which describes the research questions, data sources, search terms, quality assessment criteria, and an overview of the article selection process performed in this review.

Afterward, based on the findings from the selected articles, the results for each of the defined research questions are presented, followed by a summary and discussion of these results for future reference, ending with some closing remarks about this review and future work.

2 Methods

The systematic review was prepared according to the Preferred Reporting Items for Systematic Reviews and Meta-Analyses (PRISMA) statement [9,10].

This section starts with a presentation of the research questions that need to be answered, the data sources that provided the necessary information, and the search terms that were used. This is followed by a discussion of the quality assessment criteria used to restrict the number of results.

Finally, the results of the article selection process are presented, starting with the search and duplicate removal, followed by the abstract and full-text screening, and concluding with the extraction of the included articles.

2.1 Research Questions

As mentioned in the Introduction, the goal of this review is to discover the value and impact that PPM currently has on organizations and decision-makers, in addition to the techniques being used in its implementation, serving as a reference for the development of future PPM projects.

As a means to support this, three research questions were defined and presented in Table 1, the first focusing on the mechanisms and techniques currently employed in the implementation of PPM, with a particular interest in predictive techniques (e.g. ML algorithms) and their integration with existing PM solutions.

The second question centers on the impact of PPM on the decision-making process of stakeholders, as they are provided with insights on the future state of a given process and can therefore act to change its outcome in their favour, which should allow them to follow a more proactive approach during the process management, rather than a reactive one.

Finally, to complement the previous question, the third question aims to determine the impact that a more proactive approach to process enhancement produces in the performance and costs, both monetary and temporal, of organisations and their business processes, through the application of PPM.

Table 1. Research questions

ID	Question
RQ1	What mechanisms are being used in the implementation of Predictive Process Mining?
RQ2	How is Predictive Process Mining supporting stakeholders and their decision-making?
RQ3	How is Predictive Process Mining impacting the performance and costs of organizations?

2.2 Data Sources

Having defined the research questions, the data sources from which to extract information on each question must be defined and, according to [11], it is advisable to maintain a small number of comprehensible data sources.

As such, given their reputation and comprehensive database of articles, the data sources presented in Table 2 were chosen for this review.

Table 2. Data sources

Database	URL
IEEE Xplore	https://ieeexplore.ieee.org
ScienceDirect	https://www.sciencedirect.com
ACM	https://dl.acm.org

It should be noted that, given their size, the data sources may overlap and duplicate articles may have to be removed, as anticipated by PRISMA.

2.3 Search Terms

Following the data sources definition, search terms have to be defined in order to find information related to the research questions.

This was accomplished by identifying the common domains in each of the research questions and then registering search terms and possible synonyms in each of these domains.

These terms are then combined through the use of boolean operators (i.e. OR between search terms within a domain, AND between domains) to formulate the query string used to search articles in the data sources.

The identified domains and respective search terms are presented in Table 3.

Table 3. Domains and keywords

ID	Domain	Keywords
D1	Process Mining	Process Mining, Workflow Mining, Process Monitoring, Workflow Monitoring
D2	Predictive Techniques	Predictive Techniques, Machine Learning, Deep Learning
D3	Performance	Operating Performance, Operating Efficiency, Operating Costs, Business Performance, Business Efficiency, Business Costs
D4	Decision Making	Decision Making, Decision Makers, Stakeholders

When defining the query string based on the search terms presented above, some testing was performed on the target data sources and it was concluded that trying to find articles that covered the four domains simultaneously led to a small result set, with only 11 articles in total.

This may be due to the nature of the questions: RQ1 is more technical in nature, unlike RQ2 and RQ3 which are more qualitative. As a result, the methods and, therefore, the articles that cover each of the questions may be different.

As such, in order to obtain a larger and, possibly, more diverse result set, while still capturing the 11 articles identified during testing, it was defined that technical (i.e. RQ1) and qualitative articles (i.e. RQ2 and RQ3) would be searched separately and combined to obtain the final result set.

At the Domain level, the query string should then result in the following: D1 AND (D2 OR (D3 AND D4)), with D1 (i.e., PPM) being the common domain to every question, D2 being related to the technical questions, and D3 and D4 related to the qualitative.

From this perspective, replacing the domains in the query string presented above with the respective keywords results in the query strings presented in Table 4, which were applied to the title, abstract, and keywords.

Table 4. Query strings

Data source	Query string
IEEE Xplore and ACM	("Process Mining" OR "Workflow Mining" OR "Process Monitor*" OR "Workflow Monitor*") AND (("Predictive Technique*" OR "Machine Learning" OR "Deep Learning" OR "Natural Language Processing") OR (("*Performance" OR "*Efficiency" OR "*Costs") AND ("Decision Mak*" OR "Stakeholder*")))
ScienceDirect	("Process Mining" OR "Process Monitoring") AND (("Predictive Technique" OR "Machine Learning" OR "Deep Learning") OR (("Performance" OR "Efficiency") AND ("Decision Making")))

It is important to note that although each data source supports similar boolean operators to combine the data sources, some adaptations had to be made between each source (e.g. ScienceDirect supports up to 8 boolean operators and no wildcards (*) in the query string, so it had to be simplified).

On the other hand, since IEEE Xplore and ACM support wildcards, they have been used to shorten the query string while still covering the keywords included in Table 3 (e.g. "* Performance" covers both "Operating Performance" and "Business Performance") and other combinations that were not thought of.

2.4 Quality Assessment

In this type of review, despite the query string being written in a way that is specific and tries to restrict the number of articles returned from a search, the number of results is usually significant, in this case, 411 articles.

In order to manage this, additional inclusion and exclusion criteria are defined to better control the range of the search and the quality of the articles, while also supporting the decisions made during the screening process. In the end, this should result in a more refined and relevant set of articles to include in the discussion of each research question.

These criteria range from simple conditions such as language and time intervals to more specific conditions that aim to restrict the results to the topics at hand.

In this review, only articles written in English and in the last 5 years are included, besides being peer reviewed and focusing on the topics being studied: fully tested PPM applications and the impact they have on business performance and the decision-making process.

The full set of inclusion and exclusion criteria defined for this systematic review are presented in Tables 5 and 6.

Table 5. Inclusion criteria

ID	Criteria
IC1	The source describes a significant contribution to the field of Process Mining
IC2	The source describes a technique or methodology to implement predictive mechanisms in Process Mining
IC3	The source describes how predictive techniques are integrated with Process Mining solutions
IC4	The source describes how Predictive Process Mining impacts the performance and costs of organizations
IC5	The source describes how Predictive Process Mining supports stakeholders in the decision-making process
IC6	The source describes how Predictive Process Mining changes the design of business processes

Table 6. Exclusion criteria

ID	Criteria
EC1	The source is over 5 years old (published before 2018)
EC2	The source is not written in English
EC3	The source is not peer-reviewed
EC4	The source focuses on a system that is not fully developed and tested
EC5	The source does not focus on the process enhancement phase of Process Mining
EC6	The source focuses on fault/fraud detection
EC7	The source focuses on computer vision techniques
EC8	The source focuses on real-time process monitoring

2.5 Data Extraction

Having defined what, where, and how to seek information on the research questions at hand, all that remains is to execute the search and submit the obtained results to a screening process that filters relevant articles, based on the defined criteria, to be included in the review.

This process starts with the data extraction from each data source, by using the defined query strings on the title, abstract, and keywords of the articles.

In addition, criteria such as language and time interval are also applied, as well as including peer-reviewed and accessible articles only, as they are possible to apply beforehand and help minimize the number of results.

The number of articles obtained from each of the data sources is presented in Table 7.

Table 7. Search results

Data source	No. results
IEEE Xplore	135 articles
ScienceDirect	115 articles
ACM	161 articles
Total	410 articles (1 duplicate removed)

Compared to the initial 11 articles found during testing, the refined query string resulted in 410 articles, which should make up for a more varied set, although many irrelevant articles may have also been included and need to be filtered in the subsequent screenings.

Having obtained the initial set of articles, they were submitted to the abstract screening, where the title and abstract of each article were read and the article was kept if it appeared to help provide an answer to any of the research questions, always taking into account the quality assessment criteria defined previously.

To support the screening process, the software Covidence was used, as it allows the user to upload the extracted articles and perform the entire screening process in a platform with team collaboration and other helpful features (i.e. dynamic PRISMA flowchart).

During the abstract screening, 344 articles were classified as irrelevant and 66 as relevant, the latter moving on to the full-text screening, where the entire article was read and classified as relevant or not for the review, based on the defined criteria.

From the 66 articles that were submitted to the full-text screening, 42 were classified as irrelevant and 24 as relevant, making up the final set of articles to be included in the review.

During this entire process, notes were taken about each article, identifying the main topics presented in each of them and which questions each of them answers.

Likewise, the reasons to exclude each article were also registered, taking into account the inclusion and exclusion criteria, ranging from articles that did not focus on PM at all to articles that did not focus on predictive techniques or focused on topics outside of the scope of this review (e.g. conformance checking, real-time monitoring).

A summary of the excluded articles during full-text screening is presented in Table 8.

Table 8. Excluded articles during full-text screening

Exclusion reason	No. articles
Does not focus on Process Mining	24 articles
Does not focus on Predictive Techniques	15 articles
Focuses on Conformance Checking	2 articles
Focuses on Real Time Process Monitoring	1 articles
Total	42 articles

Finally, a summary of the described process is presented in the diagram of Fig. 1, which follows the flowchart described in PRISMA.

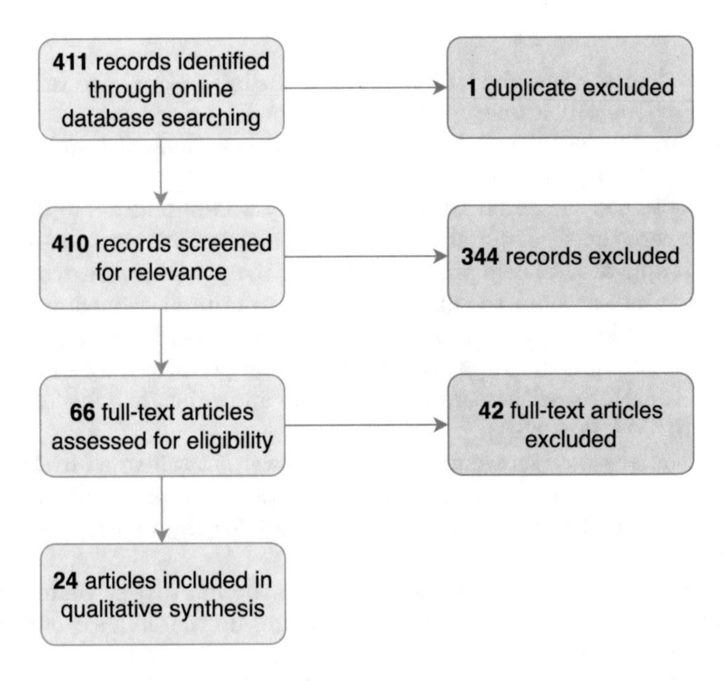

Fig. 1. PRISMA Flowchart

3 Results

The following section describes the results from the screening process detailed in the previous section, identifying for each research question the articles and the respective information that helped answer it.

3.1 RQ1: What Mechanisms Are Used in the Implementation of Predictive Process Mining?

Of the 24 retrieved articles, 22 focus on the mechanisms being employed in the implementation of PPM, ranging from predictive techniques (e.g. ML) to data representations and optimisation methods.

Article [12] identifies the three most common use cases in PPM: next-event, outcome, and suffix prediction, while also being the article that sheds the most light on suffix prediction, the most complex and least studied problem.

The article proposes a Deep Learning (DL) model whose architecture consists of embedding, sequential, and generator layers, with the first one embedding the prefix information (i.e., past activities and respective timestamps), the second one extracting knowledge from the prefix, and the last one generating the suffix based on the obtained knowledge.

It is important to note that the sequential layer consists in a DL model that is interchangeable, as the study compares 7 different algorithms to determine the best for suffix prediction, with Long Short Term Memory (LSTM), Autoencoding Generative Adversarial Networks (AE-GAN) and Bidirectional Encoder Representations from Transformers (BERT) being the most effective and some of the more prominent algorithms in this review.

In the end, it was concluded that no algorithm was better in every scenario, a conclusion that is carried over to the other studies included in this review.

The studies show that the effectiveness of a predictive algorithm depends on factors such as process length, variability (i.e. variety of possible activities), and the stage of completion in which the prediction is made (i.e. closer or farther from the outcome).

As a result of this review, a diverse set of predictive techniques currently used in PPM has been identified. When developing new PPM use cases, these techniques should be applied, evaluated, and compared, as a means of determining the best one for the characteristics of each use case. To keep it brief, only the most prominent and effective algorithms are included in this section.

The most frequent ensemble algorithm in the included studies is XGBoost, which is applied in 9 of the 22 articles: [13–21]. The ensemble nature of this algorithm makes it a robust and effective solution for most use cases, usually performing better with processes that are short in length and variability.

Considering the sequential nature of business processes and the resulting event logs, these can be represented either in a sequential or aggregated manner as the input for the predictive model. In other words, the process can be represented as a sequence of events and respective information (e.g. activity, timestamp, other relevant information), based on control-flow data and event payloads, or with aggregated information about its execution, including process metrics (e.g. number of completed activities, total execution time) and the results from its execution (e.g. a shopping cart and respective details).

In next-event and outcome prediction use cases, depending on the complexity of the process, either representation is usually effective, with the application of both being demonstrated in the included studies.

However, in the case of suffix prediction, sequential representation is advised [12]. In suffix prediction, the problem is treated, by analogy, as a text completion problem, where the start of the sentence is provided, and the algorithm completes the rest.

As use cases get more complex, a hybrid approach could also be followed, for an even richer representation of the process data, integrating both the activities and respective timestamps, in a sequential representation, with data about the case that is not specific to a given activity, including process metrics and contextual data (e.g. number of open checkouts, in the context of a supermarket process). Nevertheless, similar to the predictive techniques, no single representation is better in every scenario, as each use case has its own characteristics and available data [22].

Although XGBoost is capable of handling any of the mentioned data representations, based on the included studies, it usually fares relatively better with an aggregated representation of a process, since there are more specialized DL models to deal with sequential data.

The most frequent of these models is LSTM, applied in 10 of the 22 articles: [5,12,20,21,23–28].

Compared to XGBoost, LSTMs and other sequence DL models are generally more adequate for a sequential data representation due to their capacity to model a sequence of events so that each pattern can be assumed to be dependent on previous ones.

This characteristic makes them the most popular models in fields such as Natural Language Processing (NLP), where phrases are treated as a sequence of words or characters. Some models make use of more complex techniques such as attention mechanisms to further improve their effectiveness on these types of problems even further.

It should be pointed out that this model is often paired with other models, such as Convolutional Neural Networks (CNN) [26], or composed in an ensemble, resulting in more robust models such as Bidirectional LSTMs (Bi-LSTM) and Attention-based Bi-LSTMs (Att-Bi-LSTM) [24].

The use of CNNs is more common in fields such as Computer Vision and Image Processing, where data is usually represented in multidimensional matrices. Its application in PPM may not be obvious at first, but the article [12] proposes a possible data representation that makes the benefits of using CNNs more clear.

Considering a suffix prediction use case, where the training dataset consists of ongoing processes, the data may be represented in a three-dimensional matrix (P $\times$ E $\times$ D), where P is the number of example processes included in the dataset, E is the number of possible events in the process, and D is the number of columns necessary to represent the data on a given event (e.g., activity, timestamp). As a result, the model receives the list of events (i.e., prefix) and the data related to each of them as input, represented in a two-dimensional matrix.

By analogy, this two-dimensional matrix can be treated as an image and be processed as an image would be in a computer vision use case. Depending on the use case, the output can either be a single array with the activity and timestamp of the next event or the outcome, or it could also be a new two-dimensional array, with the predicted information about the suffix (i.e., the activity and timestamp of each future event).

Despite LSTMs being effective in most cases and the most commonly applied DL technique in PPM, studies have been conducted to find even more robust techniques, with BERT and GAN showing the most interesting results.

On the one hand, BERT is proposed in [13] as a transfer learning approach to next-event and outcome prediction, taking advantage of BERT being a pre-trained Transformer model and requiring less training time to obtain a robust and effective model. In addition, the Attention mechanism is used to obtain the long-term dependencies between activities more effectively.

On the other hand, GAN is proposed in [29] as an alternative to the more common LSTM models, where it was concluded that GANs tend to generalise better than LSTMs and do not appear to suffer from accuracy fluctuations based on the prefix length (i.e. number of included events in the prefix). It also concludes that the generator layer is able to learn the input distribution to generate predictions close to the ground truth, eliminating the need for a large number of training instances.

In addition to the accuracy fluctuations that may occur based on the prefix length, the article [30] warns of the need to monitor the temporal stability of the predictive models. In other words, if multiple predictions are performed during the duration of a process execution, it must be guaranteed that the predictions are stable and do not change every time new data appears, since it can lead to a lack of trust from the stakeholders.

However, no matter how robust and effective the algorithms are, optimisation methods are always important to register and apply whenever a given technique is not delivering the expected or necessary performance.

One such optimisation method is proposed in [31] as the usage of Word2Vec, a popular NLP technique that is capable of mapping words in a vector space, instead of more typical representations such as one-hot encoding, according to the similarity and relatedness found in the training dataset. The article proposes the encoding of processes that are represented in a sequential manner using Word2Vec, essentially treating events as words and taking advantage of the similarity comparison that this representation allows, to enhance the performance of the predictive models.

Another optimisation method is proposed in [32] as the use of genetic algorithms to optimise the hyperparameters of the predictive algorithms. This is a common optimisation method in the field of DL, given the range of algorithms, architectures, and hyperparameters to choose from, with a given configuration emerging in the end as the best for the task at hand.

Finally, to alleviate the problem of high variability and, especially, process length, several optimisation methods have been proposed, known as Trace Bucketing techniques, which consist in dividing the prefix into subsets of events that are easier to analyse and may produce better predictions. At runtime, the best bucket to use in a given prediction is determined, and the respective model is then selected to predict.

The simplest of these techniques consists in segmenting the prefix into smaller subsets of activities with a fixed length (e.g. sets of 3 activities), via sliding window, as described in the articles [16,21,29]. The authors of [21] note that, by applying this method, it is assumed that the next event or outcome, depending on the use case, is closely related to the last N activities, with N being the length of the set.

On the other hand, clustering is also suggested as a possible technique to divide the prefix into smaller subsets of events, as described in the articles [16,20].

Finally, it is important to note the warning provided by the authors of article [16] about Concept Drift, a common phenomenon with data that spans long periods of time and consists of the variation of statistical properties of variables over time, which results in less accurate predictions as time passes. As a countermeasure to combat this issue, the authors suggest that models are trained frequently and with recent data in the training dataset, so that predictive mechanisms accompany the changes.

3.2 RQ2: How Is Predictive Process Mining Supporting Stakeholders and Decision Makers?

Of the 24 articles retrieved, 6 focus on the value and support that PPM provides to stakeholders and decision-makers, allowing them to take a more proactive approach to their daily operations.

In the article [19], the author states that processes in the real-world business environment are complex and need to be monitored to obtain the earliest possible prediction of the process outcome as it is an essential factor for better customer experience and mitigation of business loss.

PPM emerged as a tool that provides businesses with timely predictive information and constructive feedback for the stakeholders to take proactive corrective actions to reduce process execution risk (e.g. exceeding time limit). An example of such actions could be the adjustment of the priority of activities [14,19,27].

The authors of article [12] state that the predictions resulting from PPM use cases provide valuable input for planning and resource allocation. In addition, they add that the prediction of the remaining execution time (i.e., outcome prediction) can be used to prioritise process instances to achieve service-level goals.

Currently, model explainability is a common topic in the research field of artificial intelligence, as more and more ML algorithms present a black-box nature, whose predictions are hard to understand. When it comes to PPM, the article [15] describes how model explainability can be leveraged by stakeholders to identify the main contributing variables for a process with a turnaround time longer than expected, which allows for a quick and more confident intervention in the process to reduce that time.

Finally, in an interview conducted in article [33], with the objective of determining how organisations perceive the value of PM, 17 stakeholders and decision-makers from the industry were asked about the impact and benefits of PM to their business and how PPM allows them to make the right decisions, at the right time.

In this interview, the interviewees identify three main benefits of using PPM: (i) production of end-to-end process visualisations and prediction of key performance indicators; (ii) data-driven and evidence-based decision-making; (iii) implementation of timely interventions and enhancements based on detected problems.

In addition, they also discuss the impact that PPM has on the performance of their operations. However, that discussion will be analysed in more detail in the next question.

3.3 RQ3: How Is Predictive Process Mining Impacting the Performance and Costs of Organisations?

Of the 24 articles retrieved, 8 focus on the impact that PPM has on the performance and costs of organisations, based on more proactive actions taken by stakeholders.

One of the most common performance indicators that PPM is reported to impact is customer satisfaction, with the article [34] stating that predictions on the future events of a given process, such as the customer journey path, allow them to produce personalised recommendations based on user behaviour, which, in turn, optimises KPIs such as shopping cart conversion rate, in addition to the aforementioned customer satisfaction.

Moreover, the article [17] states that increased efficiency, as a result of corrective actions taken based on the predicted outcome of a given process, is positively correlated with increased customer satisfaction.

On the other hand, as mentioned in the previous question, PPM has emerged as a key tool for decision-makers, as it provides constructive feedback for process improvement [19], allowing proactive corrective actions to be taken. In turn, process execution risks, such as exceeding the time limit, are reduced by adjusting the priority of activities and resource planning and allocation [14,26,27].

Furthermore, as the explainability of the model becomes more prevalent, identifying the key contributor variables to a given bottleneck in a process will also allow stakeholders to take not only more timely but also more precise actions, as they will know exactly where to intervene, maximising the value in return [15].

Finally, referring to the interview conducted in the article [33], the interviewees were asked about the impact of PPM on the performance of their operations.

Overall, they agree with the observations made above, stating that timely interventions upon problems in a process lead to improved process efficiency, monetary gains (e.g. costs), and non-monetary gains (e.g. customer satisfaction).

One of the interviewees reported that PPM has allowed his organisation to work much faster on making the necessary changes to improve operations. Moreover, it permitted PM to not only be used to analyse the processes but to actively enhance and improve upon them.

Another interviewee stated that PPM allowed his organisation to create value for the purchase-to-pay process in their business by acting on time to changes based on alerts provided by the predictive algorithms.

4 Discussion

The following section summarises the findings obtained from the review, while also providing some discussion on the subjects at hand to determine the current state of the field.

4.1 Use Cases

The most common PPM use cases are focused on the next-event, outcome, and suffix prediction [12], visually represented in the diagram in Fig. 2.

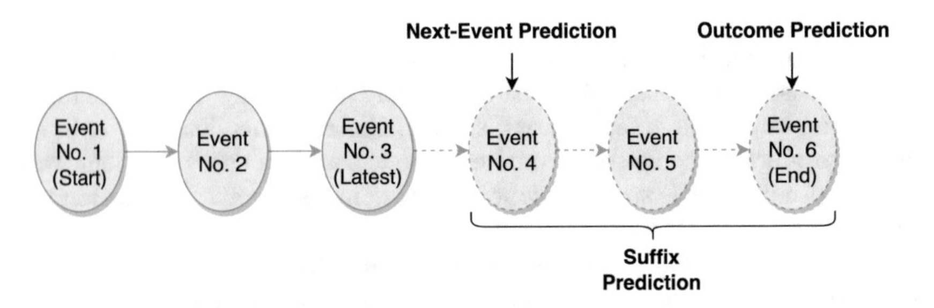

Fig. 2. Most common Predictive Process Mining use cases: next-event, outcome, and suffix prediction.

The next-event prediction task looks at the immediate future of the process execution by predicting the properties of the next event (e.g. timestamp, event type).

On the other hand, the outcome prediction task is concerned with what happens at the end of the process execution by predicting properties of the case (e.g. the acceptance of a loan application or prediction of customer churn).

Finally, the suffix prediction task attempts to construct the whole process execution by predicting properties of all future events (i.e., sequence generation).

Note that the output of these use cases may be a timestamp, activity, or both. For example, when predicting the outcome of a given process execution, the timestamp of the last event can be predicted, allowing the estimated time remaining to be calculated based on that, or the activity associated with the last event can be predicted (e.g. predict if a user will finalise the purchase of a given product), with some works describing approaches to predict both simultaneously.

4.2 Predictive Techniques

Of the articles included in the review, several of them conducted studies to find the best algorithms for a given use case and the characteristics of the problem. However, it was concluded that there is no "silver bullet" algorithm that is the best for every scenario and use case because their performance depends on factors such as process length, variability (i.e., variety of possible activities), and stage of completion (i.e., closer or farther from the outcome), among others.

From the review, four techniques stood out as the most prominent and effective: XGBoost, LSTM, BERT, and GAN.

XGBoost surges as the more traditional ML technique in the set. XGBoost is an ensemble algorithm that uses decision trees as base learners and is designed to be highly efficient, scalable, and have low computational requirements, demonstrating notable results in the majority of use cases, although it usually fares better with processes that have lower variability and overall complexity compared to sequence DL models, which tend to overfit in these cases.

However, as the complexity of processes and use cases increases (e.g. suffix prediction), sequence DL models have emerged as an appropriate technique to apply in these circumstances, combined with sequential data, due to their capacity to model a sequence of events so that each pattern can be assumed to be dependent on previous ones.

From this family of DL algorithms, LSTM is the most commonly used, usually combined with other algorithms, such as CNNs, or composed in an ensemble, resulting in robust models such as Bi-LSTM and Att-Bi-LSTM. These algorithms are designed to handle the problem of long-term dependencies in sequential data, which makes them more suitable for a sequential data representation and more complex processes.

However, despite the effectiveness of LSTMs in most cases, algorithms such as BERT and GANs have been found to bring more benefits, such as shorter training times, smaller training dataset requirements, and better scalability for more complex processes.

By following a transfer learning approach, BERT can be used to obtain a robust and effective model with significantly fewer training examples and time, while also taking advantage of the benefits provided by the Transformer architecture and the usage of an Attention mechanism to more effectively obtain the long-term dependencies between activities.

Furthermore, GANs have proven to be a robust and scalable alternative to LSTM models, as they generalise better and do not appear to suffer from accuracy fluctuations based on prefix length and variability. GANs are composed of a generator network that is trained to create new data samples similar to the real data, in contrast to a discriminator network that is trained to distinguish between real and generated data. In this way, the generator is able to learn the input distribution to generate predictions close to the ground truth, eliminating the need for a large number of training instances.

Overall, based on the literature, XGBoost seems to be the most effective model for most cases, having low computational requirements and providing notable performance in less complex use cases, in which DL models tend to overfit, while still being scalable enough to provide good results in more complex scenarios.

However, as the complexity of processes and use cases grow (e.g. suffix prediction), the usage of sequence DL models becomes more viable, with LSTM models generally performing well, despite BERT and GAN emerging as more sophisticated and robust alternatives for the most complex use cases.

Nevertheless, the prevalence of these algorithms in the literature should not discourage experimentation with other algorithms during the implementation of PPM use cases. Every use case has different characteristics, and, as stated above, a "silver bullet" algorithm does not exist for PPM.

In addition, issues such as overfitting, temporal stability, and accuracy fluctuations are difficult to predict and can happen with any predictive model, which further encourages the testing and evaluation of various algorithms and techniques.

4.3 Data Representation

Given the sequential nature of business processes, they can be represented in a sequential or aggregated manner.

The sequential representation of a process should consist of a list of events and respective information (e.g. activity, timestamp, and other relevant information), based on control-flow data and event payloads.

In contrast, the aggregated representation of a process may include information about its execution, such as process metrics (e.g. number of completed activities, total execution time), contextual data, and the results from its execution (e.g. a shopping cart and respective details).

Note that in a sequential representation, each event is included and its information is represented. Among others, this information may be the respective activity, timestamp, or even both, the latter resulting in a 2D matrix, which, as registered above, may benefit from the use of a CNN to process the matrix as if it was an image in a Computer Vision use case.

However, as the complexity of processes grows, following a single representation may limit the performance of the predictive model. For example, an aggregated representation does not include the order and timestamps in which the activities were performed, whereas sequential data does not include contextual information about the process.

Instead, a hybrid approach can be followed, providing all the available information about the process, integrating both the activities and respective timestamps, in a sequential representation, with data about the case that is not specific to a given activity, including process metrics and contextual data, resulting in richer data for the predictive model to work with.

In suffix prediction use cases, the use of sequential representation is recommended in the literature, given the sequential nature and representation of the output itself. However, given the benefits that a hybrid representation may bring, in theory, it would be interesting to apply it in implementing future suffix prediction cases.

Nevertheless, similar to the selection of predictive techniques, each use case has its characteristics, and a careful analysis of the available data is necessary to determine the right representation to be followed. Whenever possible, to support this decision, a benchmark should also be made between an aggregated, sequential, and hybrid approach.

4.4 Optimization Methods

From the review, several optimisation methods have been registered and should be considered whenever a given solution is not delivering the expected performance.

Below, a summary of the methods found during the review is presented, organised by the problem they address.

Enhancement of Sequential Data Representation - taking inspiration from the techniques applied in the field of NLP, Word2Vec has been proposed as a possible technique to apply in sequential data representation of processes, treating the sequence of events as if they were a sequence of words.

By applying Word2Vec to business processes, it is expected that benefits such as word similarity and relatedness, obtained during the training phase of Word2Vec, translate to the events and can be used to compare similar or related events, allowing the performance of predictive models to be improved.

Optimization of the Hyperparameters and Model Configuration - the use of genetic algorithms to optimise the hyperparameters of the predictive algorithms is a common and well-documented optimisation method in the field of DL, given the range of algorithms, architectures, and hyperparameters that are usually available in the field.

Nevertheless, the application of this method should be carefully considered. Its application should produce a model that, according to the defined criteria, is the best for the task at hand. However, genetic algorithms have to be designed and developed almost from scratch for the specific set of algorithms and tasks at hand, which requires some effort and, in some cases, might not justify the time spent developing them.

High Process Complexity - one of the biggest challenges in PPM concerns the complexity of business processes, both in length (e.g. number of activities in a given execution) and variability (e.g. number of distinct activities).

To alleviate this problem, the literature proposes a set of optimisation methods, known as Trace Bucketing techniques, which consist of dividing the prefix into subsets of events, designated buckets, that are easier to analyse and may produce better predictions. At runtime, the best bucket to use in a given prediction is determined, and the respective model is then selected to predict.

The most common methods found during the review were: clustering and fixed prefix length, the latter consisting of segmenting the prefix into smaller subsets of activities with a fixed length (e.g. sets of 3 activities), via a sliding window.

It must be noted that when applying either of these methods, it is assumed that the next event or outcome, depending on the use case, is closely related to a given subset of activities, which may not be accurate and may actually result in worse performance.

On the other hand, GANs have proven to generalise well and do not suffer from accuracy fluctuations based on the prefix length, while also being effective with processes with high variability, which may eliminate the need and risk of applying these optimisation techniques.

Concept Drift - this is a common phenomenon with data that spans long periods of time and consists of the variation of statistical properties of variables over time, which results in less accurate predictions as time passes.

In the literature, as a countermeasure to combat this issue, it is suggested that models are trained frequently and with recent data in the training dataset, in order for the changes to be accompanied by the predictive models.

4.5 Business Impact

Based on the results of the review, it is clear that some decision-makers are already taking advantage of PPM to optimise their business processes and the overall performance of the organisation.

The main benefits that were found about PPM include end-to-end visualisations of the business process and the prediction of key performance indicators, enabling data-driven and evidence-based decision-making, as well as the implementation of timely interventions and enhancements (e.g. resource reallocation, activity prioritisation) based on detected problems (e.g. processes that will probably not meet the deadline), allowing the outcome to be steered in the organisation's favour.

As a result of these interventions, organisations are able to work faster, business processes become more efficient, and there are monetary (e.g. costs) and non-monetary gains (e.g. customer satisfaction). Ultimately, this leads to greater value for the organisation and its clients, which is the goal of PM and what organisations should be striving for, instead of using PM as a strictly business intelligence and analytical tool.

Nevertheless, during this review, it was found that a relatively low number of articles addressed the value and impact of PPM in organisations compared to the technical articles, which comprised around 90% of the included articles. For example, only one article details an interview with stakeholders about the subject.

Furthermore, the claims found in these articles were generally not supported by any significant values or evidence, further limiting the perceived value of PPM for stakeholders interested in adopting it.

This highlights the need to demonstrate the value and effectiveness of PPM through the development and evaluation of more real-life use cases, resulting in more evidence of the impact of PPM on operational performance and costs in an organisation. Ultimately, this should lead to increased interest and adoption of PPM.

5 Conclusion

The motivation behind this systematic review was the mostly unproven value of PPM for stakeholders and organisations, which, in addition to the relatively low number of articles related to PPM as of 2019, compared to the superset of articles on PM, creates the need for more work to be performed in the field.

As a means of supporting this work, this systematic review aimed to identify the current state of PPM, the mechanisms being applied in the development of PPM use cases, and the perceived value of stakeholders and organisations that already use PPM in their daily work.

The review started with 411 scientific articles, extracted from the defined data sources, of which 24 were included and analysed, after the screening process.

This analysis allowed the most common use cases of PPM to be identified: next-event, outcome, and suffix prediction, in addition to the state-of-the-art predictive techniques being applied in their implementation, from XGBoost to sequential deep learning (DL) models, such as LSTM, BERT, and GAN.

The articles also provided insights on possible representations/embeddings of data from business processes, according to the type, complexity, and techniques used in the use case.

To complement this, some optimisation methods have also been identified, such as hyperparameter optimisation and trace bucketing, while also raising awareness of common problems in PPM, such as concept drift and temporal stability.

Finally, although limited, the perceived value of PPM from stakeholders, based on the articles included in the review, has also been summarised for future reference.

Nevertheless, given the rapid emergence of new mechanisms and the limited information on the perceived value of PPM from stakeholders and organisations, it is imperative that researchers and developers continue supporting the growth of the field in the future, and the results of this systematic review should hopefully provide a reference point for this work.

As mentioned above, more real-life use cases need to be developed and studied, measuring the impact of PPM on operational performance and costs, in addition to more interviews being conducted with stakeholders, in order to understand their needs and goals for PPM.

In the end, the hope is that these efforts will help to further validate the effectiveness of PPM in the eyes of organisations and ultimately lead to increased adoption.

Acknowledgment. This work was supported by the Ministry for Science, Technology and Higher Education funded by National Funds through the Portuguese Fundação para a Ciência e a Tecnologia (FCT) under the R&D Units Project Scope 10.54499/UIDB/00760/2020 (https://doi.org/10.54499/UIDP/00760/2020), 10.54499/UIDP/00760/2020 (https://doi.org/10.54499/UIDB/00760/2020).

References

1. Consulting, F.: Trends in process improvement and data execution - how organizations are improving processes and turning process data into real-time action, vol. 1 (2022)
2. van der Aalst, W., et al.: Process mining manifesto. In: Daniel, F., Barkaoui, K., Dustdar, S. (eds.) BPM 2011. LNBIP, vol. 99, pp. 169–194. Springer, Heidelberg (2012). https://doi.org/10.1007/978-3-642-28108-2_19
3. van der Aalst, W., Weijters, A.J.M.M.: Process mining: a research Agenda. Comput. Ind. **53**, 231–244 (2004)
4. Pishgar, M., et al.: A process mining- deep learning approach to predict survival in a cohort of hospitalized COVID-19 patients. BMC Med. Inf. Decis. Making **22**, 1–16 (2022)
5. Theis, J., Galanter, W.L., Boyd, A.D., Darabi, H.: Improving the in-hospital mortality prediction of diabetes ICU patients using a process mining/deep learning architecture. IEEE J. Biomed. Health Inf. **26**, 388–399 (2022)
6. Fernandes, M., Corchado, J.M., Marreiros, G.: Machine learning techniques applied to mechanical fault diagnosis and fault prognosis in the context of real industrial manufacturing use-cases: a systematic literature review. Appl. Intell. **52**, 14246 (2022)
7. Chiu, T., Wang, Y., Vasarhelyi, M.: The automation of financial statement fraud detection: a framework using process mining. J. Forensic Invest. Account. **108** (2020)
8. dos Santos Garcia, C., et al.: Process mining techniques and applications - a systematic mapping study. Expert Syst. Appl. **133**, 260–295 (2019)
9. Moher, D., Liberati, A., Tetzlaff, J., Altman, D.G.: Preferred reporting items for systematic reviews and meta-analyses: the prisma statement. BMJ **339**, 332–336 (2009)
10. Page, M.J., et al.: The prisma 2020 statement: an updated guideline for reporting systematic reviews. BMJ **372**, 3 (2021)
11. Paré, G., Trudel, M.-C., Jaana, M., Kitsiou, S.: Synthesizing information systems knowledge: a typology of literature reviews. Inf. Manage. **52**, 183–199 (2015)
12. Ketykó, I., Mannhardt, F., Hassani, M., van Dongen, B.F.: What averages do not tell: predicting real life processes with sequential deep learning. In: Proceedings of the 37th ACM/SIGAPP Symposium on Applied Computing, pp. 1128–1131 (2022)
13. Chen, H., Fang, X., Fang, H.: Multi-task prediction method of business process based on BERT and transfer learning. Knowl.-Based Syst. **254**, 109603 (2022). https://doi.org/10.1016/j.knosys.2022.109603
14. Sun, X., Hou, W., Ying, Y., Yu, D.: Remaining time prediction of business processes based on multilayer machine learning. In: 2020 IEEE International Conference on Web Services (ICWS), pp. 554–558 (2020)

15. Toh, J.X., Wong, K.J., Agarwal, S., Zhang, X., Lu, J.J.: Improving operation efficiency through predicting credit card application turnaround time with index-based encoding. In: Companion Proceedings of the Web Conference 2022, pp. 615–620 (2022)
16. Teinemaa, I., Dumas, M., Rosa, M.L., Maggi, F.M.: Outcome-oriented predictive process monitoring: review and benchmark. ACM Trans. Knowl. Discovery Data **13**, 1–57 (2019)
17. Ogunbiyi, N., Basukoski, A., Chaussalet, T.: Incorporating spatial context into remaining-time predictive process monitoring. In: Proceedings of the 36th Annual ACM Symposium on Applied Computing, pp. 535–542 (2021)
18. Chen, L., Klasky, H.B.: Six machine-learning methods for predicting hospital-stay duration for patients with sepsis: a comparative study. In: SoutheastCon 2022, pp. 302–309 (2022)
19. Tariq, Z., Charles, D., McClean, S., McChesney, I., Taylor, P.: Proactive business process mining for end-state prediction using trace features. In: 2021 IEEE SmartWorld, Ubiquitous Intelligence & Computing, Advanced & Trusted Computing, Scalable Computing & Communications, Internet of People and Smart City Innovation (SmartWorld/SCALCOM/UIC/ATC/IOP/SCI), pp. 647–652 (2021)
20. Verenich, I., Dumas, M., Rosa, M.L., Maggi, F.M., Teinemaa, I.: Survey and cross-benchmark comparison of remaining time prediction methods in business process monitoring. ACM Trans. Intell. Syst. Technol. **10**, 1–34 (2019)
21. Tama, B.A., Comuzzi, M., Ko, J.: An empirical investigation of different classifiers, encoding, and ensemble schemes for next event prediction using business process event logs. ACM Trans. Intell. Syst. Technol. **11**, 1–34 (2020)
22. Francescomarino, C.D., Ghidini, C.: Predictive process monitoring. In: van der Aalst, W.M.P., Carmona, J. (eds.) Process Mining Handbook. Lecture Notes in Business Information Processing, vol. 448, pp. 320–346. Springer, Cham (2022).https://doi.org/10.1007/978-3-031-08848-3_10
23. Venugopal, I., Tollich, J., Fairbank, M., Scherp, A.: A comparison of deep-learning methods for analysing and predicting business processes. In: 2021 International Joint Conference on Neural Networks (IJCNN), pp. 1–8 (2021)
24. Wang, J., Yu, D., Liu, C., Sun, X.: Outcome-oriented predictive process monitoring with attention-based bidirectional lstm neural networks. In: 2019 IEEE International Conference on Web Services (ICWS), pp. 360–367 (2019)
25. Lee, W.L.J., Parra, D., Munoz-Gama, J., Sepúlveda, M.: Predicting process behavior meets factorization machines. Expert Syst. Appl. **112**, 87–98 (2018)
26. Xia, C., Xing, M., Ye, Y., He, S.: A process mining framework based on deep learning feature fusion. In: 2022 41st Chinese Control Conference (CCC), pp. 7412–7418 (2022)
27. Heinrich, K., Zschech, P., Janiesch, C., Bonin, M.: Process data properties matter: introducing gated convolutional neural networks (GCNN) and key-value-predict attention networks (KVP) for next event prediction with deep learning. Decis. Support Syst. **143**, 113494 (2021)
28. Hanga, K.M., Kovalchuk, Y., Gaber, M.M.: A graph-based approach to interpreting recurrent neural networks in process mining. IEEE Access **8**, 172923–172938 (2020)
29. Taymouri, F., Rosa, M.L., Erfani, S., Bozorgi, Z.D., Verenich, I.: Predictive business process monitoring via generative adversarial nets: The case of next event prediction. In: Business Process Management: 18th International Conference, BPM 2020, Seville, Spain, September 13–18, 2020, Proceedings 18, pp. 237–256 (2020)
30. Teinemaa, I., Dumas, M., Leontjeva, A., Maggi, F.M.: Temporal stability in predictive process monitoring. Data Min. Knowl. Disc. **32**, 1306–1338 (2018)

31. Junior, S.B., Ceravolo, P., Damiani, E., Omori, N.J., Tavares, G.M.: Anomaly detection on event logs with a scarcity of labels. In: 2020 2nd International Conference on Process Mining (ICPM), pp. 161–168 (2020)
32. Francescomarino, C.D., et al.: Genetic algorithms for hyperparameter optimization in predictive business process monitoring. Inf. Syst. **74**, 67–83 (2018)
33. Badakhshan, P., Wurm, B., Grisold, T., Geyer-Klingeberg, J., Mendling, J., vom Brocke, J.: Creating business value with process mining. J. Strateg. Inf. Syst. **31**, 101745 (2022)
34. Terragni, A., Hassani, M.: Optimizing customer journey using process mining and sequence-aware recommendation. In: Proceedings of the 34th ACM/SIGAPP Symposium on Applied Computing, pp. 57–65 (2019)

Digital Media Doctorates: Trends, Methodologies and Contributions

Francisca Rocha Lourenço(✉), Lorena Sousa, Oksana Tymoshchuk, Fábio Ferreira, Mónica Silva, Lídia Oliveira, and Nelson Zagalo

University of Aveiro, 3810-193 Aveiro, Portugal
franciscalourenco@ua.pt

Abstract. This study aimed to provide an overview of the doctoral theses carried out in the field of Digital Media over the last six years. Two doctoral programs linked to the DigiMedia Research Centre were considered: the Doctorate in Multimedia in Education and the Doctorate in Information and Communication on Digital Platforms.

A total of 46 theses were collected from the Institutional Repositories and analyzed considering their title, abstract, keywords, objectives, adopted methodology, type of funding, scientific publications and dissemination actions that each doctoral student carried out during their journey.

The results of this analysis show that the most used methodologies were case studies and research-based design; and the most used instruments are observation, surveys, interviews and focus groups. These theses, final scientific-technological products such as digital platforms, mobile applications, models, and prototypes have been built in various areas of study. The information gathered was systematized and recorded in a spreadsheet and uploaded into Microsoft Power BI software for presentation of the results and interactive data visualization on this Research Centre's Digital Observatory website.

With these findings, we hope to contribute to generating knowledge in the field of Digital Media and provide insights for future research and development in this area. Furthermore, the study results can inform decision-making processes for funding and resource allocation within the DigiMedia Center and guide the development of new doctoral programs and curricula.

Keywords: Ph.D. Theses · Digital Media · Multimedia in Education · Information and Communication in Digital Platforms

1 Introduction

This article is the result of cooperation between the Digital Media Observatory - an integral part of the DigiMedia Research Centre of the Department of Communication and Art at the University of Aveiro - and a programme called Students@DigiMedia. Students@DigiMedia is an annual program that aims to integrate students into teams and projects developed by researchers from the same department.

Á. Rocha et al. (Eds.): WorldCIST 2024, LNNS 987, pp. 379–388, 2024.
https://doi.org/10.1007/978-3-031-60221-4_36

The present analysis was realized during the academic year of 2022/2023 and seeks to show a compilation of data extracted from doctorate theses conducted at DigiMedia between the years 2018 to 2023 (until June). The objective is to gain insight into the primary subjects explored within the doctoral programmes affiliated with Information and Communication on Digital Platforms and Education in Multimedia. Additionally, the analysis examines indicators such as methodologies applied and instruments, to identify any specific trends in these areas. To show the data, the Business Intelligence software Power BI was used to visualize the information obtained from the data analysis.

2 Theoretical Framework

Pursuing a PhD represents the highest level of formal education, serving as the starting point for academic careers and a prerequisite for independent research endeavours [1, 2]. Undertaking a PhD involves acquiring specialized knowledge within a particular field or discipline and assimilating the practices of knowledge production in that area [3]. A PhD thesis is a transformative journey that spans several years of intensive research [4]. This arduous process moulds doctoral students into experts in their academic knowledge.

The Salzburg II recommendations for PhD programmes in Europe mention that the primary outcomes of doctoral education are the graduates themselves and "their contribution to society through the knowledge, competences and skills acquired during research, as well as awareness of and openness to other disciplines" (EUA, 2010, p. 4)[1].

Given the demanding nature of developing new researchers, research centres that promote doctoral programs assume a crucial role as "gatekeepers of the field" [5], determining who will shape the future production of knowledge. These centres play a pivotal role in ensuring the implementation of high-quality practices and processes that nurture the growth and potential of aspiring scholars [4, 6].

Moreover, the research conducted during PhD programs can potentially impact addressing global challenges significantly. Recent studies have revealed that doctoral students are vital drivers of universities' research output, with one-third of research publications attributed to their contributions [7]. However, further investigation is warranted to fully comprehend the extent of PhD students' contribution to scientific development [1, 8]. Additionally, understanding the influence of financial support, including research grants, on researchers' decision-making, behaviour, and long-term research output throughout their careers is crucial [1, 9, 10].

One of the challenges faced in PhD education pertains to evaluating doctoral theses. These comprehensive studies encompass vast amounts of information and data gathered through diverse methodological approaches and within different contexts, thereby posing difficulties in analysis. The evaluation process necessitates considering various indicators, ranging from the contribution and literature review to the approach and methodology, as well as the analysis, results, and presentation [11]. Furthermore, a holistic assessment of the thesis quality is essential to comprehend its broader implications for academia and society.

[1] https://eua.eu/resources/publications/615:salzburg-ii-%E2%80%93-recommendations.html#:~:text=9789078997221-,Download,-Share

Research centres need to be equipped with a comprehensive understanding of the research outcomes generated by PhD students to enhance the quality of doctoral programs. This broader perspective enables them to identify areas of strength and areas that require improvement, thereby fostering continuous progress and excellence.

Given the substantial amount of data involved in PhD research, it becomes imperative to develop tools that facilitate in-depth analysis of the methodological processes, research findings, and the broader impacts of doctoral research.

Leveraging the power of Business Intelligence (BI) tools, such as Power BI, can provide valuable insights by identifying patterns and trends within the vast corpus of doctoral theses [12]. These advanced BI tools offer the advantage of creating interactive and visually appealing data visualisations, making insights more accessible and understandable to a broader audience, including researchers, administrators, and funders [13].

By employing such BI-driven tools, research centres can make informed strategic decisions based on a comprehensive understanding of the methodological rigour, findings, and potential societal impact of the research conducted by PhD students.

Considering the increasing use of tools and instruments designed for individual and automated creation and construction, as mentioned above, Information Visualisation (InfoVis) becomes a concept that involves idealizing what we want to show to the public, in terms of the data obtained. The field of visualisation study explores various methods and strategies to effectively communicate information in a manner that facilitates users' comprehension and interpretation of data [14]. Research findings have demonstrated using information visualisation techniques enhances the data analysis compared to relying just on statistical analysis methods [15].

The visual representation should be tailored to suit the intended audience, which can vary from individuals with limited knowledge to those with extensive expertise [16]. In this particular instance, the focus is on understanding the manner in which scientific information is conveyed.

3 Method

This study was part of the Digital Media Observatory that aims to create digital data collection, compilation, processing, and a visualization system to understand the complexity, disseminate information, and promote studies on digital media in Portugal and other countries.

This study aimed to create an overview of the themes, objectives, methodologies, and results of the doctoral theses developed in Digital Media studies in the last five years. For this purpose, the project team considered the two Doctoral Programs (Doctorate) linked to DigiMedia: the Doctoral Program in Multimedia in Education (University of Aveiro) and the Doctoral Program in Information and Communication in Digital Platforms (a consortium between the University of Aveiro and the University of Porto).

The team collected these theses from the Institutional Repository of the University of Aveiro (RIA) and the information system SIGARRA of the University of Porto to achieve this goal. Forty-six theses that were concluded between 2018 and 2023 were collected and analysed considering the title, abstract, keywords, methodology adopted as the instruments, approach, target audience, and territorial scope of the research. Besides this, we analysed the objectives of each project and the type of funding for each thesis. Finally, we have analysed the scientific publications and dissemination actions that each doctoral student throughout their academic journey.

All the gathered information was systematized and recorded in a spreadsheet. Subsequently, the team utilized the Microsoft Power BI software to effectively display the results and provide users with an interactive visualization of the data on the website of the Digital Media Observatory.

4 Results

In total, 46 theses were analyzed: 25 from the Doctoral Program in Multimedia in Education (MEdu) and 21 from the Doctoral Program in Information and Communication in Digital Platforms (ICPD). In 2018, 11 theses were concluded. In 2019, this number was nine, followed by eight in 2020, five in 2021, eight in 2022 and five in 2023, as can be visualized in Fig. 1.

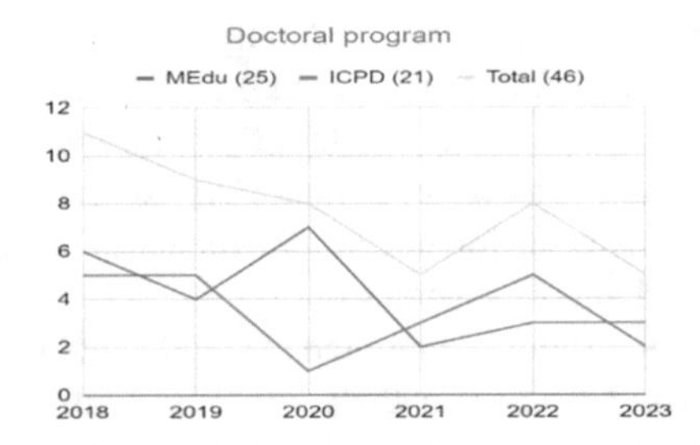

Fig. 1. Number of theses analysed under the two doctoral programmes.

The results indicate that out of the theses analyzed 1 had only one supervisor, while 25 had both a supervisor and co-supervisors. Forty supervisors and 14 co-supervisors were affiliated with the University of Aveiro in Portugal Additionally, six supervisors and three co-supervisors were from the University of Porto, also in Portugal. On supervisor and one co-supervisor were affiliated with other Portuguese universities, such as the Aberta University and the University of Lisboa. Furthermore, six thesis co-supervisors represented three Brazilian universities (Federal University of Maranhão, Federal University of Rio Grande do Norte (UFRN), and State University of Rio Grande do Norte (UERN), and only one co-supervisor was from the University of York in England.

It is worth noting that half of the theses (50%) received some kind of funding, such as funding from the Foundation for Science and Technology, CAPES Foundation (Brazil), and COMPETE 2020. Figure 2 provides a visual representation of the funding sources received by these theses, and it is interesting to observe that some of these theses benefited from multiple funding sources.

When examining the methodologies employed in the theses, it was found that the most prevalent methodology type was mixed methodologies, adopted by 60.7% of the

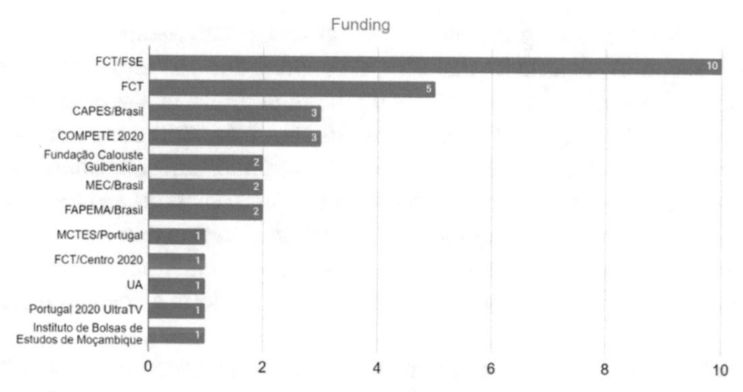

Fig. 2. Funding sources received by these theses.

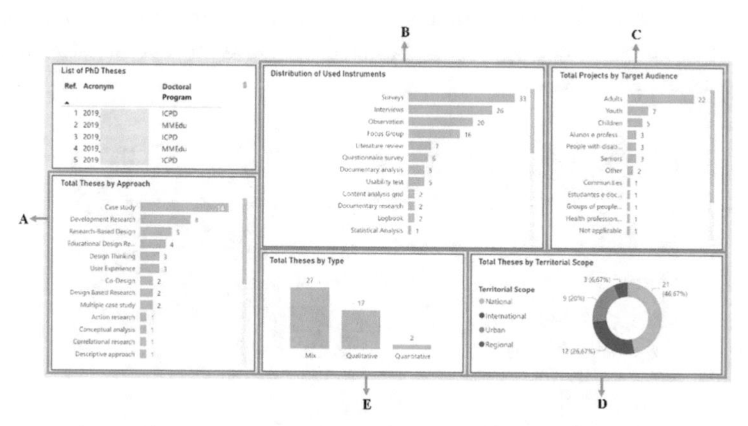

Fig. 3. Power BI Dashboard with analysis of methodological procedures.

theses. Qualitative methods were utilized in 35.7% of the theses, while quantitative methods were employed only in 3.6% (Fig. 3 - E).

Furthermore, a detailed analysis of the methodology approach revealed that: the Case Study had 14 theses; the Development Research and eight theses and with Design-Based Research, in third place, with five theses. Other methodology approaches were co-design (2), user experience (2), and design thinking (2) (Fig. 3 - A).

Still, on the methodology side, it was found that the most commonly used instruments in the doctoral theses analyzed were questionnaires (used in 33 of the theses) and interviews (used in 26 of the theses). This was followed by observation (20) and focus groups (16). Other methods used to a lesser extent included: systematic literature review, usability testing, document analysis, accessibility testing, analysis of usage logs, educational guides, and workshops (Fig. 3 - B).

Additionally, the data collected on the methodology considers the target audience studied and the territorial scope. The target audience that most theses focused on was Adults, with a percentage of 47.8%, followed by Youth in general, with a percentage of 15.2% of theses that took this audience into account for the development of the research (Fig. 3 – C).

Finally, it is worth noting that the territorial scope with the highest incidence was the national scope. A significant majority, accounting for 46.7% of the theses carried out research at the national level, indicating the importance of studying the dynamics and trends within the country. Furthermore, 26.1% of the theses explored research at an international level, shedding light on the global implications and connections of the topic. Another 20.0% of the theses focused their research on the urban level, highlighting the significance of understanding the specific urban contexts and their impact on the subject matter. And, finally, 6.7% of the theses were regional in scope, emphasizing the importance of analyzing the dynamics and challenges within a specific region (Fig. 3 – D).

The analysis of the theses revealed the recurring presence of keywords such as ICT (Information and Communication Technologies), Communication, Digital Platform and Inclusion, indicating their relevance and significance in the field (Fig. 4). These findings shed light on the current trends and priorities within the realm of information and communication technologies, emphasizing the need for continued research and development in these areas.

Fig. 4. Word cloud with the PhD Theses keywords.

During their research, the PhD students developed an impressive total of 41 scientific-technological products. Among these diverse products, the most notable were Models (10), Prototypes (6), Kits of best practice guidelines (5), and Digital platforms (5). These products showcase the students' dedication and expertise in their fields, highlighting their innovative contributions to the scientific and technological community (Fig. 5).

After analyzing the collected theses, the team discovered that the PhD students produced 227 scientific publications related to their studies. These publications included 91 conference papers and 92 articles in peer-reviewed scientific journals, which were particularly notable. Additionally, the doctoral students were involved in publishing 21 book chapters and six books, showcasing the breadth and depth of their research contributions (Fig. 6 - B).

Out of the 227 publications, more than 35,2% (80 articles) were indexed in scientific databases such as Scopus and Web of Science, which indicates that the scientific community recognized their contributions (Fig. 6 – A).

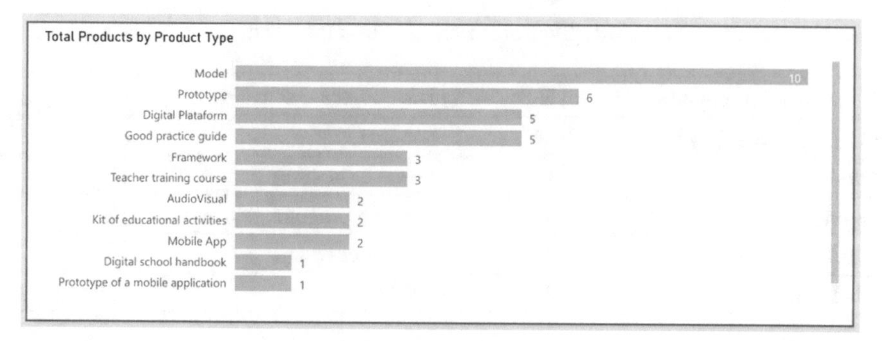

Fig. 5. Type of Scientific-Technological Products produced by the PhD students.

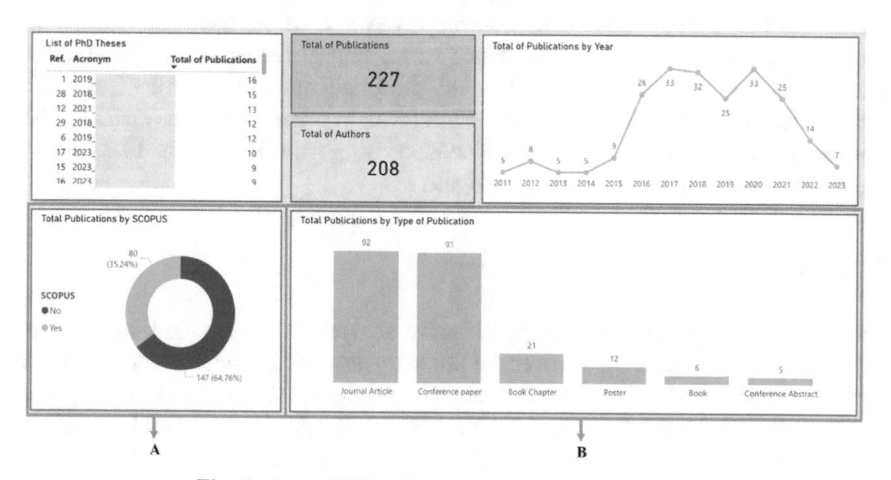

Fig. 6. Power BI Dashboard with publication results.

It is essential to consider that the theses were recently defended, and publishing scientific articles can be time-consuming. Therefore, it is highly likely that the number of publications related to these studies will increase in the future. Based on the collected data, it is evident that the PhD students effectively presented their research findings and actively engaged in scientific discussions (Table 1).

Based on the collected data, it is evident that the doctoral students presented the results of their research and actively participated in scientific discussions. The analyzed theses mention that PhD students participated in 103 dissemination activities, primarily through oral presentations and posters at scientific conferences. These dissemination activities served as a platform for doctoral students to showcase their groundbreaking research findings, exchange valuable insights, and foster collaborative networks within the scientific community.

However, some examples of the impact of PhD students' research have been documented. For instance, several theses have received national or international recognition. One example is the thesis titled "Digital Technologies to Support the Development of Mathematical Reasoning in Pupils with autism spectrum disorder," which was awarded

Table 1. Type of research findings by activities.

Dissemination activities	N	Dissemination activities	N
Conference presentation	59	Seminar	3
Poster	17	Demo Exhibition	2
Tertulia	8	Training programme	2
Sessions	5	Multiplier Event	2
Workshop	5	Total	103

the European "EDF-Oracle e-Accessibility Scholarship" in 2017/2018. Another notable example is the "Happy" mobile app developed as part of the thesis on "Cancer prevention mediated by smartphones," which was a finalist for various national awards. These include the PT Inovação 2017 award, the Challenge 2017 award - Health from Santa Casa da Misericórdia de Lisboa, the innovation prize Prémio Inovação PT 2017 - Academia Altice Labs category, and Finalist of Born from Knowledge Ideas 2017 by Agência Nacional de Inovação SA, Portugal.

5 Conclusion

This study sought to provide a comprehensive analysis of the doctoral theses carried out in the area of Digital Media over the last six years, with a special focus on the two doctoral programs: MEdu and ICPD, linked to the DigiMedia, to provide a comprehensive overview of the trends, themes, results and methodologies prevalent in these areas of research, as well as providing insights for improving the postgraduate offer at the university.

The theses covered a wide range of topics and employed various methodologies, focusing on mixed methodologies. The research conducted by the PhD students has resulted in numerous scientific-technological products and publications, demonstrating their significant contributions to the field. The doctoral students actively engaged in dissemination activities, presenting their research findings at conferences and contributing to scientific discussions. Overall, the findings highlight the importance of research centres like DigiMedia in fostering the growth and potential of aspiring scholars and advancing knowledge in digital media.

Advanced BI tools like Power BI have proven instrumental in unveiling patterns and trends within this corpus of research, allowing for in-depth and insightful visual representations. The main limitation of this study was the fact that it only focused on theses completed within two doctoral programs in the field of Digital Media. We therefore recommend expanding the analysis to the national level to include a wider variety of doctoral programs.

One of the main limitations of this study was that theses used in the analysis were sourced from different databases belonging to two universities. Another limitation arose from the unavailability of some of the more recent theses not being online. The team had to conduct extensive searches and contact PhD students directly to obtain the necessary information. The researchers encountered difficulties collecting data regarding the impact of the thesis projects. The "impact" dimension focuses on a project's long-term consequences. Since the students from the analyzed PhD Theses have recently completed their studies, understanding their research contributions' true value and significance is more complex.

Looking ahead, the team is committed to continuing this analysis and keeping it up to date, by including data from new theses as they become available. Moreover, it is of great interest to expand the scope of this research beyond the current universities and explore other research centres.

Acknowledgement. This work is financially supported by national funds through FCT – Foundation for Science and Technology, I.P., under the project UIDB/05460/2020.

References

1. Hortaa, H., Cattaneo, M., Meoli, M.: PhD funding as a determinant of PhD and career research performance. Stud. High. Educ.. High. Educ. **43**(3), 542–570 (2018)
2. McKenna, S., Quinn, L., Vorster, J.A.: Mapping the field of higher education research using PhD examination reports. High. Educ. Res. Dev. **37**(3), 579–892 (2018)
3. Dall'Alba, G., Barnacle, R.: An ontological turn for higher education. Stud. High. Educ. **32**(6), 679–691 (2007)
4. Holbrook, A., Bourke, S., Fairbain, H., Lovat, T.: The focus and substance of formative comment provided by PhD examiners. Stud. High. Educ. **39**(6), 983–1000 (2014)
5. Mowbray, S., Halse, C.: The purpose of the PhD: theorising the skills acquired by students. High. Educ. Res. Dev. **29**(6), 653–664 (2010)
6. Borkowski, N.: Changing our thinking about assessment at the doctoral level. In: Makin, P.L., Borkowski, N.A. (eds.) The Assessment of Doctoral Education: Emerging Criteria and New Models for Improving Outcomes, pp. 11–52. Stylus Publishing (2006)
7. Belavy, D.L., Owen, P.J., Livingston, P.M.: Do successful PhD outcomes reflect the research environment rather than academic ability? PLoS ONE **15**(8), e0236327 (2020)
8. Walker, S.H., Ouellette, V., Ridde, V.: How can PhD research contribute to the goal health research agenda? Dev. Pract.Pract. **16**(6), 617–622 (2006)
9. Bloch, C., Graversen, E.K., Pedersen, H.S.: Competitive research grants and their impact on career performance. Minerva **52**, 77–96 (2014)
10. Welch, E.W., Feeney, M.K.: Technology in government: how organizational culture mediates information and communication technology outcomes. Gov. Inf. Q. **31**(4), 506–512 (2014)
11. Holbrook, A.: Examining PhD and research masters theses. Assess. Eval. High. Educ.Eval. High. Educ. **38**(4), 407–416 (2013)
12. Mariani, M.M., Machado, I., Magrelli, V., Dwivedi, Y.K.: Artificial intelligence in innovation research: a systematic review, conceptual framework, and future direction. Technovation **122**, 102623 (2023)
13. Bai, X., Li, J.: The best configuration of collaborative knowledge innovation from the perspective of artificial intelligence. Knowl. Manag. Res. Pract., 1–13 (2020)
14. Shen, H., et al.: Information visualisation methods and techniques: state-of-the-art and future direction. J. Ind. Inf. Integr.Integr. **16**, 1–17 (2019)

15. Horttanainen, P., Virrantaus, K.: Uncertainty evaluation of terrain analysis results by simulation and visualisation. In: 12th International Conference on Geoinformatics – Geospatial Information Research; Bridging the Pacific and Atlantic, pp. 473–480 (2004)
16. Hepworth, K.: Big data visualization: promises & pitfalls. Commun. Des. Q. **4**(4), 7–19 (2016)

Information Economy for the Fitness of Intelligent Algorithms

Miguel Ángel Jiménez García[1(✉)] and Richard de Jesús Gil Herrera[2]

[1] Universidad Americana de Europa (UNADE), Cancun, Mexico
miguelbass2009@gmail.com
[2] Universidad Internacional de La Rioja (UNIR), Logroño, Spain
richard.dejesu@unir.net

Abstract. In today's increasingly AI-driven world, there is a growing tendency to rely too heavily on generative AI and technical tools while neglecting the importance of proper data analysis. This scenario often leads to models that are overfitted and redundant processes. Therefore, data science needs to have clear goals and strategies in place. Especially when the low barrier to entry results in fewer feasibility studies, which, unfortunately, often produce poor data results. It is crucial to Regularize AI, assess practicality, and select techniques carefully to execute successful projects. Lack of sufficient data can result in bias and false positives. Thus, each case requires a careful analysis to choose the appropriate methods, algorithms, and frameworks. This paper emphasizes the importance of training intelligent models by exploring the potential of generative AI to enable models to learn problem-solving from existing knowledge.

The text delves into the complexities of global data exchange through theoretical frameworks such as GDPR to guide data privacy and cybersecurity. It suggests solutions like VPNs and the significance of data quality and infrastructure through discussions on data lakes, big data, and data pipelines. There is also an emphasis on various model training techniques and common challenges in neural network training.

The paper also underscores the importance of ongoing analysis and adaptation in a fast-paced world. It explores data management solutions and values user input and feedback. Suggestions include efficient practices and AI recommendations via data economy, standards, and regulations to reduce pre-work and analysis time. Innovations such as IoT, distributed computing, and AI-friendly databases are crucial for model training. It attempts to show how analysis and pre-work lay the foundation for continuous improvement by providing examples of modulated architectures and emphasizing the link between research and product value.

In conclusion, technology advancements and developer contributions have propelled AI progress. This exercise strives to advocate for tailored solutions while anticipating AI's integration with richer user inputs and intelligent analysis in robust systems.

Keywords: Artificial Intelligence (AI) · Model training · AI regularization · Generative AI · Innovation

Á. Rocha et al. (Eds.): WorldCIST 2024, LNNS 987, pp. 389–401, 2024.
https://doi.org/10.1007/978-3-031-60221-4_37

1 Introduction

1.1 Problem Statement

With the recent advancements of AI, especially with generative AI, there is an over-utilization of technical services and tools that leverage its capabilities. As a result, most products that implement intelligent algorithms need to be thoroughly analyzed to determine the state of the data and its subsequent processed outputs. An over-fitting of the models with redundant processes that do not add value to the problem at hand is an increasingly common consequence of the lack of data analysis and refinement.

As with any other discipline, data science must have a clear goal, process, and results. With the lower entry barrier to data analytics and business intelligence, plus the enhancement and further development of Artificial Intelligence Models and tools, several projects are run without feasibility and applicability studies. The results are poor data forced fed into a generic algorithm hoping to understand the problem better through a recommendation of insights that jump to conclusions with the same ease as the project's conception (Geng, et al., 2019).

The regularization of AI, its practical implementation, and the techniques to determine its value must be subject to continuous revision before, during, and after lead generation. Effectively doing so can refine models that fulfill the original questions and problems (Charilaou & Battat, 2022).

The contrary can happen when more information is necessary to leverage AI's true capabilities properly. For example, a small sample that is not statistically significant can yield false positives, bias, or other non-valuable results. The general recommendation is to analyze each case for a particular method, algorithm, or framework that will best suit the goals and objectives of the stakeholders.

1.2 Justification

There are example applications that implement artificial intelligence for the sole reason it is possible. When studying a new topic, such as neural networks or deep learning, it is undoubtedly acceptable to experiment and advance an individual's knowledge (Davenport & Ronanki, 2020). Still, when the project tries to solve a real-world problem relying only on the promise that artificial intelligence will resolve the issue and find optimizations and contextual recommendations, it is recommended to proceed with performing due diligence (Niemi, 2017).

Planning and adjusting often lead to success when performing a project, compared to cases when execution with limited or no pre-work. When applying innovative methods and technologies, there is an inmate risk that the case could behave unexpectedly, as there are few historical instances, or even that the method's immaturity prevents it from being considered trustworthy.

The quality and quantity of the data are another dimension of the project that needs proper care. Especially when the complexity and criticality of its components are uncertain, the generated outputs could derive the task into sub-projects or consolidate its findings as the basis for future developments.

The importance of continuous data validation and experiment verification can lead a team of data scientists to uncover valuable insights that perfect the underlying processes of a system, leading to significant improvements.

1.3 General Objectives

Intelligent models' proper training and fitness are the backbone of the technological advancements that have recently taken the world by storm. Data science does the heavy lifting to clean, arrange, sort, and format the information for consumption by a model. The result is a conveniently formatted output that answers a business concern or user inquiry or addresses a system dependency.

The paper's primary objective is to effectively achieve the expected results that bring the most relevant information to the requesting party.

With the past wave of digitalization, several industries, governments, and the public rely increasingly on technology to perform daily tasks, conduct business, and do other personal paperwork. With the increased usage of digital assets, there is a demand for updated regulations and laws that protect the user's privacy and affirm the validity of the integrity of data sources (Zech, 2016). The usage of these data sources has proven to be of great value for the evolution of various services through the development of APIs (Application Programming Interfaces). Some of these services have become a standard for their processes, such as user registration and authentication through third-party accounts (Grzonkowski, Corcoran, & Coughlin, 2011). These services allow cross-platform connectivity and the automation of more robust services capable of lead generation. Data science and artificial intelligence make their automation in a dynamic environment possible.

Intelligent algorithms provide a wide range of solutions that can address general problems to offer sophisticated answers to complex topics in dynamic environments. The sense that AI would achieve the status of General Artificial Intelligence is quickly becoming less of a theoretical concept but a practical one. Emerging technologies and trends like generative AI, where new information appears from training data, push the boundaries of what algorithmic logic can achieve. Another objective of this study is to determine the feasibility of correcting the problem statement to accommodate the best solution and implementation. In other words, how can AI educate itself on what problems are worth solving based on the available knowledge?

In an evolving, fast-paced world where solutions appear at the same time new problems arise, how can AI models evolve and remain relevant for their original purpose? Continuous improvement via constant refinement and problem analysis is the standard approach for training intelligent models until they reach acceptable performance (Houde, et al., 2020). While this is an algorithmic approach and, thus, can be automated, is it optimal for most models? Reducing their fitness resources and procedures is practical and required as better hardware and state-of-the-art algorithms and models continue their optimization (Dudek-Dyduch, Tadeusiewicz, & Horzyk, 2009). The race for better technology and innovative applications continues with every new company and framework dedicated to its development. What does the future of AI look like?

2 Theoretical Framework

Efforts such as the General Data Protection Regulation (GDPR) aim to protect the privacy and cybersecurity of information technologies consuming and producing data. As the world connects more, country lines and borders dissipate, with data exchange coming in and out of several routes. According to Reinhard Eliger, the most intense transborder data exchange happens in personnel departments, financial companies, marketing agencies, the tourism industry, eCommerce, and the public sector of all the involved entities in the conversation (Schwartz, 1994).

As international bodies and governments redefine and implement new policies, stricter regulations hinder open communication in certain areas. For example, China applies different data via various methods such as Web and DNS server IP addresses, keywords, and DNS redirections (Zittrain & Edelman, 2003). The impact of these measures in the global scenario is hard to determine as for every imposed mechanism, there is a functional and reliable workaround that can circumvent its effectiveness. In the case of VPNs, proxies and specialized software to mask or emulate the origin of a connection are popular tools to access, consume, and generate information in restricted zones.

2.1 Subtopics

Data lakes, big data, fast data and other architectural concepts are solutions to the high demand for reliable and available data sources. Their adaption and standardization are crucial for developing and adopting intelligent products and services (Miloslavskaya & Tolstoy, 2016). Without the proper investment in the analysis of its intake process, the outputs could be different from the best possible result. The data pipeline that feeds an algorithm is as important as the trained model that generates and presents users with valuable results (Wang, Yun, Goldwasser, Vaikuntanathan, & Zaharia, 2017).

Training predictive models based on a regression algorithm is a standard process requiring two datasets to fit a given model. One dataset has rows for teaching the model about the actor's behavior. In contrast, the other has data to validate the model's predictions and effectively score its performance. Once reviews show that data is valid and formatted to the specifications of its intended application, the feeding step can occur until a satisfactory score appears.

Neural Networks present a different challenge for their training and fine-tuning, as a more laborious and experimental set of observations will uncover the necessary discoveries to find the hyper-parameter combination with its highest performance. There are ways to simulate full training, leading to a reduction in time searching for its optimal values. Still, in practice, cross-validation of more configurations is ideal before accepting a model as properly fit.

Data models at an enterprise level can reach numbers that require a more sophisticated infrastructure and platform to sustain its data and model training, deployment, and consumption. Hadoop presented an in-memory-based approach for handling and manipulating data, which later produced the desired analysis outputs. Scaling became an issue as more data entered the system to the point of surpassing the hardware's capabilities (Qiu, Wu, Ding, Xu, & Feng, 2016). Apache Spark leverages distributed computing following Hadoop's MapReduce methods, solving the scaling issue without compromising performance. The application allows larger data sets for AI model training, ranging from exploratory testing to explanatory analytics and tuning machine learning pipelines (Salloum, Dautov, Dautov, Peng, & Huang, 2016).

Data science's evolution about the possibilities that machine learning can now deliver is intrinsically related to the evolving requirements of digital industries that capture users' attention with more innovative features. With every new data visualization, service, or convenient application of an AI-powered model, a more specialized demand will arise from a now-resolved problem to a newly found one. Resolving new issues and questions is the driving force behind data scientists, AI engineers, and their company's stakeholders, each striving to innovate to match the fast pace of AI development (Zheng & Casari, 2018).

2.2 State of the Art

New methodologies and frameworks like data ops are emerging to reduce the end-to-end analysis process by automating data collection, validation, and verification (Munappy, Mattos, Bosch, Olsson, & Dakkak, 2020). As Big Data's 3V characteristics (variety, volume, and velocity) play a more significant role in modern applications' data requirements, the necessity of optimized platforms and infrastructures that can service an ever-increasing demand for inputs has become a primary objective for providers and consumers alike.

Indeed, most successful companies build their business on something other than technology alone but in a sustainable model that delivers goods, services, and overall value to customers. Once a model proves its effectiveness, competition will try to imitate it to benefit from the recently discovered market. The advantage of the incumbents relies not just on its gained market share but on its ability to generate leads based on the gathered information (Casado & Younas, 2015).

While technology is a competitive advantage, it is neither a guarantee nor a one-sized solution that can create, save, and expand a company. Analyzing quantitative and qualitative data and uncovering relationships and behaviors are the angular blocks of effective decision-making based on facts and data.

When it comes to the acquisition and storage of data, performance, and reduced storage space are critical when managing data that grows exponentially. Popular frameworks like Hadoop and Spark solve these issues while providing a good developer experience.

Models can produce results from mathematical combinations that represent an expected reality; in this sense, a set of inputs can educate a model with the necessary information to make judgments and classify, score, group, or relate a question to a known answer. More advanced models can formulate entirely new solutions without a direct query to its training material. Generative AI is the next phase in the evolution of artificial intelligence and its practical implementation. Today, it is possible to answer questions from one format and generate answers in a variety of different formats like text-to-images (DALLE-2 model), text-to-3D images (Dreamfusion model) (Lowe, 1991), images-to-text (Flamingo model) (Alayrac, et al., 2022), text-to-video (Phenaki model) (Villegas, et al., 2022), text to audio (AudioLM model), text to text (ChatGPT) (Gozalo-Brizuela & Garrido-Merchan, 2023), text-to-code (Codex model) (Chen, et al.,

2021), text-to-scientific-text (Galactica model) (Taylor, et al., 2022), text-to-algorithm (AlphaTensor model) (Lin, 2023).

Some model's architecture specializes domains due to their data processing algorithms. For example, Generative Adversarial Networks are highly effective at processing and generating images thanks to their discriminator implementation and noise reduction methods (Wang, Chen, Yang, Bi, & Yu, 2020). A recent application of generative models is for their output to train discriminatory models on world scenarios that diminish bias and allow them to adapt dynamically to new environments or data behaviors (Foster, 2022). Figure 1 shows the models according to these world scenarios.

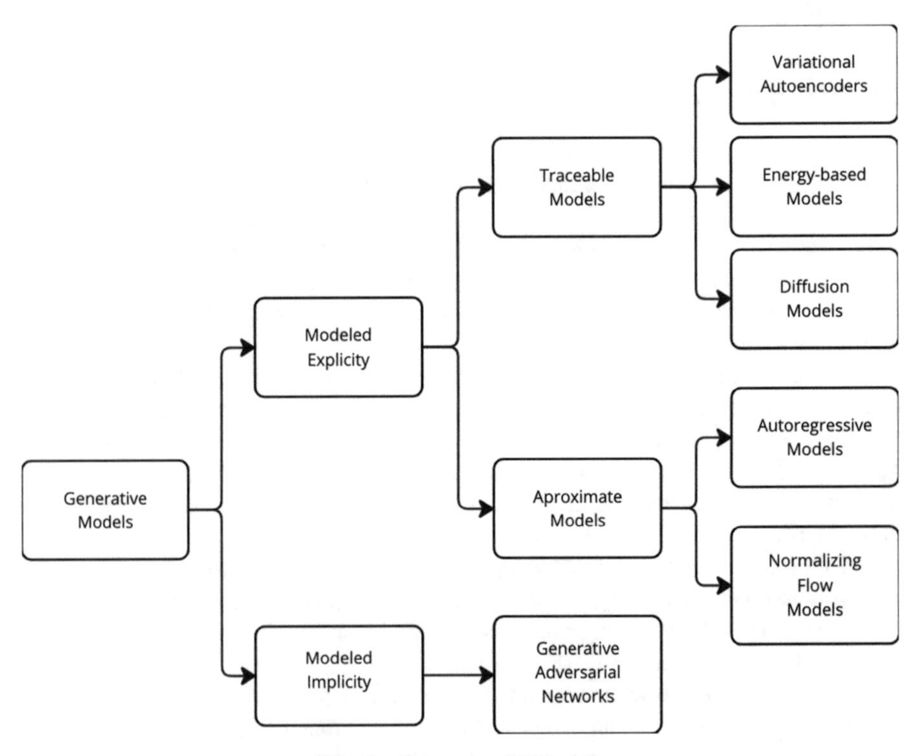

Fig. 1. Generative AI Models

The evolution and development of AI algorithms relate to the problems they are trying to solve. We can only imagine computer science taking off and branching into multiple disciplines with the demand for digital solutions that facilitate, automate, and innovate known problems or traditional processes. Artificial intelligence grew similarly, giving birth to machine learning, neural networks, deep learning, and other specializations. Thanks to the accelerated hardware development with greater processing power and storage mechanisms, the boundaries between cost-effective and technically feasible become less of a concern when executing projects with greater scope. The popularity of distributed architectures based on the cloud makes it accessible for more developers and companies to embark on their AI-based journeys. This results in quick understanding and improvement of AI products and services, including backing academic research that pushes state-of-the-art models and algorithms.

3 Proposed Solution

A data economy where standards and regulations protect their users' and owners' privacy and integrity is the premise behind most governments and technological giants that lead the development of data-based products and services (Regulation, G. D. P., 2018). Standards for personal identification, labeling of physical assets, and managing online activities and transactions are among the top priorities for an effective structure to generate, consume, and govern data worldwide (Schwartz, 1994).

While different strategies show promising results, a general recommendation is still to be made (Bygrave, 2014). Not all use cases are the same, and not all formats or solutions apply to every possible scenario. Most industry practices for data management rely on integrity, integration, and accessibility, but these basic concepts now need to be updated to account for scalability, consistency, and economical processing. Data lakes and other big data architectures provide a practical implementation for today's data inquiries. However, the matter is that deep learning and different intelligent algorithms still require significant pre-work for it to become functional (Fang, 2015).

By looking at the most performant models, a series of practices that effectively diminish the amount of pre-work and analysis time can originate a new recommendation and later formal method for AI data consumption. User data uploaded to social networks, eCommerce sites, financial services, and public services contribute to the wealth explosion of tech giants and many emerging companies. With the virtually unlimited supply of fresh data at their disposal and with rapidly evolving technology, its analysis produces insights and inputs into recommendation systems, business intelligence engines, generative models, and many more innovative data products (Sakr, Liu, Batista, & Alomari, 2011).

The mechanisms by which databases, frames, and series go through the data ingestion pipelines are usually highly refined and adapted to fit a use case. Some cases lead to new paradigms, such as graph databases that accommodate connected data models, on which the system naturally adds and queries data (Robinson, Webber, & Eifrem, 2015). In the same sense, data analytics via the continuous optimization of the models can serve as the basis for new paradigms that reduce the amount of user effort in designing and managing data assets.

IoT, distributed computing in the cloud, and the emergence of databases designed for AI consumption (DB4AI) are among the most widespread technology innovations to support and improve the state-of-the-art of Artificial Intelligence products (Li, Zhou, & Cao, 2021).

4 Methods and Results

The analysis, studies, and pre-work needed for planning and implementing a project do not guarantee a flawless design and implementation; on the contrary, they prepare the groundwork for the ongoing continuous improvement and optimization of said project.

The change in mindset from a finite project to a living product that requires ongoing support can help a team succeed in its selected problem space.

With a clear definition of a product-led development process, the question becomes: What is the goal for the current development iteration? More analysis will be required to inspect and adapt the results, their state, and the encompassing product health; the sense of perfection changes its purpose to the product's usefulness and the manageability of its resources.

For example, a web product that follows a data-first approach with high user interaction that aims to solve the relationships of user purchases with their environment, including intelligent profile building and an unbiased recommendation system, can be supported with a modulated architecture that hosts semi-independent databases with different paradigms to facilitate the process and transport of data inputs to other parts of the system.

Another example is a conservative design for a library system that intends to serve a medium-sized private school library. The system requirements and specifications can rely on a relational database with a backend engine that includes APIs and simple platform services.

The results from small, medium, and enterprise projects in the biotech industry show that despite the inclusion or exclusion of artificial intelligence in their product's design, the value and appreciation of its outputs correlate directly with the amount of analysis and adaptation to meet the problem's objective resolution.

The small project aims to determine the most profitable way to segment customer markets by their purchase history. The staff includes a project lead and a cross-functional team of five developers with data analysis and software engineering experience. The initial investigation shows that a simple graph displaying customer purchase behaviors grouped by available properties can help generate hypotheses with basic arrangements. The results assessment shows which groups share the most sales related to properties, such as age, gender, income, and preferences.

As a stretch goal, the team can automate its research by standardizing a data format and ingestion mechanism. The process can include machine learning algorithms that dynamically output segmentation and business strategy leads. The first tasks were simple enough for simple digital tools, with an almost complete disregard for intelligent algorithms.

The medium project supposes product development opportunities within a product catalog where customers provide feedback. The objective is to classify which ideas have the most potential concerning return on investment. This time, the team comprises one project lead, eleven data scientists, software engineers, and business analysts.

As with the previous exercise, the recurrent theme among data-led activities is the pre-work required in the collection, cleaning, analysis, and interpretation phases. Once the first results and their applicable implementations become available, machine learning is more likely to be helpful in driving outcomes. While AI often solves more complex problems that rely on a known area space, it is still a valuable tool that needs prior assessment before deciding on its inclusion and development, especially in small to medium projects.

The enterprise project consists of wrapping the previous results into a system capable of analyzing and generating market insights that drive customer satisfaction and business prosperity. With a higher scope and a more significant business impact, the team configuration is now: a project manager, product manager, two cross-functional teams, and one enterprise architect.

The approach is now to deliver further automized insights incrementally to become a platform-level service that multiple internal and external actors consume. The demand for AI is now a requirement to meet the dynamic nature of user data across different products, marketing campaigns, and organizations.

In any case, while products and services are in the research phase, User input and feedback are not just welcome but become mandatory to successfully deliver a meaningful product that serves its user's needs rather than just their wishes and demands. In the case of intelligent algorithms and data ingestion pipelines, the most elegant designs share a clear problem statement and an exhaustive analysis process that includes multiple cycles of research and development as well as open communication channels with their user base and subject matter experts.

The project contains a GitHub repository (CSV_Dataset_Generator) with a dataset generator and a Jupyter Notebook hosted in Kaggle (New Marketing Campaign) with the unsupervised classification model powered by AI recommendations, including market analysis, data enrichment, and campaign recommendations.

The first step is to generate more records based on a sample. The process involves analyzing the data's structure and inherent behavior regarding its fields and relationships. Once the operation finishes, it's configured to take in a row and generate several similar records based on the user's input. When the dataset is complete, the output is a comma-separated value file that the AI model can consume.

When the initial analysis begins, results are only available for the ones the team decides to pursue or the ones stakeholders suggest as potential success routes. By running a generative AI prompt with the proper questions, new data will show a series of viable options for the team and business sponsors to explore and decide on actions. In the exercise, four customer clusters have their respective analysis details and subsequent insights. These serve as guidelines to continue elaborating the marketing strategy.

The team must employ data analysis tools, frameworks, and services to explore such options. Kaggle hosts a Machine-learning environment in the cloud with the option to publish a Jupyter notebook containing the project's dataset and executable code (Kaggle, 2023).

The data relates to a marketing campaign where customer data shows their purchase activity and how they relate to previous marketing programs. The problem statement is to find hidden insights that can help craft a strategy for classifying customers' behavior towards their preferred products to deliver semi-personalized advertisements with a higher conversion rate to finalize a sale.

The initial step is to review, clean, and prepare the data for consumption by a given model. The data is clean with a high level of pre-work, so only one row was unsuitable for this exercise.

Generative AI analysis, such as ChatGPT, was part of the process for supporting the analysis, coding, and subsequent post-process of the original outputs.

As a first step, it's essential to determine a set of questions and hypotheses that focus the team on achieving an expected outcome and drafting a research strategy. The marketing team believes the data is sufficient to categorize customers with similar traits, such as spending habits, favorite products, and socio-economic situations.

Leveraging the KMeans algorithm helps the team to quickly and effectively cluster customers based on their features. The first question's answer is the output of an unsupervised machine-learning technique with a visual representation in a scatter plot.

With customer groupings ready, the data science team can determine which customer segment is the most profitable for the company, along with a breakdown of the purchases by product category. The image below provides an easy understanding of spending activity across the company's product families per group.

The following business question is, which channels do customers prefer for making purchases? And its follow-up question, which marketing campaign was most successful per group?

When drafting a new marketing strategy, the data science team can provide insights into the relationships of certain product sales per customer group. The image below shows a bar graph showing the average wine sales and the marketing campaign's effectiveness. While this is a good starting point for most data science and machine learning projects, more opportunities are feasible to achieve from a technical perspective. While the exercise proves that Artificial intelligence may not be necessary for every case, it is undoubtedly possible to assess the need to expand the scope and capabilities of the product or service with leads automatically available via AI-backed models.

The project's business sponsors can emit an educated decision leaning on the proof of concept's results to determine if users' expectations are reasonable and the potential business opportunities for employing intelligent models for drafting and expanding marketing campaigns.

The following approaches are suggestions for the use case in question:

1. KMeans: This algorithm partitions data into 'K' clusters, which in our case was 5. The process minimizes the sum of squared distances between each data point and its cluster centroid.
2. MPLClassifier: The Multi-Layer Perceptron Classifier employs multiple layers to classify data. Each one consists of neurons that compute inputs to generate expected outcomes.
3. GMM: The Gaussian Mixture Model assumes that the data originates from a mixture of unknown parameters. The process is a series of mean and covariance estimations to match each component to an overall distribution.

Once each model generates an output, it can then pass into a review, comparison, and experimentation phase to answer more detailed business questions. For example, what is each group's average spending per product category? How many purchases did each marketing campaign produce? What are the spending habits per group? What are their favorite channels? Are there any correlations with group profile data? What other statistics can factor in a targeted potential marketing campaign? All of these questions can have a scoring assignment with simulated data, which later can be run using real-time marketing metrics. The strategic advantage of having a live or near-live analysis of emerging situations is a competitive ace for most of today's tech giants.

For real-time analytics and business agility to be operation-ready, there must be an infrastructure and data-ingestion pipeline in place that supports and allows for quick and effective decision-making with the lowest risk possible. An information economy with solid foundations in business agility, accuracy, and data integrity becomes more justifiable when investments multiply their returns in the short term. The next step is to leverage experiments and prototype data to advocate for a data platform that serves the organization's vision of becoming a data-first, intelligent, and agile company.

A team of analysts can then continue elaborating small, targeted ads that validate or reject several marketing hypotheses. From a business perspective, more data-based decisions with quality results that allow for dynamic content employment is a safer route to support and expand potential opportunities. For the exercise in question, it is of higher value to iterate with small project cases and assess the situation before committing to the next course of action.

The technology and person-hours for continually generating and analyzing data are less expensive than traditional long-term plans or cutting-edge tools and services that have yet to prove their effectiveness in real-world scenarios.

5 Conclusions

The accelerated pace at which technology advances and the subsequent democratization of information has led to gigantic breakthroughs in designing and implementing AI-based algorithms. New products and services that leverage Artificial Intelligence appear at the same rate as new bleeding-edge developments materialize and are publicly available.

Historically speaking, this is an unprecedented time when more developers contribute to the progress of the AI landscape.

While different AI products and services become readily adaptable to a wide range of problems and ideas, it is not always the best option to implement a sophisticated solution to topics of simple nature. The general rule is to go with an elegant solution that solves one problem and solves it in an optimized way, generally in its most basic and straightforward manner.

More applications can appear using said technology as resources become cheaper and easier to obtain. As complex problems become more accessible, new challenges might appear as the next wave of issues to be addressed by innovative solutions.

The future of AI can interrelate with modern user interfaces where data is submitted more naturally. In contrast, users provide richer inputs that contain not just concrete questions but emotions and other meta-data.

The future of Artificial intelligence can depart in many sub-areas but converge back at the point of intelligent analysis and optimized integration into a more robust system.

References

Alayrac, J., et al.: Flamingo: a visual language model for few-shot learning. Adv. Neural Inf. Process. Syst. 23716–23736 (2022)

Bygrave, L.: Data privacy law: an international perspective (2014)

Casado, R., Younas, M.: Emerging trends and technologies in big data processing. Concurr. Comput. Pract. Exp. 2078–2091 (2015)
Charilaou, P., Battat, R.: Machine learning models and over-fitting considerations. World J. Gastroenterol. **28**(5), 605 (2022)
Chen, M., et al.: Evaluating large language models trained on code. arXiv preprint (2021)
Davenport, T., Ronanki, R.: An empirical overview of nonlinearity and overfitting in machine learning using COVID-19 data. Chaos, Solitons Fractals 139 (2020)
Dudek-Dyduch, E., Tadeusiewicz, R., Horzyk, A.: Neural network adaptation process effectiveness dependent of constant training data availability. Neurocomputing **72**(13–15), 3138–3149 (2009)
Fang, H.: Managing data lakes in big data era: What's a data lake and why has it became popular in data management ecosystem. In: IEEE International Conference on Cyber Technology in Automation, Control, and Intelligent Systems (CYBER), pp. 820–824 (2015)
Foster, D.: Generative Deep Learning. O'Reilly Media, Inc. (2022)
Geng, D., Zhang, C., Xia, C., Xia, X., Liu, Q., Fu, X.: Big data-based improved data acquisition and storage system for designing industrial data platform. IEEE Access **7**, 44574–44582 (2019)
Gozalo-Brizuela, R., Garrido-Merchan, E.: ChatGPT is not all you need. A state of the art review of large generative AI models. arXiv preprint (2023)
Grzonkowski, S., Corcoran, P., Coughlin, T.: Security analysis of authentication protocols for next-generation mobile and CE cloud services. In: 2011 IEEE International Conference on Consumer Electronics-Berlin (ICCE-Berlin), pp. 83–87 (2011)
Houde, S., et al.: Business (mis) use cases of generative ai. arXiv preprint arXiv:2003.07679 (2020)
Kaggle. Contact. Kaggle (2023, September 22). https://www.kaggle.com/contact#/datasets/license
Li, G., Zhou, X., Cao, L.: AI meets database: AI4DB and DB4AI. In: Proceedings of the 2021 International Conference on Management of Data, pp. 2859–2866 (2021)
Lin, H.: Standing on the shoulders of AI giants. Computer **56**(01), 97–101 (2023)
Liu, N., Li, S., Du, Y., Torralba, A., Tenenbaum, J.: Compositional visual generation with composable diffusion models. In: European Conference on Computer Vision, pp. 423–439. Springer, Cham (2022). https://doi.org/10.1007/978-3-031-19790-1_26
Lowe, D.: Fitting parameterized three-dimensional models to images. In: IEEE Transactions on Pattern Analysis and Machine Intelligence, pp. 441–450 (1991)
Miloslavskaya, N., Tolstoy, A.: Big data, fast data and data lake concepts. Procedia Comput. Sci. **88**, 300–305 (2016)
Munappy, A., Mattos, D., Bosch, J., Olsson, H., Dakkak, A.: ad-hoc data analytics to dataops. In: Proceedings of the International Conference on Software and System Processes, pp. 165–174 (2020)
Niemi, A.: Digital lead generation and nurturing: a holistic approach (2017)
Peng, Y., Nagata, H.: An empirical overview of nonlinearity and overfitting in machine learning using COVID-19 data. Chaos Solitons Fractals 139 (2020)
Perreault, Jr., W., Leigh, L.: Reliability of nominal data based on qualitative judgments. J. Market. Res. 135–148 (1989)
Qiu, J., Wu, Q., Ding, G., Xu, Y., Feng, S.: A survey of machine learning for big data processing. EURASIP J. Adv. Signal Process. 1–16 (2016)
RegulationGeneral Data Protection Regulation. General Data Protection Regulation (GDPR). Intersoft Consulting (2018)
Remane, G., Hanelt, A., Tesch, J., Kolbe, L.: The business model pattern database—a tool for systematic business model innovation. Int. J. Innov. Manag. (2017)
Robinson, I., Webber, J., Eifrem, E.: Graph Databases: New Opportunities for Connected Data. O'Reilly Media, Inc. (2015)

Sakr, S., Liu, A., Batista, D., Alomari, M.: A survey of large-scale data management approaches in cloud environments. IEEE Commun. Surv. Tutor. **13**, 311–336 (2011)

Salloum, S., Dautov, R., Dautov, X., Peng, P., Huang, J.: Big data analytics on Apache Spark. Int. J. Data Sci. Analyt. **1**(3–4), 145–164 (2016)

Schwartz, P.: European data protection law and restrictions on international data flows. Iowa L. Rev. **80**, 471 (1994)

Shen, Y., et al.: ChatGPT and other large language models are double-edged swords. Radiology **307**(2) (2023)

Taylor, R., et al.: Galactica: a large language model for science. arXiv preprint (2022)

Villegas, R., et al.: Phenaki: variable length video generation from open domain textual description. arXiv preprint (2022)

Wang, F., Yun, C., Goldwasser, S., Vaikuntanathan, V., Zaharia, M.: Splinter: practical private queries on public data. In: 14th USENIX Symposium on Networked Systems Design and Implementation, pp. 299–313 (2017)

Wang, L., Chen, W., Yang, W., Bi, F., Yu, F.: State-of-the-art review on image synthesis with generative adversarial networks. IEEE Access **8**, 63514–63537 (2020)

Zech, H.: A legal framework for a data economy in the European Digital Single Market: rights to use data. J. Intellect. Prop. Law Pract. **11**(6), 460–470 (2016)

Zheng, A., Casari, A.: Feature Engineering for Machine Learning: Principles and Techniques for Data Scientists. O'Reilly Media, Inc. (2018)

Zittrain, J., Edelman, B. (2003). Internet filtering in China. IEEE Internet Comput. **7**(2), 70–77 (2003)

Towards Transparent Governance: Unifying City Councils Decision-Making Data Processing and Visualization

Kristýna Zaklová(✉), Jiří Hynek, and Tomáš Hruška

Faculty of Information Technology, Brno University of Technology, Božetěchova 1/2, 612 66 Brno, Czech Republic
{zaklova,hynek,hruska}@fit.vut.cz

Abstract. The local open government data holds undiscovered potential for wider application. Municipal policies and representatives' decisions have a direct impact on citizens' daily lives and the public should be aware of them. Nevertheless, the data related to city council activities may not always be easily searchable, and if available, the varying formats complicate further extraction and processing. This paper contributes to the academic discourse by proposing a unified approach to processing and storing the data from the council's decision-making. The paper outlines a generic data model designed for broad application across municipalities. A data set within this model can be automatically processed and stored in the database. The solution presents an information system with an user-friendly interface that visualize the data in the form of dashboards, enhancing accessibility for the general public. This paper contains practical example of how the data model can be applied to real votes of the council in the City of Brno, which is the second largest city in the Czech Republic.

Keywords: City Council · Decision-making · Local Politics · Open Government Data · Data Visualization

1 Introduction

Interest in open government data (OGD) has been growing significantly in recent years [4,20,22]. OGD originated as a combination of open data and government data. It has to meet the requirements of free access, use, modification and further sharing [18]; and it must be produced or commissioned from the public bodies. Enacting OGD can be undertaken through various approaches, such as utilizing the five institutional dimensions [21] or employing alternative methodologies [11].

It should also be considered that publishing OGD can bring both benefits and risks [10]. It appears that the adoption of OGD has a positive impact on transparency, economy, and the development of innovation [21]. Furthermore, governments need to ensure that OGD is published in a way that facilitates the creation of both free and commercial products [23]. Examples of civic engagement

Á. Rocha et al. (Eds.): WorldCIST 2024, LNNS 987, pp. 402–411, 2024.
https://doi.org/10.1007/978-3-031-60221-4_38

projects are given in the study [8]. Other factors in OGD involve the assessment of data quality [9] and the level of openness, typically evaluated through the 5 Star Open Data[1] scale.

There is considerably less information on OGD initiatives at the local level (e.g. the US city of Chicago [8] or two Swedish municipalities [12]). Studies on municipalities show that the pace of data opening is lower than for larger initiatives such as national-level platforms. This paper focuses on city councils, whose authority varies based on country-specific legislation. Usually, a city council consists of representatives periodically elected by the citizens, significantly impacting daily lives of the residents through their decision-making. Municipalities have an interest in transparency and providing public access to their data, even though it is not legally mandated for all of them.

The technical challenges faced by OGD at local and higher levels involve custom datasets, formats and standards [22]. Regarding formats for policy-related data, the Open Knowledge Foundation has launched its initiative to create an infrastructure for the electoral process [17]. However, the data from representative decision-making still needs to be taken into consideration at national and transnational levels.

In the Czech Republic, there are 14 regions and over 6,200 municipalities [5]. According to acts No. 128/2000 and 129/2000 of the Czech Collection of Laws, each must have its own municipal council, but there is not yet a legal obligation to publish meeting minutes or voting records. The Czech OGD standards are available on the Portal for Providers[2]. In the Czech Republic, open formal standards (originally *otevřené formální normy*)[3] are used to unify formats and support data interoperability. So far they only cover some of the datasets.

We aim to design a versatile data model applicable to any municipality. Considering the number of existing municipalities, our initial phase will produce results applicable to the Czech Republic. The expected outputs are the unified model for the city council decision-making data and the information system for the presentation of this dataset with emphasis on the semantics of the data. The results will be practically tested on the City of Brno.

2 Current Form of Available Data and Its Visualizations

Data availability varies significantly based on local legislation and the technical sophistication of the country or region. The authors [3] dealt with OGD at the local level and confirmed the prevalence of semi-structured or non-structured formats (mainly HTML and PDF), suggesting that municipalities perceive this form of publication as open, even though it contradicts OGD principles. Individual municipalities typically have data about their council presented on their official websites. The relevant subsection of the website usually contains a list of council members, meeting minutes and voting records (sometimes directly

[1] https://5stardata.info/en/.
[2] https://data.gov.cz/english/.
[3] https://data.gov.cz/ofn/.

in the minutes). No summary statistics are available, items cannot be filtered by advanced criteria, voting records cannot be sorted by category, etc. Only a minority of municipalities have own dashboards (e.g. Victoria City Council[4]).

The voting records, which are included in a dataset, are commonly located in municipal data portals or national open data catalogs. This pattern is also noticeable in the Czech Republic, where a few municipalities publish their datasets related to council decision-making on the National Open Data Catalog[5]. The dominant formats are JSON, CSV or XML, with scheme and attributes varying by author definition. A single record in the dataset usually represents either a single voting topic (also referred to as a motion or issue in the terminology; e.g. dataset from Cambridge City Council[6]) or a single roll call by a councilor referring to the voting topic (e.g. dataset from Dallas City Council[7]). In the second case, redundant information can be noticed.

Previous research in the field of visualizing voting records usually centers around the data sourced from high-level institutions governing the entire country, as these institutions commonly provide available data. The subsequent projects were chosen due to their close thematic alignment:

- *CivisAnalysis* [2] uses open data from the Brazilian Chamber of Deputies and in the form of the web-based system provides e.g. a visualization of roll call results as n-dimensional space.
- Open-source project *LegisLatio* tool [16] was created with the objectives of visualization of legislative roll call vote data from Ecuadorian parliament and offering better understanding of the voting of multi-party legislative bodies.
- As the new *Voteview.com* [1] was created a new online open-source platform that brings together roll call voting in the United States Congress. Project founder K. Poole also proposed a simple geometric model of voting in which points represent individual legislators and pairs of points represent roll calls [19].
- As a part of the project *Paralemeter.org*, an analysis of the linguistic production paired with the metadata from the perspective of communication and political studies was carried out [13]. The initiative focuses on the Slovene, Croatian and Bosnian parliaments. However, the project currently includes visualization of several municipalities (e.g., the capital city Ljubljana). The general concept of the project is the closest to this work.

Another possible use of voting data is demonstrated by a study [6] that linked information from voting advice applications to votes in parliament. The expected voting results were compared with the actual results of roll calls. With the proposed technique, it is possible to detect legislators who vote differently

[4] https://opendata.victoria.ca/pages/mayor-and-council.

[5] https://data.gov.cz/datasets.

[6] https://data.cambridgema.gov/General-Government/City-Council-Roll-Call-Votes-2018-2019/3g59-fvq4.

[7] https://www.dallasopendata.com/Services/Dallas-City-Council-Voting-Record/ts5d-gdq6.

than they promised before the election. Such visualizations have great potential as it provides a tool for overseeing the institutions, increasing their transparency and bringing information closer to the general public.

Shortcomings in these projects often stem from inadequate targeting of users, with a tendency to favor experts over the general public. The use of OGD primarily for academic purposes is confirmed by the authors [14], who also identified a lack of educational events and activities to promote interest in OGD among young people. Furthermore, the practical deployment of these projects is frequently hindered by challenges such as a lack of data availability over an extended period and the absence of mechanisms for adding new records.

3 Solution Requirements

An analysis was conducted, including own research into the issue, an evaluation of the current situation, and the examination of data provided by selected municipalities. Additional information on the practical functioning of the municipality was provided by the staff of the Brno City Municipality[8]. Based on the analysis, the requirements for public deployment of the developed solution and the future operation of the application were gathered. It includes the addition of new records and a longer-term vision of the project. The tool also aims to educate the general public and increase their interest in local politics. To achieve this, three main objectives have been identified.

The first objective is the unification of the data model. Standardization of the same or similar datasets across public sector entities is one of the OGD challenges, according to the source [4]. The model should include as little redundancy as possible and contain mandatory basic and optional additional attributes. Further, it should be generic enough to be applicable to any municipality. Thus, iterative adjustments can be expected. All processed datasets will be published as OGD under the model with the possibility of their further use.

The second objective focuses on the cooperation with specific municipalities. In order to understand political engagement of citizens at the local policy level, it is necessary to understand their motivations. The authors of [15] conclude that institutional rules and levels of citizen engagement are related, with greater participation when citizen engagement is supported. If a municipality makes its data available, it is expected to do so as their own decision, and they are willing to support further processing and an interest in increasing citizen engagement.

The third objective emphasizes the cooperation with the general public. As described by [7], it is important to discuss the challenges of citizen participation and whether investment in the open government movement will increase citizen participation. The emphasis needs to shift to supporting public reflection on pressing issues. It is also necessary to interact with the target users and to regularly evaluate how they perceive the tool, its user-friendliness, clarity of the visualizations and the added value.

[8] Its Data department, respectively Data.Brno team, which publishes the city's data on the municipal data portal—https://datahub.brno.cz/.

4 Data Processing and Proposed Data Model

From the design perspective, it was important to understand how the process of collecting and processing data from the decision-making of the representatives is carried out. According to the diagram (Fig. 1), the life cycle of data processing has four phases that correspond to the Ubaldi's OGD value chain [22]:

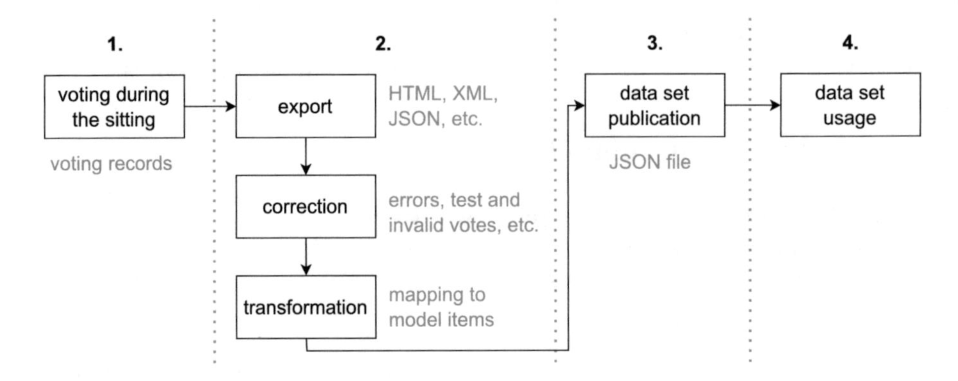

Fig. 1. Data processing flowchart. The individual phases correspond to: data generation (1.); data collection, aggregation and processing (2.); data distribution and delivery (3.); and final data use (4.).

1. **Data Generation**: the council members meet regularly, discussing voting topics according to a previously known agenda. Voting takes place on individual issues, and a voting record is taken from each vote. Larger cities usually have a voting system and voting machines. The systems can have various parameters to a certain extent, so their data outputs typically do not have the same format.
2. **Data Collection, Aggregation and Processing**: the first two phases are provided by the voting system in larger municipalities. Then, the export of the voting records is carried out, usually as a set of records (e.g. HTML files) or as a single dataset (e.g. XML file). This is followed by data processing, which ideally involves validation of the records (removal of errors, test and invalid votes, etc.), and transformation of the data into a generic model, which will be described below.
3. **Data Distribution and Delivery**: the dataset is published within the application itself to transparently provide a source of visualizations and enable further use. It can also be published in a data portal at the municipal or state level.
4. **Final Data Use**: the data can be used in academic, government or commercial sphere. One possible further use of the dataset will be demonstrated in Sect. 5, which introduces the developed information system.

The generic data model was designed as a JSON schema. It is available under CC BY 4.0 DEED license[9]. It has four separate sub-parts: basic data on municipal self-government; political entities (parties or their coalitions); representatives (members of the council); and councils, i.e. information about the councils for terms, including individual meetings and voting topic with votes.

The individual voting records contain either the summary numbers of the individual voting options (thus the votes of the representatives are anonymous), or the roll call votes of all representatives. Sometimes both options can occur within the same council—in a "secret vote" the votes of councilors are anonymous. The proposed data model takes into account all these variants.

5 Results

The designed system has a client-server architecture which is shown in more detail in Fig. 2. The data model described in Sect. 4 defines the structure of the JSON file that serves as input data for the system. The data is processed automatically and stored in a MySQL relational database. During the dataset transformation, new information is derived (e.g. voting statistics stored for better system performance).

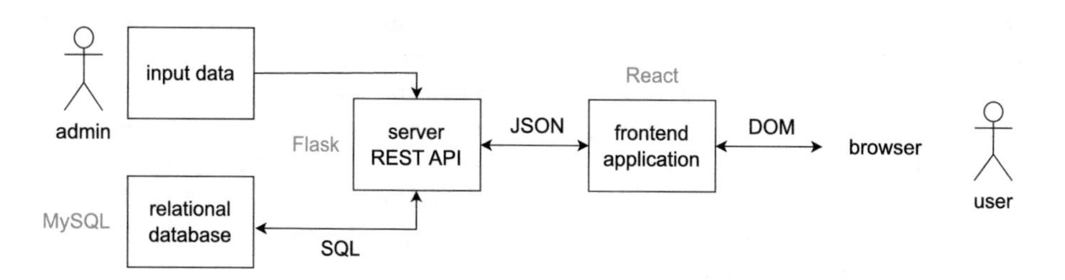

Fig. 2. A diagram of the system architecture including the technologies used and the target users.

The front-end is divided into several views (Fig. 3). It allows displaying basic information about the elected term, political entities, representatives, meetings and votes. For all of the above, it is always possible to display a detail with one specific item. The UI elements vary depending on the type of item selected, but they all have interactivity in common—either allowing to view more detailed information or quickly navigate between views. The dominant chart type is the horizontal stacked bar chart, because of its clarity and size.

More detailed analysis is also available. Councilor's attendance is monitored and it is measured as the ratio of valid votes in which the councilor participated to all valid votes, considering the duration of their mandate. Comparison of two councilors allows to determine the percentage of agreement between them.

[9] https://github.com/zastupko/data-model.

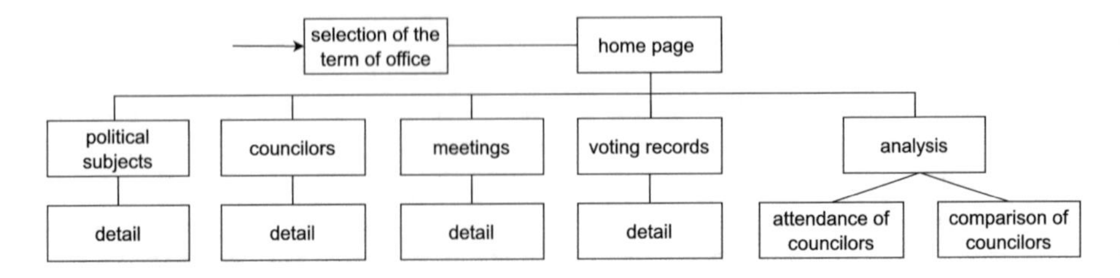

Fig. 3. Hierarchy between views in the client part of the application. For clarity some connections between sub-pages have been omitted in the diagram.

The view enables to browse individual meetings and displays matching and non-matching voting records. When comparing members of one party, it is possible to see if any of them deviate from the behavior of the others; when comparing members of different parties, it is possible to see how they agree or not.

6 Application of Results: The City of Brno

The second largest city in the Czech Republic, with a population of less than half a million, the City of Brno, was the first partner of the project. It provides data from its City Council (originally *Zastupitelstvo města Brna*) with 55 councilors as a further step in its effort to introduce more transparency and share the data from its activities. The created dataset includes the last two terms of the office (2018–2022 and 2022–2026), currently 51 meetings and 4,962 voting topics. The correctness of the created dataset was verified against the meeting minutes.

The created system was deployed within the data portal of the City of Brno[10]. Its first version was released before the municipal elections in autumn 2022 and brought the citizens an overview of the activities of their 8th City Council (Fig. 4). The developed solution has demonstrated its usability in more than a year of operation. Now, it includes a new 9th City Council and new records are added periodically.

Cooperation with Brno helped to treat edge cases, which included, for example, replacements of councilors during the term of office, changing the names of councilors, withdrawing from a political subject or introducing secret votes. To evaluate the satisfaction of Brno citizens, a questionnaire survey was used in which 43 men and 45 women participated. More than half of the respondents rated the tool positively—83 % said they understood what the tool does; 76 % rated the orientation on the website as easy; and 80 % rated the website as clear. In addition, suggestions were provided for further enhancements and improvements that we plan to address. Through this collaboration, the citizens of Brno have gained the tool for overseeing their representatives. They have embraced this positively and expressed interest in its continued utilization.

[10] https://www.brno.cz/zastupitelstvo-analyza.

Fig. 4. Statistics of the mayoress of the City of Brno in the previous term of office. The view shows the length of the mandate, the percentage of voting participation, her political affiliation, summary of voting and attendance. Interactive graphs allow more detailed information to be displayed.

7 Discussion

The main advantage of this work is its uniqueness—an effort to increase transparency of city councils and centralize its data in one place, offer open datasets from different municipalities in the same format, and provide unified visualizations for the general public. The usage scenario has only been conducted on one municipality, but the created results should be generic enough to be applicable to different municipalities, which is currently tested with cooperation to other Czech partner cities. It is necessary to verify the suitability of the proposed data processing scheme and the data model for these municipalities. The aim is to establish a standard that would be applicable in the Czech Republic and could be gradually extended in single municipalities. It would also be advantageous to include foreign city councils.

We are aware that the expansion of the project depends on municipalities' willingness to cooperate. But we believe in an upward trend, as can be observed in OGD in general, and we are ready for negotiations to establish new partnerships. Transforming data into the created data model remains a challenge, as pre-processing must accommodate the diverse data formats from each municipality. Nevertheless, there is potential to develop a universal tool capable at least partially automate this part of the process. The use of AI is suggested, namely for automatic vote records classification, knowledge mining, pattern searching, etc. This would contribute not only to a greater understanding of the source data, but also to deeper insights for finding appropriate approaches and solutions.

8 Conclusion

This paper described the problem of availability of OGD from city councils and the research works focused on visualization of such data. In the effort for enhanced accessibility of such information, this work introduces two key concepts: the generic model for city council decision-making data and the information system providing clear visualizations for the general public. The model is included in the process of collecting and processing data from the decision-making of the representatives, which was also presented. The application of results was conducted in the city of Brno, and the insights and experiences gathered thus far have proven advantageous for collaboration with other Czech cities. The attained results showcase the feasibility of implementing this concept, with upcoming steps encompassing its expansion throughout the Czech Republic.

Acknowledgments. This work was supported by project Smart information technology for a resilient society, FIT-S-23-8209, funded by Brno University of Technology.

References

1. Boche, A., Lewis, J.B., Rudkin, A., Sonnet, L.: The new Voteview.com: preserving and continuing Keith Poole's infrastructure for scholars, students and observers of congress. Publ. Choice **176**(1), 17–32 (2018). https://doi.org/10.1007/s11127-018-0546-0
2. Borja, F.G., Freitas, C.M.D.S.: CivisAnalysis: interactive visualization for exploring roll call data and representatives' voting behaviour. In: 28th SIBGRAPI Conference on Graphics, Patterns and Images, pp. 257–264. IEEE, Brazil (2015). https://doi.org/10.1109/SIBGRAPI.2015.34
3. Corrêa, A.S., Paula, E.C., Corrêa, P.L.P., Silva, F.S.C.: Transparency and open government data: a wide national assessment of data openness in Brazilian local governments. In: Transforming Government: People, Process and Policy, vol. 11, pp. 58–78. Emerald (2017). https://doi.org/10.1108/TG-12-2015-0052
4. Chlapek, D., Kučera, J., Nečaský, M., Kubáň, M.: Open data and PSI in the Czech Republic. In: ePSIplatform Topic Report No. 2014/03 (2014)
5. Czech Statistical Office—Small Lexicon of Municipalities of the Czech Republic—2021. https://www.czso.cz/csu/czso/maly-lexikon-obci-ceske-republiky-2021. Accessed 25 Oct 2023
6. Etter, V., Herzen, J., Grossglauser, M., Thiran, P.: Mining democracy. In: Proceedings of the Second ACM Conference on Online Social Networks, pp. 1–12. ACM, New York (2014). https://doi.org/10.1145/2660460.2660476
7. Evans, A.M., Campos, A.: Open government initiatives: challenges of citizen participation. J. Policy Anal. Manage. **32**(1), 172–185 (2013)
8. Kassen, M.: A promising phenomenon of open data: a case study of the Chicago open data project. In: Government Information Quarterly, vol. 30, pp. 508–513. Elsevier BV (2013). https://doi.org/10.1016/j.giq.2013.05.012
9. Kučera, J., Chlapek, D., Nečaský, M.: Open government data catalogs: current approaches and quality perspective. In: Kő, A., Leitner, C., Leitold, H., Prosser, A. (eds.) EGOVIS/EDEM 2013. LNCS, vol. 8061, pp. 152–166. Springer, Heidelberg (2013). https://doi.org/10.1007/978-3-642-40160-2_13

10. Kučera, J., Chlapek, D.: Benefits and risks of open government data. J. Syst. Int. **5**, 30–41 (2014). https://doi.org/10.20470/jsi.v5i1.185
11. Kučera, J., Chlapek, D., Klímek, J., Nečaský, M.: Methodologies and best practices for open data publication. In: DATESO, pp. 52–64. CEUR workshop proceedings (2015)
12. Lassinantti, J., Bergvall-Kareborn, B., Stahlbrost, A.: Shaping local open data initiatives: politics and implications. J. Theor. Appl. Electron. Commer. Res. **9**, 17–33 (2014). https://doi.org/10.4067/S0718-18762014000200003
13. Ljubešić, N., Fišer, D., Erjavec, T., Dobranić, F.: The parlameter corpus of contemporary slovene parliamentary proceedings. In: Language Technologies & Digital Humanities, pp. 162–167. Ljubljana (2018)
14. Lněnička, M., Nikiforova, A., Saxena, S., Singh, P.: Investigation into the adoption of open government data among students: the behavioural intention-based comparative analysis of three countries. Aslib J. Inf. Manag. **74**, 549–567 (2022). https://doi.org/10.1108/ajim-08-2021-0249
15. Lowndes, V., Pratchett, L., Stoker, G.: Local political participation: the impact of rules-in-use. Public Adm. **84**, 539–561 (2006). https://doi.org/10.1111/j.1467-9299.2006.00601.x
16. Méndez, G.G., Moreno, O., Mendoza, P.: LegisLatio: a visualization tool for legislative roll-call vote data. In: Proceedings of the 15th International Symposium on Visual Information Communication and Interaction, pp. 1–8. ACM, New York (2022). https://doi.org/10.1145/3554944.3554957
17. Open Knowledge Foundation—Digital public infrastructure for electoral processes. https://okfn.org/en/projects/digital-public-infrastructure-for-electoral-processes/. Accessed 8 Nov 2023
18. Open Knowledge Foundation—Open definition. https://opendefinition.org/okd. Accessed 4 Nov 2023
19. Poole, K.T.: Spatial Models of Parliamentary Voting, 1st edn. Cambridge University Press, New York, USA (2005)
20. Ren, G.J., Glissmann, S.: Identifying information assets for open data: the role of business architecture and information quality. In: 14th 2012 International Conference on Commerce and Enterprise Computing, pp. 94–100. IEEE, Hangzhou (2012). https://doi.org/10.1109/CEC.2012.23
21. Safarov, I.: Institutional dimensions of open government data implementation: evidence from the Netherlands, Sweden, and the UK. Public Perform. Manage. Rev. **42**, 305–328 (2019). https://doi.org/10.1080/15309576.2018.1438296
22. Ubaldi, B.: Open government data: towards empirical analysis of open government data initiatives. In: OECD Working Papers on Public Governance (2013). https://doi.org/10.1787/5k46bj4f03s7-en
23. Zeleti, F.A., Ojo, A., Curry, E.: Exploring the economic value of open government data. Gov. Inf. Q. **33**, 535–551 (2016). https://doi.org/10.1016/j.giq.2016.01.008

Exploring ChatGPT Prompt Engineering for Business Process Models Semantic Quality Improvement

Sarah Ayad(✉) and Fatimah AlSayoud

Faculty of Computer Studies, Arab Open University, Dammam, Saudi Arabia
{s.ayad,f.alsayoud}@arabou.edu.sa

Abstract. Business process modeling aims to enhance the comprehension of company processes, enabling decision-makers to attain strategic objectives. Nevertheless, inexperienced systems analysts who lack domain knowledge may produce low-quality models, thereby impacting the overall effectiveness of the modeling process. Large language models are bringing about a revolutionary transformation in the application of knowledge within various domains. This paper presents preliminary experimental findings on the potential application of Large language models in Business Process Modeling semantic quality improvement, specifically focusing on the extent to which these technologies can aid the modeler by suggesting improvements. In our study, we delve into the knowledge of GPT-4, a powerful language model. We aim to investigate its capabilities by employing various combinations of prompts, incorporating proposed textual syntax, and integrating contextual domain knowledge. Our objective is to leverage these approaches to improve the quality of business process models. Our findings indicate that the knowledge generated by GPT-4 is predominantly generic, encompassing ambiguous and general concepts that extend beyond the specific domain. However, when we apply our proposed prompting techniques, we observe a notable improvement in the specificity and comprehensiveness of the results. The utilization of these prompts helps to refine the generated knowledge, leading to more specific and comprehensive outcomes that align closely with the intended domain.

Keywords: Business Process Modelling · Semantic Quality · Prompt Engineering

1 Introduction

Business process modeling aims to visually represent, analyze, and optimize the processes that are involved in the operations of a business. The goal is to understand and improve the way that work is done within an organization, in order to increase efficiency, reduce costs, and improve overall performance. When the modeler is not a domain expert or does not have access to a full domain knowledge resource, the BPM will be incomplete. Thus it will not fully represent all

Á. Rocha et al. (Eds.): WorldCIST 2024, LNNS 987, pp. 412–422, 2024.
https://doi.org/10.1007/978-3-031-60221-4_39

of the activities, inputs, outputs, and other elements of the process that it is intended to model. We propose to assist the modeling activity with a quality centered approach that aims to exploit LLMS. Recent research concentrated on how to use NLP to create BPMs by the extraction of process models from natural language texts, and the other way around [2,3,8]. However, there has been a lack of research into utilizing LLMs to enhance existing BPMs. Despite the significant potential that they hold for enhancing the completeness of BPMs through the provision of both tacit and explicit knowledge, there has been limited exploration of the use of these knowledge sources. In order to create accurate, clear and somehow complete BPM, the modeler should have enough application domain knowledge that can be extracted from several sources such as user's requirements statements, domain expertise or existing models related to the same problem or to similar problems. Researchers, usually, rely on domain ontlologies as application domain knowledge but there is no proof that they are incomplete and that they encompass all the relevant terms and concepts within the domain. This lack of coverage presents a limitation in each approach, as it hinders the ability to fully capture and represent the domain's knowledge. To solve such limitation, we aim to exploit GPT-4 knowledge. Our approach do not ask GPT-4 to create a visual representation, instead it needs to suggest missing concepts. Thus to help the modeler improve his model. The GPT-4 may need to refine it's suggestions based on feedback from the modeler.

2 Background and Related Work

This section provides a review of the most relevant research on Business process modeling and LLMs. The state of the art in this field can be broadly categorized into two related areas: the extraction of process models from natural language texts, and the generation of natural language texts from process models.

In the last decade, there has been growing attention given to the issue of generating models from texts. Numerous approaches have been put forth to address this problem, utilizing Natural Language Processing (NLP) techniques. B. Maqbool et al. [12] explored NLP techniques for automatically generating BPMN models from textual requirements. Their systematic review of 36 studies from 2010 to 2018 indicates that while NLP tools simplify BPMN model generation, they are insufficient for real-time systems in industries. Sholiq, S et al. [16] address the challenge of converting textual requirements into BPMN diagrams by proposing a two-stage method involving NLP analysis and informal mapping rules for diagram generation. Existing methods struggle with completeness and handling complexity. Ferreira, R et al. [7] propose a semi-automatic method using NLP to extract process models from organizational texts. They define mapping rules and develop a prototype for process element identification. Moreover, Van der Aa, H. et al. [1] introduce a tailored NLP approach to extract declarative process models from natural language texts. Their automated solution effectively captures complex process behavior, offering an alternative to manual extraction methods. Bellan, P. et al. [4] explores using pre-trained language models for extracting business process information. They find the approach feasible for many aspects but note challenges with control flow relations. In the second category, Leopold et al. [11] investigates the challenges surrounding the comprehension and utilization of process models, highlighting the common issue where only business analysts possess in-depth understanding while lacking domain expertise. They proposed an approach that involves an automatic transformation of BPMN process models into natural language texts, employing a blend of linguistic methods and graph decomposition techniques. Moreover, the utilization of natural language extends to formally expressing business semantics, as demonstrated by the application of the OMG standard, Semantics of Business Vocabulary and Business Rules (SBVR). This standard leverages natural language to articulate intricate business rules [15]. The advantages of verbalization in understanding and validating models are explored from various angles. Addressing the necessity for natural language input in validating conceptual models, prior work [14] suggests natural language generation as a suitable remedy.

3 Methodology

The term knowledge refers to the part of the world investigated by a specific discipline and that includes a specific taxonomy, vocabulary, concepts, theories, research methods, and standards of justification (P. A. Alexander 1992). Our approach uses domain knowledge to improve quality of BPMs by exploiting the domain knowledge provided by LLMs. There are numerous complex graphical notations available for business process modeling, but for the sake of clear communication and consistency, we have chosen to use BPMN, which is widely recognized and utilized.

In the first step, the approach consists in exporting the BPM into an XML file in order to export all concepts ranging from flow objects to artefacts. We sort the list of concepts by type that will eventually serve as input into our AI tool. The second phase involves prompt engineering, employing three distinct techniques: Zero-shot, Few-shot, and Chain of Thought prompting [6,18]. Following this, the third phase is about enhancing our BPM by adding the recommended activities. Finally, the evaluation of the BP model is conducted using a domain ontology and employing semantic quality metrics (Fig. 1).

3.1 Data Extraction

The process begins by taking the BP model exported in XML, XPDL, or XMI as input. From this, we extract Lanes, pools, and activities. To facilitate this extraction, we developed a custom module using Python code that specifically identifies the lanes connected to each pool and the activities linked to each lane. Consequently, the data extracted adheres to a distinct .csv format, detailing the associations between pools and their respective lanes, as well as between lanes and their corresponding activities.

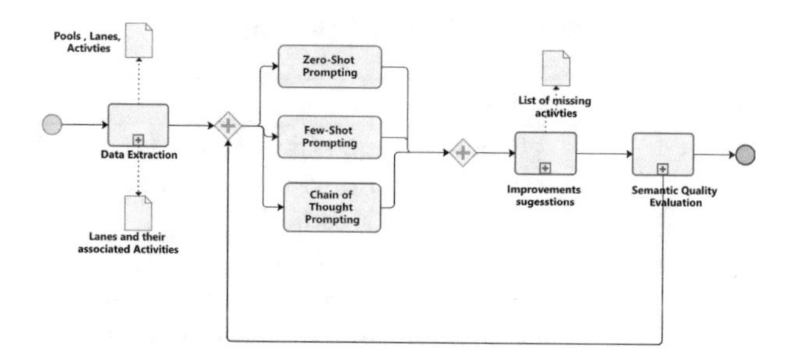

Fig. 1. Our approach

3.2 Prompt Engineering

In the second phase of our methodology, we delve into Prompt engineering, a crucial step in refining and enhancing business process models.

Prompt engineering enables the optimization of BPMs using pre-trained LMs without fine-tuning, facilitating their effective use in BPM research tasks like predictive process monitoring and text-based process extraction. In machine learning and NLP, a prompt comprises four key elements: instruction, context, input data, and output indicator, where the latter defines the expected format of the model's response, such as suggesting activities in BPMN format.

In our study, we applied three distinctive prompting techniques: Zero-shot prompting [9] , few-shot prompting [5], and Chain-of-Thought (CoT) Few-Shot Prompting [19]. Through the deployment of these varied approaches, we encompass various levels of granularity and complexity inherent in BPM tasks and processes. Each technique serves a distinct purpose, addressing varying complexities and requirements within the context of business process modeling. Utilizing these diverse prompting techniques becomes essential for assessing the comprehensiveness, ambiguity, and clarity of the suggested enhancements.

Zero Prompting. In the zero-shot prompting technique, we directly feed either the lanes, activities, activities associated with each lane, or the combination of lanes and activities extracted from the .xpdl file into ChatGPT. Then, we instruct ChatGPT to generate additional activity suggestions based on this input. During our experiments, we tested four versions of prompts for each entity type. Three are summarize in Table 1.

Few-Shot Prompting. In few-shot prompting, we provide ChatGPT with a limited set of details, such as data from other lanes, activities linked to those lanes, or specific task examples. This additional context aids ChatGPT in understanding the context of our BP model and acquiring domain knowledge. Consequently, its responses are more precise compared to zero-shot prompting, despite the restricted quantity of examples provided.

Table 1. Zero shot prompting

Input Data	Instruction
No	Please list potential BPMN activity names for ***current Lane*** within a ***current Pool*** (list only Activity names).
Activities for each Lanes	Given the BPMN Activities for each Lanes in the attached file. Please list additional activity names for ***current Lane*** in a ***current Pool*** (list only Activity names).
Lanes and Activities	Given the BPMN lanes and Activities in the attached files. Please list additional activity names for ***current Lane*** in a ***current Pool*** (list only Activity names)

As outlined in Sect. 3.1, we have retrieved the associated activities for each lane, which will be utilized in the few-shot prompting technique. For instance, let L_1 represent the lane, and A_1^1, A_1^2 , A_1^3, ... denote the list of activities linked to L_1.

Similarly to Zero-shot experience, we tested four versions of few shot prompting for each entity type. Two of them summarize in Table 2.

Table 2. Few Prompting techniques

Input Data	Instruction
Activities for each Lanes	Considering the BPMN activitiesA_1^1 and A_1^2 and ... for L_1, and referring to the Activities for each Lanes in the attached file . Please suggest more BPMN activity names suitable for L_1 within a $Pool_1$ context, considering the provided activities for each lane.
Lanes and Activities	Considering the BPMN activities A_1^1 and A_1^2 and ... for L_1, and referring to the lanes and Activities in the attached file .Please suggest more BPMN activity names suitable for L_1 within a $Pool_1$ context, based on the provided activities, lanes, and their interactions. (list only Activity names)

Chain-of-Thought (CoT) Few-Shot Prompting. In our research, chain of thought prompting helps structure prompts in a step-by-step way, guiding ChatGPT through a series of connected steps to generate accurate responses. While few-shot prompting involved using examples from Pools, Lanes, and activities extracted from the BP model, chain of thought prompting we we assist in methodically identifying associated, supplementary, and expanded tasks carried out by each Lane [13].

3.3 Improvements Suggestions

The quality enhancement process involves offering analysts a series of concepts suggestions to elevate the quality of their models. Our method introduces textual guidelines that serves as recommendations. We suggest missing tasks to the analysts to enhance the models. However, enhancing quality is a gradual process, necessitating an iterative and step-by-step approach.

The proposed activities are provided as textual guidelines and cannot be incorporated into the BP model automatically, yet. This step uses the domain knowledge to generate improvement actions. This means that the completeness and even the relevance of these guides rely partly on the prompting results.

3.4 Quality Evaluation

The fourth step involves the assessment of the semantic integrity within BPMs. The semantic quality measures the degree of correspondence between the model and the domain and it is related to both completeness and validity of the models; here the BP models [10]. The primary aim of semantic quality is to enhance the alignment between the model and the domain it represents. However, this alignment cannot be directly established or verified. The framework outlined in [10] encompasses two key semantic objectives: validity and completeness. *Validity* implies that all statements within the model are accurate and pertinent to the problem and *Completeness* entails that the model encompasses all statements that are both accurate and relevant to the problem domain. In this paper, we will evaluate the completeness of the BP models.

To appraise this quality, we used pre-defined quality deficiencies including instances of incompleteness and ambiguity [17]. These deficiencies stem from modeling decisions that generate models failing to encompass the intended requirements or possessing limited expressiveness. Such models contribute to inadequate systems, either due to their incomplete nature or the potential misunderstandings arising during their implementation.

To assess the completeness of our models, we adopt the methodology proposed by Si-said et al. [17], which relies on domain knowledge. They utilizes application domain knowledge extracted from ontologies and represented through semantic constraints to evaluate the semantic quality of BPMs. We used the tool introduced by Cherfi et al., and applied the Model- richness metric.

The richness of a model is a reflection of its ability to comprehensively cover the domain it represents. This entails evaluating whether the model encompasses all necessary steps and includes every required element. To assess the richness of the model, Cherfi et al. introduced Model Activity Richness metric.

Model Activity Richness (MAR) $= \frac{X}{Y}$

where: X : number of activities presented in the model and Y : number of actions in the enriched ontology

4 Validation

To illustrate our approach we consider an initial BP model modelled in BPMN. It's worth mentioning that for this purpose, we utilized a medical domain ontology and a hospitalization process scenario to gain valuable insights from the experience. Due to space limitations, the Business process model can be found via this link : https://github.com/fgs-fgs/BusinessModelPrompt/tree/main. We developed a prototype implementing the proposed approach for BP model quality evaluation and improvement.

Data Extraction and Prompt Engineering. Utilizing our developed tool, we conducted data extraction and implemented the three prompting techniques. The prototype generated distinct activity lists corresponding to each prompting approach. The .cvs file containing the prompting results is conveniently located in the same folder for reference. The selection of activities to be added to our BPM was chosen, focusing on essential additions identified through our process. Application to our example, Table 3 is an application of the Chain-of-Thought (CoT) Few-Shot prompting technique applied on our case study (Figs. 2, 3 and 4).

Input Data: Activities

- In a hospital setting, doctors play a crucial role in patient care. One of the primary BPMN activities they perform is "Conduct Patient Assessment", which involves understanding the patient's medical condition.
- *A: This assessment is crucial as it forms the basis for further medical actions.*
- *Following the assessment*, another key BPMN activity doctors often engage in is "Prescribe Medication". This activity ensures that the patient receives the necessary treatment for their condition.
- *A: Medication prescription is essential as it directly impacts the patient's recovery process.*
- Now, referring to the attached file which outlines the BPMN Activities for various healthcare professionals,
- Please suggest more BPMN activity names suitable for doctors within a hospital context (list only Activity names):

Input Data: Activities for each Lanes

- In a hospital setting, doctors play a crucial role in patient care. One of the primary BPMN activities they perform is "Conduct Patient Assessment", which involves understanding the patient's medical condition.
- *A: This assessment is crucial as it forms the basis for further medical actions.*
- Following the assessment, another key BPMN activity doctors often engage in is "Prescribe Medication". This activity ensures that the patient receives the necessary treatment for their condition.
- *A: Medication prescription is essential as it directly impacts the patient's recovery process.*
- Now, referring to the attached file which outlines the BPMN Activities for each Lanes for various healthcare professionals,
- Please suggest more BPMN activity names suitable for doctors within a hospital context (list only Activity names)::

Fig. 2. Chain of thoughts stage (inputs: Activities, Activities for each Lames)

Input Data: Lanes and Activities

- In a hospital setting, doctors play a crucial role in patient care. One of the primary BPMN activities they perform is "Conduct Patient Assessment", which involves understanding the patient's medical condition.
- *A: This assessment is crucial as it forms the basis for further medical actions.*
- Following the assessment, another key BPMN activity doctors often engage in is "Prescribe Medication". This activity ensures that the patient receives the necessary treatment for their condition.
- *A: Medication prescription is essential as it directly impacts the patient's recovery process.*
- Now, referring to the attached file which outlines the BPMN Lanes and Activities for various healthcare professionals,
- Please suggest more BPMN activity names suitable for doctors within a hospital context (list only Activity names):

Fig. 3. Chain of thoughts stage (inputs: Activities and Lames)

Quality Evaluation. Quality measurement procedures are employed to quantify the level of quality. Our objective is to assess the original BPM and the new one using both syntactic and semantic metrics [10]. Syntactic analysis focuses on the number of activities in the model (NA), the number of activities per Lane (NA/L), while semantic metrics (MAR) gauge the correspondence between model information and the modeled domain.

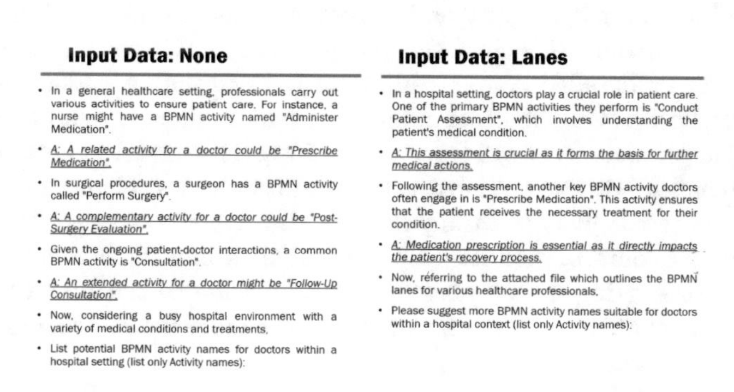

Fig. 4. Chain of thoughts stage(inputs: None, Lanes)

Table 3. Model activity richness metric application.

Technique	Input Data	NA	Model Activity Richness
Zero shot	No	30	0.12
	Lanes	19	0.07
	Activities	19	0.07
	Activities for each Lanes	16	0.06
	Lanes and Activities	19	0.07
Few shot	No	18	0.07
	Lanes	14	0.05
	Activities	12	0.04
	Activities for each Lanes	9	0.03
	Lanes and Activities	14	0.05
Chain of thoughts	No	14	0.05
	Lanes	11	0.04
	Activities	14	0.05
	Activities for each Lanes	9	0.03
	Lanes and Activities	14	0.05

The Model-Richness metric's low value confirms that the model does not cover a sufficient part of the domain. The model can be enriched by adding new activities suggested by our AI. It will increase the number of activities i.e. it will increase the model richness' metric value. Note that the ontology is used only to evaluate the quality of the model.

The quantity of activities resulting from our prompting techniques follows an expected trend: the zero-shot technique generate a higher number than few-shot, and few-shot prompts generate more activities than Chain of Thought. This is due to the increasing specificity and accuracy in the prompting, resulting in a more focused outcome. The zero-shot technique operates without specific context examples or detailed context, allowing it to explore a wider spectrum of potential activities. On the other hand, few-shot prompting uses limited, specific examples or context to guide its suggestions. This targeted approach might lead to a smaller but more refined selection of activities, aligning closely with the provided context but potentially limiting the diversity of suggestions. Therefore, the higher count of activities from few shot prompting compared to chain of thoughts prompting in our research might be due to the broader, context-independent exploration of the former versus the more focused, context-specific nature of the latter.

Finally, our evaluation incorporates the Model Activity Richness (MAR), which relies on the activities within our ontology. A higher MAR signifies a richer model, although it doesn't necessarily guarantee increased accuracy or precision. Hence, our enhancement approach is iterative, allowing modelers flexibility in incorporating or removing activities from the suggestions until achieving a comprehensive model that aligns with their requirements.

5 Conclusion

The paper introduces a significant step in semantics-based quality assessment and enhancement through LLMs. Notably, our approach's diversity lies in its independence from specific domains or notations, promising substantial support for BP modelers by automating prompts, even for non-experts. The provided quality metrics offer quantified domain coverage evaluation, yet the approach encounters limitations due to potential incompleteness in the existing domain ontology compared to the extensive propositions by LLMs. Future research will center on enriching the domain ontology utilizing LLMs. Simultaneously, we are steering efforts towards enhancing the semantic quality of Business Process Models by integrating LLMs, introducing supplementary elements like resources, gateways, events, and actors. Moreover, there's a necessity for increased validation through real-life case studies involving practitioners.

Acknowledgement. Author would like to thank Arab Open University for supporting this work.

References

1. van der Aa, H., Di Ciccio, C., Leopold, H., Reijers, H.A.: Extracting declarative process models from natural language. In: Giorgini, P., Weber, B. (eds.) CAiSE 2019. LNCS, vol. 11483, pp. 365–382. Springer, Cham (2019). https://doi.org/10.1007/978-3-030-21290-2_23
2. de Almeida Bordignon, A.C., Thom, L.H., Silva, T.S., Dani, V.S., Fantinato, M., Ferreira, R.C.B.: Natural language processing in business process identification and modeling: a systematic literature review. In: Proceedings of the XIV Brazilian Symposium on Information Systems, pp. 1–8 (2018)
3. Bellan, P., Dragoni, M., Ghidini, C.: Process extraction from text: state of the art and challenges for the future. arXiv preprint arXiv:2110.03754 (2021)
4. Bellan, P., Dragoni, M., Ghidini, C.: Extracting business process entities and relations from text using pre-trained language models and in-context learning. In: Almeida, J.P.A., Karastoyanova, D., Guizzardi, G., Montali, M., Maggi, F.M., Fonseca, C.M. (eds.) EDOC 2022. LNCS, vol. 13585, pp. 182–199. Springer, Cham (2022). https://doi.org/10.1007/978-3-031-17604-3_11
5. Brown, T., et al.: Language models are few-shot learners. In: Advances in Neural Information Processing Systems, vol. 33, pp. 1877–1901 (2020)
6. DAIR.AI: Basics of prompting (2023). Accessed 31 Oct 2023
7. Ferreira, R.C.B., Thom, L.H., Fantinato, M.: A semi-automatic approach to identify business process elements in natural language texts. In: International Conference on Enterprise Information Systems, vol. 2, pp. 250–261. SCITEPRESS (2017)
8. Honkisz, K., Kluza, K., Wiśniewski, P.: A concept for generating business process models from natural language description. In: Liu, W., Giunchiglia, F., Yang, B. (eds.) KSEM 2018. LNCS (LNAI), vol. 11061, pp. 91–103. Springer, Cham (2018). https://doi.org/10.1007/978-3-319-99365-2_8
9. Kojima, T., Gu, S.S., Reid, M., Matsuo, Y., Iwasawa, Y.: Large language models are zero-shot reasoners. In: Advances in Neural Information Processing Systems, vol. 35, pp. 22199–22213 (2022)
10. Krogstie, J., Lindland, O.I., Sindre, G.: Defining quality aspects for conceptual models. In: Information System Concepts: Towards a Consolidation of Views, pp. 216–231 (1995)
11. Leopold, H., Mendling, J., Polyvyanyy, A.: Generating natural language texts from business process models. In: Ralyté, J., Franch, X., Brinkkemper, S., Wrycza, S. (eds.) CAiSE 2012. LNCS, vol. 7328, pp. 64–79. Springer, Heidelberg (2012). https://doi.org/10.1007/978-3-642-31095-9_5
12. Maqbool, B., et al.: A comprehensive investigation of BPMN models generation from textual requirements—techniques, tools and trends. In: Kim, K.J., Baek, N. (eds.) ICISA 2018. LNEE, vol. 514, pp. 543–557. Springer, Singapore (2019). https://doi.org/10.1007/978-981-13-1056-0_54
13. Narang, S., Raffel, C., Lee, K., Roberts, A., Fiedel, N., Malkan, K.: Wt5?! Training text-to-text models to explain their predictions. arXiv preprint arXiv:2004.14546 (2020)
14. Rolland, C., Proix, C.: A natural language approach for requirements engineering. In: Loucopoulos, P. (ed.) CAiSE 1992. LNCS, vol. 593, pp. 257–277. Springer, Heidelberg (1992). https://doi.org/10.1007/BFb0035136
15. SBVR, O.: Semantics of business vocabulary and business rules (SBVR), version 1.0 (2008)

16. Sholiq, S., Sarno, R., Astuti, E.S.: Generating BPMN diagram from textual requirements. J. King Saud University-Comput. Inf. Sci. **34**(10), 10079–10093 (2022)
17. Si-Said Cherfi, S., Ayad, S., Comyn-Wattiau, I.: Improving business process model quality using domain ontologies. J. Data Semant. **2**, 75–87 (2013)
18. Wang, J., et al.: Prompt engineering for healthcare: methodologies and applications. arXiv preprint arXiv:2304.14670 (2023)
19. Wei, J., et al.: Chain-of-thought prompting elicits reasoning in large language models. In: Advances in Neural Information Processing Systems, vol. 35, pp. 24824–24837 (2022)

Industrial Data Sharing Ecosystems: An Innovative Value Chain Traceability Platform Based in Data Spaces

Josué Freitas[1,2], Cristóvão Sousa[1,2](✉), Carla Pereira[1,2], Pedro Pinto[1,2], Ricardo Ferreira[1,2], and Rui Diogo[1]

[1] INESC TEC, Campus da FEUP, Rua Dr. Roberto Frias, 4200-465 Porto, Portugal
cristovao.sousa@inesctec.pt
[2] ESTG, Polytechnic of Porto, Rua do Curral, Casa do Curral, 4610-156 Felgueiras, Portugal

Abstract. Considering the great challenge of implementing digital tools to improve collaboration in the value chain and promote the adoption of circularity strategies, as is the case with digital traceability tools and digital product passports. This paper presents an innovative proposal for implementing an industrial data sharing ecosystem, namely an architecture and platform for digital traceability between entities based on Data Spaces. To validate our proposal, a use case scenario was implemented as part of the BioShoes4All project.

Keywords: Digital Product Passport · Digital Traceability · Information Sharing · International Data Spaces

1 Introduction

The concept of product traceability in manufacturing is gaining energy in the context of the digital transformation processes towards circular economy [1]. More concretely, traceability might be seen as the main drive to adopt emergent circular management practices such as the digital product passport [2–4], which represents a circular-oriented and human-centered view on traceability data. This new model of product data organisation and inter-exchange, adds new digital transformation challenges/barriers to organisations, related to: i) interoperability [5]; ii) data security and ownership [6,7]; iii) "siloed" data storage; iv) shortage of technological resources and digital skills [8]; v) data quality as well as common schemes and vocabularies. The implementation of information sharing strategies between companies allow to obtain knowledge to support more informed decisions, identifying opportunities, anticipating challenges, reducing risks and difficulties. On the other hand, without a more comprehensive knowledge of the flow of products and materials, both upstream and downstream of the organization, the decision-making process towards designing more sustainable, durable and reusable products, is difficult. For this reason, information sharing in inter-company environments has been an area of great interest and concern over the last few decades [9].

Á. Rocha et al. (Eds.): WorldCIST 2024, LNNS 987, pp. 423–432, 2024.
https://doi.org/10.1007/978-3-031-60221-4_40

In this context, and from the point of view of small and medium-sized companies, it is important to decode and systematize how to implement digital tools to enhance value chain collaboration to promote the adoption of circularity strategies, in particular the creation and sharing of product passports. At the same time hoe to control on data sovereignty and ensure data security. To pave the path to the DPP, the specification of traceability digital tools to collect and share data across value chain is fundamental. To address this challenge a data-space based architecture was adopted to share traceability data among entities, which might be used it to populated each DPP. The approach is grounded of sharing data-sets rather than highly-structured business documents, fostering a distributed collaborative environment and overcoming the silos traps.

Aware of the need, the European Commission (EC) has reinforced the promotion of interoperability strategies [10] and reference architectures [11] so that the implementation of an industrial data sharing ecosystem becomes possible. The EC initiatives aims at facilitating information sharing, collaboration and innovation in the industrial sector, through a reference architecture that mitigate the technological challenges aforementioned, while reassuring *stakeholders* concerns regarding security, trust and quality of information.

Lastly, technology companies that offer platforms and services management for SMEs, also face several challenges when adapt to new technological innovation. These companies need to develop flexible and adaptable solutions to the specific needs of each company, however, there lack of capacity to embrace such technological challenges.

In pursuit of overcoming the challenges presented, this work discusses a technological solution, accessible to small and medium-sized companies, regardless of their technological resources, making it possible for them to participate in a data ecosystem for traceability. More specifically, this work's main objectives were to design a technological solution based on the data ecosystem paradigm and which enables the secure sharing of traceability data between companies in a value chain, safeguarding data sovereignty through contracts and usage policies.

2 Scaling Digital Traceability: Concepts

This section presents the concepts of traceability, Digital Product Passport (DPP), Data Spaces (DS) and International Data Spaces (IDS), focusing on how they relate to ensuring the digital traceability of a product throughout the value chain.

2.1 Traceability and Digital Product Passport: What Relation?

Traceability is a concept that aims to identify a product in order to obtain information about it at any point in its life cycle [1]. Information can only be obtained by recording data that characterises the structure of the product and the movements from the beginning to the end of each process by the entities involved. By implementing traceability systems, it is possible to trace the path of a product, identifying the places where it has been processed, stored, transported and sold. In [1], the four fundamental pillars for creating a traceability system are presented (*Four Pillar Traceability Framework*): Product Identification, Data to Trace, Product Routing, and Traceability Tools.

DPP, boosted by a recent initiative of the European Commission [12], is a concept that aims to provide detailed information about a product throughout its entire life cycle, from its production to its end of use. It is a tool that will make it possible to track and monitor relevant information about a product, such as its composition, the origin of materials, manufacturing processes, use, maintenance, recycling and repair instructions, among others [12]. The main goal of DPP is to promote sustainability in manufacturing, allowing manufacturers to make more informed decisions about the use of materials during the product's life cycle [13,14]. It can also be used to improve transparency between manufacturers and consumers [4]. In practical terms, the DPP will consist of a unique identifier on the product label [15], which will give access to all the data added by the participants along the value chain. Consumers will have access to detailed information about a product before they buy it, allowing them to make an informed choice, thus promoting more sustainable products.

Thus, considering the concept and objectives of traceability, we conclude that traceability is a basis for the implementation of DPP, since it consists of a detailed and reliable record of information along a product's value chain, sharing the same objective as DPP, the monitoring of the product throughout its life cycle. Both, traceability and DPP, requires the sharing of information between companies. However, although this information sharing has significant benefits, it raises problems related to security, trust and data quality. Due to these difficulties and the lack of a reference architecture for this problem, there is still a great deal of resistance on the part of entrepreneurs and *stakeholders* to sharing their companies' internal information [5,9]. It is therefore essential to adopt a decentralised model [16]. In other words, there won't be just one organisation responsible for storing and managing all the information. Instead, the aim is for the information to be distributed in a decentralised network of participants, which allows the information to be accessed and updated whenever necessary [17]. With this decentralised model it is possible to increase data security and infrastructure. As the information is not centralised in one place, there is no single point vulnerable to attacks and failures [18]. An example of a decentralised model for sharing data in a secure ecosystem is the Data Space. Thus, Data Spaces, the concept presented below, can play a key role in the creation and implementation of DPP.

2.2 Data Spaces and International Data Spaces

A Data Space consists of an environment where data that comes from different sources, but is related to each other, is stored and shared for different purposes. This includes an infrastructure and tools that enable this sharing in a unified and interoperable way [19]. This concept has been explored as a way of sharing information related to the product life cycle, and has been addressed by the

European Commission with the DPP. The idea of connecting the Data Space concept with the DPP is to create an environment where all participants can use an infrastructure that allows the exchange of relevant information about a specific product, such as its characteristics and its operations and transformations along its value chain.

The Data Spaces approach is advantageous when faced with a problem that does not depend on synchronous data exchange, but rather on the availability of data in a given ecosystem, where those who share information in the Data Space environment maintain ownership of the data, and access to this data occurs asynchronously by another entity through a data request to the ecosystem [20]. This concept is known as a data MarketPlace and due to its information exchange characteristics, it allows circularity and traceability strategies to be maximised. In this sense, the use of a Data Space presupposes a distributed approach, i.e. non-centralised, where the origins of the data always remain with its author. The data is kept in its place of origin, and the distributed platform acts as an intermediary that allows access to and interaction with this data, without the need to move or copy it to a centralised location [20].

In order to guarantee interoperability and standardisation in the exchange of information, Data Spaces have communication mechanisms and protocols aimed at facilitating the exchange of information between participants.

The International Data Spaces is a proposed reference architecture for implementing Data Spaces, with the main goal of serving as a standardised component for secure data exchange in inter-company contexts. This architecture provides a set of mechanisms to guarantee security and interoperability in the exchange of data between different companies and systems [21].

So, considering the basic requirements associated with product traceability and DPP, the use of an IDS ecosystem is a potential candidate for implementing a solution for sharing more private information. Participants need to follow certain rules that ensure data interoperability and security. In addition, each participant is duly identified and certified, which guarantees trust in the exchange of information. Through IDS it is possible to guarantee the security of connections between trusted participants due to the possibility of defining access control and the use of shared data [21].

The IDS architecture is distributed and decentralised, which means, in practical terms, that each company needs to implement its own IDS connector. The IDS connector is the main component of the IDS and is used by all participants to be able to connect with other companies via the IDS ecosystem. Only through the connector is it possible to guarantee the sharing of data with the available forms of security and policies. Within the scope of this architecture, four fundamental entities are considered for the performance of the ecosystem [21]: 1) Provider - IDS participant who makes data available, which is then accessed by Consumers. As they share data, they define the access and utilisation rules associated with the sharing; 2) Consumer - IDS participant who consumes the data made available by the Providers; 3) Broker - Acts as an intermediary that manages the data information available in the IDS, it is responsible for providing and enabling data sharing; 4) Certification Authority - Responsible for the certification of IDS connectors, through defined standards. All the data shared between the IDS connectors follows a data model made up of a set of entities, as shown in the Fig. 1.

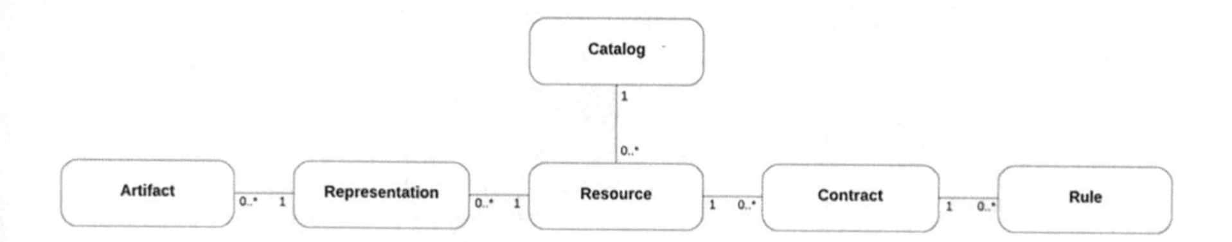

Fig. 1. IDS resource data model

In conclusion, IDS can play an essential role in the implementation of product traceability, since it is a data-sharing platform that offers a secure and reliable environment for the exchange of information between organisations.

3 Conceptual and Technological Architectures

Based on the concepts presented in the previous section, we now present an architecture, based on IDS, capable of guaranteeing interoperability between companies, while always guaranteeing the security and sovereignty of the data between the company supplying and consuming the information through the contracts established between the interested parties. The IDS reference architecture already has several certified and widely known implementations. One of the best known and most widely used is the *Fraunhofer* 5 implementation. This implementation already has the following components [21]: IDS *Connector*, *Broker* and *Dynamic Attribute Provisioning Service* (DAPS) and tools necessary for the operation of the data ecosystem (Fig. 2).

The *Broker* IDS acts as an intermediary in the data ecosystem, allowing the parties involved to liaise and exchange information in a controlled and secure manner. It is responsible for providing participants with metadata on the data made available by other participants in the ecosystem. This reinforces the idea of a *Marketplace*, where participants can consult the data they need, facilitating data sharing [21].

In the context of IDS, DAPS is a fundamental component that plays an important role in ensuring security and access control. The aim is to verify the authenticity of participants in the IDS ecosystem and provide all the functionalities related to identity management [21].

The purpose of the traceability platform created is to connect to the IDS *Connector* in order to share the traceability records created, as well as to obtain traceability records from other companies. This platform uses a database to store all the information from the traceability records.

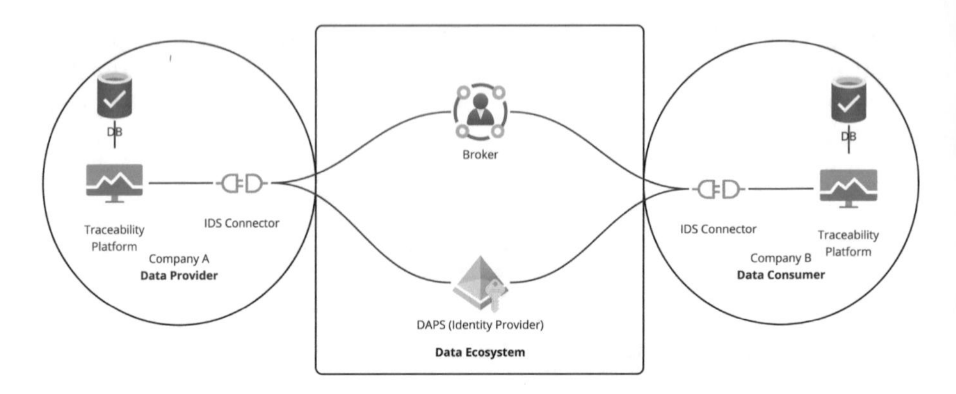

Fig. 2. Conceptual architecture

It is important to emphasise that a company can simultaneously play the role of data provider and data consumer.

In technological terms, the following technologies were used to develop the platform: Vue.js and Quasar, Spring Boot, Mongo DB. Vue.js 6 is a JavaScript framework for building user interfaces. It is a highly flexible framework that allows the creation of reusable components, facilitating the construction of interactive and scalable web applications. Quasar7, for its part, is a Vue.js-based *Framework* that offers a set of components, plugins and generic functionalities available to speed up the development of web applications. It offers a set of tools that make it possible to create responsive interfaces, manage layout and much more. Spring Boot 8 is a Java application development framework that simplifies the creation and configuration process. It is based on Spring and offers a number of features that allow you to quickly create robust and scalable applications. It also includes a built-in server, thus eliminating the need to set up a separate server. It has an ecosystem capable of modifying modules and libraries, which makes it possible to integrate various systems. Due to the need to make the data model as dynamic as possible, Mongo DB9 was used. Mongo DB is a non-relational, document-orientated database that offers a flexible and scalable approach to storing data. Documents are stored in key-value format, using the JavaScript *Object Notation* (JSON) format.

4 Bioshoes4all Demonstration Scenario

In order to validate the platform developed, the use case scenario "Record traceability" of the BioShoes4All project[1] was considered. The BioShoes4All project aims to promote the sustainable transition of the footwear industry in Portugal by adopting circular economy and sustainable bioeconomy practices. In addition to processes, technology plays a crucial and preponderant role in enabling companies to embark on this journey of transformation. It is essential to develop

[1] https://bioshoesforall.pt.

digital solutions and platforms that make it possible to valorise the main types of production waste in the footwear value chain, as well as to start valorising post-consumer products, thus increasing circularity in the production process. Likethis, the diagram below describes the overall operation of the platform using the Bioshoes4all project as use case (Fig. 3).

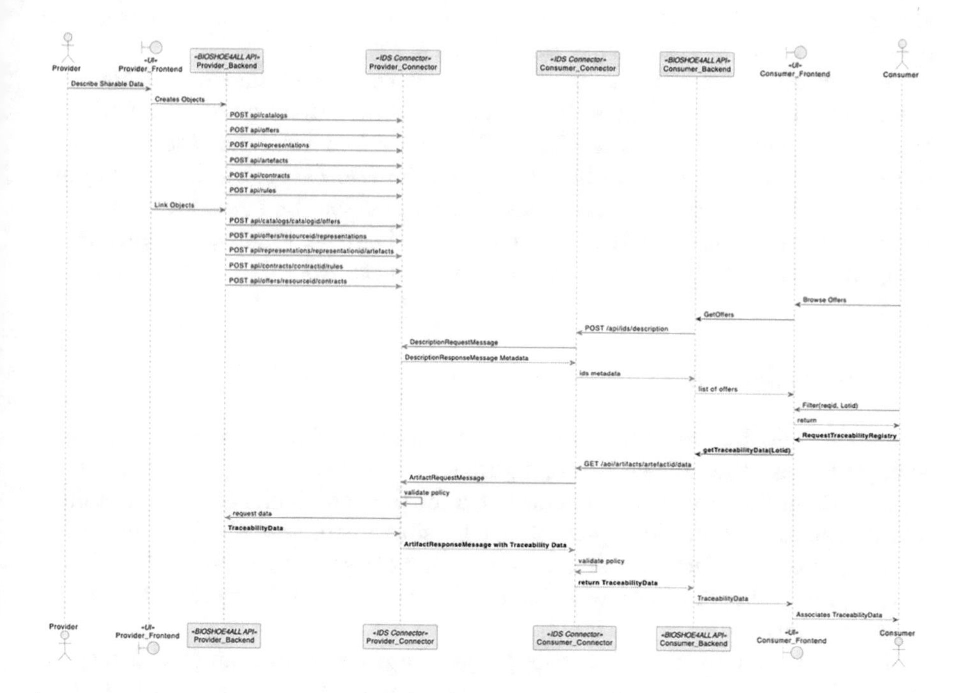

Fig. 3. Traceability sequence of interactions

As shown in figure above, the company providing the information will have to interact with the Frontend in order to insert and define shoes models from which it will share traceability records in the future. In addition, the Backend receives the data and make it available in its own IDS.

The use of the IDS must follow the *Rest API* provided by the IDS, so initially the *Catalogs* associated with the *Resource* must be created. Then the *Offer*, also known as the *Resource,* must be created. The *Offer,* as mentioned above, has *Representations* and *Artifacts* associated with it, and initially these concepts are only instantiated. Finally, there are the rules and policies associated with the *Resource*, these policies will be associated with the *Rules*, which in turn will be associated with the *Contract.*

Once the model has been created, traceability records associated with the model can be registered. In IDS, *Representation* and *Artifact* must be created to represent the traceability record directly associated with the model. On the side of the company that wants to consume the data, the search in the Frontend application communicates with the Backend, which must have the *Uniform Resource Locator* (URL) of the *IDS Connector* associated to the traceability records requested.

Using the URL, the Backend must use the associated *IDS Connector* to request the offers made available by the *IDS Connector* to which the given URL corresponds. All the offers returned will be shown to the user so that they can choose and filter the traceability record they want to view.

By choosing a specific record, the Backend will communicate with the *IDS Connector* in order to request the data of a given *Artifact* associated with the URL of the supplier company. The *IDS Connectors* communicate with each other so that this request can be made, with the *IDS Connector* of the supplier company validating the policies. If the information can be made available, it is returned to the Connector IDS that requested the data. Finally, the consuming company also checks access policies and returns the traceability record data to the user.

5 Conclusion and Future Work

The main contribution of this work was the design and development of a traceability platform that allows SMEs, regardless of their technological resources, to participate in the data ecosystem designed for the new challenges of traceability and the Digital Product Passport. It was possible to obtain a platform according to the desired architecture, capable of allowing companies to create traceability records, make these records available on the IDS and obtain records created by other companies.

It is essential to note that this project represents only an initial step in the development of traceability platforms for the possible creation of the PDP. Considering the specific case of the footwear sector, where the platform has been validated and tested, there are several challenges to be faced, such as: increasing company participation by adapting the data model created, to a model closer to companies in the footwear industry; adapting the platform to an *Enterprise Resource Planning* (ERP) in order to make this project easier to integrate, as well as the possibility of downloading the records obtained.

However, despite the limitations and complexity involved, the initiative to create an industrial data ecosystem and promote information sharing between companies is key to boosting innovation, collaboration and sustainability in the sector. The European Community has played an important role in this by supporting interoperability and the emergence of reference architectures, such as IDS, which address the issues of security, trust and quality of shared data.

Acknowledgements. This work is co-financed by Component 12 - Promotion of Sustainable Bioeconomy, integrated in the Climate Transition Dimension of the Recovery and Resilience Plan within the scope of the Recovery and Resilience Mechanism (MRR) of the European Union (EU), framed in the Next Generation EU, for the period 2021–2026, within project BioShoes4All, with reference 11.

References

1. Schuitemaker, R., Xu, X.: Product traceability in manufacturing: a technical review. Procedia CIRP **93**, 700–705 (2020)
2. Anastasiadis, F., Manikas, I., Apostolidou, I., Wahbeh, S.: The role of traceability in end-to-end circular agri-food supply chains. Ind. Mark. Manag. **104**, 196–211 (2022)
3. Giovanardi, M., Konstantinou, T., Pollo, R., Klein, T.: Internet of things for building façade traceability: a theoretical framework to enable circular economy through life-cycle information flows. J. Clean. Prod. **382**, 135261 (2023)
4. King, M.R., Timms, P.D., Mountney, S.: A proposed universal definition of a digital product passport ecosystem (DPPE): worldviews, discrete capabilities, stakeholder requirements and concerns. J. Clean. Prod. **384**, 135538 (2023)
5. Sarathy, R., Muralidhar, K.: Secure and useful data sharing. Decis. Support Syst. **42**, 204–220 (2006)
6. Moosa, H., Ali, M., Alaswad, H., Elmedany, W., Balakrishna, C.: A combined blockchain and zero-knowledge model for healthcare B2B and B2C data sharing. Arab J. Basic Appl. Sci. **30**, 179–196 (2023)
7. Richter, H.: Exposing the public interest dimension of the digital single market: public undertakings as a model for regulating data sharing. SSRN Electron. J. (2020)
8. Martens, B., de Streel, A., Graef, I., Tombal, T., Duch-Brown, N.: Business-to-business data sharing: an economic and legal analysis, July 2020
9. Lai, F., Li, D., Wang, Q., Zhao, X.: The information technology capability of third-party logistics providers: a resource-based view and empirical evidence from china. J. Supply Chain Manag. **44**, 22–38 (2008)
10. European Commission. Directorate General for Communications Networks, Content and Technology. The impact of open source software and hardware on technological independence, competitiveness and innovation in the EU economy: final study report. LU: Publications Office (2021)
11. D.-G. f. D. S. European Commission: Communication from the commission to the European Parliament, the council, the European economic and social committee and the committee of the regions - European interoperability framework - implementation strategy, March 2017
12. Götz, T., et al.: Digital product passport: the ticket to achieving a climate neutral and circular European economy?. Resreport, Cambridge Institute for Sustainability Leadership, July 2022
13. Berger, K., Baumgartner, R.J., Martin, W., Bachler, J., Preston, K., Schöggl, J.P.: Data requirements and availabilities for a digital battery passport - a value chain actor perspective, August 2022
14. Rusch, M., Schöggl, J., Baumgartner, R.J.: Application of digital technologies for sustainable product management in a circular economy: a review. Bus. Strateg. Environ. **32**, 1159–1174 (2022)
15. Paramatmuni, C., Cogswell, D.: Extending the capability of component digital threads using material passports. J. Manuf. Process. **87**, 245–259 (2023)
16. Jansen, M., Meisen, T., Plociennik, C., Berg, H., Pomp, A., Windholz, W.: Stop guessing in the dark: identified requirements for digital product passport systems. Systems **11**, 123 (2023)
17. Koppelaar, R.H., et al.: A digital product passport for critical raw materials reuse and recycling. Sustainability **15**, 1405 (2023)

18. Wang, S., Zhang, Y., Zhang, Y.: A blockchain-based framework for data sharing with fine-grained access control in decentralized storage systems. IEEE Access **6**, 38437–38450 (2018)
19. Halevy, A., Franklin, M., Maier, D.: Principles of dataspace systems. In: Proceedings of the Twenty-Fifth ACM SIGMOD-SIGACT-SIGART Symposium on Principles of Database Systems, SIGMOD/PODS06. ACM, June 2006
20. Otto, B., ten Hompel, M., Wrobel, S.: Designing Data Spaces: The Ecosystem Approach to Competitive Advantage. Springer, Cham (2022). https://doi.org/10.1007/978-3-030-93975-5
21. Otto, B., Steinbuss, S., Teuscher, A., Lohmann, S.: Ids reference architecture model (2019)

Enhancing Zero Trust Security in Edge Computing Environments: Challenges and Solutions

Fiza Ashfaq[1], Abdul Ahad[2,3], Mudassar Hussain[1], Ibraheem Shayea[3], and Ivan Miguel Pires[4](✉)

[1] Department of Computer Science, University of Management and Technology, Sialkot 51040, Pakistan
mudassar.hussain@skt.umt.edu.pk
[2] School of Software, Northwestern Polytechnical University, Xian, Shaanxi, People's Republic of China
[3] Department of Electronics and Communication Engineering, Istanbul Technical University (ITU), 34467 Istanbul, Turkey
shayea@itu.edu.tr
[4] Instituto de Telecomunicações, Escola Superior de Tecnologia e Gestão de Águeda, Universidade de Aveiro, Águeda, Portugal
impires@ua.pt

Abstract. Improving Zero Trust Security in Edge Computing Environments presents unique problems and necessitates novel solutions to secure sensitive data and lessen the dangers associated with edge computing installations. This article explores the challenges and solutions for improving Zero Trust Security in edge computing environments. The distributed and heterogeneous nature of edge ecosystems, limited resources, dynamic and mobile nature, scalability and performance concerns, network latency and bandwidth constraints, and scalability and performance concerns contribute to the challenges. Careful analysis and specific security solutions are needed to achieve a solid security posture. Several ways to overcome these issues include identity and access management mechanisms, secure network segmentation, continuous monitoring and threat detection systems, and authentication and authorization technologies. Zero Trust Security must be integrated with edge computing architectures for effective deployment. Edge security gateway and firewall systems provide perimeter protection for edge settings, while cloud-based security services and threat intelligence improve threat detection and response capabilities. The article also includes case studies and recommended practices, exploring lessons gained and ideas for implementing Zero Trust Security in edge computing settings. Future directions and emerging trends in Zero Trust Security for Edge Computing are investigated, offering insights into the growing security landscape in edge settings.

Keywords: Zero Trust · Security · Edge Computing · Edge Computing Environment · challenges and solutions

Á. Rocha et al. (Eds.): WorldCIST 2024, LNNS 987, pp. 433–444, 2024.
https://doi.org/10.1007/978-3-031-60221-4_41

1 Introduction

The perimeter-based network security method is being challenged by the contemporary security idea known as "Zero Trust Security." Implementing Zero Trust Security becomes even more critical in the age of edge computing, as computing resources are split among several edge devices and settings.

Edge computing describes a decentralized architecture where data processing and storage occur closer to the data source, enabling quicker reaction times and lowering dependency on centralized cloud resources. However, because edge computing is spread, additional security threats and difficulties must be solved. The concept of edge computing has evolved to meet the increasing need for low-latency applications and real-time data processing. Edge computing allows quicker reaction times and lessens network congestion by bringing processing resources closer to the data source. To maintain the privacy, accuracy, and accessibility of data and services, however, these edge computing environments' dispersed and dynamic nature poses substantial security concerns that need to be resolved [2].

Edge computing environments increasingly adopt Zero Trust Security as a more reliable and adaptable security method. This approach, which uses the "never trust, always verify" ethos, aims to address specific problems and implement customized solutions. However, the dynamic nature of edge settings necessitates a continuous assessment of each entity's credibility to build ongoing authentication and authorization processes. The wide variety of edge devices and their varied security capabilities provide another difficulty. Some gadgets could have little processing power or basic security mechanisms, which makes them more susceptible to assaults. Effective edge computing security solutions must balance security demands and resource-constrained devices' limitations. Additionally, because edge computing is spread, there is a problem with safeguarding data transmission and storage across several edge nodes. Data protection at rest and in transit requires encryption, secure communication protocols, and decentralized critical management systems [10].

Real-time threat detection and micro-segmentation, along with identity and access management systems, are implemented to improve Zero Trust Security in edge computing settings. Threat intelligence and security analytics can provide insights into new risks and enable preventative security actions. Platforms for information sharing and collaborative security can help develop collective defense tactics. Figure 1 shows a zero-trust security model in an IoT cloud service architecture diagram, with a controller managing access control and security policies, two edge agent servers acting as intermediaries, and three user terminals connected to each server. The network operates under a zero-trust security framework, requiring strict identity verification for accessing resources on a private network.

Organizations may strengthen the security posture of their edge computing environments by adopting the Zero Trust Security principles and putting these solutions in place. By doing this, they can use edge computing's advantages while

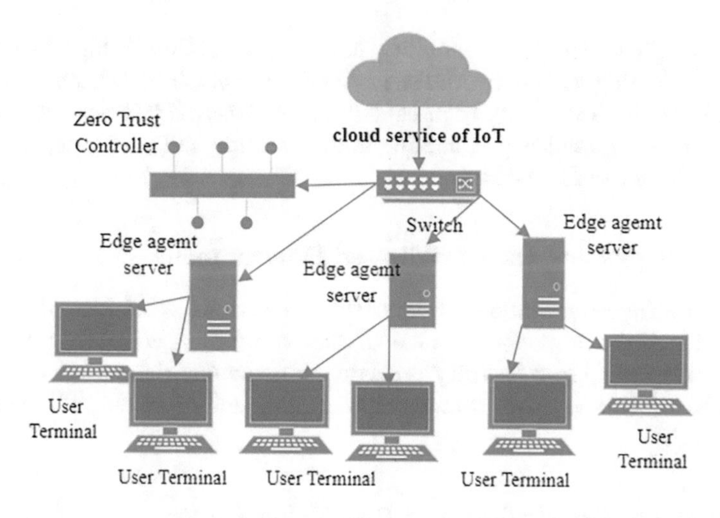

Fig. 1. Zero trust in edge computing environment architecture

minimizing the dangers of distributed computing and preserving the privacy, integrity, and accessibility of their data and services [14].

This paper addresses the unique challenges and proposes novel solutions for improving Zero Trust Security in Edge Computing Environments. The paper aims to secure sensitive data and reduce risks inherent in edge computing installations. It explores the difficulties of implementing Zero Trust Security given edge ecosystems' distributed and heterogeneous nature, edge devices' limited resources and processing power, dynamic and mobile environments, scalability and performance concerns, and network latency and bandwidth constraints.

The benefit of improving zero trust security in edge computing settings is that it provides specialized and efficient solutions to handle the special problems that arise in these dispersed computing environments. This paper highlights the importance of integrating Zero Trust Security with edge computing architectures, highlighting the role of Zero Trust frameworks and technologies. The paper also discusses the role of edge security gateways, firewalls, cloud-based security services, and threat intelligence in perimeter protection and threat detection. It also provides case studies and best practices for implementations of real-world Zero Trust Security solutions. The paper also anticipates future trends and trends in Zero Trust Security for edge computing, offering insights into the evolving security landscape.

2 Challenges of Implementing Zero Trust Security in Edge Computing Environments

This section discusses the challenges organizations face in implementing the Zero Trust model in edge computing due to its diverse, dynamic, and resource-constrained nature. Key challenges include dealing with heterogeneous environments, managing edge device mobility, ensuring scalability in high-demand scenarios, and dealing with network latency and bandwidth limitations. These factors highlight the complexity of integrating robust Zero Trust Security measures in edge computing landscapes, making understanding and navigating these challenges crucial for organizations.

2.1 Heterogeneous and Distributed Environment

Edge computing environments frequently use various platforms, devices, and technologies and are dispersed. Due to this variability, adopting uniform and standardized Zero Trust Security procedures in the environment is challenging. It becomes difficult to ensure seamless integration and interoperability of security systems [11].

2.2 Resource Constraints and Processing Power

The resources available to edge devices, such as their processing speed, memory, and storage capacity, are often constrained. Implementing adequate security measures while considering these resource limitations might be challenging. The restricted resources of edge devices may be taxed by the cryptographic operations, authentication procedures, and continual monitoring necessary for Zero Trust Security.

2.3 Mobile and Dynamic Edge Devices

Edge devices are movable and dynamic in edge computing environments and regularly join, depart, or relocate inside the network. It is challenging to guarantee ongoing authentication and authorization of these devices based on their shifting contexts and locations. While maintaining a constant security posture, zero trust security measures must be flexible enough to handle the mobility of edge devices safely [5].

2.4 Scalability and Effectiveness

Edge computing settings often have numerous dispersed devices, requiring scalable, high-performance security solutions. As the number of devices increases, administration, authentication, and authorization procedures may become resource-intensive, affecting system performance. Figure 2 illustrates Zero Trust Security, a cybersecurity approach emphasizing verifying all requests before trusting any entity. The model's circular diagram highlights the importance of strict control and visibility over network components, scrutinizing every access request, rigorous authentication, and maintaining control and visibility of all network components.

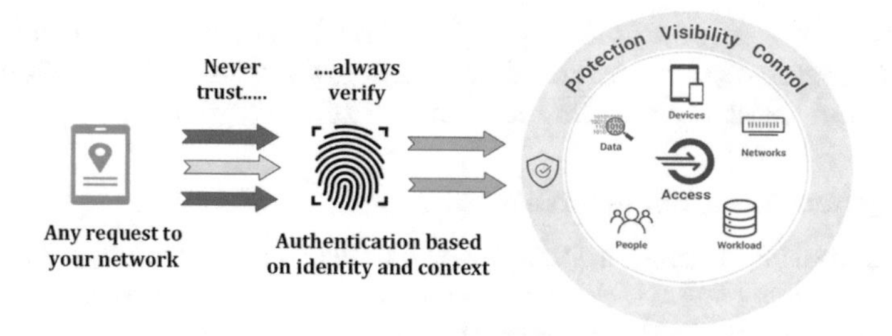

Fig. 2. Zero Test Security

2.5 Limitations on Bandwidth and Network Latency

The connection between edge devices and central security components may experience network latency and capacity limitations because edge computing environments operate on scattered networks. Continuous authentication and permission checks are only two examples of zero-trust security measures that might incur extra network overhead and reduce system responsiveness and effectiveness. For implementation to be successful, minimizing latency and maximizing network capacity consumption are essential [1].

In order to overcome these obstacles, security needs, resource limitations, and performance concerns must all be carefully balanced. Utilizing lightweight cryptographic algorithms, effective authentication protocols, and enhanced network communication techniques are just a few examples of technology and solutions organizations must use. For successful Zero Trust Security methods to be developed and implemented in edge computing settings, cooperation between security specialists, edge device makers, and network providers is crucial.

3 Solutions to Strengthen Zero Trust Security in Edge Computing Environments

This section discusses the Zero Trust model for enhancing security in edge computing systems. It highlights strategies like robust identity and access management, data protection, and encryption. It also emphasizes secure network segmentation, continuous threat detection, ongoing monitoring, device lifecycle management, and policy-based security enforcement. These measures ensure resilience against cyber threats, maintain data integrity, and enhance organizations' defensive posture in the dynamic edge computing landscape.

3.1 Identification and Access Control

A reliable identity and access management (IAM) solution must be implemented for Zero Trust Security in edge computing. Robust authentication techniques, such as multi-factor authentication and granular access restrictions, are required to guarantee that only authorized people and devices may access resources at the edge. IAM systems offer centralized administration of user identities and access privileges and assist in upholding the concept of least privilege [13].

3.2 Segmenting a Secure Network

In edge computing settings, segmenting the network into isolated, secure areas is crucial for reducing the attack surface and controlling possible security breaches. Network segmentation aids in the creation of distinct security domains based on levels of trust, where access across segments is rigorously regulated and tracked. By doing this, threats are stopped in their tracks, and the effect of any hacked devices is lessened.

3.3 Threat Detection and Ongoing Monitoring

Security hazards are identified and immediately mitigated thanks to ongoing monitoring and threat detection systems. In order to spot abnormalities, invasions, or unauthorized access attempts, this requires watching network traffic, device behavior, and user activity. It is possible to quickly identify potential risks by utilizing advanced security analytics, machine learning, and anomaly detection approaches [15].

3.4 Data Protection and Encryption

Effective encryption techniques must be implemented for both data in transit and at rest to safeguard sensitive data in edge computing settings. Data encryption guarantees that unauthorized parties cannot read or use it, even if it is intercepted or hacked. To stop unauthorized data access, leakage, or manipulation, data protection methods, including data loss prevention (DLP) approaches, can be used [3].

3.5 Lifecycle Management and Secure Device Provisioning

Throughout their entire existence, edge devices must be kept secure. It includes secure device provisioning, in which gadgets are safely connected to the network and their integrity and validity are checked. Continuous device lifecycle management often entails updating, managing patches, and checking for vulnerabilities. The integrity and reliability of edge devices can be increased using secure boot procedures and remote attestation systems [12].

3.6 Instruments for Authentication and Authorization

Zero Trust Security is based on implementing robust procedures for authentication and authorization. It calls for protocols like OAuth, OpenID Connect, or SAML for safe authentication and building trust between entities. Granular access regulations can be enforced based on user roles, device attributes, or contextual considerations using role-based access control (RBAC) and attribute-based access control (ABAC).

3.7 Security Enforcement Based on Policy

Policy-based security enforcement ensures consistent use of security precautions in edge computing environments. It involves developing and implementing policies based on company needs, legal constraints, and risk analyses. Automated policy enforcement systems can dynamically deploy security measures [9]. Organizations can manage risks related to Zero Trust Security by improving their edge computing environments. Regular security evaluations, testing, and upgrades are necessary for the solutions to be successful and resilient.

4 Integration of Zero Trust Security with Edge Computing Architectures

This section explores the integration of Zero Trust Security into edge computing architectures, a crucial aspect of modern cybersecurity. It highlights the need for robust security frameworks, leveraging advanced technologies like Zero Trust frameworks, software-defined networking, and micro-segmentation. It also discusses the components of this integration, such as Zero Trust frameworks, networking, micro-segmentation, and edge security gateways and firewalls. The section emphasizes the importance of threat intelligence and cloud-based security services in enhancing overall security.

4.1 Systems and Technologies with Zero Trust

Zero trust frameworks and technologies are the critical building blocks to deploy zero trust security in edge computing architectures. These frameworks offer a set of guiding principles, best practices, and architectural elements for developing and putting into practice zero trust security. One such framework is the Zero Trust Architecture (ZTA) [4]. Thanks to technologies like software-defined perimeter (SDP) and Zero Trust Network Access (ZTNA), access is allowed depending on user identification, device health, and contextual variables. These technologies offer granular access control and authentication techniques.

4.2 Networking and Micro-segmentation that is Specified by Software

Microsegmentation and software-defined networking (SDN) are essential for establishing Zero Trust Security in edge computing systems. SDN makes virtualized network segments and logical separation possible by providing dynamic network setup and control. Microsegmentation expands on this idea by segmenting the network into smaller sections to provide granular control and resource separation. By containing possible breaches or assaults, this strategy aids in limiting lateral network movement [16].

4.3 Solutions for the Edge Security Gateway and Firewall

Solutions for firewalls and edge security gateways are crucial to integrating edge computing architectures with Zero Trust Security. By providing perimeter protection at the edge, these solutions act as the first line of defense against dangers from the outside. Firewalls and edge security gateways enact access regulations, monitor network activity, and weed out illicit or illegal activity. From many forms of threats, including DDoS, malware, and intrusion attempts, they assist in defending the edge infrastructure and linked devices.

4.4 Intelligence About Threats and Cloud-Based Security Services

Cloud-based security services and threat intelligence are essential to improve Zero Trust Security in edge computing architectures. These services offer scalable and centralized security management, real-time monitoring, automatic response systems, and enhanced threat detection capabilities. Threat intelligence feeds provide the latest information about potential attacks and weaknesses, enabling proactive security actions and incident response [8]. Integrating edge computing architectures with Zero Trust Security using a multi-layered strategy, including Zero Trust frameworks, software-defined networking, edge security gateways, and cloud-based security services, is crucial for ensuring robust security solutions. Figure 3 illustrates an edge-to-cloud computing architecture that enhances system performance by distributing computing resources across cloud, fog, and edge layers, allowing IoT devices to efficiently process data locally.

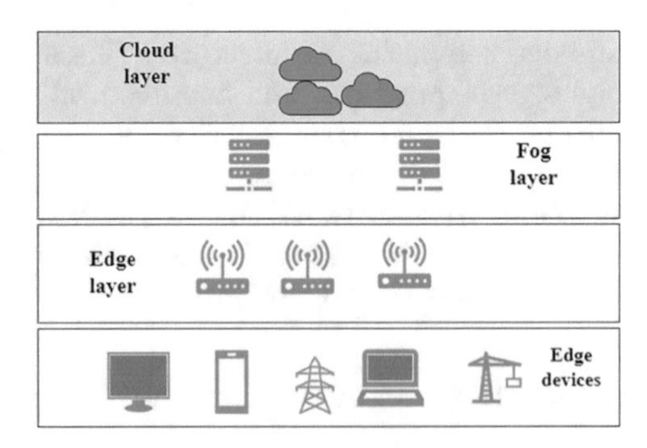

Fig. 3. Edge-to-cloud architecture layers

5 Case Studies and Recommended Methods

This study explores the implementation of Zero Trust Security frameworks in edge computing environments, including smart cities, industrial IoT, and remote healthcare. It provides real-world case studies, insights, and recommendations for organizations aiming to adopt this approach. The study emphasizes the need for a resilient and secure framework in edge computing systems' dynamic and complex nature, serving as a crucial guide for organizations navigating this advanced technological landscape.

5.1 Implementing Zero Trust Security in Edge Computing Use Cases

In this part, organizations that have effectively implemented Zero Trust Security in their edge computing environments are studied through case studies from the actual world. It emphasizes the difficulties these use cases encountered, the solutions chosen, and the results obtained. These case studies include essential information on how Zero Trust Security may be successfully used in many edge computing contexts, including smart cities, industrial IoT, and remote healthcare [6].

5.2 Observations and Current Issues

The difficulties and lessons gained in implementing zero trust security in edge computing settings are unique. The real-world difficulties businesses encountered during the implementation process are covered in this part, along with the priceless insights businesses learned. Some of these difficulties are integrating legacy systems, controlling intricate network topologies, meeting compliance standards, and managing the dynamic nature of edge devices and workloads. Organizations may better prepare for their Zero Trust Security projects by comprehending these difficulties and taking lessons from actual experiences.

5.3 Recommendations for the Adoption of Zero Trust Security

The suggestions for organizations wishing to implement Zero Trust Security in their edge computing environments based on the case studies and lessons gained must address various topics, including organizational alignment, technology choice, strategy and planning, and continuous maintenance. Conducting a thorough risk assessment, developing clear access control rules, deploying strong identity and access management systems, routinely upgrading security controls, and fostering a culture of security awareness and accountability are some critical suggestions that may be made [7].

Organizations may improve the effectiveness of their deployment of Zero Trust Security in edge computing settings by studying case studies, comprehending the difficulties that organizations confront, and adhering to the advised best practices. In the special setting of edge computing, these observations and suggestions reduce risks, strengthen security posture, and guarantee the effective implementation of zero-trust concepts. With the changing environment of edge computing and the emergence of new threats, it is crucial to evaluate and modify security measures continuously.

6 Zero Trust Security for Edge Computing: Future Directions and Emerging Trends

The demand for security products specifically designed for edge settings is expected to rise as edge computing develops. Future research will focus on creating edge-native security frameworks and tools that can interact with dispersed and heterogeneous edge architectures. Automation and orchestration will be crucial for implementing and managing Zero Trust Security in edge computing deployments. Intelligent automation systems and solutions will be created to dynamically enforce security policies, authenticate and authorize edge devices, and quickly identify and address security problems.

Integrating artificial intelligence (AI) and machine learning (ML) techniques will improve the capabilities of Zero Trust Security in edge computing settings. These technologies will enable large-scale edge data analysis, abnormal behavior detection, and the identification of new threats. These technologies will make possible predictive security measures, adaptive access restrictions, and proactive threat intelligence.

Privacy protection is increasingly essential in edge computing, as it involves processing sensitive data close to the edge. Future efforts will focus on integrating privacy-protecting methods into edge computing's Zero Trust Security frameworks, including differential privacy, secure multi-party computing, and federated learning.

Blockchain technology's decentralized and immutable characteristics will be used to improve security and trust in edge settings. Blockchain-based identity management, transaction validation, and distributed consensus techniques may be added to edge computing installations. Compliance with industry-specific rules and standards is vital for edge computing in various sectors. Collaboration and standardization initiatives will ensure interoperability and compatibility among edge platforms and security solutions. The future of Zero Trust Security will focus on addressing the unique challenges and demands of edge settings.

7 Conclusions

Improving zero-trust security in edge computing environments is challenging due to heterogeneous environments, resource scarcity, dynamic devices, scalability issues, network latency, and bandwidth restrictions. Robust authentication procedures, efficient identity and access management, secure network segmentation, ongoing monitoring, threat detection systems, data encryption, secure device provisioning, and data protection strategies can be implemented to address these issues. Policy-based security enforcement is crucial for uniform security measures. Integration of Zero Trust Security with edge computing architectures includes software-defined networking, micro-segmentation, edge security gateways, firewall solutions, cloud-based security services, and threat intelligence.

Future developments in Zero Trust Security for edge computing include edge-native solutions, automation, AI, machine learning, privacy-preserving methods, blockchain integration, compliance, regulatory considerations, and collaboration. Organizations can enhance security in dispersed computing environments by addressing issues and implementing recommended solutions, offering specialized and practical solutions to address the security challenges.

Acknowledgement. This work is funded by FCT/MEC through national funds and co-funded by FEDER—PT2020 partnership agreement under the project **UIDB/50008/2020**. This article is based upon work from COST Action CA21118 - Platform Work Inclusion Living Lab (P-WILL), supported by COST (European Cooperation in Science and Technology). More information on www.cost.eu.

References

1. Alagappan, A., Venkatachary, S.K., Andrews, L.J.B.: Augmenting zero trust network architecture to enhance security in virtual power plants. Energy Rep. **8**, 1309–1320 (2022)
2. Ali, B., Hijjawi, S., Campbell, L.H., Gregory, M.A., Li, S.: A maturity framework for zero-trust security in multiaccess edge computing. Secur. Commun. Netw. **2022** (2022)
3. Cheng, R., Chen, S., Han, B.: Towards zero-trust security for the metaverse. IEEE Commun. Mag. (2023)
4. Federici, F., Martintoni, D., Senni, V.: A zero-trust architecture for remote access in industrial IoT infrastructures. Electronics **12**(3), 566 (2023)
5. He, Y., Huang, D., Chen, L., Ni, Y., Ma, X.: A survey on zero trust architecture: challenges and future trends. Wirel. Commun. Mob. Comput. **2022** (2022)
6. Manan, A., Min, Z., Mahmoudi, C., Formicola, V.: Extending 5G services with zero trust security pillars: a modular approach. In: 2022 IEEE/ACS 19th International Conference on Computer Systems and Applications (AICCSA), pp. 1–6. IEEE (2022)
7. Mohammed, S., Rangu, S.: To secure the cloud application using a novel efficient deep learning-based forensic framework. J. Interconnection Netw. 2350008 (2023)
8. Phiayura, P., Teerakanok, S.: A comprehensive framework for migrating to zero trust architecture. IEEE Access **11**, 19487–19511 (2023)
9. Ray, P.P.: Web3: a comprehensive review on background, technologies, applications, zero-trust architectures, challenges and future directions. In: Internet of Things and Cyber-Physical Systems (2023)
10. Saleem, M., Warsi, M., Islam, S.: Secure information processing for multimedia forensics using zero-trust security model for large scale data analytics in SAAS cloud computing environment. J. Inf. Secur. Appl. **72**, 103389 (2023)
11. Sangaraju, V.V., Hargiss, K.: Zero trust security and multifactor authentication in fog computing environment. Available at SSRN 4472055
12. Sethi, P.S., Jain, A.: Edge computing. IN: Future Connected Technologies: Growing Convergence and Security Implications, p. 162 (2023)
13. Sharma, J., Kim, D., Lee, A., Seo, D.: On differential privacy-based framework for enhancing user data privacy in mobile edge computing environment. IEEE Access **9**, 38107–38118 (2021)
14. Sharma, R., Chan, C.A., Leckie, C.: Probabilistic distributed intrusion detection for zero-trust multi-access edge computing. In: NOMS 2023-2023 IEEE/IFIP Network Operations and Management Symposium, pp. 1–9. IEEE (2023)

15. Wang, J., Xu, J., Li, J., Xu, C., Xie, H.: Research on trusted security protection method for edge computing environment (2023)
16. Yang, Y., Bai, F., Yu, Z., Shen, T., Liu, Y., Gong, B.: An anonymous and supervisory cross-chain privacy protection protocol for zero-trust IoT application. ACM Trans. Sens. Netw. (2023)

Blockchain Adoption in Education with Enhancing Data Privacy

Khadeejah Abdullah[(✉)], Kassem Saleh, and Paul Manuel

Kuwait University, Kuwait, Kuwait
lkhadeejah.abdullah@ku.edu.kw

Abstract. Blockchain technology adoption has influenced several fields, including healthcare, financial, and supply chain systems. Lately, this technology has expanded its application in education because of its unique attributes, which are decentralization, immutability, consensus, and security. Despite the bright side of centralized credentialing and implementation, there are some challenges that need to be addressed. Thus, a comprehensive literature review of the reasons that blockchain adoption in education field is considered necessary and effective. To do so, we investigate and explore the benefits and gaps of decentralized credentialing systems that are utilized to securely issue, verify, and share digital credentials such as grades, degrees, diplomas, and certificates through simulation of blockchain and smart contract embedded logic. In this paper, we focus on two major themes: adoption of blockchain in education and enhancing the data privacy on blockchain through means of consensus algorithm. We express and analyze the results conducted as experimental simulations by stakeholders which are university administrations, professors, and students, when combining the two themes to propose a solution in terms of trust and adoption to enhance the privacy and security of student learning data and maintain individual privacy rights. we have discussed the methodology which is Design Science Research (DSR) and evaluated the results to emphasize on the significance of blockchain in education. We believe there is future work in enhancing the logic and validation of smart contract through means of hashing and cryptography.

Keywords: Blockchain · Smart Contract · Consensus · Hash · Distributed Ledger · Timestamp · Immutability · P2P

1 Introduction

A whitepaper [18] was published by Satoshi Nakamoto in 2008 revealing Bitcoin and its underlying technology, named blockchain: Blockchain [1, 19] is referred as a distributed database that chronologically store a chain of data sealed into blocks that run in a secure immutable manner, known as ledger. New blocks are appended to the end of the chain, in which each block hold reference to the previous block content using SHA256 hash algorithm [13], ensuring immutability and compactness of block. This startling concept has attracted a lot of attention from the financial industries. Following the financial institutions, different fields such as healthcare, software industries, and education began

Á. Rocha et al. (Eds.): WorldCIST 2024, LNNS 987, pp. 445–455, 2024.
https://doi.org/10.1007/978-3-031-60221-4_42

to introduce and apply this technology. Blockchain is a technology that eliminate the presence of centralized authority as all blockchain operations and transactions must be stored securely on a decentralized distributed ledger, as a result, blockchain consist of different layers, and each layer comprise of its unique functionality and objective such as infrastructure, consensus algorithm, smart contract, and application. In financial institutions, to this date, processing of transactions traditionally utilizes intermediation or third-party collaboration. However, collapse of the investment bank Bear Sterns in 2008 [17] has established the necessity and reliability of distributed digital transaction. This is the starting point towards blockchain which is built on proven hash signatures and digital cryptographic algorithms. Blockchain [16] can maintain the history of all transactions by means of timestamp and asymmetric cryptography. Another notable institution that adopted the blockchain technology is education [1]. The future of learning will be upgraded using blockchain technology in terms of data privacy and usability. By means of this technology, teachers, university administrators and students create academic records in a chronological list of events in real time and store the records in an immutable ledger with high degree of visibility and transparency.

The content of the ledger is synchronized across peer-to-peer (P2P) network as a distributed ledger. There are three main types of blockchains, which are public (not permissioned), private (permissioned), and consortium blockchain (hybrid) [18].

1.1 Problem Statement

A vast amount of sensitive data about students is collected and stored in educational institutions in centralized data centers, including personal information, academic records, and financial data. Any failure in privacy and security on the students' data management will be a disaster to the education institutions.

1.2 Research Hypothesis

Blockchain applied in education can revolutionize the learning experience and the data management of students' data. It will protect the data privacy of students and teachers while allowing participants to reach a consensus by creating an immutable record in the distributed ledger.

1.3 Research Objectives

- Study the impacts of privacy and safety of student's data in transit under a blockchain environment by simulation.
- Assess how blockchain can overcome the current limitations in the education system.

1.4 Research Outcomes

- Enhanced data privacy of student data and related documents.
- Optimized processes and transactions under the blockchain environment.

2 Literature Review

The evolution of blockchain began with cryptocurrency and has paved the way toward developing smart contracts in areas such as healthcare [10], supply chain, real-estate management etc. The emphasis on blockchain adoption is due to its ability to build a trusted and decentralized environment. Hence, the higher education sector is a potential user for implementing blockchain technology because of its capability to give control to stakeholders to validate learning records and identity management [1–5]. For instance, different institutions can decide with which other institutions to share data to avoid falsified grades or certificates. Additionally, the removal and lack of need for a trusted third party can introduce distributed ledgers that improve smart-contract based protocols throughout multiple levels of administration which is automatically enforced for the higher education field with the ability to ease processes while mitigating probability of error.

A global blockchain-based higher education credit platform, named EduCTX, is based on the concept of the European Credit Transfer and Accumulation System (ECTS), which offer a globally unified viewpoint for students and higher education institutions (HEIs) involving potential stakeholders, such as companies and organizations [2]. Basically, it presents a prototype implementation of the environment based on the open-source Ark Blockchain platform as a distributed peer-to-peer network that process, manage, and control ECTX tokens, which represent student credits that they gain for completed courses. EduCTX utilizes the benefits of blockchain, for instance, a decentralized architecture, security, immutability, integrity, transparency, anonymity, and longevity, to create a global grading system using proof of concept prototype. The adoption and implementation of EduCTX faces challenges in terms of data privacy due to the fact that students' academic records are sensitive and have complicated management regulations [2].

The process in which a next block is generated by solving a complex cryptographic puzzle using lot of computational power is known as Proof of Work (PoW) consensus [5], while Byzantine Fault Tolerance is an algorithm objective is to safeguard the system from failure by employing collective decision making (correct and faulty nodes) to limit the influence of the faulty nodes, and it is based on Byzantine Generals' problem [11, 18]. There is a common alternative to PoW, which is Proof of Stake (PoS) that Etherum have shifted to because this type of consensus rather than invest on expensive hardware, it employs validators that invest in the coins of the system by locking up them as stakes, and a validator chooses to generate a new block based on its economic state in the network. The PoS mechanism requires validators to lock up their assets in a smart contract, hence, blockchains utilizing PoS mechanism must ensure their smart contracts are secure. The consequences of dishonesty in PoS and PoW are similar, as validators lose staked assets in PoS, whereas, in PoW, they lose money spent on hardware and power.

3 Methodology

3.1 Design Science Research

In this paper, we select the Design Science Research (DSR) problem solving paradigm that aim to improve human knowledge by following the creation of innovative artifacts, DSR seek to create artifacts to enhance technology and science knowledge for a problem in an environment which lead to results that include both design knowledge and newly designed artifacts providing better understanding in the design theory of how the artifact can enhance the relevant concept and context [12]. The artifacts are designed as sequence diagrams with activity diagram explaining the algorithm on building blockchain for education.

3.2 Network Design Description

In education, we can apply blockchain following a benchmark and best practices as the amount of encryption is Asymmetric-key algorithm and hash function across every node, privacy of data is secured in every block due to the fact every node consist of private key and public address, and when a node takes part in any transaction, only public address get shared, consensus algorithm is embedded to ensure identity verification, where we can utilize legal framework ISO27001 in conjunction with GDPR to protect Personal Identification Information (PII) following a smart contract as a function of blockchain which is validated and implemented then shared across a peer-to-peer network to form distributed ledger technology. The smart contract policy is established with cryptography as a cybersecurity control method, and it provides a powerful capability of code execution for embedding business logic on the blockchain. The smart contract could be directed toward three main classifications of blockchains, such as public, private, and permissioned [14].

The artifact design or diagram emphasizes the role of smart contract in the contract layer as an algorithm and mechanism to provide a robust framework for managing student enrollments, exam results, and certifications on the blockchain which allow for seamless payment and certification process [3].The smart contract for education is a self-executing agreement with predefined conditions, and the contract can optimize learning experience with accepting payments in certain cryptocurrency that students can hold in their digital wallet. We can introduce smart contracts to the educational institutions that provide transparency. We can leverage the verifiable records of grades and achievements of students by complying with regulations frameworks such as, ISO27002 and GDPR/DPA for data privacy and security.

The Figs. (1, 2, 3 and 4) illustrate the flow of student processes chronologically throughout the phases of their academic life, for instance, when the student graduate from high school and apply to a program in university. After graduation, student decide to apply for a job at an employer, or the student decide to continue studying at a higher education program. In all cases, a blockchain is utilized to store student data securely with integration with application blockchain firewall to validate compliance with regulation frameworks, such as ISO27001 and GDPR/DPA. Student data are decentralized and available over the permissioned blockchain that communicate directly with universities amongst each other, and for employment, facilitating transfer of student data reliably and securely.

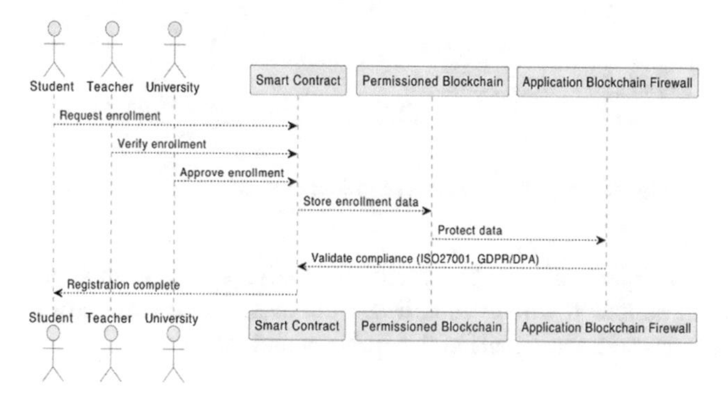

Fig. 1. Sequence diagram of high-school student application to university

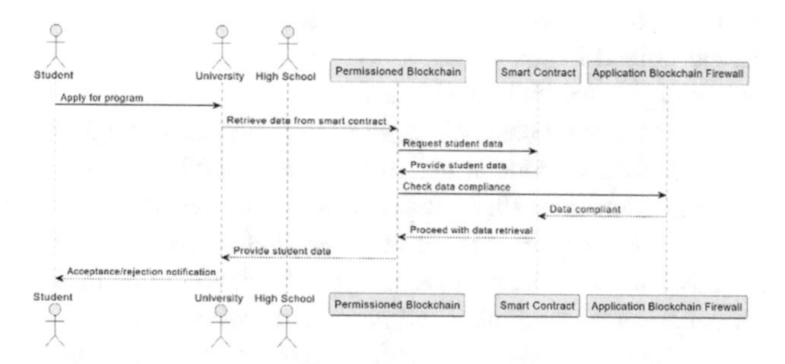

Fig. 2. Sequence diagram of student registration to university program

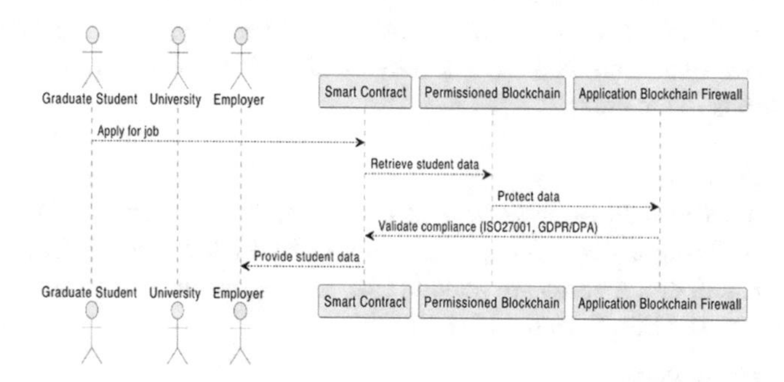

Fig. 3. Sequence diagram of graduate student application for employment

Figures (1, 2, 3 and 4) showcase how flow of activities is happening when we implement blockchain technology in adherence with school processes and employment. The actors are student, university, teacher, and employer, they undergo validation and verification procedure when smart contract request for data retrieved in the blockchain.

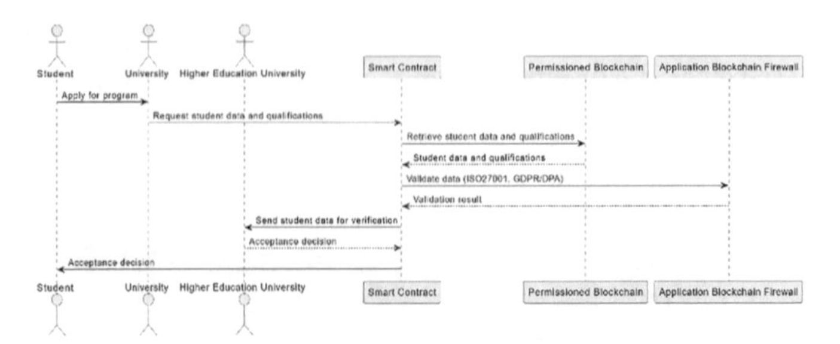

Fig. 4. Sequence diagram of graduate student application for

And the blockchain is protected by specific firewall-based blockchain to ensure traffic in transit is not tampered with or modified. Moreover, the data are cross-checked with regulatory frameworks for enhanced data privacy and integrity.

3.3 Construction

Simulating a blockchain in education involves creating a simplified version to demonstrate how a decentralized ledger could be used for tasks like storing student records securely. In this example, we will provide a simple Python simulation of a blockchain for educational records using a basic block structure and hashing mechanism.

In addition to blockchain code simulation, we add another security layer and that is Blockchain Application firewall, it inspects traffic and instills rules and policies to direct the incoming and outgoing traffic between nodes where security is optimized and monitored, this allows for best-practice cyber hygiene in controlling the way this permissioned blockchain transmit data and perform transactions.

This smart contract allows users to transfer tokens between addresses, and the token balances are stored on the blockchain. This is a simple example, and real-world smart contracts can be significantly more complex depending on the use case. To simulate the interaction with this smart contract, we would typically deploy it on a blockchain testnet (like Ropsten for Ethereum) and use a development environment or a tool like Remix to send transactions to the contract.

Consensus Algorithm:
The algorithm outlines the basic operations performed by each function in the provided Python code, offering a step-by-step explanation of how the blockchain is initialized, blocks are created, and records are added [13]. The algorithm is represented in a flow activity diagram, see Fig. 5. Here is the algorithm [3]:

1. **Initialize Blockchain:**

   – Create an empty list for the blockchain (**chain**).
   – Create an empty list for the current records (**current_records**).
   – Create the genesis block using - the **create_block** function with a predefined previous hash.

2. **Create Block Function:**
   - Input: **previous_hash** (hash of the previous block).
   - Create a new block with the following attributes:
     - **index**: Incremented index based on the length of the blockchain.
     - **timestamp**: Current timestamp.
     - **records**: List of records from the current block.
     - **previous_hash**: Hash of the previous block.
     - Reset the **current_records** list.
     - Append the new block to the blockchain.
     - Return the created block.
3. **Create Record Function:**
   - Input: student_id, course, and grade.
   - Create a new record with the given attributes.
   - Append the record to the **current_records** list.
   - Return the created record.
4. **Hash Function:**
   - Input: **block** (a block from the blockchain).
   - Convert the block to a JSON string and encode it.
   - Calculate the SHA-256 hash of the encoded string.
   - Return the hash.
5. **Get Last Block Function:**
   - Return the last block in the blockchain.

The Blockchain class represents the entire blockchain. And Each block contains an index, timestamp, a list of educational records, and the hash of the previous block. The create_record method adds an educational record to the list of current records. The create_block method creates a new block with the current records and adds it to the chain. The hash method generates the SHA-256 hash of a given block [20].

This is a basic illustration, and a real-world implementation for educational records, it would require additional features, security considerations, and likely the use of a specialized blockchain platform. This simulation serves as a conceptual starting point for understanding the principles of blockchain in an educational context.

Simulating a smart contract in a blockchain involves creating a simplified version to demonstrate how the contract can be deployed and interacted with. Below is a basic example of a smart contract written in Solidity, the programming language for Ethereum smart contracts. This example represents a simple token contract component [14].

- **SimpleToken** is a basic smart contract representing a token system.
- The **balanceOf** mapping stores the token balances for each address.
- The **Transfer** event is emitted whenever tokens are transferred.
- The **constructor** initializes the token supply with the address deploying the contract.
- The **transfer** function allows users to transfer tokens to other addresses.

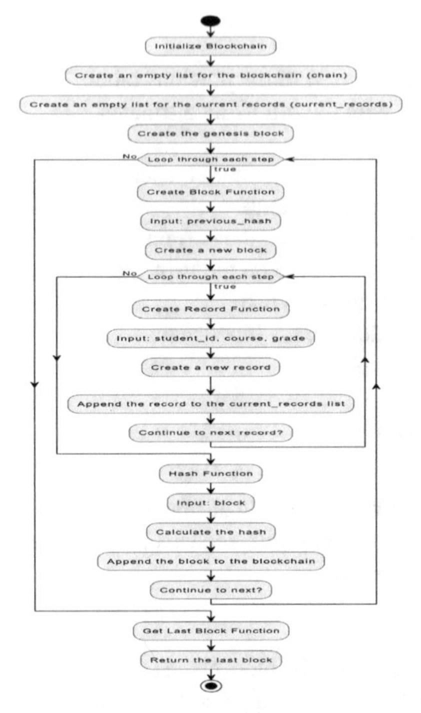

Fig. 5. Activity diagram of blockchain algorithm

3.4 Results

In a blockchain simulation for education, the results would typically be a series of blocks containing educational records. The simulation could be visualized or analyzed to demonstrate how a decentralized ledger could store and secure student records. Below is a hypothetical finding of what the results might look like after simulation:

The first block is the genesis block, containing no educational records and a previous hash of "1". The second block contains a record for student 123 who took the Math course and received a grade of A. The hash of the previous block is used to link the blocks. The third block contains a record for student 456 who took the History course and received a grade of B. The hash of the second block is used as the previous hash.

In a real-world scenario, each block would be cryptographically linked to the previous one, ensuring the integrity and immutability of the educational records. The simulation provides a visual representation of how a blockchain could be used to securely store and link educational data over time.

The findings found in these simulations are that which relate to how blockchain operate and their classes with respect to functions and objects representing the education processes such as transferring grades and records in conjunction with tokenized operations that are a result of smart contract, as a result, we obtained a proposed solution to overcome current centralized systems such as security concerns that involve transparency and immutability, hence, we can validate the hypothesis that blockchain in education does improve security and performance.

PWC reports that 24% of executives are initiating investments in blockchain to enhance the education system [15]. In 2020, Europe held a significant market share of 27.3% in the blockchain in the education sector. Projections indicate that educational institutions will adopt blockchain technology at a compound annual growth rate (CAGR) of 16.0% by 2026. According to JRC science hub, the implementation of blockchain in the education sector has the potential to reduce operational costs by over 5%. Despite the promising outlook, a survey by Gartner reveals that only 1% of Chief Information Officers (CIOs) have indicated any form of blockchain adoption within their educational organizations [15].

3.5 Discussion

Applying blockchain technology with smart contracts in education holds the promise of significantly improving both performance and security within academic systems. By leveraging the decentralized and tamper-resistant nature of blockchain, educational institutions can establish a robust framework for managing student records, course data, and administrative processes. The use of smart contracts automates various tasks, streamlining workflows, and reducing the reliance on manual intervention. This not only enhances overall system performance by minimizing processing delays but also significantly reduces the risk of human errors in data entry and record-keeping. The transparency and immutability of blockchain contribute to enhanced security, ensuring the integrity of academic records and protecting sensitive information. Additionally, the decentralized nature of blockchain mitigates the risk of a single point of failure, providing a resilient and secure environment for academic data. While challenges such as integration complexity and regulatory compliance need careful consideration, the potential for better performance and heightened security in education through blockchain and smart contracts is a compelling prospect that merits further exploration and implementation.

3.6 Evaluation

When applying a smart contract in a blockchain for enhancing data security, the results can be demonstrated through various improvements in the handling and storage of sensitive information. Here is a conceptual representation of the results:

Immutability:
Smart contracts on a blockchain provide data immutability. Once data is recorded, it cannot be altered or deleted. And as a result, protection is enhanced against unauthorized tampering or modification of critical information [6].

Decentralization:
Blockchain operates on a decentralized network of nodes, reducing the risk of a single point of failure or malicious attacks. And as a result, resilience is increased and security optimized due to the absence of a central authority that could be exploited [7].

Transparency:
Smart contracts facilitate transparent and auditable transactions. All interactions are transparent with the contract and are recorded on the blockchain and can be publicly verified. Hence, accountability and traceability of data access and modifications are improved [8].

Access Control:
Smart contracts can implement sophisticated access control mechanisms, ensuring that only authorized parties can interact with specific functions or data. Strengthening security by restricting access to sensitive data and operations [9].

Automation:
Smart contracts automatically execute predefined rules without the need for intermediaries. The trustless execution reduces the reliance on third parties or vendors. This allows to minimize trust issues and less probability of human error which increase efficiency in data [15].

Cryptography:
Blockchain relies on cryptographic techniques to secure transactions and ensure data confidentiality and integrity. Therefore, data security is enhanced through robust cryptographic procedures, protecting sensitive information from unauthorized access [11].

It's important to emphasize that the effectiveness of data security measures in a blockchain depends on the specific implementation, configuration, and the underlying blockchain platform. Additionally, ongoing monitoring, testing, and updates are essential to address emerging security challenges.

4 Conclusion

The integration of blockchain technology and smart contracts in education offers a transformative approach to securing and validating school-related items. By leveraging the immutability, transparency, and automation capabilities of these technologies, educational institutions can create a more secure, efficient, and trustworthy system for managing academic data. While challenges exist, the potential benefits make blockchain and smart contracts a promising avenue for enhancing security and validation in education. Ongoing research, collaboration, and technological advancements will play a crucial role in further refining and optimizing these solutions for widespread adoption in the education sector.

Acknowledgement. I would like to express my gratitude to the Kuwait Foundation for the Advancement of Sciences (KFAS) for their generous support and funding, which were crucial to the completion of this research.

References

1. Raimundo, R., Rosário, A.: Blockchain System in the Higher Education. MDPI, 16 Mar (2021). www.mdpi.com/2254-9625/11/1/21
2. EduCTX: A blockchain-based Higher Education Credit Platform [Internet]. IEEE. [cited date not available]. https://ieeexplore.ieee.org/abstract/document/8247166
3. Solidity Academy. Revolutionizing Education with Smart Contracts. Medium. 2023 Jun 29. medium.com/@solidity101/revolutionizing-education-with-smart-contracts-742e570a414f
4. Alammary, A., Alhazmi, S., Almasri, M., Gillani, S.: Blockchain-based applications in education: a systematic review. Appl. Sci. **9**(12), 2400 (2019). https://doi.org/10.3390/app9122400
5. Kholishotulaila, S., Laila, K., Angga, A.L.: Benefits provided by blockchain technology in the field of education. Blockchain Front. Technol. **1**(2), 74–83 (2022). https://doi.org/10.34306/bfront.v1i2.59
6. Llambias, G., González, L., Ruggia, R.: Blockchain interoperability: a feature-based classification framework and challenges ahead. Clei Electronic J. **25**(3) (2023). https://doi.org/10.19153/cleiej.25.3.4
7. Fan, S., Min, T., Xiao, W., Cai, W.: Altruistic and profit-oriented: making sense of roles in web3 community from an airdrop perspective (2023). https://doi.org/10.1145/3544548.3581173
8. Gulen, K., Karaagac, A.: Agricultural food supply chain with blockchain technology: a review on Turkey. J. Global Strategic Manag. (2024). https://doi.org/10.20460/jgsm.2023.314
9. Almajed, H., Almogren, A.: Simple and effective secure group communications in dynamic wireless sensor networks. Sensors. **19**(8), 1909 (2019). https://doi.org/10.3390/s19081909
10. Barbaria, S., Mont, M., Ghadafi, E., Mahjoubi, H., Rahmouni. H.: Leveraging patient information sharing using blockchain-based distributed networks. IEEE Access. **10**, 106334–106351 (2022). https://doi.org/10.1109/access.2022.3206046
11. Yang, Z., Salman, T., Jain, R., Pietro, R.: Decentralization using quantum blockchain: a theoretical analysis. IEEE Trans. Quantum Eng. **3**, 1–16 (2022). https://doi.org/10.1109/tqe.2022.3207111
12. Peffers, K., Tuunanen, T., Rothenberger, M., Chatterjee, S.: A design science research methodology for information systems research. J. Manag. Inf. Syst. **24**(3), 45–77 (2007). https://doi.org/10.2753/mis0742-1222240302
13. Dutta, S., Saini, K.: Blockchain implementation using python. Adv. Syst. Analy. Softw. Eng. High-Performance Comput., 123–136 (2022). https://doi.org/10.4018/978-1-7998-8367-8.ch006
14. Pierro, G., Tonelli, R., Marchesi, M.: An organized repository of Ethereum smart contracts' source codes and metrics. Future Internet. **12**(11), 197 (2020). https://doi.org/10.3390/fi12110197
15. Essential Blockchain in Education Statistics in 2023, Zipdo. ZipDo, Sep 28 (2023)
16. Johar, S., Ahmad, N., Asher, W., Cruickshank, H., Durrani, A.: Research and applied perspective to blockchain technology: a comprehensive survey. Appl. Sci. **11**(14), 6252 (2021). https://doi.org/10.3390/app11146252
17. Grund, S., Nomm, N., Walch, F.: Liquidity in resolution: comparing frameworks for liquidity provision across jurisdictions. SSRN Electron. J. (2020). https://doi.org/10.2139/ssrn.3742257
18. Nakamoto, N.: Centralised bitcoin: a secure and high-performance electronic cash system. SSRN Electron. J. (2017). https://doi.org/10.2139/ssrn.3065723
19. Kurniawan, R., Oktaviani, D.: Characteristics of blockchain technology in educational development. blockchain Front. Technol. **1**(2), 23–28 (2022). https://doi.org/10.34306/bfront.v1i2.41
20. Tang, Q.: Towards using blockchain technology to prevent diploma fraud (2021)

Boats Imagery Classification Using Deep Learning

Dumitru Abrudan[1], Ana-Maria Drăgulinescu[2(✉)], and Nicolae Vizireanu[2]

[1] Doctoral School of Electronics, Telecommunications and Information Technology, National University of Science and Technology Politehnica Bucharest, 060042 Bucharest, Romania

[2] Telecommunications Department, National University of Science and Technology Politehnica Bucharest, 060042 Bucharest, Romania

ana.dragulinescu@upb.ro

Abstract. Sea transportation is the cheapest way to ship goods all over the world. This type of transportation involves legal and illegal transportation. As an example of illegal activities, we can exemplify drug transportation, immigrants transportation etc. Compared to road transportation where all the routes are connected and monitored by cameras installed in different places, in sea transportation this cannot be possible. Other methods are used for ship monitoring such as satellites. Due to fact that satellite imagery requires lots of resources and high costs, alternative methods can be used. An alternative method in areas close to coastal sites are RGB cameras placed in specific zones suitable for anchoring. In this paper, we proposed to identify different type of images with boats taken in the water. To this goal, we trained a convolutional neural network (CNN) on a dataset comprising images with different types of boats such as bowrider, deck, dinghy, ponton, runabout, vela. Due to fact that at the time of image acquisition noise can be inserted in images, mathematical morphology is used to remove this noise. We obtained 98.27% accuracy using our proposed method for boat classification.

Keywords: Mathematical Morphology · Image Processing · Deep Learning · Maritime applications · Surveillance

1 Introduction

Illegal activity detection is a frequent issue in coastal border areas. Criminal organizations specialized in activities as contraband of drugs, immigrant's traffic, illegal cargo movement etc. are constantly using these areas for their illegal purpose. Due to fact that these offenses affect the economy and the security of the citizens, the governments were forced to take immediate measures to stop and prevent further illegal activities given their major impact on economy, livelihoods, and environment [4].

The vehicles that operate by water can be classified in two types of vehicles: i) Surface vehicle (SV), ii) Underwater vehicle (UWV). SV are also categorized according to their utility, as mentioned by [13], such as:

Á. Rocha et al. (Eds.): WorldCIST 2024, LNNS 987, pp. 456–465, 2024.
https://doi.org/10.1007/978-3-031-60221-4_43

- Commercial. All forms of commercial vehicles that are dedicated to commerce, trade and used for industrial purposes are defined as being commercial vessels.
- Passenger and public vessels. In this category, there are included all vessels dedicated to passenger transport such as: ferries, water taxis, river cruises and so on.
- Private Vessels. Small to medium-sized vessels owned by a concern, organization, or a single person.
- Defense. Owned by countries as part of its military capabilities. These vessels range from frigates, warships, small patrol vessels and so on.
- Special Purpose Vessels. These vessels are dedicated for exploration or research. They are owned by government or institutions.

Taking into consideration that our method is based on images captured on the surface, the area of interest are private vessels. In the literature, different techniques were implemented to identify or to locate ships that run in all oceans, seas, rivers, lakes etc. Location or identification can be made using signals generated by vessel devices or RGB images.

The authors in [12] present a device for boat tracking called Distress Alert Transmitter, installed on fishing boats and used in case of emergency (medical, fire etc.). In case of emergency, the boat position is obtained through Global Positioning System (GPS). The important feature of this device is low cost, in-built GPS, signal transmission for 24 h with 5 min frequency. In addition, in the emergency center, data are transmitted with regards to vessel ID. This kind of device is suitable only in case of emergency situations. Another safety measure for crew to counteract terrorism or piracy is presented by [3], Ship Security Alert Systems. In case of emergency, a silent alarm is activated. A report which contains location information's is sent to specified locations. Part of these locations are the closest national authorities and/or military or law-enforcement forces. According to [11], an important part of the process which allows ships to transmit rescue information to authorities is the International Mobile Satellite Organization (INMARSAT/IMSO). For operation, INMARSAT uses satellite telecommunications. Also, beside collaboration with International Maritime Organization, there is also a collaboration with International Civil Aviation Organization.

The authors in [11] studied the system called automatic identification system (AIS) which consists in a transporter installed on a vessel. When the transporter is in function, it transmits data on a VHF channel which can be received by other vessels. The most important data transmitted are name of the vessel and course, longitude, latitude, speed, type of ship, length. A main issue to this kind of system is the switching off possibility, which could lead to potential illegal activities on the sea. Therefore, our paper proposes a non-AIS-based solution for boat recognition and include the following specific contributions:

- Creating a CNN for boat classification.
- Testing the CNN on images not affected by noise.
- Creating a mathematical morphology filter for boat images affected by noise.
- Testing the CNN on images affected by noise.

The paper is organized as follows: Sect. 2 describes the architecture of the CNN and the mathematical morphological filters. Next, Sect. 3 describes the proposed algorithm used for boat classification along with the obtained results and the discussions. Finally, in Sect. 4, the conclusions and the future improvements to be taken into consideration are presented.

2 Related Work

CNNs are used to carry out various image processing tasks and can fulfill with success the task of boats classification as mentioned by [8].

In [3], the authors propose an algorithm named Search for Unidentified Maritime Objects (SUMO) for vessel detection using satellite images. These images are captured from satellites such as RADARSAT, ERS and JERS. The images provided for processing are known as Synthetic Aperture Radar (SAR) images. This algorithm is a pixel-based Constant False Alarm Rate detector and is used for captured images supplied by radar. Sea conditions affect the SAR images generating high noise. In the process of vessel detection, due to a high level of noise, a large number of false alarms is generated and the accuracy for detection is lower. The authors conclude that, using the proposed method using SAR images, one can detect objects in the sea with a size between 1 m - which is the size of a buoy - to 450 m - which is the size of a super tanker.

In [15], the authors proposed a method for ship detection which included two modules: a pre-training module used for feature extraction and a fine-tuning module for the feature extraction. The images used in the process of feature extraction are SAR images. This method detects three types of commercial vessels: cargo, tanker and container. The accuracy rate for vessel classification using this method ranges between 75.90% and 91.70%.

In [9], the authors used a database of 31.455 images with six types of commercial vessels: cargo ship, ore carrier, fishing boat, bulk cargo carrier, passenger ship and container ship. These images were acquired by the monitoring cameras placed in an area close to the coastline. The work presents a comparison for the ship detection methods using the following detectors: YOLO v2, SSD (Single-Shot Detector) and R-CNN (Region-based Convolutional Neural Network), the latter using a region of interest during pooling, training, and sampling to improve accuracy and training used by [7]. R-CNNs have some drawbacks because they generate positive and negative samples during the process of network training. This leads to slowing down the speed of the algorithm. In addition, performance of the network is limited due to fact that, for each region of interest, features are extracted. Nevertheless, the best results for ship classification using YOLO v2, SSD and R-CNN detectors were obtained by R-CNN detector. R-CNN detector outperforms YOLO with 16.22% and SSD with 12.58%. A comparison was made between YOLO v2 network and region-based methods and the authors concluded that YOLO v2 was faster. The accuracy classification for different types of object ranges between 71.90% and 90.40% [6].

Christopher et al. [2] used an algorithm to detect illegal fishing boats based on lighting features arising from boats. The images acquired were taken in low light conditions. This system generates good results when the lunar illuminance is low, but in case of full moon conditions, it generates a large number of false detections. Moreover, a problem for misdetection can occur in case of boats which shut down the lighting bulbs. Due to the possibility of interference occurrence such as salt and pepper noise known also as impulse noise, it is necessary to remove this noise before using the images in a trained CNN for recognition.

The authors in [1,14] used mathematical morphology (MM) for image noise removal. The processed images with morphology operators are not affecting the essential shape characteristics. MM is a nonlinear filter, and it is suitable to be used for impulse noise removal. To restore an image corrupted with noise an open-close sequence filter is used. To test the efficiency for the same image, other filters are tested such as: median filter, center weighted median filter and noise adaptive median filter. From all these tested filters, the highest performance was obtained by the open-close sequence filter. To restore an image by removing the haze effect, in [5] convolution operation together with morphological operations was used. Haze effect is an atmospheric phenomenon which affect the quality of the captured images due to the presence of smoke, dust and other dry particularities suspended. Nonetheless, a convolution operator is linear while a morphological operator is, as we mentioned above, non-linear. The morphological operators used to build the network layers for image de-hazing are erosion and dilation. The structuring element used for erosion and dilation process has the shape of a rectangle with the size of 4×4. The output of the network is obtained using a sigmoid activation function.

Quantitative evaluation for image de-hazing is expressed in terms of peak signal-to-noise ratio (PSNR). Using a network, in which is used an opening operator followed by a closing operator the haze from images is removed, a high rate of PSNR value is obtained, being a superior in comparison with other methods (such as a closing network, or an opening network) for haze removal. For an image with resolution of $M \times N$ pixels, the PSNR can be computed based on Eq. (1).

$$PSNR(dB) = 10 \cdot lg \frac{255^2 \cdot M \cdot N}{\sum_{x=0}^{M-1} \sum_{y=0}^{N-1} [f'(x,y) - f(x,y)]^2} \tag{1}$$

3 Algorithm Description, Results and Discussions

The proposed method is based on the workflow in Fig. 1. The proposed algorithm was implemented using MATLAB R2018b. To train the CNN for boat type recognition, we used the dataset generated by [10] which consists in a set of 5523 RGB images. We extracted six types of boat classification and the images associated to a boat type were stored in corresponding folders. All the images from [16] dataset is unique and are the result of a hand-made cropping operation. The initial dataset consisted in about 800 images, but using data augmentation techniques (such as rotation, flipping, translation, padding, mirror), their number increased to 5520 images.

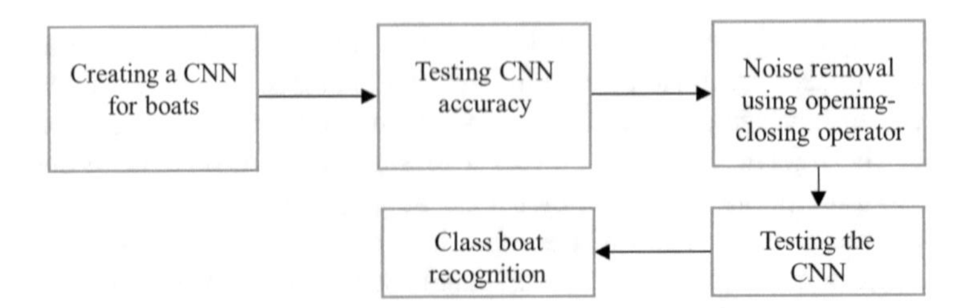

Fig. 1. Flowchart of the proposed algorithm

The classes of boats are as follow: bow rider, deck, dinghy, pontoon, runabout, vela. For each class, there are 920 images with the boat type which will be used in the process of training and validation (with a 75% of images used for training and 25% of images used for validation). Due to fact that image dataset did not have the same dimension, all the images were resized to a size of 227 × 227 pixels. The architecture of the proposed CNN for boat type recognition is presented in Fig. 2.

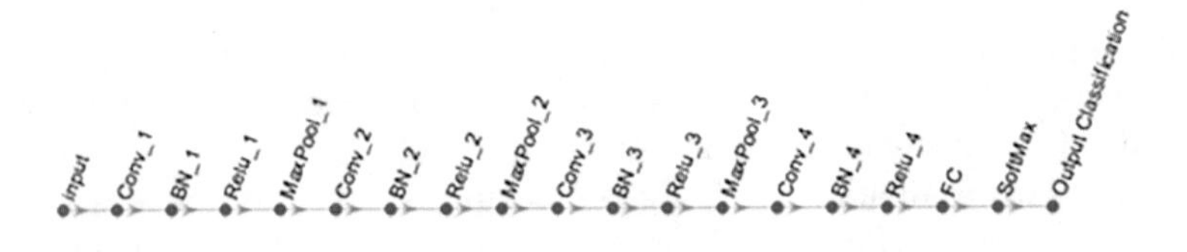

Fig. 2. CNN architecture for boat type recognition

The architecture of the CNN is composed of 5 convolution layers. In Table 1, for each layer, the values for the filter size and filters number are presented.

A convolutional layer received an image as input. The output is a new image which has in composition a defined number of channels. These channels are known as feature maps. The pooling layer is used to subsample the feature maps. The purpose of this layer is to replace the output generated by the convolutional layer with a summary statistic of the neighboring outputs. The most used types

Table 1. Layers values used in CNN

Layer no	Filter Size	Number of Filters
1	3	8
2	3	16
3	3	32
4	3	68
5	3	128

of pooling layers are average pooling, global average pooling, max pooling and min pooling. The pooling layer used in our CNN is max pooling. This operation computes the largest or maximum value from the region of the feature map which is covered by the selected filter.

The fully connected layer served as a classifier and was placed at the end of the CNN architecture. At this stage, each neuron is connected with all neurons in the previous layer.

The optimization algorithm used is Adam. At this time, it is the most used and, in comparison with other optimization algorithms, Adam is easy to be implemented, is faster, requires minimum tuning and does not require a large amount of memory to run. In the CNN training process, there were set the following parameters: learning rate equal to 0.001 and number of epochs were set to 2, 4 and 6. The number of epochs influenced the accuracy rate of the CNN as is presented in Table 2.

Table 2. Accuracy rate vs epochs number

Epoch number	Accuracy rate
2	88.25%
4	95.57%
6	98.27%

In our proposed method, the number of epochs took in consideration is equal to 6 due to the highest accuracy rate obtained during training and validation. Due to fact that number of epochs in matter of accuracy do not register significant values if there are more than 6, a bigger value for epoch was not took into consideration. Next, we compared the result of the classification for an image affected by noise and another which was not. In Fig. 3 (left), the boat type was recognized correctly which meant that the CNN was proper to be used for boat classification. In the case where noise is presented at the time of image acquisition, CNN cannot correctly recognize the boat type as can be seen in Fig. 3 (middle). To remove the noise from images and avoid misclassification occurrence, we used mathematical morphology filters. In Fig. 3 (right), the effect of a morphology filter called opening was presented. After noise removal, the CNN correctly recognizes the boat type.

Different morphology filters were used: dilation, erosion, opening and closing as in Table 3, where A denotes the original image and B denotes the structuring element. Opening operator involves two consecutive operations: dilation followed by erosion. The closing operator involves two consecutive operations: erosion followed by dilation.

The opening operator used in on our images affected by noise, has the effect to smooth the contours of object, so the image can be recognized by the CNN. The closing operator has the effect to smooth sections of contours. Due to fact that noise is presented in images in small zones, the opening operator used is

Fig. 3. Correct boat type (left), misclassification of boat type (middle) and correct recognition after noise removal.

removing enough noise from our images affected by noise, in a matter that our trained CNN can correctly recognize the correct recognition of boat type. Closing operator is suitable to restore connectivity between other objects from images which are closer to each other. Table 3 presents the efficiency of the proposed filters filers.

Table 3. Filters for noise removal efficiency

Filter used	Equation	Efficiency
Dilation operator	$A \oplus B = \{x \vert [(\hat{B})_x \cap A] \subseteq A\}$	Not efficient
Erosion operator	$A \ominus B = \{x \vert (B)_x \subseteq A\}$	Not efficient
Opening operator	$A \circ B = (A \ominus B) \oplus B$	Efficient. The boat type was recognized
Closing operator	$A \bullet B = (A \oplus B) \ominus B$	Not efficient

In Table 4, a confusion matrix to represent the CNN classification accuracy in the case when morphology operator called opening is used in the process of images noise removal.

Table 4. Confusion matrix. BD - Bowrider Deck, DEB - Deck boat, DIB - Dinghy boat, PB - Ponton boat, RB - Runabout boat, VB - Vela boat

		PREDICTION					
		BD	DEB	DIB	PB	RB	VB
ACTUAL	BD	**892 (97%)**	28 (3%)	0 (0%)	0 (0%)	0 (0%)	0 (0%)
	DEB	28 (3%)	**892 (97%)**	0 (0%)	0 (0%)	0 (0%)	0 (0%)
	DIB	0 (0%)	0 (0%)	**911 (99%)**	9 (1%)	0 (0%)	0 (0%)
	PB	0 (0%)	0 (0%)	9 (1%)	**911 (99%)**	0 (0%)	0 (0%)
	RB	0 (0%)	0 (0%)	0 (0%)	0 (0%)	**920 (100%)**	0 (0%)
	VB	0 (0%)	0 (0%)	0 (0%)	0 (0%)	0 (0%)	**920 (100%)**

Table 5 highlights the differences and similarities between our work and other approaches in the literature in terms of target boats identified, techniques employed and performances. The reference metric used to compare the systems was the accuracy.

Table 5. Comparison with other works (OW - Our work, BD - Bowrider Deck, DEB - Deck boat, DIB - Dinghy boat, PB - Ponton boat, RB - Runabout boat, VB - Vela boat)

Ref	Target boat	Input	Approach	Accuracy [%]	Disadv
[3]	boats with length between 1 to 450 m	Satellite images	SUMO	78.00%	Images are affected by sea conditions
[15]	cargo, tanker, container	SAR images	SSD	75.90-91.70%	Images are affected by weather conditions
[9]	cargo ship, ore carrier, fishing boat, bulk cargo carrier, passenger ship container ship	Real-world video segments	Yolo v2	77.51%	Images collection require a large space for capturing
			SSD	77.66%	
			R-CNN	92.40%	
[2]	illegal fishing boats	Visible Infrared Imaging	sharpness of spike	99.32%	Images are affected by weather and light conditions
OW	BD, DEB, DIB, PB, RB, VB	RGB	CNN	98.27%	Images can be affected by noise during acquisition

One can observe that in our paper we approached more types of boats with respect to other works and we classified them based on CNN with second highest accuracy after [2]. An important advantage of our work resides in the efficiency of opening filter used. If an image is affected by noise, using this filter will result in a correct boat recognition.

4 Conclusions

In this work, we presented an algorithm that used a CNN for boat type class recognition. Because the CNN generates a misclassification of boat types when images are affected by noise, a morphological noise filter called opening was used. The accuracy of the CNN without the filter is 15.07% whereas employing the filter we obtained an accuracy of 98.27%. We compared our results article [8] whose method accuracy ranges between 75.90% and 91.70% but their employed images are not affected by noise. For future works, we propose to recognize boat types in images captured in infrared. At this time our proposed method is suitable for images captured during the daytime.

Aknowledgement. This work was supported by a grant from the National Program for Research of the National Association of Technical Universities - GNAC ARUT 2023, contract no. 21/2023.

References

1. Abrudan, D., Drăgulinescu, A.M., Preda, R.O., Vizireanu, N.: Fuel burn reduction in commercial aviation using mathematical morphology. In: Vladescu, M., Cristea, I., Tamas, R.D. (eds.) Advanced Topics in Optoelectronics, Microelectronics, and Nanotechnologies XI. SPIE, March 2023. https://doi.org/10.1117/12.2643287, http://dx.doi.org/10.1117/12.2643287
2. Elvidge, C., Zhizhin, M., Baugh, K., Hsu, F.C.: Automatic boat identification system for viirs low light imaging data. Remote Sens. **7**(3), 3020–3036 (2015)
3. Greidanus, H., Alvarez, M., Santamaria, C., Thoorens, F.X., Kourti, N., Argentieri, P.: The sumo ship detector algorithm for satellite radar images. Remote Sens. **9**(3), 246 (2017)
4. Kuemlangan, B., et al.: Enforcement approaches against illegal fishing in national fisheries legislation. Marine Policy **149**, 105514 (2023)
5. Mondal, R., Dey, M.S., Chanda, B.: Image restoration by learning morphological opening-closing network. Math. Morphol. Theor. Appl. **4**(1), 87–107 (2020)
6. Redmon, J., Farhadi, A.: Yolo9000: better, faster, stronger. In: 2017 IEEE Conference on Computer Vision and Pattern Recognition (CVPR), IEEE, July 2017. https://doi.org/10.1109/cvpr.2017.690, http://dx.doi.org/10.1109/CVPR.2017.690
7. Ren, S., He, K., Girshick, R., Sun, J.: Faster r-cnn: towards real-time object detection with region proposal networks. IEEE Trans. Pattern Anal. Mach. Intell. **39**(6), 1137–1149 (2017)
8. Scherrer, R., Aulnette, E., Quiniou, T., Kasarherou, J., Kolb, P., Selmaoui-Folcher, N.: Boat detection in marina using time-delay analysis and deep learning. Int. J. Data Warehousing Min. **18**(2), 1–16 (2022)
9. Shao, Z., Wu, W., Wang, Z., Du, W., Li, C.: Seaships: a large-scale precisely annotated dataset for ship detection. IEEE Trans. Multimedia **20**(10), 2593–2604 (2018)
10. Spagnolo, P., Filieri, F., Distante, C., Mazzeo, P.L., D'Ambrosio, P.: A new annotated dataset for boat detection and re-identification. In: 2019 16th IEEE International Conference on Advanced Video and Signal Based Surveillance (AVSS), IEEE, September 2019. https://doi.org/10.1109/avss.2019.8909831, http://dx.doi.org/10.1109/AVSS.2019.8909831
11. Syed, M.A.B., Ahmed, I.: Multi model LSTM architecture for track association based on automatic identification system data (2023). https://doi.org/10.48550/ARXIV.2304.01491, https://arxiv.org/abs/2304.01491
12. Vanparia, .P., Ghodasara, Y.: : Review paper on to study and enhance coastal security system using GIS/GPS Tool. Int. J. Comput. Appl. Inf. Technol. (2012)

13. Xing, B., Zhang, L., Liu, Z., Sheng, H., Bi, F., Xu, J.: The study of fishing vessel behavior identification based on AIS data: a case study of the east china sea. J. Marine Sci. Eng. **11**(5), 1093 (2023)
14. Ze-Feng, D., Zhou-Ping, Y., You-Lun, X.: High probability impulse noise-removing algorithm based on mathematical morphology. IEEE Sig. Process. Lett. **14**(1), 31–34 (2007)
15. Zhang, C., et al.: Evaluation and improvement of generalization performance of SAR ship recognition algorithms. IEEE J. Sel. Top. Appl. Earth Observ. Remote Sens. **15**, 9311–9326 (2022)

The Importance of a Framework for the Implementation of Technologies Supporting Talent Management

Helena Rodrigues Ferreira[1] (✉), Arnaldo Santos[1,2], and Henrique S. Mamede[1,2]

[1] Universidade Aberta Lisboa, Lisbon, Portugal
helena.ferr09@gmail.com, {arnaldo.Santos,jose.mamede}@uab.pt
[2] INESC TEC, Porto, Portugal

Abstract. The speed and scale of technological change are raising concerns about the extent to which new technologies will radically transform workplaces. Competition for the best talent is being intensified, and talent management requires new approaches and innovative strategies for developing talent based on corporate culture and its unique properties. By implementing and adopting technology in Human Resources Management (HRM), organizations create a digital employee lifecycle that spans from the initial Hiring Process to encompassing areas such as Performance Management, Learning and Development until the Offboarding, shaping a Talent Management journey. Despite the implementation of technologies being a continuous practice observed in numerous organizations, there are still challenges. The HRM technological market has become massive, and concerns arise about adopting these technologies' costs, practicality, and purpose. Because of that, designing strategies for implementing technologies in HRM, specifically in talent management, is hard to overview. In this context, this document aims to present the necessity and significance in developing a framework that aggregates the implementation process of technologies in talent management supported by Design Science Research (DSR). The holistic perspective of the forthcoming framework consolidates insights into business challenges and their correlation with technology selection, technological capabilities, implementation procedures, as well as anticipated metrics and their impact.

Keywords: Workplace Transformation · Talent Management · Human Resources Management · Technology Implementation · Technology Adoption

1 Introduction

The speed and scale of current technological change are raising concerns about the extent to which new technologies will radically transform workplaces or displace workers altogether. Additionally, the profound impact of COVID-19 accelerated those changes, generating what we usually call New Ways of Working (NWoW) [1].

Human Resources (HR) and Information Technology (IT) experts are vital in designing and developing systems that empower organizations to embrace NWoW while fostering the necessary culture for talent to thrive. By adopting technology in Human Resources

Á. Rocha et al. (Eds.): WorldCIST 2024, LNNS 987, pp. 466–472, 2024.
https://doi.org/10.1007/978-3-031-60221-4_44

Management (HRM), organizations foster an enriched digital employee experience that begins at the hiring process and extends to performance management and development [2], until the offboarding. Thanks to the collective efforts of scholars and IT, substantial progress has been made in understanding how these technologies have changed the world of work by inspiring the adoption of new talent management initiatives and bringing up new types of human capital. However, technology evolves continuously, and new technologies are undoubtedly emerging. Researchers will continuously set out to critically assess their impacts [3].

In addition, despite implementing technologies in talent management being a continuous practice observed in numerous organizations, there are challenges. The "HR tech market" becomes more massive daily with new Artificial Intelligence (AI) driven platforms being announced. Some raise cautionary fags over the cost justifiability of different technologies, tools and applications. Others sound alarms over their practical implications. Still, others ask whose needs are being served and for what purposes they adopt or integrate a particular technology tool or application [4]. A study by HR advisory and research firm Josh Bersin Co. [5] found that 42% of organizations rated their technology implementations in talent management as needing to be entirely successful two years after implementation. Another central concern to HR and IT is how to bolster employee and organizational resilience to disruption from new technologies [6], given that adopting new technologies in the organization may lead to some resistance at group and individual levels [7]. Because of that, employee technology adoption remains a crucial factor for many organizations [8]. Besides, designing strategies for implementing technologies in talent management within organizations is hard to overview, especially when some organizations fail to realize the anticipated benefits from their technological investments.

This phenomenon will only be reduced by an increased understanding of aligning with the organization's strategy and effectively coordinating IT implementation [7]. Since talent management is commonly recognized as a decisive factor of success in organizations with a high demand for skilled individuals [9], along with the increasing recognition of technology as vital in this domain, this research holds significant importance and relevance.

Based on this, the research problem lies in the widespread adoption of technology in talent management; many organizations struggle to achieve successful implementation and adoption and fail to recognize its impact on their operations and outcomes.

In this document, we first present the frame of this research and the research problem (this section). It is followed by Sect. 2, where we present a preliminary literature review with conceptual theories on the key definitions. We conclude the document in Sect. 3 with the future research methodology and final remarks.

2 Research Background

We argue that a way forward for organizations to face the uncertainties and complexities posed by Digital Transformation (DX) and the NWoW in the workplace is by shifting towards a learning organization [10]. The learning organization concept flourished in the 1990s, stimulated by Peter M. Senge's The Fifth Discipline [11]. Senge defined the learning organization as one where people continually expand their capacity to create the desired results, new and expansive thinking patterns are nurtured, where collective aspiration is set free, and where people continually learn how to learn [11]. "Learning organization" is still an appropriate and widely used term to describe an organization that prioritizes and values continuous learning and improvement. However, considering the new workplace demands and rapid changes, this represents an opportunity to reframe the learning organization concept. Besides the disciplines that promote learnability, a learning organization should also learn and adjust to a new context [10]. The transformation of work practices is now characterized by increased flexibility, digitization, virtualization, and mediation. This shift-like work fundamentally challenges the traditional "formal" bureaucratic organizing logic, emphasizing flexibility, adaptability, and dynamism [1]. Suppose previously organizations followed a job-based approach to talent management - that involves using a well-established jobs architecture to define job levels, grades, career paths, spans of control, the criteria for career progression and compensation based on job value - more recently. In that case, organizations must rethink their operating model [12].

In recent years, the concept of talent management has been distinguished as new, capturing the interest of most researchers [13]. According to Armstrong [14], talent management is an integrated series of activities to ensure that an organization attracts, retains, motivates, and develops the potential and talented people it needs now and in the future. To "win the talent battle", an organization must make talent management a top business priority. To recruit and keep the right individuals, the organization must constantly develop and improve the employee value proposition [15]. Within NWoW, talent management must give employees increased autonomy and flexibility regarding when and where they work [16]. Competition for the best talent is being intensified, and talent management requires new approaches and innovative strategies for developing talent based on corporate culture and its unique properties [9].

Technology supporting talent management plays a crucial role. Since the 1990s, the terminology for HRM technologies has changed regularly as technological innovations have advanced. The Human Resource Information Systems (HRIS) established centralized systems responsible for gathering, storing, and handling HRM information. Their goal was to enhance cost-effectiveness and efficiency by encouraging process uniformity through automation. The rise of cloud-based IT and the notion of Talent Management, has led to specialized technology claiming to develop and implement talent management 'systems'. These systems aim to identify investment and retention targets, establish a unified accessible talent database, and structured communication, criteria, and development processes [17]. Nowadays, organizations increasingly depend on technologies to recruit, select, and development their workforces [15]. Integrating AI in technology supporting talent management, is at the center of its development. IT must ensure that the new technologies are free from bias because such technologies are more likely to be accepted and provide a fairer and more positive workplace environment if legal, ethical, and employee rights are adequately addressed [15]. The research proposed will focus on technologies that support the field of talent management with some examples including Skills Architecture Platforms - used to understand the workforce and to inform decisions about where to build, borrow or buy talent [18], - Learning Management Systems (LMS)

- used to track, record, and report learning data [19], - Learning Experience Platform (LXP) - used to complement the LMS and deliver a personalized learning journey, - and Talent Experience Platform (TXP) - connects learning to talent management, skills management, career growth opportunities and insightful analytics [20].

To achieve successful outcomes, accurate implementation of technologies supporting talent management is an imperative. The study of technological implementations and adoption within organizations has been part of academic research for decades. Holahan [21] defined implementation effectiveness as the "human connection" between an organization's decision to adopt a new technology and realizing a return on its investment. According to their study, implementation effectiveness is related to the successful adoption of the technology by the users, and it depends on the climate of implementation, organizational receptivity toward change, and innovation values fit. Several other theories specifically address the adoption of technology. The Unified Theory of Acceptance and Use of Technology (UTAUT), for example, identifies three effects arising from determinants of behavioral intention: performance expectancy, effort expectancy, and social influence [22].

In today's dynamic landscape, where technology advances rapidly, talent strategies are intricately linked with the implementation of technologies. Technology adoption and strategy renewal must emerge in parallel and inform one another. An organization cannot devise a new strategy without assessing the real potential of new technologies and its ability to acquire the necessary skills and resources. Conversely, it cannot adopt every new piece of digital technology without a strategic plan to leverage it [23]. Technological implementation is critical to why some organizations outperform others, as even a well-formulated strategy cannot guarantee success until it is effectively implemented [24].

3 Conclusion

By utilizing Design Science Research (DSR), the future research aims to understand the key factors organizations should consider when implementing and adopting technology that supports talent management.

Our strategy merges the IT and HR domains, aiming to develop a comprehensive framework for organizations. This framework enables the evaluation, selection, and implementation of technologies into talent management operations, maximizing their potential to enhance organizational performance and facilitate workplace transformation. In this research, successful implementation is conceptually defined as a process that starts before the technology implementation. This conceptualization encompasses understanding both the defined business challenges and priorities within organizations, along with the anticipated workplace changes. It also considers how these elements are interconnected with technology selection and technological capabilities. The strategic approach favors 'capabilities' over 'features,' emphasizing broader functionalities to address the specific challenges. The anticipated metrics and impact resulting from the implementations and adoptions are also part of this holistic perspective. Segmenting into these three dimensions fosters an ecosystem approach for successful technology implementation and adoption, strategically guided by change management. Figure 1 presents the implementations dimensions proposed.

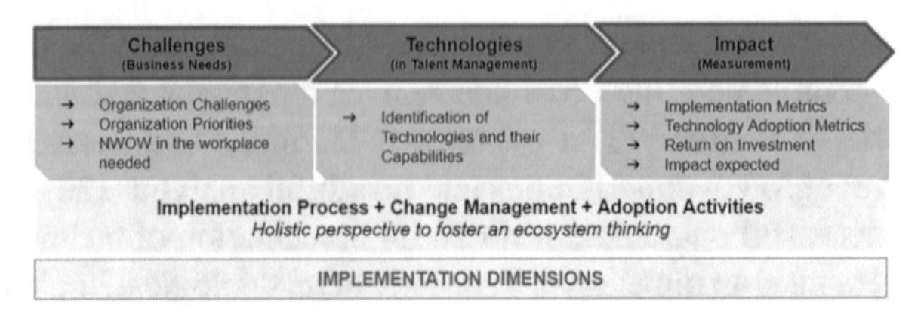

Fig. 1. Proposed Implementation Dimensions

The research methodology that will be employed is DSR due to its focus on the solution that addresses a generalized and real problem [25]. We will incorporate the process of six different activities outlined by Peffers [26] and the guidelines proposed by Hevner [27] into this research. Figure 2 presents the process DSR of the research.

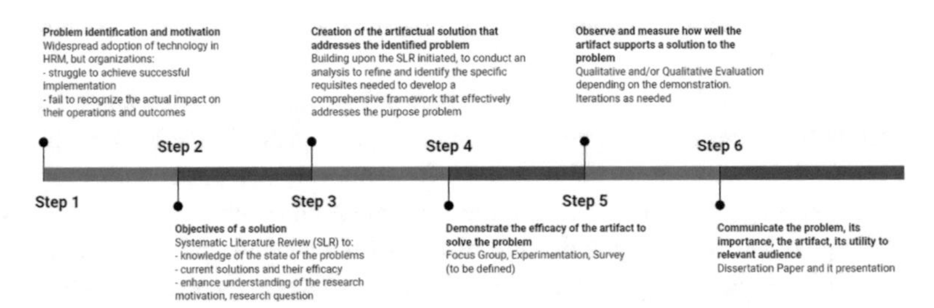

Fig. 2. Design Science Research Process (adapted from Peffers et al., 2007)

In Step 1, we atomized the problem conceptually (challenges, technology and impact) so that the solution could capture its complexity. In addition, our justification of the solution's value brings the understating of the reasoning associated with the problem [26]. By following Kitchenham's Procedures for Performing a Systematic Literature Review (SLR) [28], we will derive the objectives of the solution in Step 2. In addition, we propose conducting interviews with professionals specialized in the field to gather valuable insights regarding implementing technology in Talent Management. In Step 3, we will develop the framework that effectively tackles the identified problem and holds significance in solving a business challenge. The selection of a demonstration process remains undecided and will depend on the opportunities we create to use the framework effectively in Step 4. Considering our purpose to rigorously demonstrate the framework's utility, quality and efficacy, in Step 5, we will design and conduct a research survey to collect feedback from professional practitioners in the field. In the final step (Step 6), we will communicate the artefact, its utility and novelty, the rigor of its design, and its effectiveness to researchers and practicing professionals.

Despite the already sustained relevance of this research, specific points need a more in-depth analysis. Until now, the research has primarily concentrated on the organizational perspective and challenges, with limited exploration of the influence of employee motivations. Moreover, a deep exploration of the technologies in talent management is also needed. The concept still needs to be recognized by academic research, and special attention is needed in the SLR to ensure that the relevant information is captured. It is essential to categorize the technology based on its capabilities rather than relying on labels that may become obsolete rapidly. Finally, yet importantly, there are specific fields that, although yet to be fully defined, can be included in this research. Software architecture and its integration with HRIS are fundamental aspects to consider. By incorporating these dimensions into the research, we can better ensure the relevancy of the practical aspects of the framework and its effective implementation within the technological landscape of organizations.

References

1. Aroles, J., Cecez-Kecmanovic, D., Dale, K., Kingma, S.F., Mitev, N.: New ways of working (NWW): Workplace transformation in the digital age. Inf. Organ. **31**(4), 100378 (2021). https://doi.org/10.1016/j.infoandorg.2021.100378
2. Gartner. Gartner Identifies Top Four HR Investment Trends for 2023, June 3 (2023). https://www.gartner.com/en/newsroom/press-releases/03-06-2023-gartner-identifies-top-four-hr-investment-trends-for-2023
3. Wiblen, S., Marler, J.: The Human–Technology Interface in Talent Management and the Implications for HRM, pp. 99–116 ((2019)). https://doi.org/10.1108/S1877-636120190000023009
4. Bonk, C.J., Wiley, D.A.: Preface: Reflections on the waves of emerging learning technologies. Education Tech. Research Dev. **68**(4), 1595–1612 (2020). https://doi.org/10.1007/s11423-020-09809-x
5. Zielinski, D. Why Do So Many New HR Technology Implementations Fail? SHRM. Obtido 8 de junho de 2023, de (November 2022). https://www.shrm.org/hr-today/news/hr-magazine/winter2022/pages/why-new-hr-technology-implementations-fail.aspx
6. Trenerry, B., et al.: Preparing workplaces for digital transformation: an integrative review and framework of multi-level factors. Front. Psychol. **12** (2021). https://www.frontiersin.org/arti-cles/https://doi.org/10.3389/fpsyg.2021.620766
7. Huda, M.Q., Sa'adah, N.U.: Measurement model of information technology implementation effectiveness. In: 2014 International Conference on Cyber and IT Service Management (CITSM), pp. 21–24 (2014). https://doi.org/10.1109/CITSM.2014.7042168
8. Sanders,S.: Putting Technology into Perspective: The best thing to happen to technology in the future will not be technology. Workforce Solutions Rev. **7**(5), 20–22 (2016)
9. Sinyagina, N.Y.: New trends in HR technologies: Overview of foreign studies. In"SHS Web of Conferences, vol. 103, p. 01033 (2021). https://doi.org/10.1051/shsconf/202110301033
10. Givel, I.: TEDx Talks (Diretor). Growing through change: A How-To for leaders of learning organizations, dezembro 8 (2017). https://www.youtube.com/watch?v=D1iO2QwJYAI
11. Senge, P.M.: TheFifth Discipline: The Art and Practice of the Learning organisation, Doubleday (1990)
12. The Conference Board. Navigating to a Skills-Based Approach to Talent Development. Obtido 2 de junho de 2023, de // , 19 Maio (2021). http://www.conference-board.org/publications/skills-based-to-talent-development
13. Al Dalahmeh, M.: . Talent management: a systematic reviewOradea J. Bus. Econ. **5**, 115–123 (2020). https://doi.org/10.47535/1991ojbe102
14. Armstrong, M.: A handbook of human resource management practice, 10th ed. Kogan Page Publishing (2006). http://surl.li/mltcy
15. França, J.F., T., São Mamede, H., Pereira Barroso, J.M., Pereira Duarte dos Santos, V.M.: Artificial intelligence applied to potential assessment and talent identification in an organizational context. Heliyon **9**(4), e14694 (2023). https://doi.org/10.1016/j.heliyon.2023.e14694

16. Andrulli, R., Gerards, R.: How new ways of working during COVID-19 affect employee well-being via technostress, need for recovery, and work engagement. Comput. Hum. Behav. **139**, 107560 (2023). https://doi.org/10.1016/j.chb.2022.107560
17. Wiblen, S., Marler, J.H.: Digitalised talent management and automated talent decisions: the implications for HR professionals. Inter. J. Human Res. Manag. **32**(12), 2592–2621 (2021). https://doi.org/10.1080/09585192.2021.1886149
18. Gloat, Skills Foundation - How does it work? Obtido 8 de junho de 2023, de, junho 12 (2022). https://gloat.com/platform/skills-foundation/
19. Cooke, G.: Learning Content Management Systems Guide: LCMS vs. LMS (May 2021). https://www.elucidat.com/blog/learning-content-management-systems/ (Accessed 19 September 2022)
20. Suri, K.: TXP101:Whatisa talentexperienceplatform?(semdata). Obtido 8 dejunho de2023 (2023). https://www.cornerstoneondemand.com/resources/article/txp-101-what-talent-experience-platform/
21. Holahan, P.J., Aronson, Z.H., Jurkat, M.P., Schoorman, F.D.: Implementing computer technology: A multi-organizational test of Klein and Sorra's model. J. Eng. Tech. Manage. **21**(1), 31–50 (2004). https://doi.org/10.1016/j.jengtecman.2003.12.003
22. Momani, A.: The Unified Theory of Acceptance and Use of Technology: A New Approach in Technology Acceptance. Inter. J. Sociotechnol. Knowl. Developm. **12**, 79–98 (2020). https://doi.org/10.4018/IJSKD.2020070105
23. Bughin, J. R., Kretschmer, T., van Zeebroeck, N.: Experimentation, Learning and Stress: The Role of Digital Technologies in Strategy Change (SSRN Scholarly Paper N.o 3328421) (2019). https://doi.org/10.2139/ssrn.3328421
24. Tawse, A., Tabesh, P.: Strategy implementation: A review andan introductory framework. Euro. Manag. J. **39**, 22–33 (2021). https://doi.org/10.1016/j.emj.2020.09.005
25. Dresch, A., Lacerda, D., Antunes, Jr., J.A.V.: Design Science Research: A Method for Science and Technology Advancement (2014). https://doi.org/10.1007/978-3-319-07374-3
26. Peffers, K., Tuunanen, T., Rothenberger, M.A., Chatterjee, S.: A design science research methodology for information systems research. J. Manag. Inform. Syst. 24(3), 45–77 (2007)
27. Hevner, A.R., March, S.T., Park, J., Ram, S.: Design Science in Information Systems Research. Manag. Inf. Syst. Q. **28**(1), 75–105 (2004). https://doi.org/10.2307/25148625
28. Kitchenham, B.: Procedures for Performing Systematic Reviews Kitchenham, B., 2004. Keele, UK, Keele University, vol. 33 (2004)

Unlocking Education Together: A Meta-model Emphasizing Usability and Reuse of Open Educational Resources

Antonio Bucchiarone[1], Andrea Vázquez-Ingelmo[2(✉)], Alicia García-Holgado[2], Francisco José García-Peñalvo[2], Gianluca Schiavo[1], and Roberto Therón[2]

[1] Fondazione Bruno Kessler (FBK), Trento, Italy
{bucchiarone,gschiavo}@fbk.eu
[2] GRIAL Research Group, University of Salamanca, Salamanca, Spain
{andreavazquez,aliciagh,fgarcia,theron}@usal.es

Abstract. Open Educational Resources (OERs) represent a paradigm shift in education, offering diverse, freely accessible materials to democratize knowledge. However, challenges in usability and integration hinder their practical application. This paper addresses these issues within the innovative ENCORE (ENriching Circular use of OeR for Education) project, dedicated to optimizing OER utilization. Focusing on the direct usability challenges faced during ENCORE's development, an extension to the ENCORE meta-model is proposed. This enhancement ensures uniform representation, empowering educators to seamlessly integrate and utilize OERs in personalized learning paths and assessments. Beyond conventional classrooms, the extended meta-model fosters community awareness through collaborative material sharing. This paper serves as a call to action, emphasizing the pivotal role of OERs in reshaping education. Additionally, the paper introduces a research roadmap addressing key research areas such as personalized learning paths, generative AI applications, and community building, among others.

Keywords: Education · Open Educational Resources · Meta-Modeling · Research Roadmap

1 Introduction

In the evolving landscape of educational resources, Open Educational Resources (OERs) [1] stand out as a symbol of innovation and accessibility. These resources, encompassing a wide array of educational materials, are freely available for use and redistribution, aiming to democratize access to knowledge [2]. However, the journey from the conceptualization of OERs to their practical application in educational settings is filled with challenges, particularly in terms of usability and integration [3,4]. This paper delves into these challenges within the context of the ENCORE (ENriching Circular use of OeR for Education) project[1], a pioneering initiative aimed at optimizing the use and effectiveness of OERs.

[1] https://project-encore.eu/.

Á. Rocha et al. (Eds.): WorldCIST 2024, LNNS 987, pp. 473–482, 2024.
https://doi.org/10.1007/978-3-031-60221-4_45

The approach adopted by ENCORE is multifaceted, combining data-driven tools for educational design with comprehensive pedagogical guidelines. This strategy is designed to assist educators in crafting learning scenarios (i.e., courses, modules, etc.) that not only impart essential knowledge but also link learning outcomes directly to skills and concepts. To enhance this linkage, ENCORE utilizes ESCO, the European framework of Skills, Competences, Qualifications, and Occupations[2], as a foundational resource. By applying Natural Language Processing (NLP) techniques, the project aligns the skills detected in OERs with those listed in ESCO, creating a structured database that correlates educational resources with relevant skill sets and concepts.

During the development of the ENCORE database, we encountered significant challenges related to the *direct usability of OERs* [1,5,6]. The heterogeneity of sources and formats made it difficult to standardize and integrate these resources into existing educational frameworks (i.e., Moodle[3]).

This paper addresses the challenges by proposing an extension to the ENCORE meta-model. This enhanced meta-model is designed to overcome limitations by providing a comprehensive characterization of OERs. The objective is to enable their direct utilization in educational tools.

Our approach is rooted in the conviction that unlocking the true potential of OERs requires a systematic and user-friendly framework. This framework facilitates seamless integration and application of OERs in various educational contexts. The extended meta-model not only ensures uniform representation but also enhances the adaptability of OERs for immediate use in educational tools. Furthermore, it empowers educators to search for OERs aligned with their learning objectives, integrate them into personalized learning paths, and utilize them effectively in assessment activities.

Beyond the classroom, the extended meta-model supports learners in upskilling by providing a versatile platform for accessing high-quality educational materials. It also fosters a sense of community awareness by encouraging the sharing of materials, ultimately contributing to the creation of a collaborative and knowledge-sharing environment. In essence, this extension to the ENCORE meta-model aims to support the utilization of OERs, making them more accessible, adaptable, and impactful in the field of education.

This paper further contributes to the call to action for the educational community by inviting educators, technologists, and policymakers to reassess the pivotal role of OERs in the educational field, fostering a collaborative effort to enhance accessibility, adaptability, and effectiveness in the utilization of OERs. The emphasis lies in recognizing the evolving demands of modern pedagogy and the need for innovative tools and models that seamlessly align with these dynamic requirements.

The rest of this paper is structured as follows. Section 2 introduces the ENCORE project and its aims (in Sect. 2.1), followed by an exploration of OER characterization, including a review of previous works and schema definitions

[2] https://esco.ec.europa.eu/en/about-esco/what-esco.
[3] https://moodle.org/.

(in Sect. 2.2). Section 3 introduces an extended version of the ENCORE meta-model, building upon the initial work conducted within the ENCORE project. This extended meta-model seeks to comprehensively characterize Open Educational Resources (OERs), with a primary focus on facilitating their direct integration into educational tools. Section 4 reflects on the benefits of this approach, emphasizing the improved usability, community validation, and enhanced search capabilities of OERs, followed by Sect. 5, where we conclude with a summary of our findings and reflections on the future trajectory of OER utilization in educational contexts.

2 Background

2.1 The ENCORE Ecosystem

The ENCORE project addresses the evolving landscape of education by introducing an intelligent system designed to recommend high-quality Open Educational Resources (OERs) aligned with key skills influenced by societal macro-trends: specifically, Green, Digital, and Entrepreneurial (GDE) skills. Recognizing the challenge posed by the widespread availability of OERs, the ENCORE project emphasizes the necessity to support educators in exploring and identifying relevant OERs. The project provides educators with a set of digital tools within the ENCORE Ecosystem, called *Enablers*, offering features such as content recommendations, advanced filtering, and task automation. While these technologies empower educators to create more engaging learning experiences, it is crucial for educators to receive adequate training and resources to seamlessly integrate these systems into their teaching practices. Taking these aspects into account, the two-phase approach of the ENCORE project aims to provide valuable insights and guidelines for educators looking to incorporate such systems into their pedagogical efforts.

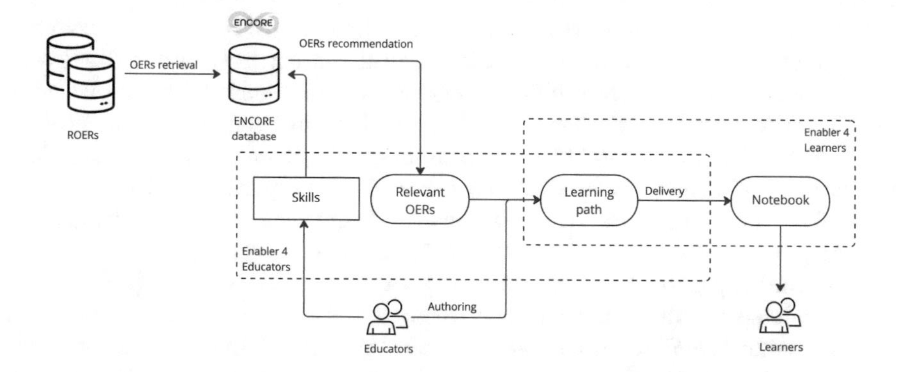

Fig. 1. The ENCORE Ecosystem.

As illustrated in Fig. 1, the ENCORE Ecosystem relies on two key enablers: the Enabler for Educators (E4E) and the Enabler for Learners (E4L). The primary goal of E4E is to assist educators in crafting learning paths [7] that leverage available OERs related to green, digital, and entrepreneurial *skills*. E4E's core service provides access to an extensive collection of OERs, enabling educators to efficiently explore relevant resources by inputting specific skills or pertinent concepts for use in learning activities. Educators have the flexibility to select relevant OERs either directly from the search list or while reviewing individual resources. This functionality empowers them to curate custom collections of OERs, proving invaluable during the *learning path* design phase. E4E features a Learning Path Editor (LPE) centered on enhancing the teachers' experience, employing graph-like visual-editing capabilities integrated into the broader framework. This design facilitates the creation of ready-to-use learning paths, striking a balance between providing useful abstraction for teachers to define paths naturally and ensuring concrete clarity for students during consumption.

The E4L allows for sharing the formulated learning paths through digital notebooks, leveraging tools such as Visual Studio Code[4]. The process begins within the LPE, where educators can download the notebook file and distribute it to learners. Upon accessing the notebook, learners are presented with an overview of the learning path crafted by the educator. Each node within the path corresponds to a distinct cell within the notebook, enhancing organization and streamlining the learning experience by integrating textual content, code snippets, and interactive elements. This structured approach facilitates an efficient and cohesive learning journey for students. Both Enablers utilize OERs as potential educational resources that educators can include in the learning paths and that can be accessed by learners.

2.2 OER Characterization and Discoverability

The Dublin Core Metadata Schema stands out in the Open Educational Resources (OERs) landscape as a pivotal structure that supports the discovery and description of electronic resources. It comprises a set of 15 metadata elements that simplify the process of representing digital content in a standardized way, thus enhancing the discoverability of resources across varied digital platforms and libraries [8,9]. This schema has been widely adopted due to its flexibility, aiding in the interoperability between different content types and metadata standards, and promoting accessibility across digital repositories [10–13].

Linked data principles have significantly influenced the OER domain by enriching metadata, which in turn facilitates the integration of educational services and mitigates the issue of metadata heterogeneity [14–16]. This approach underscores the potential for creating a web-wide, accessible network of reusable educational resources. Additionally, the use of linked data for semi-automatic classification of OERs enhances their categorization and discoverability [17]. Another significant development is the incorporation of semantic

[4] https://code.visualstudio.com/.

annotations within HTML code using microdata and schema vocabulary, which further improves the visualization and selection process for end-users, particularly aiding those with disabilities [18].

Sustainability is a key concern for OERs, with various models focusing on aspects like funding, technical infrastructure, and content adaptability, emphasizing the need for community involvement and partnerships for a holistic approach to sustainability [19].

Finally, innovations such as the introduction of efficient text file formats for assessment OERs in Learning Management Systems like Moodle, including Aiken[5], GIFT[6], and Moodle XML[7], reduce the workload for educators and facilitate a smoother integration of these resources into the educational ecosystem.

Despite these advancements, challenges persist, particularly in standardizing metadata across different educational and cultural institutions, which can affect access and use. In response, a proposed OER meta-model aims to establish a common framework for the development, sharing, and exploitation of OERs, factoring in community feedback and diverse OER types [1,3]. This model seeks to overcome the barriers of metadata heterogeneity and to foster a more unified, global approach to educational resources.

3 Meta-model Proposal

The previous meta-model laid the groundwork for characterizing Open Educational Resources (OERs) in the ENCORE project and constructing the initial version of the ENCORE database [20]. It established a schema for standardizing OER metadata, encompassing Dublin Core Metadata Element Set (DCMES) [8,9] attributes such as `title`, `subject`, `description`, and `publication_date`. For a detailed description of the preceding meta-model, please refer to [21].

In this context, the meta-model extension (Fig. 2) introduces significant enhancements, particularly focusing on community engagement and the detailed characterization of assessments, which was one of the main concerns when extracting OERs for their subsequent use in the ENCORE database.

- The Feedback class is now integrated, enabling instructors to provide ratings, purpose and comments, facilitating a richer interaction between educators and content.
- The Assessment class is a central addition, representing various evaluation mechanisms like Quiz, Case Evaluation, Fill in the Blanks or Coding Assignments.
- The Tool class encapsulates tools that execute OERs, underlining the meta-model's new emphasis on the interoperability of resources across different platforms and technologies.

[5] https://docs.moodle.org/403/en/Aiken_format.
[6] https://docs.moodle.org/403/en/GIFT_format.
[7] https://docs.moodle.org/403/en/Moodle_XML_format.

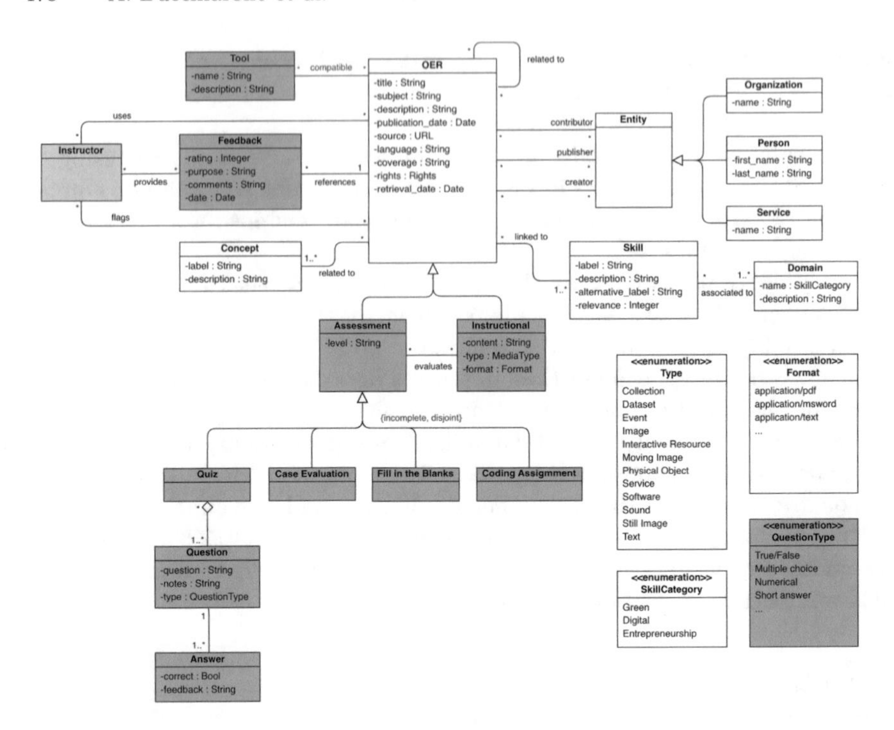

Fig. 2. OER meta-model in the context of ENCORE. Colored items represent the introduced entities.

The additions to the meta-model are justified by the imperative to validate and structure OERs consistently before their utilization in educational contexts.

The Feedback entity is connected to the concept of Instructor, an entity that was already present in a previously developed ENCORE meta-model [22], representing the connection of the OER meta-model with the ENCORE framework.

The integration of Feedback ensures that OERs can be vetted and refined based on community inputs, leading to a continuous improvement cycle, as well as an enhanced process of OER reuse, as instructors and educators can search for other educators' OER use cases in similar contexts.

This class is complemented with two additional relationships between the Instructor and the OERs:

- A relationship that accounts for the use of OERs, so it can be computed the number of times an OER has been used by an Instructor
- A relationship to conceptualize the possibility of flagging OERs as unusable or not appropriate, fostering the aforementioned refinement of OERs by the community

On the other hand, the Assessment class is a strategic enhancement that addresses specific challenges faced in the utilization of OERs-namely, the issues of scraping OER content and managing the variety of their formats. The diversity of formats in which assessments are created and disseminated has posed significant barriers to the standardization and integration of these resources into educational frameworks. The Assessment class aims to mitigate these challenges by providing a structured and consistent way to characterize assessments, ensuring they can be more readily identified, accessed, and integrated across various educational platforms. The subdivisions of this class into types like Quiz, Coding, Case Evaluation, and Image Evaluation, introduces a taxonomy (which can evolve over time, adding more types of assessment) that enables the categorization of OERs according to the nature of the learning evaluation they offer.

Additionally, each assessment activity includes a "level" attribute, which could be associated with a specific taxonomy, such as Bloom's taxonomy [23]. This "level" attribute provides a classification based on cognitive complexity, helping to categorize the assessment's difficulty or depth of understanding. The purpose of defining a class for assessment activities is to provide a structured and standardized way to represent and manage assessment-related data within an educational system.

Assessment and instructional OERs are distinct yet interconnected components of the educational process, each serving a specific purpose.

The primary purpose of assessment is to measure and evaluate the extent of a student's understanding, skills, and knowledge. Assessments provide insights into learning progress, identify areas of strength and weakness, and inform educators about the effectiveness of their teaching strategies.

Instruction, on the other hand, is focused on teaching and facilitating learning. The main goal of instruction is to impart knowledge, develop skills, and guide students in acquiring the necessary competencies. It involves planned activities and strategies to promote learning.

Instruction involves the delivery of content through lectures, discussions, hands-on activities, and other methods. It is geared towards building a student's knowledge base and fostering the development of critical thinking and problem-solving skills. The focus of instruction is on the teaching process itself, emphasizing the delivery of content, engagement, and the facilitation of learning experiences.

While assessment and instruction have different primary functions, they are interconnected in a cyclical process. Assessments inform instruction, guiding educators in refining their teaching methods to better meet the needs of students, and instruction lays the groundwork for assessments to measure learning outcomes.

Finally, the Tool class represents the practical application of OERs, acknowledging the need for resources to function seamlessly across diverse educational tools and platforms. This interoperability is fundamental for the OERs' adaptability and scalability, ensuring they remain relevant and accessible in various educational infrastructures.

4 Enhancing OERs: A Research Roadmap

The extended meta-model of the ENCORE project aligns seamlessly with our research roadmap, addressing key areas that shape the future of Open Educational Resources (OERs).

Our emphasis on the **easy creation of OERs** is underpinned by the meta-model, which provides a *standardized template for defining and sharing OERs*. This meta-model serves as a "lingua franca", fostering a common understanding and streamlining the creation process, making OERs more accessible and directly used in a learning path.

The integration of assessments related to each OER is a pivotal aspect of our roadmap. By leveraging classical taxonomies like Bloom's taxonomy [23], we empower educators to **evaluate students effectively**. The meta-model facilitates the definition of the *appropriate level and type of assessment*, ensuring alignment with educational objectives and enhancing the overall educational experience.

The focus on a shared *Data Space of OERs* and on *recommendation algorithms* underscore our commitment to **personalized learning**. The meta-model supports educators and learners in navigating and selecting the most preferred learning paths. This not only streamlines the preparation process but also enhances the consumption of OERs, tailoring educational experiences to individual needs.

Moreover, our roadmap incorporates the dynamic nature of learning paths. OERs can be exploited to provide new algorithms, *incrementally revising learning paths as it is executed by learners*. This variability in OERs enhances personalization, allowing for an **adaptive and responsive learning environment**.

Furthermore, the *exploitation of generative AI*, opens new opportunities for OER creation [24]. The ENCORE project envisions leveraging AI to suggest and create OERs, providing educators with **innovative and tailored resources** that align with their instructional objectives.

Lastly, the creation of a **community of practice**, complete with *open recognition through open badges*, is a cornerstone of our roadmap. Educators, through their contributions and experimentation with OERs, become integral parts of an open community. This collaborative environment ensures the sharing and utilization of high-quality, experimented-upon OERs, fostering a culture of continuous improvement and excellence.

5 Conclusions

This paper introduces a novel approach—an extended meta-model designed explicitly to surmount barriers arising from the diverse formats and origins of OERs. With a robust feedback system, assessment classifications, and interoperability features, the meta-model is designed not only to enhance the usability of OERs but also to ensure their relevance in dynamic educational settings.

Looking ahead, the ENCORE project's meta-model will guide the implementation of new mechanisms, supporting the creation and utilization of OERs by educators through the ENCORE enablers. Our future research line includes the introduction of evaluation criteria and scoring mechanisms for assessing OERs.

Addressing key areas such as content creation, assessment integration, personalized learning paths, dynamic content revision, generative AI integration, and community building, we aim to support the effective utilization of OERs, contributing to the advancement of educational practices that are open and innovative.

Acknowledgment. This work has been partially supported by the Erasmus+ Programme of the European Union ERASMUS-EDU-2021-PI-ALL-INNO, Project ENCORE - ENriching Circular use of OeR for Education (Reference number 101055893).

References

1. Wiley, D., Bliss, T.J., McEwen, M.: Open educational resources: a review of the literature. In: Spector, J.M., Merrill, M.D., Elen, J., Bishop, M.J. (eds.) Handbook of Research on Educational Communications and Technology, pp. 781–789. Springer, New York (2014). https://doi.org/10.1007/978-1-4614-3185-5_63
2. Hilton, J.: Open educational resources, student efficacy, and user perceptions: a synthesis of research published between 2015 and 2018. Educ. Technol. Res. Develop. **68**(3), 853–876 (2019). https://doi.org/10.1007/s11423-019-09700-4
3. Otto, D.: Adoption and diffusion of open educational resources (oer) in education: A meta-analysis of 25 oer-projects. Int. Rev. Res. Open Distrib. Learn. **20**(5), 122–140 (May 2019). https://www.irrodl.org/index.php/irrodl/article/view/4472
4. Hylén, J.: Open educational resources: Opportunities and challenges (2020)
5. Whitfield, S., Robinson, Z.: Open educational resources: the challenges of 'usability' and copyright clearance. Planet **25**(1), 51–54 (2012)
6. Tlili, A., et al.: Towards utilising emerging technologies to address the challenges of using open educational resources: a vision of the future. Educ. Technol. Res. Develop. **69**, 515–532 (2021)
7. Martorella, T., Bucchiarone, A.: Adaptive and gamified learning paths with polyglot and .net interactive. CoRR abs/2310.07314 (2023). https://doi.org/10.48550/arXiv.2310.07314
8. Kunze, J., Baker, T.: The dublin core metadata element set. Tech. rep. (2007)
9. Weibel, S., Kunze, J., Lagoze, C., Wolf, M.: Dublin core metadata for resource discovery. Tech. rep. (1998)
10. López, C., García-Peñalvo, F.J., Pernía, P.: Desarrollo de repositorios de objetos de aprendizaje a través de la reutilización de los metadatos de una colección digital: De dublin core a ims. RED, Revista de Educación a Distancia IV (2005)
11. Alasem, A.: An overview of e-government metadata standards and initiatives based on Dublin core. Electron. J. e-Govern. **7**(1), 1–10 (2009)
12. Kakali, C., et al.: Integrating dublin core metadata for cultural heritage collections using ontologies. In: International Conference on Dublin Core and Metadata Applications, pp. 128–139 (2007)
13. García-Peñalvo, F.J., Merlo-Vega, J.A., Ferreras-Fernández, T., Casaus-Peña, A., Albás-Aso, L., Atienza-Díaz, M.L.: Qualified Dublin core metadata best practices for gredos. J. Libr. Metadata **10**(1), 13–36 (2010)

14. Dietze, S., et al.: Interlinking educational resources and the web of data - a survey of challenges and approaches. Emerald Program: electronic Library and Information Systems 47 (02 2013)
15. Navarrete, R., Luján-Mora, S.: Use of linked data to enhance open educational resources. In: 2015 International Conference on Information Technology Based Higher Education and Training (ITHET), pp. 1–6 (2015)
16. Piedra, N., Chicaiza, J., López, J., Tovar, E.: Seeking open educational resources to compose massive open online courses in engineering education an approach based on linked open data. JUCS - J. Univ. Comput. Sci. **21**(5), 679–711 (2015). https://doi.org/10.3217/jucs-021-05-0679
17. Chicaiza, J., Piedra, N., López, J., Tovar, E.: Domain categorization of open educational resources based on linked data. vol. 468, pp. 15–28 (09 2014)
18. Navarrete, R., Lujan-Mora, S.: Microdata with schema vocabulary: improvement search results visualization of open educational resources. In: 2018 13th Iberian Conference on Information Systems and Technologies (CISTI), pp. 1–6 (2018)
19. Downes, S.: Models for sustainable open educational resources. Interdiscip. J. Knowl. Learn. Objects 3 (01 2007)
20. Bucchiarone, A., et al.: Designing learning paths with open educational resources: An investigation in model-driven engineering. In: 2023 18th Iberian Conference on Information Systems and Technologies (CISTI), pp. 1–7 (2023)
21. Vázquez-Ingelmo, A., García-Holgado, A., García-Peñalvo, F.J., Chiarello, F.: Usability study of a pilot database interface for consulting open educational resources in the context of the encore project. In: Zaphiris, P., Ioannou, A. (eds.) Learning and Collaboration Technologies, pp. 420–429. Springer Nature Switzerland, Cham (2023)
22. Bucchiarone, A., et al.: Towards Personalized Learning Paths to Empower Competency Development in Model Driven Engineering through the ENCORE platform (Aug 2023)
23. Krathwohl, D.R.: A revision of bloom's taxonomy: an overview. Theory Into Pract. **41**(4), 212–218 (2002)
24. García-Peñalvo, F., Vázquez-Ingelmo, A.: What do we mean by genai? a systematic mapping of the evolution, trends, and techniques involved in generative ai. International Journal of Interactive Multimedia and Artificial Intelligence In: Press(In Press), 1–10 (07/2023 9998). https://www.ijimai.org/journal/sites/default/files/2023-07/ip2023_07_006.pdf

Author Index

Á. Rocha et al. (Eds.): WorldCIST 2024, LNNS 987, pp. 483–485, 2024.
https://doi.org/10.1007/978-3-031-60221-4